Asset Management of Urban Drainage Systems

Asset Management of Urban Drainage Systems

if anything exciting happens, we've done it wrong!

Edited by

Frédéric Cherqui, François Clemens-Meyer, Franz Tscheikner-Gratl and Bert van Duin

IWA PUBLISHING

Published by **IWA Publishing**
Unit 104–105, Export Building
1 Clove Crescent
London E14 2BA, UK
Telephone: +44 (0)20 7654 5500
Fax: +44 (0)20 7654 5555
Email: publications@iwap.co.uk
Web: www.iwaponline.com

First published 2024

British Library Cataloguing in Publication Data
A CIP catalogue record for this book is available from the British Library

ISBN: 9781789063042 (paperback)
ISBN: 9781789063059 (eBook)
ISBN: 9781789063066 (ePub)

Doi: 10.2166/9781789063059

This eBook was made Open Access in June 2024.

Contents

doi: 10.2166/9781789063059_xi

Preface

This book covers a vital dimension of the infrastructure of the towns and cities that are home to an ever-growing number of people around the world – their drainage systems and, more specifically, the task of managing and maintaining these assets over time.

Management of these assets is becoming increasingly important. Infrastructure that is already ageing is being exposed to greater extremes due to a changing climate and the pressures of growing urban populations. Yet, the infrastructure needs to be preserved to ensure it delivers protection against flooding, health and environmental risks.

The methods, tools and approaches to support this management are often only thought about after design and construction. What is needed is forward-looking, future-proofed and, where appropriate, holistic systems that support the liveability of our cities to their fullest potential. This means accommodating the shift from grey to green infrastructure with its associated transformation of our urban environment.

This book responds to that need and is the first comprehensive book ever published on the asset management of rain and stormwater management facilities. This is of real note, especially when considering the extensive body of information that the book's contributors have drawn upon.

The book is particularly timely given the urgency presented by climate change. Changes in long-term rainfall patterns are overtaking the historical designs of our systems. This brings the long-term management of our urban drainage to the forefront – particularly relevant given our desire to avoid the 'excitement', alluded to in the book's title, of when the infrastructure becomes overwhelmed.

The book is ambitious in its scope. It aims to bridge the gap between developments in research and practice. It also offers a comprehensive entrance to the vast body of literature built up over the past two to three decades. In addition, it highlights areas of new developments, such as blue-green infrastructure solutions.

The book is aimed at a very broad readership: from the technician who wants to know more about tools and methods, to the researcher and student who wants an overview, to the engineer who needs to create and implement short-, medium-, and long-term strategies. It also responds to the fact that, in education, the focus is still typically on new design principles and procedures, while in practice there is an increasing need to enhance or even redesign existing systems.

With these goals in mind, the book has been developed through a collaboration between 50 authors from across four continents, spanning academics, operators and commercial companies. This combination of global input from varied disciplines combines to create a book of broad relevance.

Furthermore, it is being made available as an Open Access publication, supporting the widest possible access to the book's insights.

We face an essential need to improve the resilience of our urban areas. Drainage systems are integral to this. Furthermore, we need to maintain and operate this infrastructure for decades to come. Because this infrastructure and our operation of it will evolve over time, especially through greater use of digital technologies used in monitoring and control, our agility and openness to innovative and holistic approaches will be relied upon to ensure the resilience of our ever-expanding urban environments. The foundations presented in this book respond directly to this need.

Kala Vairavamoorthy
Executive Director, International Water Association

Acknowledgements

The real ambiguity of this book is in its subtitle: ‘If something exciting happens, we’ve done it wrong!’ On one hand, the editorial team and contributing authors hope to invoke a spectacular impact on those mustering the courage to read or even use it. On the other hand, we hope that, after the content has been digested by increasing numbers of people, the number and severity of spectacular mishaps in urban drainage practice will decrease: time will tell.

Starting as an initiative back in 2021 during one of the covid online workshops organized by the IWA/IAHR working group on Urban Drainage Asset Management (https://udam.home.blog/), within the scope of a little over two years, draft texts have been written and edited to form the core of the present work. Given the fact that there was no formal budget available (even worse, the fee for Open Access publication had to be somehow raised), we acknowledge the fact that the employers of the contributing authors allowed them to spend time on this enterprise, although a substantial amount of spare time was spent as well. This intrinsic motivation is exemplary for the mentality encountered in this professional realm and we wholeheartedly thank all contributing authors for living up to our expectations in that respect. We like to highlight the pleasant cooperation with the production team at IWA Publishing, who showed great professionalism in supporting the final editing process.

Finally, we encourage the reader to provide feedback as we don’t pretend to present an exhaustive nor faultless overview of the everchanging state-of-the-art and the looming challenges in the field of Urban Drainage Asset Management.

February 2024
The editorial team:
Frédéric Cherqui, François Clemens-Meyer, Franz Tscheikner-Gratl and Bert van Duin

List of editors and authors

Editors (and authors)

Frédéric Cherqui
INSA-LYON, UCBL Lyon 1, DEEP, F-69621, F-69622, Villeurbanne, France; WERG, SAFES, The University of Melbourne, Burnley, VIC 3121, Australia

Frédéric Cherqui is an Associate Professeur at Université Lyon 1 / INSA Lyon, and an Honorary Senior Fellow at the University of Melbourne. For more than 15 years, his research in the field of urban hydrology has focused on conceptual modelling, performance analysis, asset management and innovative metrology, applied to urban drainage networks and nature-based solutions for stormwater management. Former chairman of the Graie association, he is an ambassador for the European Climate Pact, vice-chair of the international research group on Urban Drainage Asset Management, and of the working group on nature-based solutions of the Water Europe association. He heads up the French office of the Franco-Australian Association for Research and Innovation, and is co-chair of the Novatech international conference on stormwater management.

0000-0001-8836-3970

François Clemens-Meyer
Department of Civil and Environmental Engineering Norwegian University of Science and Technology (NTNU), NO-7491 Trondheim, Norway; SkillsInMotion B.V.

François holds an MSc in hydraulic and in sanitary engineering at Delft University of Technology (TUD); while working with Witteveen + Bos and DHV consulting engineers he finished his PhD thesis (TUD) on monitoring and model calibration in urban drainage; between 2000 and 2021 he was appointed professor on urban drainage at TUD which he combined with working as head of department (Witteveen + Bos) and from 2013 onwards as senior researcher with Deltares. At present he is adjunct professor with The Norwegian University of Science and Technology in Trondheim, a position he combines with running his small company SkillsInMotion B.V. He supervised 25 PhD Students (TUD, Radboud University Nijmegen, INSA Lyon), was (co)author of >200 scientific publications (>150 Journal articles) and was (co)editor on a recent comprehensive IWA publication on applied metrology in the field of Urban Drainage.

0000-0002-5731-0582

Franz Tscheikner-Gratl
Department of Civil and Environmental Engineering Norwegian University of Science and Technology (NTNU), NO-7491 Trondheim, Norway

Franz Tscheikner-Gratl has a Masters in Civil Engineering (2011) and a PhD in Technical Sciences (2016) from the University of Innsbruck. After his PhD he was part of the Quantifying Uncertainty in Integrated Catchment Studies (QUICS) FP7 Marie Curie Initial Training Network (ITN) at TU Delft. In 2019 he joined the Water and Wastewater Group at NTNU. In 2013 he was awarded with the Eduard-Wallnöfer Prize of Appreciation of Tyrolean Industry and in 2016 with the Award of Excellence of the Austrian Federal Ministry of Science, Research and Economy. He is currently secretary of the working group for Urban Drainage Asset Management (UDAM) of the Joint committee on Urban Drainage (JCUD), where he is also a part of the leadership team. Furthermore, he is an active reviewer for several scientific journals and a member of the editorial board of Urban Water Journal.

iD 0000-0002-2545-6683

Bert van Duin
City of Calgary City & Regional Planning, Calgary, Canada; University of Calgary, Canada and University of Alberta, Canada

Bert van Duin has over 35 years of experience in the analysis, planning, design and management of urban drainage related projects across Canada. He is the Drainage Technical Lead at the City of Calgary Water Utility, guiding the evolution of Calgary's stormwater practice, coaching and mentoring staff, and performing QA/QC on drainage related topics. He also guides the Water Utility's efforts in stormwater related research, overseeing its Drainage Knowledge Program. In addition, he is adjunct professor at the Universities of Calgary and Alberta and chairs the 'Green Infrastructure for Stormwater Management' Committee under the auspices of the Canadian Standards Association Group. He is a founding member and Past President of the Alberta Low Impact Development Partnership.

iD 0000-0002-0416-1800

Authors

Maria do Céu Almeida
Laboratório Nacional de Engenharia Civil, Lisbon, Portugal

Civil Engineer, MSc in Operations Research and Systems Engineering, PhD Civil and Environmental Engineering and researcher at the Urban Water Systems Division, National Laboratory of Civil Engineering, Portugal since 1993. Research experience in the urban water systems and services, including hydraulics and biochemical processes modelling and analysis, asset management, inspection and condition assessment, monitoring, reliability and risk management, integration of blue/green components in urban areas, water-energy nexus, resilience of urban systems. Member of CEN/TC165/WG22 on sewer systems outside buildings since 1998, and of the Portuguese ISO TC 262 and TC 292 mirror committees on risk management and security and resilience.

iD 0000-0001-8488-2474

Steve Auger
Lake Simcoe Region Conservation Authority

Steve Auger, MSc, PEng., is a water resources engineer and the coordinator of stormwater management with the Lake Simcoe Region Conservation Authority (LSRCA). He has been providing leadership and coordination for various LSRCA and associated STEP-Water partnership research and innovation projects, training, and other outreach strategies to advance stormwater management (SWM) including Low Impact Development (LID) practices. Steve's experience in practice spans over a decade as a consultant before joining LSRCA in 2014. He is also a certified professional in erosion and sediment control (C.P.E.S.C.) and a certified teacher in Ontario (OCT). Steve is honoured and appreciative that LSRCA and partners (including the Ontario Ministry of Environment, Conservation and Parks) work on municipal SWM maintenance prioritization and inventory asset management is being featured in this IWA handbook.

Mahdi Bahrami
Norwegian University of Science and Technology (NTNU), Department of Civil and Environmental Engineering, Water and wastewater engineering (VA) group, Trondheim, Norway

I have a master's degree in water resources engineering from the University of Tehran, Iran. My master's thesis was on simulation and optimization of Green Infrastructure. I have worked as a researcher with Iran's universities and public water sector. Currently, I am doing a PhD at NTNU, working on asset management of Green Infrastructure. I am interested in performance assessment, maintenance issues, and inspection of Green Infrastructure.

0000-0003-1399-3008

Sylvie Barraud
INSA Lyon, DEEP, UR7429, Department of Civil Engineering and Urban Planning, 69621 Villeurbanne, France; OTHU; H2O'Lyon

On the one hand, her research focuses on i) the long-term monitoring and modelling of water and pollutant transfer in urban stormwater control measures based on retention and infiltration processes at different spatial scales (from end-of-pipe systems to local nature-based solutions) and on the other hand ii) on the development of multi-criteria decision support systems to assess their performance and sustainability.

0000-0002-9473-4034

Jean-Baptiste Besnier
Unité Gestion des Patrimoines, Service Pilotage Assainissement et GEMAPI, Direction de Cycle de l'Eau, Métropole de Lyon, France

As a young engineer with a engineering degree in Geographic Information: Spatial Analysis and Remote Sensing from the Ecole Nationale des Sciences Géographiques, I've been working on improving our knowledge of the materials and years of laying the sewer network for the Métropole de Lyon. All thanks to graph path algorithms and machine learning tools, I'm now working on decision-support tools for prioritising inspections and rehabilitation operations.

0009-0006-1744-5809

Floris Boogaard
Deltares; Spatial Transformations at Hanze University of Applied Sciences Groningen

As a senior researcher at Deltares Floris Boogaard focusses on climate adaptation and optimizing the long term efficiency of nature-based solutions. As a professor of Spatial Transformations at Hanze University of Applied Sciences in Groningen he launched tools as climatecafe.nl and climatescan.org. With over 25 years' experience in consultancy, research and management he works to make cities more resilient to climate change. Research fields include: climate adaptation, nature-based solutions, interdisciplinary monitoring, citizen science, water supply, stormwater quality and management and urban planning.

0000-0002-1434-4838

Emanuela Bosco
Eindhoven University of Technology

Emanuela Bosco is an Associate Professor in the Chair of Applied Mechanics at Eindhoven University of Technology (TU/e). Her research focuses on advancing multi-scale and multi-physics models, integrating them with dedicated experiments for model calibration and validation, and applying them across various significant applications. These include the multi-scale modelling of granular materials, fibrous materials, chemically-induced degradation of concrete, and the multi-physics degradation of historical art objects. In 2023, she was awarded an ERC Starting Grant from the European Research Council. https://www.tue.nl/en/research/researchers/emanuela-bosco

0000-0003-1985-2150

Bert Bosseler
IKT – Institute for Underground Infrastructure, Gelsenkirchen, Germany

Bert Bosseler studied civil engineering at the Ruhr University Bochum, where he completed his doctorate with distinction on the behaviour of deformed wastewater pipes. He began his career in water management at Emschergenossenschaft, one of Germany's largest water management companies. In 2000, Bosseler took over the scientific management of the IKT – Institute for Underground Infrastructure in Gelsenkirchen, Germany, a position he still holds today. In 2010, he habilitated at the University of Hanover and was awarded the Venia Legendi for 'Underground Sewer and Pipeline Construction'. Since 2012, he has also been a lecturer for pipeline maintenance and network management at the Ruhr University Bochum. As a delegate and expert at DIN, CEN and ISO, Bosseler is involved in international standards for water management and smart infrastructure. He is the lead topic spokesperson for 'Cities and Infrastructure' at the Johannes Rau Research Foundation (JRF). www.bosseler.eu

0000-0001-5413-7043

Liam Carson
University of Alberta, Canada

Liam R. Carson is an environmental health scientist at the University of Alberta, where his research is currently focused on the microbial water quality of stormwater, and how it can be safely and viably used as an alternate source water to address growing water scarcity. Liam received a Master of Science in Environmental Health Sciences at the University of Alberta (2023), where he previously received a Bachelor of Science (Hons) in Immunology and Infection (2020).

Gilles Chuzeville
Unité Gestion des Patrimoines, Service Pilotage Assainissement et GEMAPI, Direction de Cycle de l'Eau, Métropole de Lyon, France

Gilles Chuzeville is head of the wastewater and GEMAPI assets strategy team for 3 years and formerly GIS administrator for the Water Cycle Division.

Shamsuddin Daulat
Norwegian University of Science and Technology (NTNU), NO-7491 Trondheim, Norway

Shamsuddin has a Master's in Hydro Science and Engineering from TU Dresden, Germany, and a PhD in Civil Engineering from NTNU, Norway. He is currently working in Klepp Municipality as a water and wastewater engineer. Areas of interest include urban water infrastructure, machine learning and its applications in water infrastructure and decision support.

0000-0001-7385-4195

Cornelis de Haan
De Haan Water Solutions

A civil engineer, Cornelis de Haan graduated from TU Delft in 2002 in.the field of sewer management. He has 20 years of experience in sewer management. He is an independent consultant and project manager in knowledge projects for pipelines and sewer management.

https://www.linkedin.com/in/cornelis-de-haan-17981612/

Gersende Fernandes
Univ Lyon, Université Claude Bernard Lyon 1, CNRS, ENTPE, UMR 5023 LEHNA, F-69518, Vaulx-en-Velin, France

A Doctor in Environment (Soil Physics), I worked during my PhD on infiltration of stormwater and filtration of pollutants in urban soils. I developed an approach with an infiltrometer, a radar, and nano-tracers, to mimic some nano-pollutants' behaviour during infiltration. Currently, I continue working on developing this method and its application in different fields. I have a passion for scientific mediation, arts, and cultures. The social and environmental issues are also important subjects for me.

0000-0001-7419-7377

Fabian Funke
University of Innsbruck, Department of Infrastructure, Unit of Environmental Engineering, Technikerstr. 13, 6020 Innsbruck, Austria

Fabian Funke is a research associate at the University of Innsbruck and works mainly on hydrodynamic flood modelling, blue-green infrastructures and the influence of malfunctions on urban drainage systems. He studied geography at the FU Berlin with a focus on applied physical geography. https://www.uibk.ac.at/umwelttechnik/staff/index.html.de

0009-0007-7657-3509

Marcel Goerke
IKT – Institute for Underground Infrastructure, Germany

Marcel Goerke has a Master's degree in Waste and Water Engineering (environmental Engineering) at RWTH Aachen University, and is a senior Research Fellow at IKT. Since 2016, he is head of the state-approved test centre for flow measurement at IKT. He is a member of the Expert Committee 'Construction Products and Types of Construction for the Treatment and Infiltration of Precipitation Runoff Containing Mineral Oil' and 'Plants for the Limitation of Hydrocarbons in Wastewater Containing Mineral Oils with a Biodiesel, Biofuel and Ethanol Content' at the German Institute for Construction Engineering (DIBt), Berlin. www.linkedin.com/in/marcel-goerke

0000-0001-8514-6487

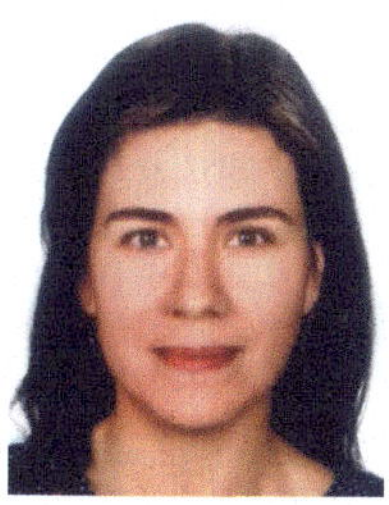

Nathalie Hernandez
STEIN Infrastructure Management, Germany

Nathalie Hernandez is a Civil Engineer, with a Master's in water systems and a PhD In Engineering from Pontificia Universidad Javeriana (Colombia). During her Master's and PhD studies, she worked on developing sewer asset management tools to support the making decision in developing countries. She has also participated in different international and Colombian events as a speaker, and is the author of different articles in international journals. Currently, she works at Stein Infrastructure Management in Germany as a Project Engineer, one of the pioneer specialized consultant companies in sewer/water asset management. Additionally, she belongs to the UDAM Committee.

0000-0001-5084-7937

Martin Gilje Jaatun
SINTEF Digital, Trondheim, Norway; University of Stavanger, Norway

Martin Gilje Jaatun is a Senior Scientist at SINTEF Digital in Trondheim, Norway. He graduated from the Norwegian Institute of Technology (NTH) in 1992, and received a PhD in critical information infrastructure security from the University of Stavanger in 2015. He is an adjunct professor at the University of Stavanger, and serves on the editorial board of several international journals. Previous positions include scientist at the Norwegian Defence Research Establishment (FFI), and Senior Lecturer in information security at the Bodø Graduate School of Business. His research interests include software security, security in cloud computing, and security of critical information infrastructures. He is vice chairman of the Cloud Computing Association, vice chair of the IEEE Technical Committee on Cloud Computing, an IEEE Computer Society Distinguished Visitor, and a Senior Member of the IEEE. http://jaatun.no/papers

0000-0001-7127-6694

Francine Kelly-Hooper
GHD Consulting Ltd

Dr Francine Kelly-Hooper is a subject matter expert (SME) in the field of stormwater management facility sediment chemistry and beneficial reuse and petroleum hydrocarbon studies. Francine completed her PhD at the University of Waterloo, Ontario, Canada in 2014. She has operated the Canada-wide stormwater management sediment survey since 2004, which currently includes over 400 sediment samples collected from over 130 facilities. Francine's ongoing collaborations with global regulators, municipalities and industry has lead to increasing support for beneficial reuse alternatives to typical waste disposal practices. Her mission is to enhance circular economies through the safe beneficial reuse of waste materials, including stormwater management facility sediment.

Karsten Paul Kerres
Institute of Smart City Engineering FH Aachen University of Applied Sciences (ISCE)

The ISCE was founded in April 2023 as an interdisciplinary institute of Aachen University of Applied Sciences. Smart City stands for holistic and sustainable development concepts in urban and rural settlement areas, including in the areas of grid-bound infrastructures, water-sensitive urban development, green-blue infrastructures, land management and digitization. Professor Dr-Ing. Karsten Kerres was appointed to the Department of Network Engineering (FB2), which deals with the planning, construction and operation of underground infrastructures (electricity, district heating, gas, water, wastewater), at FH Aachen in September 2011 and is also Chairman of the ISCE Board. As part of his scientific and advisory activities, Professor Kerres deals with the forecast-based strategic and operational maintenance of water supply networks, sewer systems and green-blue infrastructures.

0000-0003-4752-1893

Manfred Kleidorfer
University of Innsbruck, Technikerstrasse 13, 6020 Innsbruck, Austria

Manfred Kleidorfer, a distinguished professor at the University of Innsbruck, is renowned for his expertise in sustainable urban water infrastructure. His pioneering work focuses on developing innovative, nature-based solutions and efficient rainwater management strategies. With a strong background in environmental engineering, Kleidorfer excels in numerical modelling of urban drainage and supply systems, emphasizing resilience in the face of climate change and urbanization. His research contributions significantly impact integrated sewer system planning. Beyond academia, he is a passionate educator, mentoring the next generation of environmental engineers. https://www.uibk.ac.at/umwelttechnik/staff/personal/manfred_kleidorfer.html.de

0000-0002-4001-1711

Ingo Kropp
3S Consult GmbH, Germany

Diploma in Civil Engineering – Urban and transportation planning, TU Dresden, Germany. 2000 – 2006 Research Assistant at Chair of Urban Engineering, TU Dresden, Germany; 2003–2011 Founder and CEO of Baur & Kropp, Dresden, Germany. Since 2011 Associate Director and partner, Head of Asset Management, 3S Consult GmbH, Germany. Work fields and experience: Asset management for urban infrastructure networks (water, sewer, gas, heating/cooling, electricity), Development of asset management simulation software KANEW 3S. Member of working groups in DVGW and AGFW for development of German standards on asset management/rehabilitation planning. Regular participation in research projects. Work experience in consultancy projects in Germany, Europe and North America.

Jeroen Langeveld
Delft University of Technology, the Netherlands; Partners4UrbanWater, the Netherlands

Dr.Ir. Jeroen Langeveld is Associate Professor of Urban Drainage at TU Delft since 2010. In 2014 he founded the company 'Partners4UrbanWater', where he is mainly involved in (applied) research projects, such as SARS-CoV-2 monitoring in sewage. He is a very active member of the scientific community, exemplified by his position as chair of the IWA/IAHR JCUD (2017–2021) and his appointment as member of the Strategic Council of IWA in December 2021 and IWA Fellow in 2022. His research projects, covering the broad field of urban drainage and linkages to many related fields such as urban flooding, asset management, public health, ecology, modelling and monitoring, contributed both to progress the scientific state of the art in the field of urban drainage and to facilitate the uptake of new knowledge by the Dutch urban drainage sector.

ORCID 0000-0002-0170-6721

Yves Le Gat
INRAE, France – National Research Institute for Agriculture, Food and Environment, ETTIS Research Unit – Environment, Territories in Transition, Infrastructures, Societies

Yves Le Gat has been working (from 1995 until his retirement in 2023) at French National Research Institute for Agriculture, Food and Environment (INRAE), as a researcher in the field of Probabilities and Statistics applied to Water and Drainage Infrastructure Asset Management. His PhD thesis work, defended in 2009, focuses on a Linear Extension of the Yule Process applied to repeated failures of drinking water pipelines.

ORCID 0000-0002-1264-0019

David Lembcke
Lake Simcoe Region Conservation Authority

David Lembcke is the Manager of Environmental Science and Monitoring. He oversees a multidisciplinary team responsible for monitoring Lake Simcoe and its watershed and investigating issues affecting its ecological health. David's interest in urban stormwater lies in understanding and mitigating its impacts on receiving waters. This includes monitoring the function and performance of stormwater management features and has conducted a number of studies into the drivers of stormwater pond function. This has been translated into practical methodologies and resources to improve stormwater pond operation and maintenance, including the development and delivery of a municipal stormwater pond maintenance training course.

Mathieu Lepot
Un Poids Une Mesure, Lyon, France

After being a researcher for 16 years in France and the Netherlands, I left academic world late in 2023 to work in a water utility and run my R&D company. I conducted research in metrology and asset management.

Rianne Luimes
Eindhoven University of Technology, Department of the Built Environment, Chair of Applied Mechanics, Eindhoven, the Netherlands

Rianne Luimes obtained her MSc degree in Structural Design, at the Eindhoven University of Technology, the Netherlands, in 2011. After that she worked for 3 years as a structural engineer at an engineering and consultancy company. In 2014, she started her PhD research on climate-induced damage in oak museum objects at the chair of Applied Mechanics of the Department of the Built Environment of the Eindhoven University of Technology and obtained her PhD degree in 2019. From 2018 until 2022 she continued working at the chair of Applied Mechanics as a postdoctoral researcher on the condition assessment of concrete sewer pipes. Since 2022 she works as a scientific researcher at the Rijksmuseum Amsterdam and as a guest researcher at the chair of Applied Mechanics.

iD 0009-0007-9734-5337

Konstantinos Makris
Partners4UrbanWater

Konstantinos (Kostas) Makris works currently as a consultant at Partners4UrbanWater. Within his function, he cooperates with municipalities, water boards and research institutes on projects related to asset management and monitoring of urban water systems. Konstantinos obtained his PhD degree from TU Delft based on his research on material aging and acoustic inspection of plastic pipes in urban water systems. He has also earned an MSc in Water Management from TU Delft, during which he focused on inspection techniques for stormwater systems, and an MEng in Civil Engineering from UTH. Most of his research findings can be found in published articles in peer-reviewed journals.

iD 0009-0002-1275-913X

Magnus Moglia
Centre for Urban Transitions, Swinburne University of Technology, Australia

Magnus Moglia is an Associate Professor in Urban Sustainability at the Centre for Urban Transitions at Swinburne University of Technology, Australia. Magnus has published more than 70 peer-reviewed journal articles and worked closely with industry and government on a range of sustainability-related topics, including water management, infrastructure management, climate adaptation, decarbonisation strategies, and infrastructure resilience. He is a transdisciplinary researcher, focusing on how solutions that help achieve sustainability goals can be equitably and effectively promoted and implemented in cities and regions. He has a background in Physics, Mathematics, Engineering, and Environmental Management, and explores social systems to drive transformative change.

iD 0000-0002-8290-610X

Sheida Moin
Lake Simcoe Region Conservation Authority, Ontario, Canada

Dr Sheida Moin is the Stormwater Management Specialist at the Lake Simcoe Region Conservation Authority. She earned her PhD from NC State University, where she focused on near-continuous monitoring and real-time control. With experience in both private consulting and the public sector, she contributes to the improvement of stormwater management in the Lake Simcoe watershed. In her role, Dr Moin manages the watershed's stormwater asset database and oversees inspections for partner municipalities. She actively leads data analysis efforts for on-site monitoring and innovative stormwater management approaches while also sharing her knowledge through university-level teaching and training initiatives.

Marius Rokstad
Department of Civil and Environmental Engineering, Norwegian University of Science and Technology, Trondheim, Norway

Marius Møller Rokstad holds a Master's in Civil and Environmental Engineering (2012) and a PhD in Water and Environmental Engineering (2016), both obtained at the Norwegian University of Science and Technology (NTNU). After completing his PhD, Marius has worked as a consultant in Asplan Viak in Trondheim (2016–2018), and as a postdoctoral researcher at NTNU (2018–2020). In 2021 he started as an Associate Professor at the Water and Wastewater group at NTNU. His research interests include infrastructure asset management, modelling of urban water systems, water loss mitigation, smart water systems and digitalisation.

0000-0003-4650-3554

Tone Merete Muthanna
Norwegian University of Science and Technology, Trondheim, Norway; Luleå University of Technology, Luleå, Sweden

Professor in the Civil and Environmental Engineering Department, at the Norwegian University of Science and Technology, and head of the Water and Wastewater research group. Research interest and background in urban stormwater management, performance of green infrastructure in a changing climate, modelling of green infrastructure, and water quality of urban stormwater.

0000-0002-4438-2202

Norman Neumann
University of Alberta, Edmonton, Alberta, Canada

Dr Neumann is a Professor in the School of Public Health at the University of Alberta (since 2007). He received his PhD at the University of Alberta in 1999 in. the Department of Biological Sciences and subsequently assumed a postdoctoral position in the Department of Civil and Environmental Engineering. From 2002–2002, he worked as a research scientist at the National Water Research Institute with Environment Canada, but was subsequently recruited to the University of Calgary as an Assistant Professor in the Faculty of Medicine in 2002. His research centers on 'One-Water', with a particular focus on anthropogenic and zoonotic waterborne pathogens, tracking host sources of fecal pollution in the environment, bacterial ecophylogenetics, evolution of water-treatment and antibiotic resistant microbes, and public health risks associated with exposure to, or consumption of, contaminated water.

0000-0003-2708-4964

Martin Oberascher
Unit of Environmental Engineering, Department of Infrastructure Engineering, Faculty of Engineering Sciences, Universität Innsbruck Technikerstrasse 13, 6020 Innsbruck, Austria

I am a researcher in urban water management at the Unit of Environmental Engineering at the University of Innsbruck. My research interests are on information and communication technologies and applications for smart (water) cities including real-time control to achieve sustainability and to increase urban resilience against future challenges.

0000-0003-3968-4684

Aad Oomens
Strategic policy advisor, Waterboard De Dommel

Awarded his PhD from TU Delft on decision-making processes in sewer management. Working in the field of sewerage and water policy for over 30 years. Initially closely involved in the design, improvement and decision-making processes of sewer management and asset management. Directly involved in many national studies on sewerage and made many contributions to national guidelines and manuals. For the last 6–7 years, working as a strategic policy advisor at water board De Dommel from the team Policy and Innovation. In this role, also involved in the implementation of asset management at the water board.

Eric Oosterom
Stichting RIONED

Educated originally as a research chemist, in the past two decades Eric has been active in the field of urban drainage, currently as the programme manager Asset Information and Standardization. Responsible for the development, maintenance and adoption of the Dutch urban drainage semantic model (https//data.gwsw.nl).

0009-0004-3403-250X

Bardia Roghani
Norwegian University of Science and Technology, Norway and Norwegian University of Life Sciences, Norway

Bardia is a dynamic PhD graduate from the School of Civil Engineering at the University of Tehran, where he honed his expertise in urban infrastructure management. Embarking on a new academic journey in Norway since 2022, he brings over a decade of rich academic and practical experience to his endeavours. His project portfolio encompasses 'Nature-based Solution and Grey Infrastructure Asset Management', 'Integrated Water Resource Management', and 'Risk assessment for Interdependent Urban Critical Infrastructures'. Recognized with the UNESCO prize in the Blue Sky Ideas competition, he actively contributes to industrial innovation. Stay connected with Dr Roghani through his Twitter profile: https://twitter.com/BardiaRoghani

0000-0001-5435-6748

Frits Rooyackers
Eindhoven University of Technology, Eindhoven, the Netherlands

Frits, a PhD researcher at the Eindhoven University of Technology, focuses on the structural degradation of concrete sewer pipes under sulfuric acid corrosion. Proficient in applied mechanics, his work combines numerical analysis with experiments, contributing to a better understanding of sewer pipe durability challenges. Also with a background in innovative structural design for lightweight structures, parametric modelling and dynamics, Frits aims to offer practical insights for sustainable urban infrastructure, finding a balance between numerical simulations and real-world experimental validation.

Jon Røstum
Volue

Dr Jon Røstum is a Chief strategist at Volue Infrastructure responsible for R&D&I activities related to developing cloud-based solutions for water utilities. Up to 2015 he was a senior researcher at SINTEF being involved in a wide variety of projects covering the whole urban water cycle and both technical and non-technical aspects.

Irene Scheperboer
Senior Research Fellow, IKT, Arnhem, the Netherlands

Irene has obtained a PhD at the Eindhoven University of Technology on the topic of condition assessment of concrete sewer pipes through an integrated experimental-numerical approach. Thereafter, she has started as a senior research fellow at IKT – Institute for Underground Infrastructure. https://www.linkedin.com/in/irenescheperboer/

0000-0003-3959-7345

Lisa Scholten
Delft University of Technology, Delft, the Netherlands

Lisa is an Associate Professor at the Faculty of Technology, Policy and Management at TU Delft, the Netherlands, with a special interest in improving decision making and decision competence in complex systems, especially urban water systems. She holds an Engineering Degree in Water Resources Management with a minor in East Asia/China studies from TU Dresden, Germany, a PhD from ETH Zürich, Switzerland, and an executive MSc in Behavioural Science from LSE, United Kingdom.

0000-0002-4487-758X

Will Shepherd
University of Sheffield, Department of Civil and Structural Engineering, UK

Will is a Civil Engineer with twenty years of post-doctoral experience in the field of urban drainage. His main interests lie in the performance of urban drainage systems in terms of flow and quality, including asset performance modelling, hydraulic modelling, laboratory and field testing, and application of artificial intelligence in urban drainage. He is currently researching the application of autonomous robotics for inspection of urban drainage pipes. Will's contributions were supported by the UK's Engineering and Physical Sciences Research Council (EPSRC) Programme Grant EP/S016813/1. For publications and further details, see https://uk.linkedin.com/in/shepherdwill

0000-0003-4434-9442

Joakim Skjelde
COWI AS, PO Box 4220 Torgarden, N-7436 Trondheim, Norway

Joakim Skjelde is a water- and wastewater engineer at the engineering consultancy COWI, working at the department of water and municipal engineering. He holds an MSc in urban water systems from NTNU in Trondheim, with a specialization in urban drainage systems. His main research interests include, but are not limited to urban drainage asset management, hydroinformatics, hydrodynamic modelling, machine learning, and statistical and mathematical modelling. Joakim has an analytical mindset, and he is always seeking to find innovative and effective solutions on complex problems. Through his work, he strives to improve the resilience and efficiency of urban water infrastructure, by implementing cutting-edge methods. https://www.linkedin.com/in/joakim-skjelde/

Akke Suiker
Eindhoven University of Technology, Eindhoven, the Netherlands

Since 2012 Akke Suiker is Full Professor at the Eindhoven University of Technology, Department of the Built Environment, leading the group of Applied Mechanics. He obtained his MSc degree in Civil Engineering (Cum laude, 1995) and his PhD degree in Applied Mechanics (Cum Laude, 2002) at the Delft University of Technology. He had 1-year post-doc visits to the University of Massachusetts, USA and Cambridge University, UK. His research interests relate to modelling of failure and deformation of materials and multi-scale and multi-physics modeling of solids. Recent applications are 3D printing, historical museum objects, wind energy turbines, sewer systems, subsurface erosion, among others. He has published over 80 peer-reviewed journal articles in the fields of Theoretical and Applied Mechanics, Computational Mechanics, Material Science, and Applied Engineering and was elected as 'Best Teacher of the Delft University of Technology, 2009–2010'. Several of his journal articles have led to registered patents.

Rita Maria Ugarelli
SINTEF COMMUNITY, Oslo, Norway and Norwegian technical university (NTNU), Trondheim, Norway

Rita Ugarelli is Chief Scientist at SINTEF Community and adjunct professor at the Norwegian technical university in Norway. She holds a PhD in Civil Engineering with a focus on urban water Infrastructure Asset Management (IAM). Her field of expertise covers IAM applied to urban infrastructure towards sustainable maintenance and rehabilitation planning based on risk management, resilience analysis, sustainability assessment, reliability analysis, and condition assessment. Most recently, she has coordinated the H2020 STOP-IT project, she led the work package dealing with interoperability and cybersecurity in the Digital Water City project and with the development of a 'water smartness' framework in the H2020 B-WaterSmart project. Rita is Associate Editor of the Urban Water Journal, co-leader of the Water Europe vision leader team on Digital Water and member of the European Commission expert group on critical infrastructure protection (CERIS expert group).
iD 0000-0002-2096-8591

Marco van Bijnen
M. van Bijnen Advies, Drunen, the Netherlands

From 1994 to 1999 Marco was employed at Oranjewoud in Oosterhout, specialising in hydraulic modelling of sewer systems. In 2000 he started working at ARCADIS consultancy company. At ARCADIS, he focused on the field of sewer monitoring networks. From 2002 to 2015 Marco was employed at the municipality of Utrecht (the Netherlands) as an advisor on urban drainage. In 2018 he finished his PhD at Delft University of Technology. The research focussed on the impact of sewer condition on the performance of sewer systems. Since 2015 Marco has been an independent consultant in the field of urban drainage at M. van Bijnen Advies.

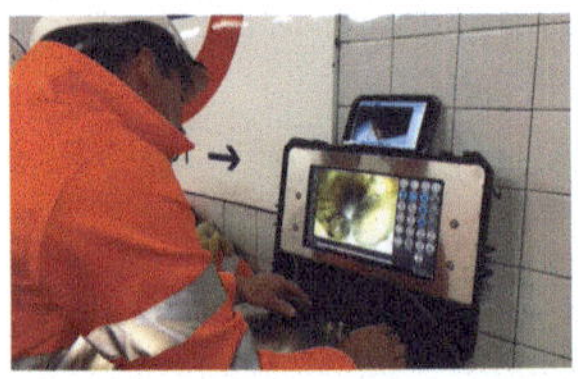

Ferry van der Valk
Founder and Managing Director Sewer Robotics, the Netherlands

I was literally born into the sewer infrastructure management industry – the third generation in my family – and I pioneered the first multi-functional robot system with UHP water jet cutting two decades ago. Sewer Robotics' mission is clear – to revolutionize the sector by developing innovative technology. At Sewer Robotics, we've seamlessly integrated HD inspection into the cleaning process developing video nozzles that upload data to the Sewer Analytics cloud. Here, AI efficiently codes defects and prioritizes maintenance and rehabilitation. Our versatile robots then step in to perform these prioritized tasks, from UHP jetting and obstructions removal to installing spot repair liners and cutting laterals, shaping the future of sewer management. Explore the future of drainage asset management at https://www.sewerrobotics.com/

Martijn van der Valk
vandervalk+degroot BV, ABC-Westland 231, 2685 DC Poeldijk, the Netherlands

Project experience regarding Dutch and international tenders for road infrastructure, urban sewer and drain water systems, dredging, offshore and oil and gas industry. Key interests are commercial and organizational projects involving civil and water related constructions, with geotechnical and cost engineering challenges.

Wouter van Riel
infralytics, Reduitlaan 33, 4814 DC, Breda, the Netherlands

Awarded his PhD at Delft University of Technology on decision-making processes for sewer system management, focussing on the role of information in a multi-actor environment. After several years at consultancy firms, Wouter became an independent consultant with his main focus on data engineering and developing and implementing data models in the electric energy and urban drainage sector.

List of Acronyms and Abbreviations

AHP	Analytical hierarchy process
AI	Artificial intelligence
AM	Asset management
AMP	Asset management plan
ANN	Artificial neural network
API	Application programming interface
ATU	Afzetting Traceer Unit (Dutch)/Deposition Tracking Unit (English)
AVMA	American Veterinary Medical Association
BPNN	Back propagation neural network
CAHI	Canadian Animal Health Institute
CAPEX	Capital expenditure
CAs	Conservation authorities
CBA	Cost–benefit analysis
CCTV	Closed-circuit television
CEA	Cost-effective analysis
CEN	European Committee for Standardization
CoF	Consequence of failure
CSO	Combined sewer overflow
CTMC	Continuous time Markov chain
CUA	Cost utility analysis
DAS	Distributed acoustic sensing
DIY	Do it yourself
DNA	Deoxyribonucleic acid
DSS	Decision support system, distributed stress sensing
DTMC	Discrete time Markov chain
DTS	Distributed temperature sensing
DW	Drinking water
DWA	Deutsche Vereinigung für Wasserwirtschaft, Abwasser und Abfall – German for 'German Association for Waste, Wastewater and Waste'
DWF	Dry weather flow
EC	European Commission

EMAR	Electro-magnetic acoustic resonance
EMI	Electro-magnetic induction
ERP	Enterprise resource planning
ERT	Electrical resistivity tomography
ESC	Environmental stress cracking
EU	European Union
FEM	Finite-element method
FELL	Focussed electro-leak location
FIB	Fecal indicator bacteria
FMEA	Failure mode and effect analysis
FMECA	Failure mode, effects and criticality analysis
FMF	Fuzzy membership function
FN	False negative
FOG	Fat, oil and grease
FP	False positive
FRP	Fibre-reinforced plastic
FSIT	Full-scale infiltration testing
GIS	Geographic information system
GPR	Ground penetrating radar
GW	Ground water
HAZOP	Hazard and operability study
HDPE	High-density polyethylene
HDTMC	Homogeneous discrete time Markov chain
HTTP	Hypertext transfer protocol
IAHR	International Association for Hydro-environment Engineering and Research
IAM	Institute of Asset Management
ICG	Internal condition grading
IPCC	Intergovernmental Panel on Climate Change
IR	Infrared
IRR	Internal rate of return
ISO	International Organization for Standardization
IWA	International Water Association
KPI	Key performance indicator
LACDPH	Los Angeles County Department of Public Health
LIDAR	Light detection and ranging
LoF	Likelihood of overall failure
LRC	Log10 reduction credits
LSRCA	Lake Simcoe Region Conservation Authority
MC	Markov chain
MCDA	Multi-criteria decision analysis
MECP	Ministry of Environment, Conservation and Parks
ML	Machine learning
MLR	Multinominal logistic regression
MSCC	Manual of Sewer Condition Classification
MTBF	Mean time between failures
MTTF	Mean time to failure
MZC	Manhole zoom camera
N	Nitrogen
NASEM	National Academies of Science, Engineering, and Medicine

NBS	Nature-Based Solutions
NHDTMC	Non-homogeneous discrete time Markov chain
NPV	Net present value
NRMMC	Natural Resource Management Ministerial Council
OECD	Organization for Economic Co-operation and Development
OF	Optical flow
P	Phosphorus
PAHs	Polycyclic aromatic hydrocarbons
PAS	Performance assessment system
PDCA	Plan/do/check/act
PDEng	Professional Doctorate in Engineering
PE	Polyethylene
PIV	Particle image velocimetry
PP	Polypropylene
PPR	Pipe penetrating radar
PTV	Particle tracking velocimetry
PVC	Polyvinylchloride
QUEST	Quantification of exfiltration from sewers with artificial tracers
RBI	Risk-based inspection
RCG	Rapid crack growth
RCM	Reliability centered maintenance
RDF	Resource description framework
RMSE	Root mean square error
RV	Risk value
SAMP	Strategic asset management plan
SCADA	Supervisory control and data acquisition
SCG	Slow crack growth
SCM	Stormwater control measure
SMART	Specific, measurable, achievable, realistic and time bound
SRM	Sewerage Rehabilitation Manual
SSET	Sewer scanner and evaluation technology
SSO	Sanitary sewer overflow
STEP	Sustainable Technologies Evaluation Program
SUDs	Sustainable urban drainage systems
SWM	Stormwater management
TDR	Time-domain reflectrometry
TFN	Triangular fuzzy number
TWDB	Texas Water Development Board
UCLA	University of California, Los Angeles
UD	Urban drainage
UDAM	Urban drainage asset management
UDO	Urban drainage ontology
URIs	Uniform resource identifiers
USDA	United States Department of Agriculture
UV	Ultraviolet
VCR	Video cassette recording
VSMOW	Vienna standard mean ocean water
WATS	Wastewater aerobic/anaerobic transformations in sewers
WDPT	Water drop penetration time

WFD	Water framework directive
WHG	Wasserhaushaltsgesetz – German for 'Water Resources Act'
WHO	World Health Organisation
WSP	Water safety plan
WWTP	Wastewater treatment plant

doi: 10.2166/9781789063059_0001

Chapter 1

Introduction to urban drainage asset management: if something exciting happens, we've done it wrong!

Frédéric Cherqui[1], François Clemens-Meyer[2], Bert van Duin[3,4,5] and Franz Tscheikner-Gratl[2]

[1]University of Lyon, INSA-LYON, Université Claude Bernard Lyon 1, DEEP, F-69621, F-69622, Villeurbanne, France
[2]NTNU – Norwegian University of Science and Technology, Trondheim, Norway
[3]City of Calgary, City & Regional Planning, Calgary, Canada
[4]University of Calgary, Calgary, Alberta
[5]University of Alberta, Edmonton, Alberta

ABSTRACT

This chapter introduces urban drainage (UD) asset management (AM) emphasizing on the scientific and practical developments over the past few decades. The book collates experiences from scientists and practitioners from around the world. A brief historical overview is provided outlining the journey from the urge for a broad application of UD systems in the 19th century born out of the awareness of public health, to the emergence of the necessity to embrace concepts as sustainability and the application of relatively new approaches that incorporate 'building with nature'. The complexity of the tasks related to AM of urban drainage systems is briefly outlined along with an overview of the content of the remaining chapters.

Keywords: Asset management, urban drainage, sustainability, public health.

1.1 INTRODUCTION

When being responsible for the well-being or proper functioning of any object, system or entity, one needs to be able to always answer at least three seemingly simple questions:

(1) What and where is the 'thing' I am responsible for?
(2) What is the current functionality of the 'thing' I'm responsible for?
(3) In what condition is the 'thing' I'm responsible for?

However simplistic they may seem, answering these three questions, when it comes to the systems deployed for urban drainage (UD), is not trivial by any standard. According to the IWA Urban Drainage Multilingual Glossary (Ellis *et al.*, 2006), *Urban Drainage* is defined as '*the science and technology dealing with the quantification, technical handling and management of all sewage sources, flows and effluents such as: surface runoff, sanitary sewage, infiltration from groundwater and other inflows into sewers, leakage from sewers into the ground (exfiltration), storm water flows and discharges, combined sewer flows and overflows and discharges from sewage treatment works; including their respective effects on the urban area and the receiving waters'.* Being able to answer the above questions *at all times* asks for substantial effort and a sound understanding of hydrology, hydraulics, water quality and public health, materials science, soil science, vegetation, ageing/deterioration processes, inspection technology, decision making, negotiation, legal aspects, guidelines, protocols, risk assessment, socio-economic aspects, policy making, organizational aspects and many more.

Next to this notion of interdisciplinarity, the stage on which the task of materializing the responsibilities as asset manager is played, is anticipated to shift substantially and constantly in the foreseeable future: most of the world population has been living in urban areas since 2007 (World Bank, https://data.worldbank.org/indicator/SP.URB.TOTL.IN.ZS) and this percentage is expected to rise in the future, adding pressure on existing systems and likely demanding expansion and adaptation. Climate change is expected to have a significant impact on the urban environment, worsening for example the 'urban heat island effect' and causing more severe droughts while also inducing increased flooding frequency and severity (Miller & Hutchins, 2017; Nuruzzaman, 2015). Apart from these important autonomous processes, the following circumstances are adding to the complexity of the tasks related to asset management (AM):

- The technical life span of the assets is longer than the time scale on which relevant autonomous processes develop (e.g. urbanization and climate), as such obscuring the possible evaluation of the relation between policy/decision making and effects obtained. In this respect it is noted that during the life span of the bespoke assets many changes in personnel involved in their management occur, limiting the accumulation of experience, unless managerial actions are taken to ensure knowledge and experiences are stored and made accessible over long periods of time (i.e., decades).
- The technical life span of the assets is also much longer than the collective memory of AM staff given the inevitable turnover rate: consistency of AM strategies will depend on successive asset manager backgrounds and a consistent AM strategy of the utilities in question.
- The existing systems are the result of decades of implementation/construction following evolving design rules and changing demands and are very likely not the most effective or efficient solution for the current desired outcomes. Older parts are not immediately updated, if at all. In practice, change can only be pursued piecemeal, due to technical and budget restrictions. As a result, adaptation management can be appropriately described as 'muddling through' (see e.g., Neumann *et al.* (2015).
- Much information on the existing systems has been lost over the years (or has become practically inaccessible); historical records are scarce, incomplete and subject to the vagaries of subjective interpretation (hampering learning from experience).
- UD infrastructure is in many respects 'entangled' with other urban infrastructure like roads, potable water supply systems, district heating, gas and power supply which can be above or below ground systems as are many components of our UD systems. Poor functioning of an element, for example a leaking potable water supply system, may well induce malfunctioning in for example a sewer system.
- Decision making is often influenced by political/societal considerations along with changing rules and regulations and, in many cases, the personal preferences of the people involved.

The realization of the complexity involved in urban drainage asset management (UDAM) has triggered (from the 1990s onward for sewer AM) many research projects on the one hand and the evolution of 'best practices' at often local/regional levels or sometimes at a national level, or within stakeholders' organizations on the other.

It must be acknowledged that local/regional differences in stormwater management approaches are inevitable due to climatological and topographic contexts; however, the fundamental principles to be applied are more generic. Notwithstanding this, the communication between these two approaches is cumbersome at best; scientific literature is (a) far from freely accessible; (b) often very specialistic and narrowly focused on details without consideration of the bigger picture or implementation realities; (c) due to jargon not readily understandable by uninitiated readers; and (d) requires extensive data for implementation that is often not available to a typical utility. Local and even national regulations, protocols and best-practice documents are often formulated in such a manner that the bureaucratic world dictates what the real world should do (with the latter refusing to do so on a regular basis), in the process largely neglecting scientific progress. In return, scientists often fail to identify and come up with solutions for issues encountered by practitioners. These challenges are compounded by language barriers or imprecise differences in terminology that hinder the diffusion of such documents. An overview of the state of the art of sewer AM can be found in Tscheikner-Gratl *et al.* (2020), and an overview of the current limits of blue-green infrastructure AM can be found in Langeveld *et al.* (2022).

The purpose of this book is to provide a comprehensive basis for scientists and practitioners in the field of UDAM, aiming to cover the main international scientific developments and practical experiences over the last 2–3 decades. The texts are written by a selection of collaborating international authors from both practice and scientific affiliations, in strong collaboration with the working group on UDAM (https://udam.home.blog/) of the IWA International Water Association) and IAHR (International Association Hydraulic Research) Joint Committee on Urban Drainage. The ambition is that the book will serve both scientists and practitioners in a balanced manner.

The remainder of this chapter will provide a brief history of the employment of UD systems and the evolution of day-to-day practice with respect to AM. Then, after defining the scope of the book and the systems boundaries applied, a brief overview of the expected challenges ahead is given. It concludes with an outline of the book illustrating the mutual coherence of the individual chapters.

1.2 A BRIEF OVERVIEW OF THE HISTORICAL DEVELOPMENT OF ASSET MANAGEMENT OF UD SYSTEMS

The challenge of adequate fresh water supply and its disposal after usage (as mentioned in ancient religious texts) is an old one. Early civilizations were built on complex systems of water management and hydraulic engineering (Mithen, 2010). It is no coincidence that the oldest advanced civilizations emerged near rivers, which enabled a permanent supply of fresh water, and were later also used for transport. The main examples are the cultures at the banks of the rivers Nile, Euphrates, Tigris, Indus, Yangtze and Huang He. An impressive example of the technical knowledge of ancient cultures is the so-called Indus Valley or Harappan civilization (ca. 3200–1900 BC) (Garbrecht, 1988). Although not the oldest example of urban water infrastructure – UD pipes for rainwater were already in use in Ur, Babylon and Habuba Kabira (ca. 4000 BC) (De Feo *et al.*, 2014; Strommenger, 1980) – it is together with the Minoan Culture (who used the first water distribution pipes in 1500 BC (Walski *et al.*, 2003)), the oldest example of well-organized and structured water infrastructure. Water supply was achieved by wells inside most of the houses; some of the houses also had flushing toilets (Büker, 2000; Garbrecht, 1988). The largest city of the Harappan civilization, Dholavira, depended solely on decentralized solutions (e.g., soak away pits) and had no centralized sewer system (De Feo *et al.*, 2014).

The notion that it takes effort to maintain the functionality of water systems is not new either: although not directly part of UD, but closely related, is the well-recorded aspect of the management of Roman water supply systems. To keep the aqueducts functioning in the Roman Empire, the so-called 'Aquarius' was responsible, a function that resembles the responsibilities of a director of a water supply company nowadays (there is also mention of privately owned enterprises in water supply in the Roman empire, see e.g. Kessener (2017). A great deal about aqueducts, their construction, operation and maintenance has been described in some detail by Frontinus (~100).

With respect to UD less has been recorded. The famous sewer system in Rome, first initiated by the Etruscans around 500 BC, was mainly meant to drain swamps and divert storm runoff. Only later, when the city became the heart of the Roman Empire, public and to a lesser extent, private latrines were connected to the sewer system. The management of the sewer system was, among a long list of other responsibilities, a task for the 'aediles' who were elected officers in the local authorities. This implies that the management of sewer systems has been regarded as a public task for more than two millennia. The next big historical development coincided with the industrial revolution in the 18th century, more specifically in England (Lucas and Walton, 2020). As early as 1835, physicians drew attention to the unhealthy aspects that evolved in the rapidly growing urban areas (see https://www.thoughtco.com/public-health-in-the-industrial-revolution-1221641), and public bodies to address this problem were established. It was however only in the second half of the 19th century that large-scale construction of sewer systems commenced, mainly initiated in response to repeated cholera epidemics and a rapidly emerging notion of the need for hygiene: finally, the relation between diseases like cholera and public exposure to wastewater was the driver to start construction on a large scale. In London, the scientific findings of Dr. Snow (1849, 1854) were instrumental for this; however, in London it took the Great Stink of 1858 to convince politicians of the need to do so. Nevertheless, not till 1875 a Public Health Act was issued providing the legal basis for local authorities to start implementing sewer systems. Major European cities adopted the development of sewer networks, often under the supervision of physicians or health specialists, in direct connection with the hygienic sanitation of the city. For example, Figure 1.1 which presents one of the first maps of the sewer system in Lyon, belongs to the personal archives of Dr. Lacassagne (Faculty of Medicine), who oversaw the construction of the network, in close collaboration with the chief engineer of the Municipal Roads Department (position created in 1854).

The development of sewer systems eventually resulted in a situation in which the presence of UD systems had become standard in even small villages across Europe by the 1950s. At the beginning of the 20th century, pipe networks became the most obvious choice for managing drinking water, stormwater, and sewage. Figure 1.2 illustrates the idea that, already in the early 19th century, underground networks were seen as the solution to many urban infrastructure challenges.

It has to be stated however, that at present, even in densely populated, rich countries not every individual property is connected to an UD system.

Treatment of wastewater prior to discharging it to surface waters started in the early 20th century, becoming standard by the 1970s. Speaking for Europe and the United States, the next big step (1980–2000) was the reduction of uncontrolled discharges of (diluted) wastewater at combined sewer overflows (CSOs) as well as the reduced emission of runoff-related pollutants (e.g., heavy metals, polycyclic aromatic hydrocarbons (PAHs), etc.). In parallel, the notion of the sustainable use of materials and the need to reduce the impact of UD systems on the natural (water) systems emerged resulting in the design of new concepts with respect to sanitation and stormwater management.

Flooding and fear of water has historically been the main driver for stormwater management. It started with the draining of swamps and marshes. In the 19th century, cities exploded in size with stormwater and sewage to be discharged as fast as possible outside the city. The last decades have seen a strong paradigm shift with the emergence of sustainable urban drainage systems (SUDS) and similar solutions with many different names (Fletcher *et al.*, 2015). These solutions were first used as storage capacity (detention basins) to reduce flooding or overflows but have since diversified in terms of composition and expectation. The stormwater paradigm shifts in the last two centuries are well summarized by Reese (2001). Stormwater control measures (SCMs) are nowadays associated with concepts such as 'pervious city', 'water-sensitive urban design' or 'sponge city'. More recently, they have been considered as 'Nature-based Solutions' (NBS) according to the definition from Cohen-Shacham *et al.* (2016): *'actions to protect, sustainably manage and restore natural or modified ecosystems, which address societal challenges (e.g., climate change, food and water security or natural disasters) effectively and adaptively, while simultaneously providing human well-being and biodiversity benefits'*. World Bank (2021) presents an exhaustive list of NBS for urban resilience. Such solutions will be discussed in greater depth in Section 1.4 (Scope, Outline and Coherence of the Book).

Figure 1.1 Map of the sewer network of Lyon, France in 1887. Colours correspond to the different shapes of pipes. The map is taken from the personal archives of Dr. Lacassagne (Bibliothèque municipale de Lyon, 429314).

1.3 CHALLENGES AHEAD IN RELATION TO UDAM

In developed countries access to sewer systems is, certainly in urban areas, nearing 100%. The existing systems are the result of a continuous process of building, rehabilitating and making extensions and small alternations to existing systems. Stormwater management systems always exist, whether designed for that purpose or not. Where specifically in stormwater management the focus of guidelines has been mainly on facilitating the creation of newly built systems, the sustained functionality of these systems requires AM over longer terms. In the course of time the demands put on UD systems have increased or at least have become defined in more detail asking for adaptation of the physical

Figure 1.2 "Le sol de Paris" – Paris's underground: this illustration was published in 1892 (Robida, 1892) and imagined Paris in the 21st century, full of networks: "tele" (television as a kind of videoconferencing), driving force, light, theatre, music, sewer, waters and even food and drinks with the "compagnie alimentaire" (food company).

systems as well as changing the manner they are operated and managed. In this respect, the following, partially overlapping challenges are identified.

1.3.1 Sustainability (in design and management!)

Sustainability can be understood as designing and managing processes and hardware in such a manner that the needs of the present are met without compromising the ability of future generations to meet theirs (*living on the interest without consuming the capital*). As any definition of sustainability this is prone to discussion as we cannot really foresee future needs. However, as far as we can see, the need for protection against flooding and waterborne diseases while protecting the environment to the highest degree possible is not likely to be abandoned in the foreseeable future. So, if we assume these needs are invariants, the meaning of sustainability related to UD boils down to working with materials and applying construction technologies that allow for the reuse of materials, striving for 'zero emission', using only renewable energy and having no or negligible effects on the natural environment. More generally, one important facet of sustainability is based on the circular economy concept (Delgado *et al.*, 2021). This comes with a potentially large number of challenges: for example, the application of recycled materials often does not match existing regulations or may turn out to be of inferior quality in the long run; an example of this is the application of recycled PVC, see Makris *et al.* (2021). Another example is the application of pavement systems that allow for infiltration of stormwater to replenish groundwater in urban areas: existing practices for the maintenance of pavement are primarily focused on traffic demands largely ignoring the specific needs of a desired minimum infiltration capacity. Specifically in this field much research has been published in the past 2–3 decades; however, the comprehensiveness of the field is such that it is not easy to keep track of every detail and distinguish between minor and major gains to be achieved when implementing a suggested solution/technology. With respect to the latter, much depends on the local situation in terms of geography and climate, legislation and available resources. In short, tailor-made solutions are likely only best from a purely technological perspective. At the same time, in practice, this may lead to inhomogeneous situations which can prove to be hard to manage.

1.3.2 Limitations set on systems due to health, environment

Notwithstanding the desire for sustainable solutions, this cannot be taken to the full limit. In the end, any man-made system will have a certain impact on the (direct) environment. If we maintain the demand to reduce the probability of the public getting exposed to contaminated water to avoid the occurrence of waterborne diseases to almost zero, this will require robust systems which could be implicitly interpreted as having been 'overdesigned'. This is not at odds with the principles of sustainability. Similar reasoning can be applied, for example, to systems to reuse stormwater or systems aiming at replenishing groundwater by infiltrating stormwater. As such, the design, management and operation of our UD systems are subject to a delicate balance.

1.3.3 Pace of urbanization

The process of urbanization is resulting in an increased pressure on existing UD systems. In many Third World countries this leads to hardly manageable situations in suburbs that are referred to as 'informal settlements' in which the availability of sanitation and/or drinking water are very poor while having a serious environmental impact as well, see for example Ndoziya *et al.* (2019). In such situations the main concern is to construct the UD systems in the first place rather than the UDAM aspects, although making proper choices in the design phase is omni-important for the sustained effectiveness of these systems. In contrast, urbanization in developed countries is less of a concern as the pace at which it develops is less extreme.

1.3.4 Multi-infrastructure management

In urban areas there are obvious (physical) relations between different types of infrastructure; many elements of UD are situated underground or have some additional degrees of functionality (e.g., parts of public parks that function as detention facilities). Furthermore, in many countries the responsibility of the management of urban infrastructure lies with a (limited) number of (municipal) authorities who may have differing objectives which makes multi-infrastructure AM a challenging task (Daulat *et al.*, 2022). To ensure effective AM these relations must be kept in mind when making decisions on (re)design, rehabilitation or maintenance activities. Although not in the main scope of this book, the importance of multi-infrastructure AM in connection with infrastructure outside the realm of UD is hard to overestimate, especially because of the increasing application of nature-based or 'blue-green solutions' for stormwater management demanding the maintenance of a mix of functionalities and the fulfillment of a multitude of objectives (e.g., water discharge, water retention, recreation, traffic, aesthetic aspects, urban heat island mitigation). Given the relative short time span in which these 'new' solutions have been applied and the knowledge pertaining to their AM needs is still developing, to the extent available, experience with the management of such infrastructures is presented.

1.3.5 Organizational challenges

In most cases municipalities are responsible for managing UD systems, in some cases waterboards are responsible for the regional transport systems (i.e., transporting sewage from various systems to a waste water treatment plant (WWTP)), in other cases waterboards bear the responsibility for the UD systems, regional system and WWTPs. Apart from these public bodies, there are also private parties that should manage their own UD systems; for example, industrial or commercial areas and individual homeowners. With respect to the latter, public bodies can put demands on the design but also on certain aspects of AM. For example, in some part of Germany it is mandatory to inspect the house connections on a regular basis to avoid contamination of groundwater. With respect to industrial or commercial zones, note that these can have a significant impact on the public system to which they normally discharge. The impact can be quantitative (i.e., the amount of runoff generated, or wastewater produced) but the composition of the water may also have a significant impact on the deterioration of public systems. This has resulted in many countries in complex permitting systems setting limits to runoff area and rate, as well as to the amount and composition of this water (runoff as well as wastewater). In some specific cases monitoring is mandatory and a system of regular site visits for inspection is enforced.

In practice, it turns out to be a challenge to coordinate between all involved parties as every party has some limitation in their system and may have specific demands; added to this is the tension that may exist between economical interest (employment versus a loss of functionality in public UD systems) and the organizational challenge in this respect is clear. Another challenge is of a more practical nature: UD system owners, be it public or private, may not have all necessary resources in-house and therefore often work with contracting firms for cleaning ditches, repairs, sewer cleaning, inspection, rehabilitation, gully pot cleaning and so on.

In some situations, mostly in situation of service delegation, this has resulted in the organizations that are ultimately responsible for UDAM lacking 'hands-on' knowledge, hampering their capability to judge the quality of the work done by contractors. These organizations face the challenge of retaining sufficient in-house knowledge (and historical perspective) to remain in control of short- to long-term objectives and be aware of the real performance of their infrastructure.

In the case of private UD system owners, the expertise, or often rather the lack of expertise, may need to govern the complexity of the system components implemented. For instance, individual homeowners typically have little to no expertise. On the other hand, the presence of a property manager may allow for more sophistication where a municipality focuses its education and enforcement efforts on this smaller subset. Specific focus should also be expended on educating tenants where the property owner may be absent, and the tenant may not have all relevant information pertaining to the intended operation of the UD system.

1.3.6 Changing climate

Obviously, climate change has a significant impact on UD systems and the manner in which they need to be managed. Increases in rain intensity and volume, occurrence of long periods of drought, extreme temperatures not only influence the functioning of these systems but also initiate changes in the demands put on them.

Many existing UD systems have been designed based on climate statistics dating back decades (and implicitly assumed to be stationary). To a certain extent, safety margins may have been applied (albeit in many cases not explicitly) to components of the original design. However, the occurrence of urban flooding seems to have increased significantly over the past decade (Feng *et al.*, 2021; Hosseinzadehtalaei *et al.*, 2021; Kourtis & Tsihrintzis 2021; Miller & Hutchins, 2017; Schreider *et al.*, 2000). Temperature rise worsens the urban heat island effect, which can to a certain extent be counteracted by more holistic management of the entire urban water cycle, aided by introducing green infrastructure that has co-benefits beyond stormwater management. At the same time, the selection of the vegetation making up these green components should already account for the anticipated change in climate. Additional effects of temperature rise include the accelerated (bio)chemical processes associated with the corrosion of materials applied in UD systems or the water quality in urban waters. Adapting (existing) UD systems to these changing conditions and demands is by no means a trivial task.

1.3.7 Data-driven asset management

Using modern technology huge amounts of (historical) data can be processed, analysed and used for proper AM. One example is the use of historical CCTV inspection data for validating ageing/deterioration models (see e.g., Caradot, 2019). The use of 'big data' is very appealing although several aspects should be kept in mind in the context of UD:

- The completeness and quality of the available data sets are not always known.
- Data are often gathered with a specific goal in mind, yielding a certain preselection that is not always properly documented. Application of such data for another goal may introduce bias.
- In many cases the time window spanned by the data sets is limited, as such not giving information on long-term processes.

Having said that, it should be acknowledged that the potential of for example AI algorithms to detect patterns in and relations between datasets is beyond question and some first applications have already been published (e.g., Caradot *et al.*, 2018; Hawari *et al.*, 2020; Hernández *et al.*, 2018 or Laakso *et al.*, 2018). An important challenge for UDAM is the provision of (large amounts of) reliable and relevant data to create a sound basis for the application of AI. Adopting (universal) data standards will facilitate the necessary data sharing between practitioners and scientists. Another important challenge concerns the black-box effect: it is crucial that deterioration (but not only) models and processes remain understandable by utility managers.

1.3.8 Interaction with overarching societal demands and desires

As UD is mostly a public responsibility, societal demands and desires and changes in regulatory requirements impact UDAM and changes in long-term decision making. An example of such a large-scale change is the transformation from the initial drive to construct UD systems (driven by public health concerns and the desire for protection against urban flooding) via environmental protection (from about 1970 onwards) to sustainability and reuse of materials (from 1990 onwards). In some cases, systems constructed more than 150 years ago are still in use, be it in an adapted manner in order to cope with the changed demands resulting from the large-scale changes in societal demands and desires. In many cases the existence of such old systems hampers the practical response to changing demands, as the cost to implement the necessary upgrades is deemed too high. For instance, see the challenges with the decision-making pertaining to transforming Paris' combined sewer system, Figure 1.3, into a system of SUDS combined with a sustainable wastewater system.

Nevertheless, such a transformation has to be made sooner or later and will most likely take decades to be put in place. Such transformation asks for careful planning of any reconstruction activities while maintaining the overall functionality during the process. On top of that, it can be expected that new demands or desires will emerge during such transformation, potentially interfering with the original objectives, effectively changing them to ever-moving targets. All in all, this implies that all professionals involved in UDAM need to be flexible in their decision making, always ensuring a degree of 'flexibility' in the system that gets implemented. The latter automatically implies that striving for 'super-optimized' solutions is pointless in the long run. Possibly, it is better to strive for robust

Figure 1.3 Old post-card edited by ND Phot, N°2174: underground Paris – sewer, sanitation department, collection system for the Sébastopol Boulevard – https://commons.wikimedia.org/wiki/File:PARIS_SOUTERRAIN_-_Les_%C3%A9gouts,_service_de_l%27assainissement_;_collecteur_du_Boulevard_S%C3%A9bastopol.jpg

and resilient infrastructure that can be applied using a wide operational envelope employing state-of-the-art (monitoring and control) technology in combination with appropriate documentation of the decisions made over the years.

1.4 SCOPE, OUTLINE AND COHERENCE OF THE BOOK

1.4.1 Some thoughts on stormwater management including SUDS and asset management

Stormwater management is fundamentally different from wastewater collection and transportation in that it includes two components, an underground conveyance or minor system which conveys runoff during the more frequent storm events and an overland conveyance or major system which conveys excess runoff when the capacity of the pipe system is exceeded. These systems are connected at the inlets/storm drains/gully pots/catch basins which may or may not be equipped with a variety of inlet control devices. Over the years, various forms of storage were added to this dual drainage system, ranging from storage provided in sags or depressions created by how roadways or parking lots are graded to rooftop storage, to a variety of underground storage options (e.g., concrete tanks or plastic chambers), to dry or wet ponds which can store the runoff generated by large areas.

In addition, over the past few decades, the notion of sustainable development has taken root in many technical/scientific fields. This undoubtedly holds true for UD as well. As early as the second half of the 1980s replenishment of ground water in urban areas and the reduction of CSO events (frequency and volume) were main drivers for diverting stormwater runoff from the classical sewer systems (be it storm sewers or combined sewer systems). Over time, the considerations were expanded covering increasingly water quality considerations (e.g., capture of sediments, nutrients, metals, hydrocarbons, pathogens) or hydromodification effects. The article from Fletcher *et al.* (2015) gives an overview of the different trajectories depending on the country.

This has resulted in a wide spectrum of hybrid systems ('grey' and 'green') aimed at intercepting rainwater and stormwater runoff at or near its source and at most transporting it (normally over short distances) to locations where natural infiltration into the ground is possible. In semi-arid regions, rainwater harvesting and stormwater capture approaches (e.g., for irrigation or toilet flushing) were added. In both cases, considerable attention was paid to integrating the design of such systems with the built environment involving (landscape) architects, ecologists, and urban planners, reflecting the multi-disciplinary nature of the field. Figure 1.4 provides some examples of integration of stormwater control measures. Practically, this has also meant that the materials used have fundamentally evolved from the more traditionally engineered materials (e.g., concrete, brick, steel, plastics) to also include vegetation, engineered 'soil' or growing media and a wide array of filter media.

This has resulted in systems where runoff is not just considered a nuisance and stormwater needs to be disposed of as quickly and efficiently as possible but rather systems that are expected to live up to a wide range of additional demands such as (non-exhaustive):

- Provide enhanced aesthetics without obtrusive odours;
- Reducing the heat island effects in urban areas;
- Provide a more or less natural environment to allow for some ecological diversity in the urban environment; and
- Provide recreational options for residents.

Figure 1.4 Example of stormwater management infrastructure integration: (a) detention basin in Bordeaux City Center, France, (b) raingardens, Meyzieu, France, (c) detention basin, Bron, France, (d) infiltration surface, Burnley, Australia, (e) place (Villeurbanne, France), and (f) tree pit, Villeurbanne, France. Photo credit: Frédéric Cherqui.

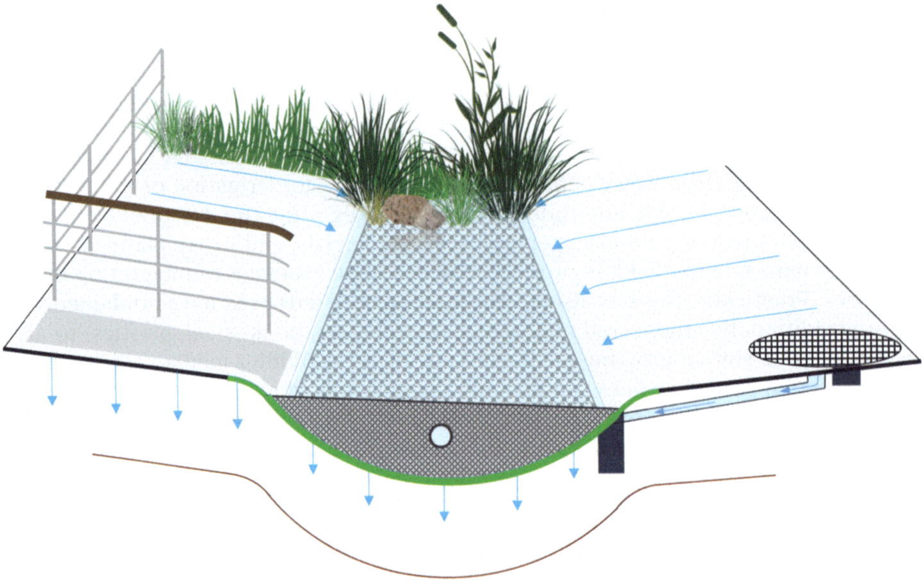

Figure 1.5 Fictitious SCMs representing the diversity of components that should be considered: soil, protection, inlet, outlet, retention volume, flow control, vegetation, runoff surface and so on (drawing by Kaminski & Vanlerberghe, 2017).

Hybrid 'grey'/'green' solutions and the need for integration within an urban landscape and functions, asked for tailor-made designs in most cases. Each asset may have specific functions and thus specific components, see Figure 1.5. Combined with the largely novel or 'experimental' state of mind of all involved and the missing plans for long-term measurements of the benefits, it did not necessarily lead to designs in which much thought was given to AM. It is only in recent years that these practices have started to mature to the extent that maintenance has been given any thought. To a certain extent a parallel can be drawn with the course of events with respect to the 'classical' solutions of UD. These systems had been constructed since roughly the 1850s while serious attention to AM only emerged in the early 1980s. For these classical systems design methods, tools used herein, and design guidelines were developed to a relatively high and detailed level. They are applied across the world based on the same sound physical principles with standard elements being covered in a wide range of courses and educational programmes. Furthermore, the resulting designs are such that processes causing deterioration of functionalities of sanitary or combined sewer systems are, although not fully known or described in detail, widely recognized and studied (for some of them in great depth: see e.g., biogenic sulphuric corrosion). Further, the choice of materials from which the classical systems are constructed is relatively limited (i.e., mostly either concrete, plastics, glass-reinforced resins, ceramics, brickwork or nodular steel). Once the notion of the need for AM was embraced, a substantial industry with standards for sewer inspection was developed. Along with quick digitization this has allowed for the development of tools with which, utilizing 'big data' amongst others inspection data, tactical planning for replacement/rehabilitation is enabled. This situation of limited uniformity, which is based on the few common objectives of public health, flood protection and environmental impact minimization, allows for bringing together scientific progress and experience obtained by practitioners in the field to come to a kind of 'state-of-the-art' and point out global weaknesses and knowledge gaps.

However, since this situation has not yet been reached for stormwater management systems, it is therefore pointless to strive for an identical approach in the book for 'classical' and 'sustainable' systems. This situation is compounded by the fact that stormwater management approaches are typically regional, reflecting the local geography and climate, and as such will be much more subject to the vagaries of for instance drought vs. monsoon conditions, or heat vs. cold conditions. Stormwater management guidelines and standards are still evolving in most jurisdictions. As a result, levels of service and expectations are not uniform and have evolved organically over the last couple of decades. Until quite recently, most guidance was focused on greenfield development with little guidance on what to accomplish in older, established communities. Retrofit programmes in older communities typically only address a single purpose, with little consideration of optimization of multiple objectives based on triple bottom line principles. As a result, many jurisdictions see a hodge-podge of infrastructure with poorly documented and not normalized expectations. In addition, a significant portion of the stormwater infrastructure may not be located within the public realm but rather on private land. This means often overlooked risks and liabilities to municipal entities where they count on this private infrastructure to be in good operation condition. A final challenge is that available performance data – even assuming that as-built/as-landscaped data are available and in good and complete condition which often will not be the case – has often not been normalized as to hydrologic/hydraulic or contaminant loadings or has been collected during relatively short-lived monitoring programmes when the system had not yet gained maturity (a counter example being the International Stormwater BMP Database – https://bmpdatabase.org or the Field Observatory for Urban Water Management OTHU – http://othu.org). A direct result of this huge diversity is that an analytical and systematic approach to AM is challenging when the desired outcomes of the intended operation of the stormwater system are poorly documented. That said, while there is this huge diversity, the unit operations and processes are still similar. AM of SCMs is at an early stage where run-to-failure is currently often the only implemented approach and decisions rely on experience and 'gut' of the asset manager. The focus is for now mostly on inspection approaches. Moving forward requires not only data and expertise regarding performance assessments or rehabilitation decisions pertaining to these assets, but it also needs strong foundations, that is, a solid AM culture and science. As such, the book focuses on the processes and operations without getting caught up in how the actual implementation thereof differs across the world.

In summary, SCMs have different specificities (i.e., compared to pipes) that deserve to be considered and which are addressed in the book in the following manner:

- SCMs are often a combination of more traditionally engineered and natural components (see Figure 1.6): each component (or their combination) delivers specific services. Each component will also require its own adapted rehabilitation needs: vegetation cannot simply be replaced but needs to grow in order to achieve its full potential. As such, stormwater control measure AM is not only about maintaining the 'engineered' parts of the asset.
- The range of services delivered by stormwater control measure is broader than 'collect and convey' (for pipes), and the importance of these services will depend on the context. For example, SCMs installed in highly visible areas of a city require landscape integration and more frequent maintenance as they contribute to the image of the city.
- SCMs are very often in the open air, and the hydraulic function is often not the main function: it could be a road (permeable), a park, a playground and so on. SCMs share public spaces, and require coordination between services (e.g., a drainage department, cleaning department, road department, park department).
- Compared to pipes, inspection should not only be conducted during dry weather conditions: it is often important to investigate the condition/performance of these assets during dry weather, wet weather, winter (snow conditions, road salt).

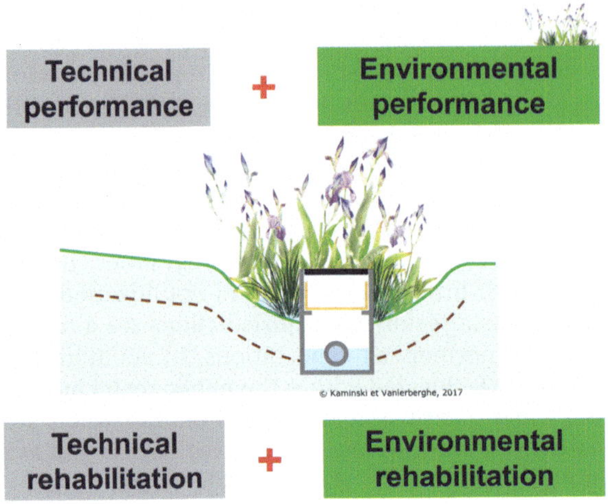

Figure 1.6 SCMs are often a combination of built and natural components which have specific performance and rehabilitation requirements (Cherqui, 2020).

The first issue to deal with is to embrace a uniform vocabulary. Since inventing new terminology is beyond the scope of this book, the suggestions made by some prominent colleagues have therefore been adopted (Fletcher *et al.*, 2015).

1.4.2 New sanitation

A somewhat different situation can be observed for what is known as 'new sanitation'. Here the focus is on the relative recent evolutions pertaining to the sanitary part of UD. The major paradigm shift strived for is to change the perspective from 'waste' to 'raw materials and energy'. The systems developed aim at reducing the consumption of flushing water and aspire to a wastewater composition that allows for advanced processing achieving a recovery of nutrients (P and N) and the possibility to extract energy from the organic components in the wastewater, by for example producing biogas.

Another driver for new sanitation is to isolate water streams holding micro-pollutants (e.g., medication). This has resulted in some experimental systems in which urine was collected in a separate system, to allow for specialized treatment. It is known that such systems suffer from ageing/deterioration processes not identified hitherto in classical systems like the clogging of transport systems with urea crystals.

Over the past 10–20 years several pilot systems (servicing up to 100–200 houses) have been built with the primary goal of demonstrating the feasibility of such concepts with an emphasis on the methods to extract nutrients and energy. Collection and transport systems in these pilot projects were either vacuum or pressure-based mechanical systems, which were not at the heart of the evaluation of the success of these projects. It is known that in several countries projects are being planned that aspire to scaling up these systems to the level of >1000 houses raising questions on the AM aspects for the long term. Although the pilots mentioned did not focus on gathering data on malfunctioning or maintenance efforts needed, some of the experience gathered in classical systems may be utilized here. What is different is the in-house part of the system: in some cases, four different waterflows should be kept separated: 'brown', 'yellow', 'grey' and 'stormwater'. This places the organization that manages the public part of the system at the mercy of the houseowner and the discipline of the inhabitants. With respect to the latter some experience can be obtained from managing classical separate systems in which wastewater and stormwater have to be kept apart.

All in all, we feel that the available information/experience with respect to new sanitation AM is too scarce to provide any meaningful review or 'state-of-the-art' impression. Therefore, the treatise on this subject is left at a summary of the (scarcely) available case studies.

1.4.3 Scope

Keeping in mind what has been said on the position of SUDS and new sanitation, and the notion that UDAM covers a wide field of different disciplines, timescales and geographical dimensions, a book of this nature must set limits on the subjects dealt with.

In terms of technical means the scope of the book is limited to all infrastructure in the urban environment that primarily serves to:

- Collect and transport wastewater from urban areas; and
- Manage stormwater in urban areas (i.e., collection, transport, detention, infiltration, sedimentation).

The classical systems will be treated with some depth and comprehensiveness, while SCMs and new sanitation, because of the relatively poor practical experience in practice will largely be treated by hand of examples or case studies. Co-benefits of SUDS, NBS and natural infrastructure will be touched on, where appropriate, to provide a comprehensive perspective to asset managers; however, given the novel nature of these considerations, only in a limited fashion. That said, the process descriptions responsible for deterioration of assets along with their characteristic time and space scales are discussed in several chapters and will be applicable to these systems as well.

In some cases, structural integrity of UD components is recognized as being a functionality; however, in this book this aspect is merely treated as a prerequisite and is therefore only addressed where appropriate.

REFERENCES

Büker D. (2000). Mensch – Kultur – Abwasser: Von der Annehmlichkeit für wenige zur Existenzfrage der Menschheit: der Umgang des Menschen mit Abwässern: ein kulturhistorischer Längsschnitt von den Anfängen bis zum Beginn des 20. Jahrhunderts (in English Man – Culture – Wastewater: From the convenience of a few to the existential question of mankind: man's treatment of wastewater: a cultural-historical longitudinal section from the beginnings to the beginning of the 20th century), Historie in der Blauen Eule, Die Blaue Eule, Essen, Germany, vol. **9**, ISBN: 978-3-89206-980-5.

Caradot N. (2019). The use of deterioration modelling to simulate sewer asset management strategies. PhD thesis, INSA Lyon, France.

Caradot N., Riechel M., Fesneau M., Hernandez N., Torres A., Sonnenberg H., Eckert E., Lengemann N., Waschnewski J. and Rouault P. (2018). Practical benchmarking of statistical and machine learning models for predicting the condition of sewer pipes in Berlin, Germany. *Journal of Hydroinformatics*, **20**, 1131–1147, https://doi.org/10.2166/hydro.2018.217

Cherqui F. (2020). The challenges of nature-based stormwater solution asset management. French 'Les défis liés à la gestion patrimoniale des techniques alternatives de gestion des eaux pluviales', workshop on nature-based solutions and hybrid water-related systems: characterisation and asset management, 16–17 November, INRAE, France. [In French] https://www6.paca.inrae.fr/recover/Actualites2/Webinaire-SFN-et-systemes-hybrides-lies-a-l-eau (last accessed 23.08.2023).

Cohen-Shacham E., Walters G., Janzen C. and Maginnis S. (2016). Nature-based Solutions to address global societal challenges, Gland, Switzerland, International Union for Conservation of Nature and Natural Resources. http://doi.org/10.2305/IUCN.CH.2016.13.en

Daulat S., Rokstad M. M., Klein-Paste A., Langeveld J. and Tscheikner-Gratl F. (2022). Challenges of integrated multi-infrastructure asset management: a review of pavement, sewer, and water distribution networks. *Structure and Infrastructure Engineering*, **20**(4), 546–565, https://doi.org/10.1080/15732479.2022.2119480

De Feo G., Antoniou G., Fardin H. F., El-Gohary F., Zheng X. Y., Reklaityte I., Butler D., Yannopoulos S. and Angelakis A. N. (2014). The historical development of sewers worldwide. *Sustainability*, **6**, 3936–3974, https://doi.org/10.3390/su6063936

Delgado A., Rodriguez D. J., Amadei C. A. and Makino M. (2021). Water in Circular Economy and Resilience. 68p., Washington, DC: World Bank. https://openknowledge.worldbank.org/handle/10986/36254

Ellis B., Chocat B., Fujita S., Rauch W. and Marsalek J. (2006). Urban Drainage: A Multilingual Glossary, June, IWA Publishing, London. 512 p. https://doi.org/10.2166/9781780402536

Feng B., Zhang Y. and Bourke R. (2021). Urbanization impacts on flood risks based on urban growth data and couples flood models. *Natural Hazards*, **106**, 613–627, https://doi.org/10.1007/s11069-020-04480-0

Fletcher T., Shuster W., Hunt W. F., Butler D., Scott A., Trowsdale S., Barraud S., Semandeni-Davies A., Bertrand-Krajewski J.-L., Mikkelsen P. S., Rivard G., Uhl M., Dagenais D. and Viklander M. (2015). SUDS, LID, BMPs, WSUD and more – the evolution and application of terminology surrounding urban drainage. *Urban Water Journal*, **12**(7), 525–542, https://doi.org/10.1080/1573062X.2014.916314

Frontinus S. J. (~100) De Aqueductibus Urbis Romae (On the water supply of Rome), a full Latin text and English and French translation can be found here: https://penelope.uchicago.edu/Thayer/E/Roman/Texts/frontinus/

Garbrecht G. (1988). 'Mensch und wasser im altertum' (in English 'Man and water in antiquity'). In: Die Wasserversorgung Antiker Städte, ed. by Frontinus Gesellschaft e.V., Geschichte der Wasserversorgung, Verlag Philipp von Zabern, Berlin, Germany, vol. **3**, chap. 1, pp. 13–40, ISBN: 3-8053-0984-8.

Hawari A., Alkadour F., Elmasry M. and Zayed T. (2020). A state-of-the-art review on condition assessment models developed for sewer pipelines. *Engineering Applications of Artificial Intelligence*, **93**, 103721, https://doi.org/10.1016/j.engappai.2020.103721

Hernández N., Caradot N., Sonnenberg H., Rouault P. and Torres A. (2018). Support tools to predict the critical structural condition of uninspected pipes for case studies of Germany and Colombia. *Water Practice and Technology*, **13**, 794–802, https://doi.org/10.2166/wpt.2018.085

Hosseinzadehtalaei P., Ishadi N. K., Tabari H. and Willems P. (2021). Climate change impact assessment on pluvial flooding using a distribution-based bias correction of regional climate model simulations. *Journal of Hydrology*, **598**, 126239, https://doi.org/10.1016/j.jhydrol.2021.126239

Kaminski H. and Vanlerberghe F. (2017). Stormwater control measures asset management – condition assessment based on failure modes (In French: Gestion patrimoniale des ouvrages de gestion des eaux pluviales – évaluation de l'état de santé basé sur l'analyse des modes de défaillance), INSA Lyon, 30 p.

Kessener H. P. M. (2017). Roman Water Distribution and Inverted Siphons – until our day. PhD thesis, Radboud University, Nijmegen, The Netherlands, ISBN 978-90-827410-0-1.

Kourtis I. M. and Tsihrintzis V. A. (2021). Adaptation of urban drainage networks to climate change: a review. *Science of the Total Environment*, **771**, 145431, https://doi.org/10.1016/j.scitotenv.2021.145431

Laakso T., Kokkonen T., Mellin I. and Vahala R. (2018). Sewer condition prediction and analysis of explanatory factors. *Water*, **10**, 1239, https://doi.org/10.3390/w10091239

Langeveld J. G., Cherqui F., Tscheikner-Gratl F., Muthanna T. M., Juarez M. F.-D., Leitão J. P., Roghani B., Kerres K., do Céu Almeida M., Werey C. and Rulleau B. (2022). Asset management for blue-green infrastructures: a scoping review. *Blue-Green Systems*, **4**(2), 272–290, https://doi.org/10.2166/bgs.2022.019

Lucas A. and Walton S. A. (2020). Technology and Change in History. ISSN: 1385-920x.

Makris K. F., Langeveld J. G. and Clemens F. H. L. R. (2021). Extensive testing on PVC sewer pipes towards identifying the factors that affect their operational lifetime. *Structure and Infrastructure Engineering*, **18**(12), 1601–1613, https://doi.org/10.1080/15732479.2021.1907601

Miller J. D. and Hutchins M. (2017) The impacts of urbanization and climate change on urban flooding and urban water quality: a review of the evidence concerning the United Kingdom. *Journal of Hydrology: Regional Studies*, **12**, 345–362, https://doi.org/10.1016/j.ejrh.2017.06.006

Mithen S. (2010). The domestication of water: water management in the ancient world and its prehistoric origins in the Jordan Valley. *Philosophical Transactions of the Royal Society A: Mathematical. Physical and Engineering Sciences*, **368**, 5249–5274, https://doi.org/10.1098/rsta.2010.0191

Ndoziya A. T., Hoko Z. and Gumindoga W. (2019). Assessment of the impact of pit latrines on groundwater contamination in Hopley Settlement, Harare, Zimbabwe. *Journal of Water, Sanitation and Hygiene for Development*, **9**(3), 464–476, https://doi.org/10.2166/washdev.2019.179

Neumann M. B., Rieckermann J., Hug T. and Gujer W. (2015). Adaptation in hindsight: dynamics and drivers shaping urban wastewater systems. *Journal of Environmental Management*, **151**, 404–415, https://doi.org/10.1016/j.jenvman.2014.12.047

Nuruzzaman M. (2015). Urban heat island: causes, effects and mitigation measures – a review. *International Journal of Environmental Monitoring and Analysis*, **3**(2), 67, https://doi.org/10.11648/j.ijema.20150302.15

Reese A. (2001). Stormwater Paradigms. Stormwater, 9p. https://www.stormh2o.com/bmps/article/13001153/stormwater-paradigms

Robida A. (1892). The Electrical Life: the 20th Century, the Illustrated Library. (*org. Title in French: La vie éléctrique: le vingtième siècle, La Librairie Illustrée,*) Paris, 266 p. Source: https://gallica.bnf.fr/ark:/12148/bpt6k101948n/f139.item

Schreider S. Y., Smith D. I. and Jakeman A. J. (2000). Climate change impacts on urban flooding. *Climatic Change*, **47**(1), 91–115, https://doi.org/10.1023/A:1005621523177

Snow J. (1849). On the mode of communication of cholera. https://johnsnow.matrix.msu.edu/work.php?id=15-78-52

Snow J. (1854). On the model of communication of cholera (2nd edition, much enlarged, includes cholera map). (https://archive.org/details/b28985266/page/8/mode/2up?view=theater)

Strommenger Eva (1980). Habuba Kabira. Eine Stadt vor 5000 Jahren (in English Habuba Kabira. A city 5000 years ago), (ed) Mainz, Philipp von Zabern, ISBN 3805304498.

Tscheikner-Gratl F., Caradot N., Cherqui F., Leitão J. P., Ahmadi M., Langeveld J. G., LeGat Y. L., Scholten L., Roghani B., Rodríguez J. P., Lepot M., Stegeman B., Heinrichsen A., Kropp I., Kerres K., Almeida M. do C., Bach P. M., Vitry M. M. de, Marques A. S., Simões N. E., Rouault P., Hernandez N., Torres A., Werey C., Rulleau B. and Clemens F. (2020). Sewer asset management – state of the art and research needs. *Urban Water Journal*, **16**, 662–675. https://doi.org/10.1080/1573062X.2020.1713382

Walski T. M., Chase D. V., Savic D. A., Grayman W., Beckwith S. and Koelle E. (2003). Advanced Water Distribution Modeling and Management, Civil and Environmental Engineering and Engineering Mechanics Faculty Publications, 18, Haestad Press, Waterbury, CT, USA. https://ecommons.udayton.edu/cee_fac_pub/18

World Bank (2021). A Catalogue of Nature-Based Solutions for Urban Resilience. World Bank Publications, The World Bank Group, 1818 H Street NW, Washington, DC 20433, US. https://openknowledge.worldbank.org/handle/10986/36507

Worldbank, https://data.worldbank.org/indicator/SP.URB.TOTL.IN.ZS, accessed on 19 July 2021.

doi: 10.2166/9781789063059_0019

Chapter 2
Asset management in a nutshell

Maria do Céu Almeida[1] and Ingo Kropp[2]

[1]National Laboratory of Civil Engineering, Lisbon, Portugal
[2]3S Consult GmbH, Garbsen, Germany

ABSTRACT

Asset management (AM) has become increasingly important in various sectors and organizations over the last few decades. This introductory chapter provides an overview of asset management to facilitate a comprehensive understanding of AM, summarizing its application to urban drainage asset management (UDAM), an asset-intensive sector. The chapter includes a short history, from general AM to UDAM, and refers to milestones, key organizations, and standards. It describes AM principles and different perspectives to approach the subject. The UDAM context is outlined, providing the general processes which can influence strategic, tactical, or operational decision levels, accounting for the specific perspectives and objectives on the subject. The main approaches applied to UDAM are broadly described, some of which incorporate a balance of costs, risks, opportunities, and performance, while others focus on specific aspects. The chapter presents different types of approaches to AM, ranging from those based on condition, reliability, and risk to those based on value, followed by those based on performance or quality of service, highlighting common aspects or overlaps between them. When appropriate, guidance to the chapters where additional information on the subject can be found is provided. Finally, challenges, limitations and opportunities are highlighted.

Keywords: Asset management, condition, cost, information management, performance, reliability, risk management, strategic planning, urban drainage.

2.1 OVERVIEW ON ASSET MANAGEMENT

Asset management (AM) is a widely used concept relevant to organizations, independent of size or type of activity (IAM, 2015). Using the term without considering the organization's internal and external context can cause misunderstanding and difficulties in its application. The relevance of AM to an organization becomes clear, especially when using a broad definition of an asset, as given by the ISO 55000 (2014) standard 'item, thing or entity that has potential or actual value to an organization', independent of the organization mission and objectives. 'Asset management involves the balancing of costs, opportunities and risks against the desired performance of assets to achieve an organizational objective' (ISO 55000, 2014).

Asset-intensive organizations such as utilities providing essential services are at the forefront of the need for AM since they rely on physical assets to provide services for society. Urban water cycle functions rely on infrastructures, especially those assets functioning as components of networks like many other urban infrastructures, including those for transportation, energy supply or telecommunications. These services have indefinite lives: as long as there is demand for it, 'these infrastructure assets are not replaced as a whole but rather are renewed piecemeal by the replacement of individual components whilst maintaining the function of the system as a whole' (Burns *et al.*, 1999). Increasing competition, and consumer, regulatory and stakeholder pressures foster the need to further advance in continuous improvement by the organizations.

ISO 55000 (2014) states four core principles for AM (Clause 2.4.2): (i) value, (ii) alignment, (iii) leadership, and (iv) assurance. 'Assets exist to provide value to the organization and its stakeholders'; the value is not limited to its intrinsic value but incorporates the contribution of the asset to the organization's objectives. These tangible or intangible elements of value involve the needs and expectations of stakeholders and other interested parties. Quality of service is an example of a dimension incorporating tangible and intangible elements.

Alignment refers to the need to have coherence between AM activities and the organization's strategic plan, ensuring the translation of organization objectives into the AM policy, strategy, objectives, plans, and activities (i.e., technical, financial, and operational). This alignment aims to connect the staff at the different sectors and levels with organization policy and objectives to understand how their contribution adds to the overall purpose.

Leadership and organizational culture are essential to good AM; this materializes by ensuring effective senior leadership and commitment across all levels of management.

Assurance includes effective monitoring and auditing of processes and outcomes to confirm the fulfillment of AM and organizational objectives and goals and ensure commitment to continual improvement (IAM, 2015; ISO 55000, 2014).

The scope of AM in an organization extends from senior management to the operational services, incorporating decision-making and activities at different levels, often assumed as the strategic, tactical and operational levels. Key elements to account for in the case of infrastructures are the risks associated with assets failure, the performance of the organization as a whole, the quality of service provided, and the cost-effectiveness of AM from short to long term (Alegre *et al.*, 2013; Vanier, 2001).

Organizational AM core activities can be implemented through an AM system composed of an integrated set of tools, including policies, plans, processes, and information systems. The requirements of this system are described in ISO 55001 (2014).

2.1.1 A short history: from general AM to UDAM

Concepts and methods for managing assets have been in use for decades. Depending on the needs and objectives, organizations adopt specific practices and terminology adapted to their core activities, products or services provided. Even within organizations, often distinct terms or approaches are adopted by different departments. AM is increasingly recognized as an aggregating and holistic discipline. Interest in adopting holistic AM principles increased in the last two decades (Gavrikova *et al.*, 2020).

Some authors position the origins of AM in the 1960s, then understood as 'an approach to maximize the value of asset portfolios throughout their life cycle' (Gavrikova *et al.*, 2020) or even before by investment managers and insurance companies (Haldane, 2014; Schneider *et al.*, 2006). The term 'asset management' associated with physical assets emerged in the 1980s by private and public organizations stimulated by pressing needs to better manage safety, levels of service and costs (IAM, 2015). In some countries and sectors, deregulation, and privatization of infrastructures (e.g., energy, transport, water sectors) during the 1980s and 1990s fostered the use of methods of asset and risk management to face new challenges (Schneider *et al.*, 2006; Too & Tay, 2008).

Developments in the following decades were significant. Some countries, including Australia, Canada, New Zealand, and the UK, have been at the forefront of the evolution of infrastructure AM (Jones *et al.*, 2014). In the UK, following the privatization of water services in England and Wales, requirements to companies included the development of AM plans, initially basically investment plans (OFWAT & DEFRA, 2006). This trend in the water and other sectors in different countries has driven the development of an AM standard approach, a publicly available specification, for the optimized management of physical assets, PAS 55 – Asset Management (PAS 55-1, 2008), first published in 2004 aimed at supporting the development of a sound AM plan. PAS 55 provides a common and complete structure to develop an AM plan, indicating the steps needed to be covered during the asset's life cycle to meet the demands of shareholders, regulators, and communities. PAS 55 outlines what needs to be done, not how to do it. The 2008 revision of PAS 55 brought clarification and improvements to facilitate its application in practice.

Following PAS 55 publication, work within ISO converged on the series of ISO 5500X standards currently in use. Somehow, these global efforts had a similar aim as those leading to the ISO 31000 (2018) and companion standards on risk management: to propose a common broad framework and terminology for all applications. The main difference between the ISO 5500X standards and PAS 55 are in their scope: PAS 55 focuses on physical assets, while ISO 5500X applies to any asset. PAS 55 and ISO 55000 are milestones for disseminating a common language and understanding of AM.

The AM system requirements described in ISO 55000 (2014) include the context of the organization in question, its leadership, planning, support, operations, performance evaluation, and improvement. Benefits from AM highlighted in these standards are improved financial performance; informed asset investment decisions considering an effective balance between costs, risks, opportunities, and performance; better risk management; better quality of services and products; demonstrated compliance with legal, statutory, and regulatory requirements; enhanced reputation; and, improved organizational sustainability, efficiency, and effectiveness.

Currently, the use of the term AM is widespread. The understanding and concept can still vary from the long-used meaning in financial services to the notion adopted by software promoters when labelling their products to support AM (Too & Tay, 2008). However, the concept involving an integrated, whole-life, risk-based management high-level approach has gained broad acceptance.

Within the broad reach of the relevant literature, infrastructure AM encompasses a wide range of asset-intensive sectors. Managing physical assets is a requirement for any organization to be sustainable, especially in capital-intensive and asset-intensive sectors (e.g., urban water, transport infrastructures, energy supply), where maintenance costs and asset depreciation are extremely high. The pressure-to-cost efficiency applies to most organizations; when managing infrastructure and providing essential services to society, it is necessary to find a balance between management costs, resources for investment in upgrading deteriorating assets, and service and regulatory requirements (Gavrikova *et al.*, 2020; Schneider *et al.*, 2006).

The concept of strategic AM in the context of infrastructure is a multidimensional approach combining organizational planning and decision levels, strategic, tactical and operational, each with its planning cycle, with systematic and coordinated activities and practices. The latter broadly incorporates competence in management, engineering, and information within the organization to better manage its assets using an assessment framework combining three dimensions of analysis performance, risk, and cost. This concept was aggregated in a scheme illustrated graphically in Figure 2.1 based on Alegre *et al.* (2006).

Within urban water systems, as in other infrastructures, the construction of water supply and drainage infrastructure financed, by and large, by public capital, allowed improvements in public health and conditions for economic development. Numerous systems were built during the 1900s, peaking in the second half of the last century. The insufficient maintenance and renewal of these assets resulted in the gradual deterioration of the large, complex, and expensive infrastructure leading to the realization of the management needs and limited resources available for that purpose (Burns *et al.*, 1999; Jones *et al.*, 2014; Tafuri & Selvakumar, 2002; Vanier, 2001). As in other urban water systems, urban drainage AM faces challenges: complex systems, interacting with and interdependent on other urban infrastructure systems, functioning under dynamic conditions and prone to rapid changes, with most components underground (i.e., out-of-sight, out-of-mind) and with long service life. While system components deteriorate over time, under the action of internal (e.g., in-pipe) and external factors, maintenance and rehabilitation rates are low. In the management of urban water systems, needs and complexity differ depending on asset condition and service life. An organization managing

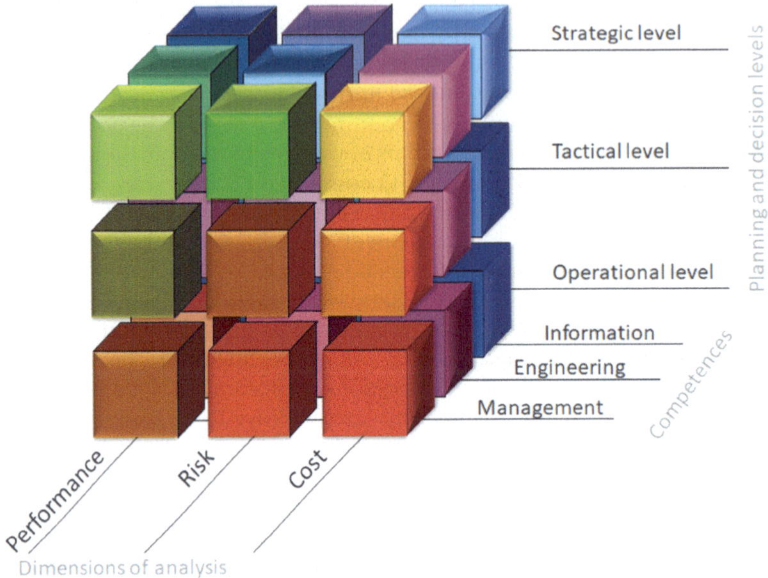

Figure 2.1 Scheme for infrastructure AM concept as proposed by Alegre *et al.* (2006).

infrastructure whose components have average service lives above the design expected live or severe deterioration faces very different challenges compared to another having relatively new assets.

A considerable body of knowledge and experience exists in the management of urban drainage infrastructure assets, including comprehensive approaches (CAs) in core activities including data management, performance assessment, operation and maintenance activities, condition assessment of assets, rehabilitation techniques, and decision support tools. Increasing pressure to improve performance and limited resources drives a shift toward AM in the sector.

The standardization work under the CEN umbrella provides an overview of the core activities in urban drainage, especially the work of the CEN/TC 165 (Technical Committee 165) on wastewater engineering, created in 1988 to develop 'functional standards, standards for performance and installation in the field of wastewater engineering for systems and components. Standards for design, calculation, construction, commissioning, operation, and maintenance in the field of wastewater engineering, from the point of origin (...) up to the point of disposal, including treatment plants and use of treated wastewater for purposes other than agricultural irrigation.' (https://standards.cencenelec.eu/dyn/www/f?p=CEN:6) The scope of the Working Group WG 22 – Drain and Sewer Systems Outside Buildings, published standards include functional requirements, performance requirements, planning, design, construction, rehabilitation, operation and maintenance. The European Standard EN 752 (2017), Drain and Sewer Systems Outside Buildings – Sewer System Management, specifying the objectives for drain and sewer systems outside buildings, was first published in 2008. 'This European Standard specifies the objectives for drain and sewer systems outside buildings. It specifies the functional requirements for achieving these objectives and the principles for strategic and policy activities relating to planning, design, installation, operation, maintenance, and rehabilitation. It is applicable to drain and sewer systems from the point where wastewater leaves a building, roof drainage system, or paved area, to the point where it is discharged into a wastewater treatment plant or receiving water body. Drains and sewers below buildings are included provided that they do not form part of the drainage system for the building' (EN 752, 2017). The term 'wastewater' is used in a broad sense also referring to stormwater.

The standard EN 752 (2017) focuses on integrated sewer system management, is aimed at senior utilities managers and sets performance standards to support decision making. The fundamental objectives for urban drainage systems are public health, occupational safety and health, environmental protection, and sustainability. CEN standards on tactical and operational activities have three groups: investigation and assessment; design and construction; management and control.

In line with the ISO 55000X series, the ISO 24516-3 (2017) Guidelines for the Management of Assets of Water Supply and Wastewater Systems – Part 3: Wastewater Collection Networks adopts the framework of the EN 752 (2017) integrated management planning, objectives, requirements, and criteria providing a framework for the design, construction, rehabilitation, maintenance and operation of drain and sewer systems outside buildings. The reader is referred to Chapter 3 for further information on guidelines and regulations.

The urban drainage sector is aligned and converging on adopting common AM language and good practices. Infrastructure services have a predominant monopolistic nature and tend to operate within a more formal regulatory framework. Consumer awareness of quality-of-service issues and increasing costs to access these services add a challenging dimension to the sector. External challenges such as the need to efficiently use resources, accounting for circularity and rising costs, and exposure to natural risks increase the pressure on this sector. Climate change effects strongly impact urban infrastructure, and drainage is no exception.

The actual service life of urban drainage assets tends to exceed the design service life, with many sewers in service for over 100 years. Urban drainage system AM decisions over time have a long-lasting impact, but a myriad of approaches and tools are available for areas comprising inspection techniques, condition assessment, deterioration models, data management, performance assessment systems, risk-based approaches, cost–benefit analysis (CBA), rehabilitation techniques or multi-criteria decision support (Tscheikner-Gratl *et al.*, 2019).

Unlike controlled manufacturing processes or visible structures and infrastructure, drainage systems often are hidden, their assessment is limited, and the service life of components is vastly exceeded. The value given to drainage infrastructure and the priority to act in the face of limited resources and competition with other more visible demands is unquestionably minor. The perception of the risks associated with a drainage system's failure is low.

The urban drainage system has relevant interactions and interfaces with the other urban systems, with some degree of cross-interdependencies (Table 2.1, adapted from Almeida *et al.* (2020)). Newer paradigms evolving since the 1980s relevant to nature-based solutions and hybrid systems (Fletcher *et al.*, 2015; Kapetas & Fenner, 2020; Langeveld *et al.*, 2022) require increasing levels of involvement and interaction of interested parties (e.g., stakeholders such as utilities, municipalities, regulators, environment agencies, communities, land planners, energy suppliers, mobility, and traffic networks managers). Integrating blue-green infrastructure as part of urban drainage adds challenges to urban drainage asset managers opening the way to shared AM and raising opportunities for stronger cooperation and coordination between urban system managers. Adopting the principles of ISO 5500X facilitates the alignment between services (GIOC, 2021).

Incorporation of interdependencies and interactions in the general management structure of utilities and alignment of AM plans facilitates the coordination between services, creating opportunities to improve the quality of service and reduction of efforts. A good example is the Canadian CSA W211 (2021) standard about the management of stormwater systems, which provides an overview of organizational components including the interactions with the organization's AM planning. This standard provides a framework identifying a set of comprehensive processes a utility can incorporate for effective stormwater systems management.

2.2 GENERAL PROCESSES

2.2.1 Overview

AM processes can be described and pictured in different ways. There are multiple points of view. Internal or external stakeholders have different perspectives to look at the subject. The Institute of Asset Management (theIAM.org), an organization aiming to bring together all kinds of people dealing

Table 2.1 Interdependencies matrix for Lisbon urban services and infrastructures.

Service| Component	Service/Infrastructure Failure	Exposure to	Potential Derived Risks and Cascading Effects
Electricity transport and distribution: Substations, overhead and underground	Damage, collapse, interruption of energy supply	R: flood W: storms	Urban drainage: failures of pumping and control systems; WWTP: failures of electromechanical elements and control systems
Communications: network and operational canters	Damage, collapse, interruption of communications	R: runoff, flood W: storms	Effects on several urban services depending on communications
Urban water cycle: wastewater and rainwater systems	Sewer systems: limited conveyance capacity, high street runoff (level and velocity), CSO Pumping stations: electrical or mechanical failures due to flooding (pumping capacity and CSO), damage to components, high inflow WWTP: lower treatment efficiency, CSO due to excessive flows; lower treatment efficiency and corrosion of infrastructures by saltwater intrusion	SLR; R: high inflows, runoff, flood	Mobility (road, rail): disturbance and interruptions, flooding of underground infrastructures (metro, train, parking, tunnels); Wastes: overturn, dragging and damage to containers; Electricity: damage to equipment and lines; Receiving water pollution Recreational uses affected
Waste collection | Cleaning, containers	Container damage, displacement and overturn	R: runoff, flood W: storms	Urban drainage: obstruction of components and surface flows
Mobility | Roadways and railways	Runoff, flooding and windstorm: disruption, interruption of mobility functions	SLR; R: runoff, flood; W: storms	Several urban services can be affected by cascading effects if maintenance or repair tasks are required during failures
Green and blue infrastructure and urban equipment	Collapse of trees Damage and collapse	R: runoff, flood W: storms	Urban drainage: components obstruction (e.g., inlets, sewers); electrical energy, Communication: damage to equipment and lines; mobility (road, rail): traffic disturbances and interruptions

R: rain; W: wind; SLR: sea level rise; WWTP: wastewater treatment plants.
Source: Adapted from Almeida *et al.* (2020).

with AM topics, published helpful guidelines and other technical literature. The conceptual model for AM published in IAM (2015) was adapted generally for UDAM (see Figure 2.2).

As in other types of organizations, urban drainage utilities should develop their organization strategic plan, considering the requirements and expectations of the various stakeholders. Private, public, and industrial customers produce sewage, and they expect a continuous and safe drainage service of wastewater and stormwater.

The legislation and regulations form the statutory framework for the operators. Today, multiple international, national, and regional laws and regulations must be complied with. Technical standards support the implementation of and compliance with these rules.

The natural environment is part of the Institute of Asset Management (IAM) external context and requires the involvement of additional stakeholders. Climate change is one of the biggest challenges we face nowadays. We must re-examine existing philosophies and systems and either enhance them or substitute them with more contemporary techniques.

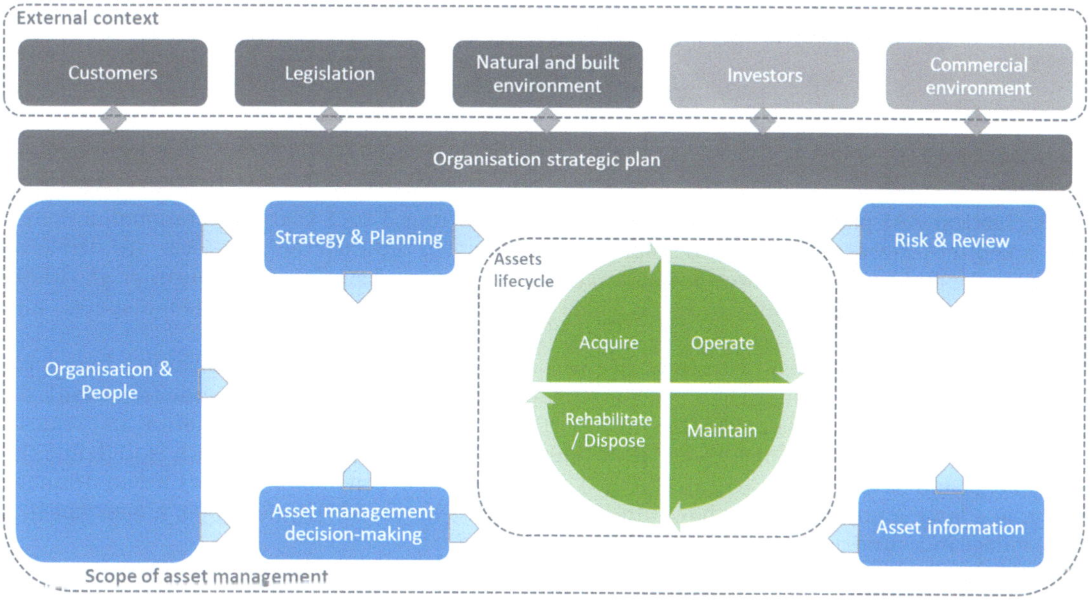

Figure 2.2 Adapted IAM concept of AM model.

Urban drainage systems are usually publicly managed. In specific situations, the property and management can be private. Different solutions involving private partners started over the last decade. Investors, who provide money to construct new or improve existing urban drainage systems, expect a return on their investments. However, these services are often subject to regulation considering the monopolist nature of these businesses. Financing opportunities for the urban drainage system operator differ based on the commercial environment. The sector environment is also about running a business without significant unplanned disturbances, for example, because of sewer collapses or overflows.

AM is co-responsible for implementing the strategic plan and comprises six subject groups:

(1) Strategy and planning.
(2) Decision making.
(3) Life-cycle delivery.
(4) Asset information.
(5) Organization and people.
(6) Risk and review.

The order above does not reflect any importance or priority between the various subjects. Figure 2.2 shows the interaction between the different areas. Thus, the entire AM process is not a direct step-by-step waterfall approach. Establishing AM in a systematic way usually starts with many small existing AM 'pieces' that need to be connected. Using an agile approach at the beginning, learning, and understanding what is there already and what is missing will quickly lead to a straightforward methodology after a while.

2.2.2 Strategy and planning

AM comprises different scales in time and space (see Figure 2.3). The strategic perspective focuses on the long-term, typically above five years, and the entire urban drainage system managed by the

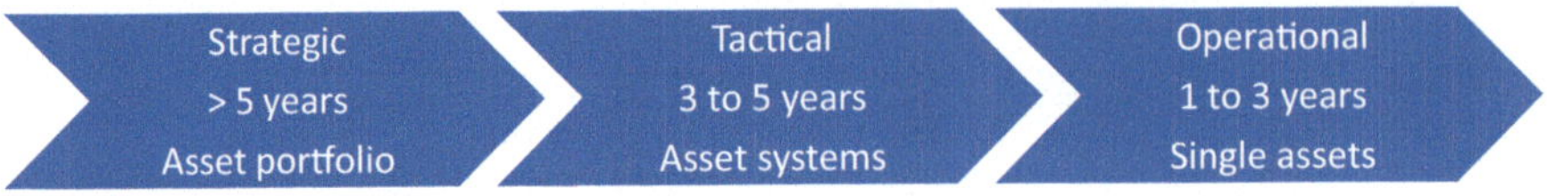

Figure 2.3 Typical time frame for different AM levels.

utility. Tactical AM planning is usually for the mid-term from 3 to 5 years. Operational activities have a shorter-term perspective. It is important to acknowledge that these services are delivered continuously, and analysis requires both a broad system view and single assets (Burns *et al.*, 1999). At the operational level, actions can be on single or sets of assets, following a system diagnosis and decision making.

An AM policy in alignment with the objectives and requirements of the organizational strategic plan set by senior utility management, accounting for the external context, forms the framework for developing a strategic AM plan (SAMP). The SAMP should be 'appropriate to the nature and scale of the organization's assets and operations' (ISO 55001, 2014). An alternative approach is to define the AM strategy and develop the AM at the tactical level. The management of a complex urban drainage system in a large municipality can have other requirements for an AM strategy than a manager in a small village in a rural area.

An AM plan deals with:

- the organizational objectives,
- the current and future performance of the systems,
- future demands reasonably foreseen,
- organization resources (e.g., budgets, human and technical resources),
- needs for assets rehabilitation and construction,
- incorporating new technologies and methods,
- support investment needs.

An AM plan should specify future capability and performance requirements for urban drainage systems. Objectives should be specific, measurable, achievable, realistic and time-bound (SMART) as described by Doran (1981). The plan should describe how to reach the objectives in the development and improvement processes. It should be documented and communicated within the organization and external parties and reviewed and updated regularly. Another substantial part of the plan is the demand analysis or the urban drainage system capacity requirements, including forecasting the future amount of wastewater or runoff from private, public, and industrial sources. The other – maybe even more important – aspect is the amount and frequency of rainfall events in the future, which determines the required capacity of stormwater and combined pipes as well as temporary stormwater detention. A level of acceptable risk is assumed when designing the systems and selecting the design return period (e.g., 10- or 20-year design rainfall return period). Today, there is broad recognition of the need to consider runoff and minimize flooding risks (Bertrand-Krajewski, 2021). Stormwater should not be mixed with wastewater and transported to a WWTP or provoke combined sewer overflows. The conversion to separate systems without undue inflows is a challenge requiring utilities, municipalities, and users to act together to improve infiltration, retention, and surface flow conveyance, with impacts on AM by more than one department (see Section 2.4). Combined sewer overflows happen but reduction of frequency and volumes should be targeted as much as possible.

The potential of new technologies and methods is under investigation (e.g., NBS) to overcome the limitations of traditional systems alone. Especially for larger projects, but not limited to these, the development of business cases within a SAMP can justify the required investments. It is worth mentioning that investments in urban drainage systems do not have to pay off economically in the short term. The benefits of investments should be assessed in the long run, considering avoided costs due to avoided flooding and pollution and co-benefits for quality of life and environment. A business case should emphasize on environmental performance and sustainable measures.

AM plans at the tactical level are the next step to achieve the strategic AM objectives. Tactical activities detail is required including capital investments, operational actions, maintenance activities, required resources (funding, personnel, material, equipment), specialized skills and knowledge, and risk assessment to reach objectives.

2.2.3 Decision making

Each asset or asset system requires decision making at various stages, depending on the organization (see also Chapter 7). Questions asset managers deal with in the AM decision-making processes include:

- When do we need a new or rehabilitate an existing asset or component?
- Which are the asset type and capacity requirements for fulfilling the requested function?
- Where should it be installed?
- How can the investment be financed?
- What is the service life of the asset?
- Can we extend the service life by doing regular maintenance?
- What are the periods for maintenance?
- Can the asset fulfil its function also in 10 or 20 years?

Regarding capital expenditures (CAPEX) CBA is a typical method for evaluating the benefits of an investment project. Evaluation criteria include net present value (NPV), internal rate of return (IRR) or payback period (the time until the investment pays off). CBA focus solely on monetized factors. Cost-effectiveness analysis (CEA) and cost-utility analysis (CUA) are alternative methods accounting for other factors without monetization.

Dealing with new assets (greenfield planning) and capital investment decision-making is straightforward. For instance, urban drainage for a new housing area can involve assessing different solutions. For existing assets (e.g., brownfield planning or infill development), it is harder to decide when to rehabilitate. Even when plenty of information on the condition (and its development) is available, there are still uncertainties about the probability of failure of the assets leading to unprecise know-how on future maintenance requirements and respective costs. It is easier to replace an existing component because a technical standard has changed, or a failure has already occurred.

Throughout the life of an asset, operation and maintenance activities allow for keeping the performance of the asset on a required or accepted level. Inspections, tests, and monitoring provide information on the asset's (actual) condition and performance. For instance, Germany has regulations for regular inspections of the whole wastewater network (e.g., SüwVO Abw, 2013), but this is not the case in all countries. If inspection is just for part of the assets done, statistics or AI/ML mathematical models can help to predict the condition of uninspected assets of the same type. Decisions about preventive maintenance regimes (time-based, condition-based, usage/duty-based) are required to keep the assets running. Other applicable methods include failure mode and effects analysis (FMEA), reliability-centered maintenance (RCM) and risk-based inspection (RBI).

The cost-based optimal time (see Figure 2.4) for replacing a single asset or component does not lead automatically to the decision to replace that asset. The challenge for the management team is to aggregate results for individual assets accounting for system functioning (e.g., a subsystem) and the needs for the entire system (replacing pipes, pumps, or equipment in a WWTP) and to evaluate these options to make the right decision with the resources available.

Another aspect not to be overlooked is resource planning when selecting the asset strategy. Physical asset maintenance requires appropriate labor resources and specific equipment and tools. Personnel requirements need to consider all planning tasks. When these resources are insufficient internally, an alternative is to use external support. The organization must evaluate and balance the right proportion between internal and external resources.

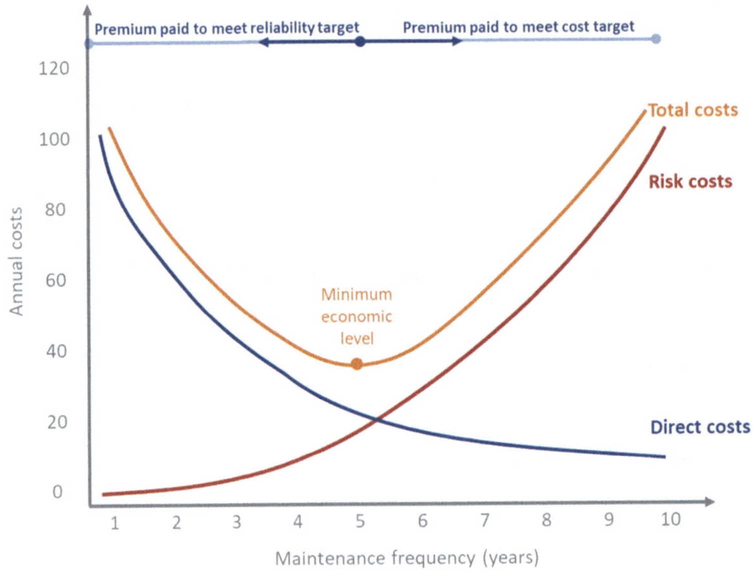

Figure 2.4 Cost-based optimal frequency of planned maintenance.

Financial and technical information on urban drainage systems is often stored independently. Enterprise resource planning (ERP) systems can combine technical and financial data but often only for above-ground assets. Usually, the storage of pipe network data is in GIS, sometimes with interfaces to financial systems (e.g., ERP). However, frequently there are different perspectives on technical and financial data. For instance, after the installation of a new sewer pipe or rehabilitation of an existing one, information is stored in GIS as an independent item; in ERP systems, investments in a given year are often not stored as single investments for every single asset, but as a sum of all investments (for the same type of asset). The latter is sufficient for the correct calculation of depreciation and the update of book values in total. The problem with this procedure is that the asset manager cannot directly link all costs to a single asset as part of life-cycle assessments. Depending on the condition and degradation, there is always the question of whether the depreciation time is in line with the service life of the asset or not. Therefore, a conservative approach uses much shorter depreciation times than the expected service life to be on the safe side.

In addition to investments, all other costs should be documented and assigned to a specific asset, if possible. Operational costs include labor, energy, chemicals, and plant and equipment. In emergencies, costs associated with third-party can exist (e.g., social, environmental, safety, and reputational).

If total costs are assigned to a single asset, the cost-optimal or best time for rehabilitation could be determined. It estimates the present value for the total based on investment and cumulative operational costs over the lifetime (see Figure 2.5). In general, it turns out that economically it is more favorable to replace an asset later than too early.

Finally, risk management must deal with the requirements and expectations of the stakeholders. Examples of requirements and expectations can be found in Table 2.2.

2.2.4 Life-cycle delivery

Various life-cycle delivery subjects support the implementation of strategic and tactical AM planning activities while ensuring system functioning and the lifetime of an asset or system component (see Figure 2.6).

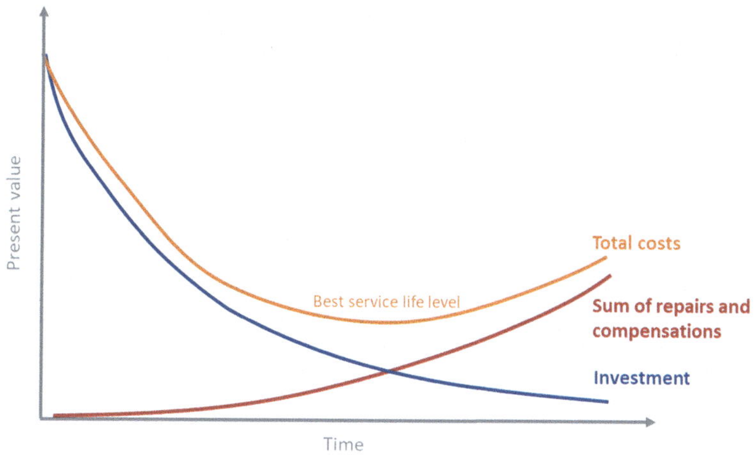

Figure 2.5 Determination of optimal service life.

Table 2.2 Examples of requirements and expectations for different stakeholders.

Stakeholder	Requirements and Expectations
Customers	Secure operation of urban drainage system Stable tariffs
Investors	High return on investment
Local communities, neighborhoods, third parties	No (unplanned) disturbances
Regulators/ Governments	Meeting all legislation and technical standards
Senior management	Meeting objectives of AM in line with all stakeholders
Owners	Maintain a good state of the urban drainage system
Employees	Safe workplace A good work–life balance Good salaries
Vendors/suppliers, consultants	A good relationship with the utility Sufficient prices

Figure 2.6 Life-cycle phases of an individual asset or system component.

The legislation provides the legal framework, with the support of standards, for the construction, operation, maintenance, and disposal of systems or assets (see Chapter 3). Applicable legislation and relevant technical standards should be identified and used. A good practice is compiling these documents in a known location, ensuring their regular review and update.

The AM planning process and its implementation facilitate the timely action to rehabilitate the infrastructure and its components. Experience shows that good project management in all phases,

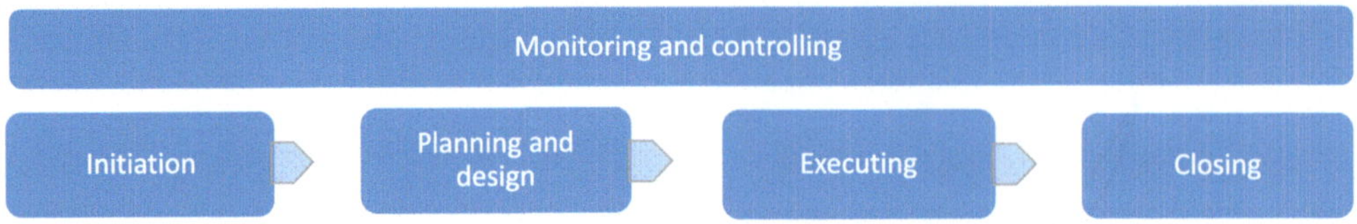

Figure 2.7 Project management steps.

from conception to construction, is essential for the asset's successful operation and maintenance. Continuous monitoring should ensure the required quality (see Figure 2.7). Software tools can support people involved in these processes. The digital documentation of activities contributes to the traceability and quality of management processes.

The design and management of urban drainage systems can benefit from adopting the RAMS system engineering perspective, composed of the following aims:

- Reliability: an asset should be able to perform a specific function; it can be design reliability or operational reliability.
- Availability: the capability of the asset to keep a functioning state in a specific environment.
- Maintainability: the ability to be timely maintained (including servicing, inspection, repair, or modification).
- Safety: the ability not to harm people, the environment, or any assets during a complete life cycle.

Once built, operation and maintenance of the assets are required. During the service life of the assets, management should ensure the achievement of functional requirements, defined service levels and health, safety, reliability, and other quality of service requirements. The objective for the maintenance of assets is to prevent or mitigate the deterioration of the assets to keep their performance and manage the risk of failure (see Chapter 9 for operation, maintenance, and rehabilitation). Depending on the type of asset (e.g., pipes, manholes, pumps), periodic maintenance and inspection activities are executed during the life cycle. Thus, maintenance plans should incorporate three categories of activities:

- Inspection, testing, and monitoring: to receive information on the actual asset's condition and performance (see Chapter 4).
- Preventive maintenance: planned activity (e.g., time-based, condition-based, risk-based – see Chapter 6).
- Corrective maintenance to repair defects and damages to restore the required performance, which is often unplanned.

The network operator needs to analyze and ultimately understand the causes of damage and faults. The evaluation can benefit from the large amounts of data gathered in regular inspections and continuous monitoring. It is also a prerequisite for assessing risks arising from damages by external actions or inappropriate operations (see the examples of deterioration mechanisms in Chapter 5).

Safe operation also requires foresight about the planning and provision of necessary resources (i.e., personnel, material, equipment, tools).

The plan to respond to emergencies is an essential tool to react timely and professionally to these situations. All the means to restore network operation and minimize harm to life and the environment should be available.

The last phase of the life cycle concerns the decommissioning and proper disposal of the assets. The decision to decommission can have various reasons, including:

- non-compliance with legal or technical standards,
- the asset can no longer provide the required level of service,

- there is a lack of spare parts,
- costs for ongoing operation are no longer economic,
- the originally intended functionality is now provided by other assets.

Waste disposal must be carried out properly, that is, without environmental impact, especially with contaminated assets. In addition, recycling options for dismantled assets are increasingly under scrutiny.

2.2.5 Information management

Professional information management is critical to a successful AM. Besides physical assets, data in digital support are becoming increasingly important to facilitate the management of large volumes of data. These include:

- inventory data of the physical assets using all relevant attributes, for example, installation year, material, type, and dimensions (reflecting as-built or as-landscaped conditions),
- performance and capacities information, including hydraulic and contaminant loadings,
- spatial information,
- logical groupings, for example, pipe types, zones, periods,
- historical records of events (e.g., faults),
- historical log of operation and maintenance activities,
- historical records of the condition state,
- documents, drawings, photographs,
- cost data (CAPEX, OPEX).

In the past, data were 'stored' mainly on paper, but today most data are managed digitally in systems, for example:

- Geographic information system (GIS).
- Workforce management system.
- Enterprise resource planning system (ERP).
- Hydraulic calculation software and other engineering software for modelling and simulation.
- Emergency management systems.
- Decision support systems (DSS) tools.
- Process control systems, SCADA.
- Document management systems.

However, many urban drainage systems managers still have hybrid (analogue and digital) data management environments but with increasing digitalization levels. Moreover, utilities have also often decentralized data management (e.g., people keeping their own 'excel sheets' instead of using central databases).

All assets, underground and above-ground, should be inspected regularly to keep up-to-date information about their performance and condition state. Data generated from inspections should be documented and linked to the assets in the AM system. In some countries, for example, Germany, legislation forces urban drainage system operators to inspect their assets (pipes, manholes, service connections) every 10–15 years. Over the last few years, this has led to overall knowledge about the condition of their assets. Annual statistics can show ongoing degradation, and statistical models for estimating an asset's future condition can be developed based on these data.

As the volume of data grows, the need for data quality standards in AM increases, including tailored checking and correction methods.

Data quality requirements as detailed in standard ISO 8000-1 (2022) include:

- Accuracy: data on assets are correct, and the accuracy of measurement data.
- Completeness: data for all relevant attributes is complete.

- Consistency: the data are consistent, for example, format, value ranges, units.
- Validated: data validation with adequate procedures.
- Updated: the data describe the status of the assets, including the management of historical events.
- Uniqueness: each record applies to one component without duplicates.
- Source: the data source is known, and a level of data reliability is used, for example, in situ measurements, as-built drawings, design drawings, and simulation results, among others.

In Chapter 8, issues related to data and information management are discussed.

2.2.6 Organization and people

Nowadays, organizations face a lack of human resources in many business areas. Urban drainage system operators must take care of their human 'assets' as well. Depending on the personnel structure, the brain drain is expected to continue in the years to come because of more people retiring and taking their long-time knowledge home. As in other areas, the aging of personnel is currently an issue of concern in the water sector. Feliciano *et al.* (2016) propose a personnel aging index to support the evaluation of human resources, considering employee ages and professional categories and the time needed to transfer the know-how to new staff members.

A good AM culture needs to be implemented in the entire utility. It applies to all levels, from the executive board to the operations staff, and involves external parties and stakeholders.

The executive board must communicate the objectives for AM, considering stakeholder expectations. They should act reasonably toward the employees and motivate people within the utility to align with the various AM activities. The management must keep an overview of all activities, planning, construction, operations, maintenance, and finances to reach an overall benefit and not exclusively an optimum for one task. This broad application of AM also includes results evaluation at all levels.

Good AM is based on a clear organizational structure in the utility. For the evaluation of an existing structure, one needs to assess which departments exist and their current tasks and duties. Depending on the size of the utility, it must be decided whether critical AM tasks are assigned to different departments or are aggregated in a central department. In either case, each area or department is responsible for its AM tasks. Therefore, people from distinct departments must work together, and it might not be an easy adaptation. An organigram chart should describe the roles of each department and contribute to underlining the significance of AM in the utility. If the organizational structure needs modifications to align AM activities, leading to changes in duties and responsibilities, communication of the advantages is essential.

Good AM is also teamwork. 'Lone warriors' should realize that teamwork can be more successful. In many utilities, reactive action and applying firefighting actions are business as usual. In good AM, activities are proactive to minimize risk and increase benefits. An existing culture of 'the way things are done around here all the time' must be challenged. A clear vision and transparent and comprehensible decisions can help. If resources in a utility are limited, people are too busy with their day-to-day work and have no free time for AM-related topics. In this situation, resources should be increased; where this is not possible, one should try to integrate AM tasks into the day-to-day work.

Everybody should understand the need for collaboration, communication, and transferring information. It should be clear how AM is organized in the utility and where specific information can be derived, for example, pipe data, condition information, hydraulic simulation, and rainfall series.

Another aspect of good AM is the development of specific AM competencies as described in CSA W211 (2021) or IAM (2014a, 2014b). This can be overcome by providing internal or external training on specialized knowledge and soft skills, for example, people who have not done a risk assessment yet should receive training (https://www.epa.gov/sustainable-water-infrastructure/advanced-asset-management-workshop-materials). Even experts should know about recent developments in their field. Since people from different departments and disciplines are involved in AM, it is essential to identify existing know-how and knowledge. A first approach can be to have informal rounds in which a team member reports on his work and experiences.

A good relationship with external contractors, manufacturers, consultants, academic institutions, watershed organizations and stakeholders is also essential to AM. Required service levels and resources must be communicated and fixed. Delivery of goods, materials and work must comply with legislation and required technical standards. External consultants or service providers can bring added knowledge, particularly when familiar with the system. In some areas, the utilities gain in promoting capacity building of the utility personnel. A good working relationship with academic institutions is advantageous to improve the overall knowledge, understanding and training of future professionals and workers.

2.2.7 Risk management and resilience

A classic definition of risk involves consideration of the likelihood of an event and the associated consequences. In ISO 31073, (2022), the risk is defined in a comprehensive way as 'the effect of uncertainty on objectives,' a broad concept incorporating both negative and positive effects, but also accepting changes not limited in time as are, for instance, the effects of climate change. The set of relevant events, 'the occurrence or change of a particular set of circumstances,' is central for risk assessment. Each event should be detailed, including the risk sources, the hazards involved, the influencing risk factors (e.g., associated with people, environment, materials, and equipment), likelihood and consequences.

An event can have different causes and may lead to a range of consequences, for example, a CSO can result from heavy rainfall, inadequate capacity of a sewer, or a blockage downstream.

This standard enhances the risk management principles (see Figure 2.8) and follows with minor changes the widely known risk process steps (see Figure 2.9). The first step, establishing the context, is essential to understand the internal and external contexts and define the scope and criteria for the assessment. Risk assessment is composed of risk identification, analysis, and evaluation:

- Identification: to find and describe relevant risks, understood as the identification of the pertinent set of events.
- Analysis: to describe the features and determine the level of risk associated with each event.
- Evaluation: to compare the levels of risk obtained with the established criteria and find priorities and risk treatment needs.

The following step, risk treatment, consists of identifying, selecting, and planning implementation of the measures. The monitoring and review step implies the planning and the implementation of monitoring actions. The step of communication and consultation is transversal to all the other phases and instrumental to providing information and involving all parts of the organization and all relevant stakeholders.

Establishing the context should not be overlooked. The purpose is to understand the utility and its internal and external contexts, including the drivers and trends affecting the management. This step also includes defining the scope and approach to risk management, including methods and criteria for risk assessment. These should enable estimating each risk and selecting a unique consistent scoring or ranking system to compare risks. Examples can be found in Chapter 6.

Monitoring and review are crucial for the success of risk management as it incorporates the actions ensuring the effectiveness of the whole process by critical analysis and observation, gathering, and analyzing the information, and reporting results to ensure continuous improvement. Essential components are the register of risk events (accidents or incidents) and the periodic review and reporting. Traceability should be ensured throughout the entire process to facilitate understanding and plan's regular review.

It is recommended, but not mandatory, to establish different consequence categories, for example, financial, health and safety, environmental, reputational or a combination of these. Consequently, setting priorities and selecting risk treatment measures can focus on specific consequences. In

Figure 2.8 Risk management principles. *Source*: adapted from ISO 31000 (2018).

addition, the importance (criticality) of assets helps to rank risk since it reflects the magnitude of the potential consequences.

Risk assessment is rarely straightforward, and many uncertainties and their effects must be considered, for example:

- uncertainty of available information and data,
- uncertainty of models and parameters used for prediction and decision-making,
- lack of knowledge to understand the problems.

Dealing with risk means balancing levels of acceptable or tolerable risks with available resources and the feasibility of measures to treat the risks. In practice, barriers to action and limited resources restrain appropriate responses. Furthermore, answering the question 'Which risk level can be tolerated?' is not straightforward. Reducing the probability or the consequences of an event can be

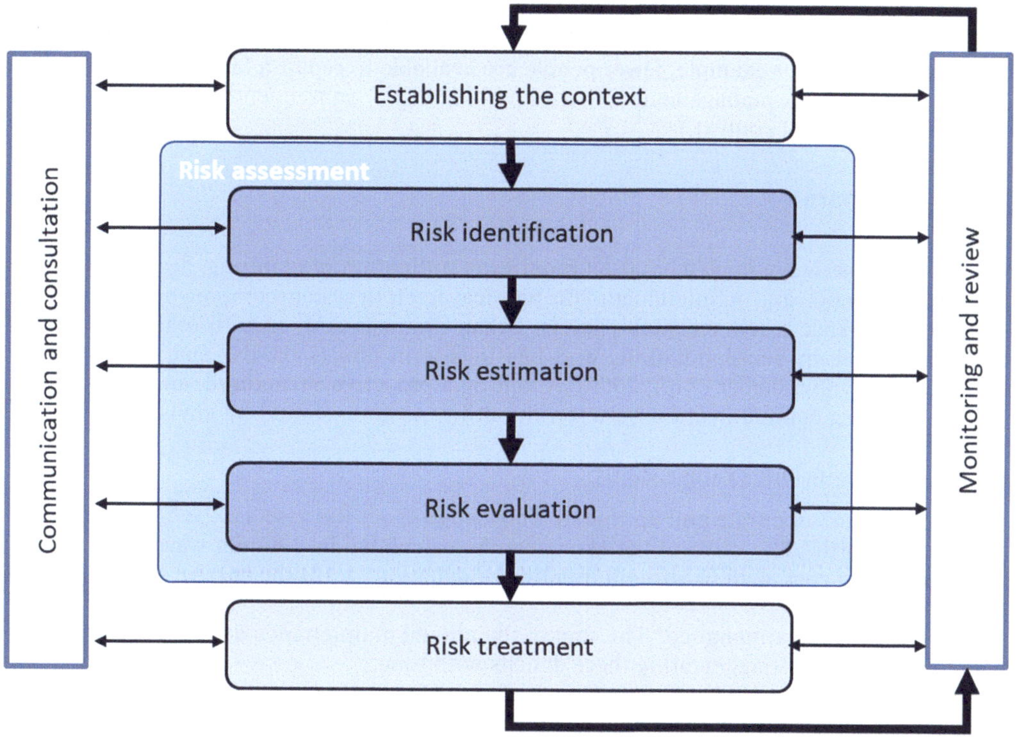

Figure 2.9 Risk management process. *Source*: adapted from ISO 31000 (2018).

undertaken in several ways, for example, for a flooding event, increasing the capacity of a sewer system or installing barriers to flood water.

Resilience is defined as the 'adaptative capacity of an organization in a complex and changing environment' by ISO 31073 (2022). Three qualities are intrinsic to this definition, following a disruptive event: (i) resisting or sustaining the functionalities of a system; (ii) restoring rapidly if affected; and adapting or improving to better respond to future events.

The fields of risk management and resilience management are connected (Aven & Thekdi, 2018). The integration of these two fields brings benefits to decision-making (Aven, 2018). Unlike risk management, resilience management can be pursued without considering potential events but allows a better understanding of cause–effect relationships and the use of resources, for instance, by estimating the likelihood of an event (Aven & Thekdi, 2018).

In case of an intense rainfall event, the resilience of an urban drainage system defines the ability to maintain service and recover its service within a specific time. The design and construction phases have a high influence on the resilience of the system but also of the served communities. In this situation, the implicit risk level of design criteria can be questionable. If the system or operating environment changes, the effects on risk and resilience need to be revised and adaptations must be considered for the urban drainage system. Examples of adaptation actions to the increasing risk of flooding are additional water-retention capacity or emergency power supply in pumping stations. Unlike wastewater or combined systems, stormwater systems encompass two outdoor subsystems; managing better the overland flow component can provide significant resilience to flooding in urban areas (see, e.g., van Duin *et al.*, 2021).

Concerning the long service life of urban drainage systems components, one activity of AM is to deal with changes in legislation and standards and availability of resources because these issues can lead to changes in risk, for example, fewer people are available to repair a failure. So, facing changes implies reviewing the risk management activities periodically or subsequently to substantial changes in the external or internal context.

2.3 MAIN APPROACHES

2.3.1 Overview

Infrastructure AM strategy implementation requires coordination of planning, design, construction, rehabilitation, operation and maintenance of the physical assets or system components to ensure system and utility performance reach acceptable levels. Urban drainage AM, as with other infrastructures, plays a decisive role in decision-making and has long-term effects on the quality of service and utility sustainability (Schneider *et al.*, 2006). Adopting a proactive, structured, and forward-looking framework, ensuring communication between the utility sectors, should be envisaged to reach the desired objectives.

Typical questions include (Vanier, 2001):

- What do I own? Accurate and up-to-date knowledge about the assets to be managed.
- What is it worth? The value of the assets can be calculated in different ways. For instance: (i) the investment cost, with or without inflation or deflation; or (ii) the current replacement value, among others.
- What is deferred maintenance? The sum of the annual maintenance deficits accounting for the compounding effects aggravating these deficits with time.
- What is the asset's condition? An asset condition is understood as how an asset performs its intended functions. Structural integrity is often the primary issue analyzed, but other functions can be evaluated, including hydraulic, environmental, and operational performance (see, e.g., EN 752, 2017 or Tscheikner-Gratl *et al.*, 2019).
- What is the remaining service life? The life expectancy of assets is central to AM. From a technical perspective, it is understood as the time remaining until the performance deteriorates to a level where a failure occurs (Marlow *et al.*, 2009).
- What do you fix first? Comprises setting intervention priorities and developing a plan after answering the previous questions. Faced with data and resource limitations, a more or less formal multi-criteria decision-making process must be selected to balance performance, risk, and cost.

Information (see Section 2.2.5 Information management) and techniques used in life-cycle processes (see Section 2.2.1 Overview) of assets are often similar. Therefore, it is usual to find similar practices to deal with, for instance, maintenance and rehabilitation management, focusing strongly on prevention replacing reactive action.

Asset and maintenance management techniques described in the literature are often interlinked, and sometimes borders are indistinct. For instance, limits are fuzzy when dealing with rehabilitation and maintenance planning of urban drainage. Another example is the need to consider the context when dealing with repair, as the techniques are used in rehabilitation and maintenance activities. As defined in EN 16323 (2014), 'rehabilitation (...is the set of...) measures for restoring or upgrading the performance of existing drain and sewer systems and includes renovation, repair and replacement', while maintenance is the 'routine work undertaken to ensure the continuing performance' of the systems.

This section is not exhaustive, and only some common practices are presented. These practices have a preventive nature contrasting with a 'run to fail' passive attitude. Perhaps the simplest way is time-based planning, where the answer to 'What and when to maintain or rehabilitate an asset' assumes the correlation between asset or system failures with service time. Even if many deterioration processes are time-dependent, many factors play a role in these processes, and other causes can decrease the performance. Therefore, despite its simplicity, this approach is not the most effective in identifying priority interventions.

Condition-based is a specific case of performance-based AM where the criteria used for decision-making are based on the condition of the assets. Information about the condition of physical assets is instrumental to all infrastructure AM practices. Condition assessment provides an understanding of asset deterioration level. It is a proxy or an indicator of the failure likelihood and a component of the risk of failure. The following approaches (reliability, risk, and value-based) explicitly combine information on condition, integrating criticality or consequences of failure.

Reliability-based AM focuses on the application of reliability methods and tools. Reliability is defined as the 'ability of a component, subsystem or system to perform a required function, under given environment and operational conditions and for a stated time' (Rausand, 2011; Rausand *et al.*, 2020; Smith, 2017). It is an integral part of the RAMS concept presented in Section 2.2.4 (Life-cycle delivery). When looking at criticality it is implicit that some components are more important as consequences of failure have a higher impact (Marlow *et al.*, 2007).

Risk-based AM practices incorporate the identification of which assets are at risk of imminent failure (likelihood of failure), of those having high consequences should they fail (often called critical assets), the use of risk management techniques to plan rehabilitation and maintenance, and plan condition monitoring to ensure continued identification of risk levels.

Value-based or value-driven AM is based on the perspective that AM decisions should be 'value-driven' instead of focused on cost. It is about ensuring the best level of value creation from the assets over the life cycle of these assets, achieved by managing adequately costs, risks, and performance.

Performance-based AM focuses on how the assets fulfill the different functions expected in alignment with utility objectives. It can be applied to sets of individual components or overall service delivered, often called service-based AM (Marlow *et al.*, 2007). The objectives defined incorporate a range of dimensions, including 'performance, risk and cost' (Alegre *et al.*, 2009).

The current trends are the approaches combining performance, risk, and cost (see Figure 2.10), as mentioned in Section 2.1 (Overview on AM). As multi-criteria problems, where many authors

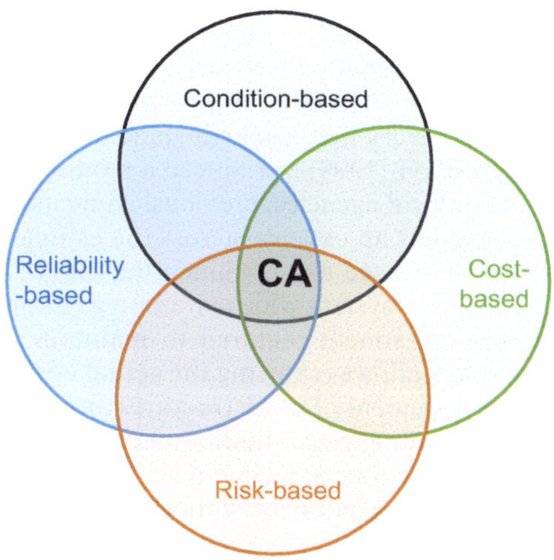

Figure 2.10 Perspectives of AM approaches and current trend of adopting a CA.

proposed the adoption of complex techniques under the vast field of multi-criteria/multi-objective decision support methods and tools (e.g., Carriço *et al.* 2021; Debères *et al.*, 2012).

For urban drainage systems, the perspectives of structural, hydraulic, environmental, and operational assessments, as in EN 752 (2017), are required to ensure the inclusion of all essential functions of these systems.

2.3.2 Condition-based AM

Network condition assessment is necessary to support investment in rehabilitation and maintenance (i.e., impacts on CAPEX and OPEX) and other related operational activities (e.g., assets inspection and systems monitoring). This information is the basis for the management practices of drainage assets. According to WRC (2023), asset condition is 'a measure of deterioration in the physical state of the asset when compared to its new state'. Figure 2.11 illustrates the uses of condition data or the derived condition rating in processes and decision-making, directly or indirectly.

The primary condition data results from inspections of network components, mainly sewers and manholes, using visual inspection or alternative inspection techniques, as detailed in Chapter 4. Other assets, such as pumping stations or overflow structures, also called vertical assets, are managed differently and periodically subject to maintenance and renewal by utilities, given their criticality (component importance), accessibility, and low numbers compared to sewers and manholes. Vertical assets are assets within a building or a facility, often composed of multiple components and frequently above ground, including pumping stations, treatment plants, or storage facilities (AWWA, 2018).

Condition-based approaches are a subtype of performance-based assessments where the performance criteria are based mainly on condition ranking or metrics. Condition assessment is typically undertaken to manage individual components, for example, using a grading system to associate a condition class to each component. For each class, intervention type and priority are previously defined, thus facilitating decision making. The reasoning behind condition assessment is that it provides an understanding of the level of asset deterioration, an indicator of failure likelihood.

Condition classes or grades, as a rule, follow a five-level ordinal scale, and are associated with typical courses of action for inspection, maintenance, and rehabilitation, for example, class or grade 1 – acceptable structural condition to 5 – collapsed or collapse imminent (WRC, 2023). In this example, grade 5 corresponds to higher intervention priority, for example, immediate action if it is a critical asset. A critical asset is a system component whose failure has a significant effect on consequences for the service and the community.

Regulatory or governmental rules may require the use of condition assessment by utilities managing water, wastewater, or stormwater systems. In the UK, utilities report on the condition and performance of their assets each year (Scotland) or every five years (England and Wales) (Marlow *et al.*, 2007). In the USA, the publication of the GASB 34 (1999) introduced a modified approach to report the value of infrastructure managed by state or local agencies, previously only using the depreciation approach, that is, the cost of an asset was spread as an expense across the estimated service life. The modified approach requires that an AM system is used and documenting that the assets are at an established minimum condition level. This calls for a system-wide condition assessment at least once every three years and estimates of the annual investment required to maintain the assets at the established condition level. The modified approach allows reporting the actual cost of maintaining infrastructure assets, encourages the proactive management of infrastructures and facilitates justifying the levels of investment, increasing prospects of funding rehabilitation needs. In Portugal, the regulator ERSAR (2023) undertakes a national quality of service assessment yearly based on a performance assessment system; the revised system includes a metric addressing the condition of sewers as part of the criteria for infrastructure sustainability. Similar practices are found in other countries.

In Chapters 4, 5 and 6, respectively, insights on the investigation of condition, deterioration processes and models, and condition-based decision making are provided.

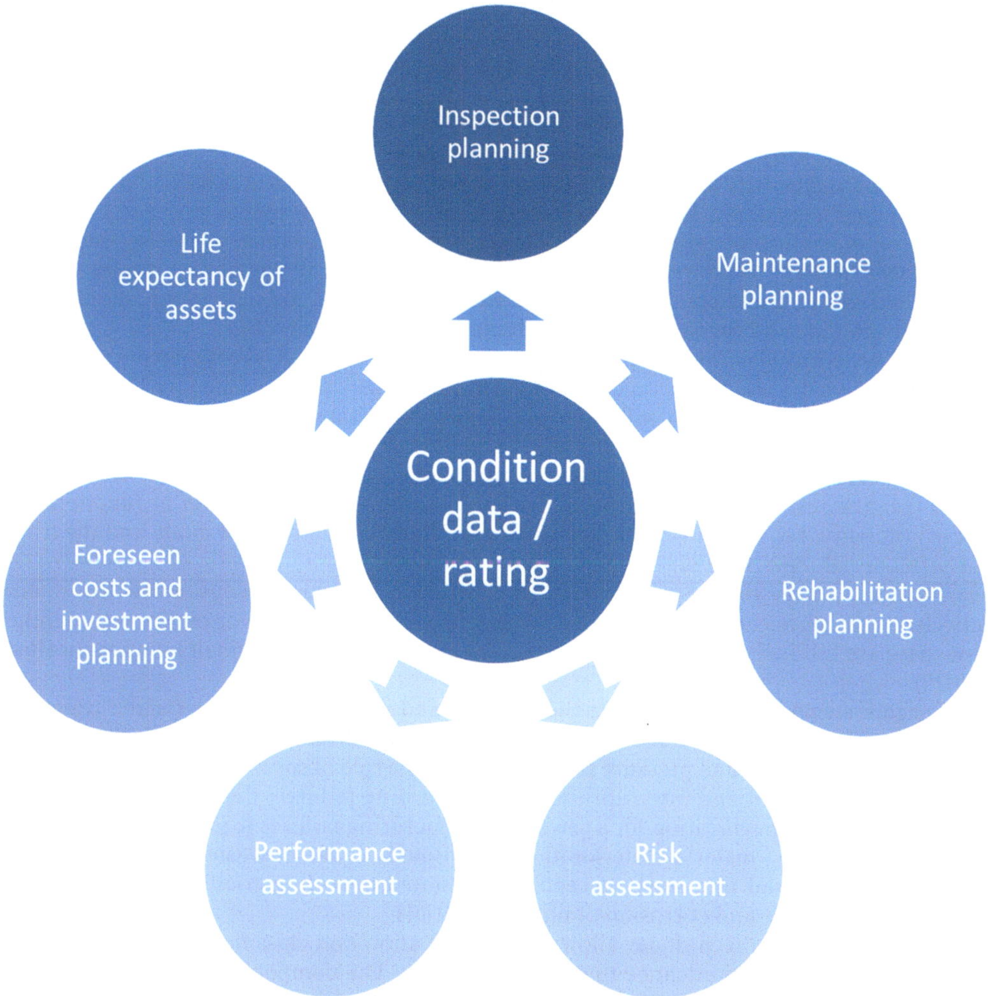

Figure 2.11 Direct or indirect uses of condition data or the derived condition rating in processes and decision making.

2.3.3 Reliability-based AM

The reliability concept associated with technological systems goes back to the beginning of the 20th century, with extensive developments after 1960, with three main branches of knowledge, hardware, software, and human reliability (Rausand *et al.*, 2020). The approach to the systems composed of several components is called system reliability analysis, as with applications to drainage systems.

The term reliability refers to the ability of a system, subsystem, or component to fulfill a function under given environmental and operational conditions for a stated period or during the design service life; it is often expressed as a probability and covers safety, serviceability, and durability (adapted from ISO 8930, 2021). To assess the reliability, it is necessary to specify the required system functions previously. System reliability is a time-dependent characteristic depending on many factors, including component maintenance and rehabilitation.

The purpose of system reliability analysis is to use specific techniques to reduce the likelihood of failures. Failure modes, root causes, and mechanisms identification are essential to act effectively and make informed decisions. Reliability techniques have wide application in risk-based approaches, as illustrated in the IEC 31010 (2019) standard on risk assessment techniques.

Two main failure types are sudden and gradual failures (e.g., degradation resulting from the wear-out of system components). The latter is the focus of condition-based AM.

Measures or indicators of reliability include the mean time to failure (MTTF), the mean time between failures (MTBF), mean residual life (MRL), failure rate and availability (proportion of time an item is functioning). These can be calculated from available data and with failure models based on statistical distributions and further used to predict failures (see Chapter 5).

Qualitative system analysis techniques are based on functional analysis of systems and interfaces. An important step is the identification of the relevant functions and the related performance criteria; once these are described, the failure events and faults ('the state of an item characterized by the inability to perform a required function') can be identified and classified (Rausand *et al.*, 2020). Some techniques, such as the FMECA (failure modes, effects, and criticality analysis), incorporate the effects (i.e., the results or consequences of a failure upon system operation or function) and the concept of criticality (i.e., a relative measure of the consequences of a failure mode and the frequency of its occurrence) (Borgovini *et al.*, 1993).

The way a system is viewed can have a structural or functional focus when the interest is, respectively, on the physical structure of the subsystems and components or on the system functions and how these are fulfilled. A function is 'what has to be done' by the system or component (Rausand *et al.*, 2020).

In drainage systems, functions include those associated with requirements for the transport and treatment of wastewater, stormwater, or surface water (e.g., for sewers, containing the fluid, conveying the fluid, resistance to loads and pressure from flow) or to the role of specific components (e.g., a sluice gate must contain the fluid and interrupt the flow). Regarding failures, these include design failures (e.g., unsuited pressure specification for a sewer), manufacturing failures (e.g., defective components), degradation failures (e.g., material deterioration with time because of wastewater flow) and misuse failures (e.g., external load in a sewer exceeding the stated operating conditions). Frequent failure mechanisms include wear and corrosion of piping materials.

Component importance techniques allow the identification and classification of components in a system according to their significance to system reliability. The component importance depends on the specified system function (Rausand *et al.*, 2020).

Using reliability techniques to support urban drainage AM has broad potential. The approaches are not disjointed. Reliability-based AM has clear overlaps with condition-based and risk-based AM. Some reliability techniques apply to all three approaches (e.g., fault tree analysis, FMECA, HAZOP). For instance, FMECA is used in the risk identification phase of the risk management process of networked infrastructures (e.g., for road AM as presented in MASC (2009). Stanić *et al.* (2014) show an application of HAZOP. Several authors undertook the identification of failure modes or failure rates to feed on reliability or risk-based approaches (e.g., Almeida *et al.* 2009; Ana & Bauwens, 2010; Davies *et al.* 2001; Laakso *et al.* 2017; Mancuso *et al.* 2016).

2.3.4 Risk-based AM

Risk-based AM considers explicitly the risks a utility manager can face when managing the utility. These include health, safety, structural, environmental, business, functional, reputation, and service-related risks.

The risk management process presented in Section 2.2.7 (Risk management and resilience) shows how the accepted broad approach is independent of the application. Development of each phase depends on the application and many contextual factors; data and information availability determine the techniques to use. In urban drainage, data may be scarce and with significant levels of uncertainty. In a narrow interpretation, a risk-based approach should include the fundamental characteristics of risk (i.e., likelihood and consequences) and incorporate the phases of the risk processes. However, in broader terms, authors considered risk-based approaches incorporating any direct or indirect measure of risk in the methodology or decision-making process, as illustrated in a comprehensive review by Tscheikner-Gratl *et al.* (2019). In the literature review provided by these authors, advancement opportunities comprise improvement of the terminology and concepts used, aligned with those accepted by risk experts, and the application of the risk process phases.

Risk analysis results are more robust if sound knowledge is behind the risk values (Aven, 2017). The description of techniques or methods should limit the subjective interpretation of users on the application, which can reduce the robustness of the results. As an example, when using a risk matrix, the scales for likelihood, consequences, and risk should be defined in a way that reduces the subjectivity by the user; it is not enough to state that a particular class of likelihood is rare and requires the expression of the corresponding range of probabilities (e.g., Patt & Schrag, 2003).

An additional perspective of risk applied to AM is the identification of the relevant risks and how they are characterized. Two broad situations are: risk events correspond to occurrences limited in time (e.g., failures, accidents) or changes in circumstances, as those associated with climate change (e.g., average temperature increase, rising sea water level). Furthermore, the purpose of risk analysis might be limited to a specific type of risk or embrace several relevant risks.

The consequences included in the analysis can comprise public health and safety, business continuity (e.g., service provision of utility or other services), financial effects (e.g., to utility, critical services, commerce, public and private property), effects on urban functions (e.g., mobility), environmental effects, damages to natural and built heritage, among others.

The reader finds more detailed information related to decision making in Chapter 7.

2.3.5 Value-based AM

An essential term in ISO 55000 (2014) is value. The standard states 'Asset management does not focus on the asset itself, but on the value that the asset can provide to the organization' and its stakeholders. This changes in a way the view from cost to value. Thus, AM decisions should be 'value driven' and not 'cost focused' as they traditionally were. The cheapest solution is not always the best one.

A central question within this AM approach is which value brings assets to specific stakeholder expectations within the decision-making process. Srinivasan and Parlikad (2017), see also Figure 2.12, describe a three-stage approach for decision making in value-based AM.

Since infrastructure in urban drainage systems has rather long service lives, the time frame for the evaluation should include a long-term perspective, together with mid- and short-term perspectives.

Typical stakeholder expectations, to mention a few again are, for example:

- safe and reliable transport of wastewater from origin to WWTP,
- safe and reliable transport of stormwater to an outfall or disposal nearby,
- operate efficiently,
- reduce costs,
- reduce risk of flooding or CSO and their effect, for example, reduce the risk for disturbances and damages,
- stable and low wastewater and stormwater rates,
- sufficient investment opportunities, preferably financed by internal measures and not external loans.

A - Establishment of the context

- Set the objectives and scope
- Define the problem statement
- Determine the time period for evaluation

B - Value mapping

- Identify the stakeholders
- Identify stakeholders' requirements and objectives
- Identify the value drivers that contribute to stakeholders' requirements
- Identify the value metrics to assess each value driver
- Determine how the asset can directly influence each of the value metrics
- Determine how the asset can indirectly influence each of the value metrics
- Determine the external factors that influence asset and value metrics
- Determine the various intervention and control options
- Identify and map the link between various factors to value generation
- Determine the factors that influence the decisions

C - Value assessment

- Identify modelling requirements
- Develop the model
- Identify the best solution
- Perform sensitivity analysis

Figure 2.12 Systematic approach to value-based infrastructure AM. *Source*: adapted from Srinivasan and Parlikad (2017).

An example of the value of assets is the decision on the alternatives to replace an old existing sanitary pipe section with a new one. The old sanitary pipe section is likely to be fully depreciated and might have a poor condition, which requires increasing inspection and maintenance activities over time. A new pipe section requires a certain amount of investment budget (including disposal of the old pipes), but the expectation of low maintenance costs in the following years. Without additional evaluation factors, a simple cost evaluation often results in keeping the old pipes as long as possible in operation. But what kind of value do the pipes bring to the value of low risk and good performance? The old pipes may have a low capacity compared to future demands and thus, the risk of sewer overflows may rise, meaning its performance value is likely to be lower. Due to its bad condition, the risk of sewer collapses increases as does the potential for operational interruptions and damages and disturbances near the location of the sewers. The increasing risk means that the risk value is therefore decreasing. Finally, because of the full depreciation, the sanitary pipes do not contribute to wastewater fees. This seems to be good at first glance. However, if wastewater fees decrease over time, there is less income to build up internal investment budgets, and the need for external loans may rise. On the other hand, the performance and risk values of the new pipe sections should be high. Based on future demand analysis, planning the capacity of the new sewer pipes should be sufficient for the upcoming x years. A new pipe is likely to be in (very) good condition with a low risk of failure and low operational costs. Due to new investments, ongoing depreciation can be used to contribute to stable fees.

The aforementioned depreciation issue also describes an inherent 'problem' many urban drainage system operators face. Depreciation periods are often lower than real technical service lives, and the challenge is to get better knowledge about actual service lives to increase depreciation periods.

2.3.6 Performance or service-based AM

Performance assessment systems are applied to support strategic and AM tactical planning from the organization's strategic and AM objectives. Performance assessment applies to responding to regulatory requirements.

Performance assessment, including service assessment aspects, is a component of the PAS 55-1 (2008) framework in the performance assessment and improvement phase. The industry recognizes performance assessment as a tool to drive efficiency and best practices (Alegre *et al.*, 2009).

The framework described in EN 752 (2017) incorporates the need to define performance requirements, considering local, regional, national, and European requirements. A four-stage process is proposed for the management of sewer systems: (i) investigation, (ii) assessment, (iii) development of a plan and (iv) implementation. The performance level is a trigger to the need for rehabilitation or upgrading. Four primary aspects of sewerage investigations and performance assessment are (EN 752, 2017; WRC, 2023):

- Structural: remaining sewer resistance to internal and external actions eventually leads to component collapses and failure to achieve the primary purpose of the sewer (i.e., to carry flow).
- Hydraulic: capacity to contain and transport flow, a failure potentially resulting in surcharge, combined sewer overflow discharges or flooding.
- Environmental: events of pollution of the environment.
- Operation and maintenance: activities to maintain the integrity of sewer components, thus ensuring the normal functioning of the sewer system and quality of service.

A dedicated performance assessment system (PAS) is an essential part of AM, allowing a more effective response to actual challenges aligned with the utility objectives (Cardoso *et al.*, 2016). The tailored performance assessment for AM is used for system diagnosis, monitoring the utility AM strategy, and planning of AM tactics and associated decision-making processes.

Almeida *et al.* (2021) present an example of a strategic utility assessment system. Figure 2.13 illustrates the IAM planning process structure proposed. The performance assessment system comprises the O-C-M-RV tree, that is, objectives, criteria (i.e., perspectives to assess each objective), metrics (i.e., measures to quantify the criteria) and reference values (i.e., range of values used to classify and judge the metric results), accounting for contextual factors (Figure 2.14). This tree structure applies to all decision levels (strategic, tactical, and operational) and facilitates the incorporation of multiple perspectives relevant to the utility.

ALA (2004) presents a risk-based approach using a simpler structure for performance assessment with objectives and criteria, aiming at evaluating wastewater systems performance subject to natural hazards and human threats. The objectives are associated with risk acceptance for 100-year and 500-year return events.

The results of applying the performance assessment system can be used to define courses of action aiming at global improvement while focusing on assets with poor scores for objectives and criteria. These courses of action are not necessarily alternatives; for instance, if the knowledge of the system and its functioning is insufficient, the way to proceed is to upgrade monitoring and data collection. If the rehabilitation needs identified are above the actual capacity of the utility, priorities need to be defined; multi-criteria methods are adequate, even if requiring a minimum level of expertise. Another frequent situation is the decision between rehabilitation techniques.

The use of performance metrics or indicators, together with multi-criteria decision support methods, has been investigated since 2000 by either looking at the whole system, components, or specific issues. Le Gauffre *et al.* (2010) presented the application for several purposes, from processing visual

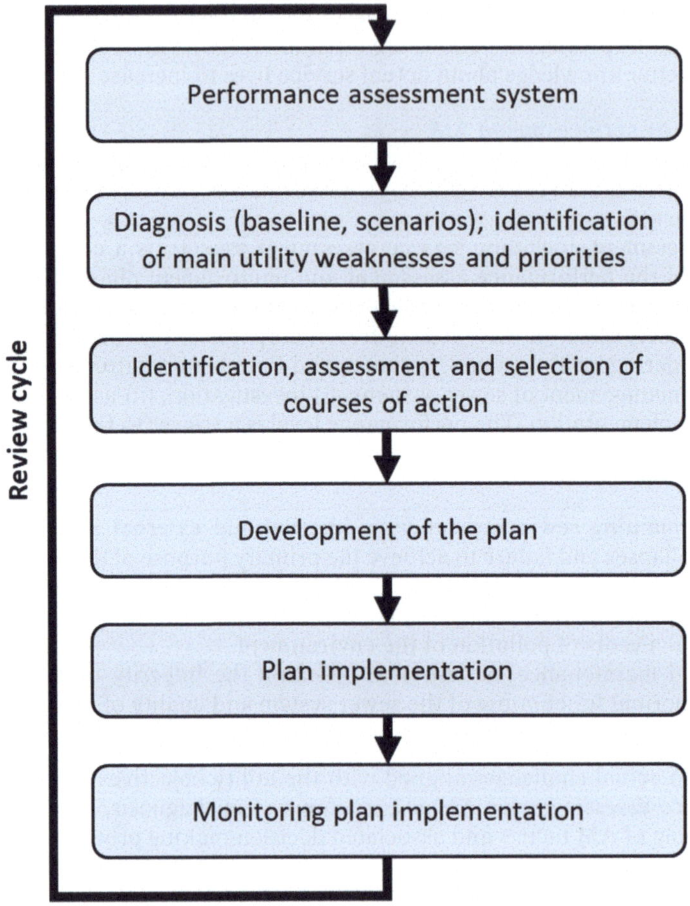

Figure 2.13 Typical structure for the IAM planning process.

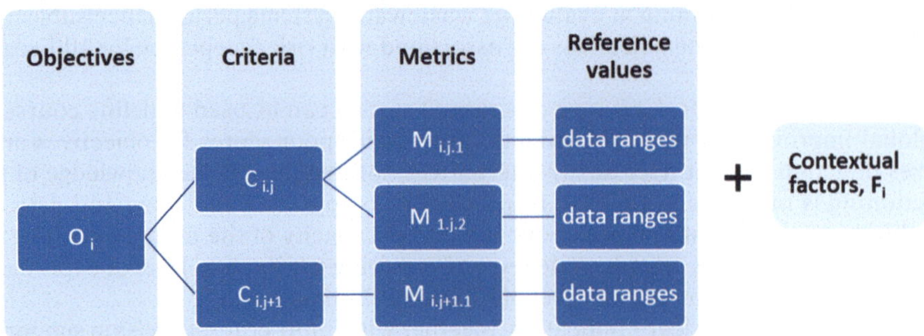

Figure 2.14 O-C-M-RV tree.

inspection results to setting rehabilitation priorities. Carriço *et al.* (2021) presented a multi-criteria decision support methodology for the selection and prioritization of rehabilitation intervention in urban water infrastructures considering performance, cost and risk and a long-term perspective.

A performance assessment system is suitable for all kinds of reporting, especially for management reviews and stakeholder meetings. Some metrics are just for information, for example, the total length of sewer pipes; others allow for measuring performance and identifying needs for change, if their performance does not meet targets. The asset manager can take all available information, on performance, risk, and cost metrics, to see if the AM objectives are achieved. Adaptation of the performance assessment system over time allows following external and internal context changes, including available information.

The reader will find more detailed information related to service-based AM in Chapter 6.

2.4 CHALLENGES, LIMITATIONS, AND OPPORTUNITIES

Adoption of AM by drainage managers brings benefits in terms of organizational knowledge and improvement of working practices, among others, globally improving the quality of service.

Current challenges to asset-intensive organizations are the increasing rehabilitation of components in poor condition or reaching the design service life and the deleterious effects of climate change. Natural hazards such as floods, landslides, and earthquakes have the potential to damage assets severely. Demographic and water consumption patterns change, resulting from pressing needs to use water more efficiently to face climate change, potentially affect services performance and need consideration in AM.

Application of existing conceptual approaches is often limited by not fully integrating the organizational and practical limitations. Most utilities have limited budgets; thus, pragmatic action is required, for instance, looking at critical components, that is, acting on those with a higher risk of failure (WRC, 2023), a wise alternative to run-to-fail, while applying this to the non-critical components.

Existing systems are combined or separate, or a combination of these two, each with advantages and disadvantages. To increase the overall sustainability of our systems, including wastewater and stormwater treatment, the rehabilitation of assets provides an opportunity. Reduction of the pollution of natural water resources is not achieved only by increasing the treatment level at treatment plants. One needs to increase the tightness of systems carrying wastewater, severely reduce the CSO discharges and ensure the separation of wastewater from rain and other non-polluted water. There is an opportunity to use nature-based solutions to reduce the buried infrastructures, where poor performance and illicit connections are difficult to detect, with potential collateral benefits. For a consistent change, the adoption of new drainage concepts and paradigms using a multi-disciplinary perspective is required (Bertrand-Krajewski, 2021).

Incorporating new drainage paradigms, namely nature-based solutions, implies the inclusion of different components and the interaction of more than one service in their management and maintenance, for example, drainage services and green areas managers, requires a clear understanding of how and by whom these assets are managed. Adapting the current AM systems to incorporate different components involves changes, but substantial difficulties are not foreseen.

Urban drainage systems are perfect examples of challenges and opportunities for sustainable development. In the Brundtland report from the World Commission on Environment and Development (WCED, 1987) sustainability was defined as 'development that meets the needs of the present without compromising the ability of future generations to meet their own needs'. A very suitable definition, also appropriate for urban drainage systems, comes from Sustainability Committee of the University of California in Los Angeles (UCLA, 2016): 'Sustainability is defined as the integration of environmental health, social equity, and economic vitality to create thriving, healthy, diverse, and resilient communities for this generation and generations to come. The sustainability practice recognizes how these issues are interconnected and requires a systems approach and an acknowledgement of complexity.' The increase in challenges because of climate change requires focusing on the environmental effects of urban drainage systems throughout the whole life cycle. Resources are limited; therefore, the design of urban drainage systems must find optimal solutions for today but with the flexibility to adapt in the future. One goal should be to reach a long service life for the assets. This can be achieved through proper planning and construction, quality assurance measures, and continued maintenance actions during the system's operation. Socioeconomic sustainability requires the operation of urban drainage systems to remain affordable for all customers.

REFERENCES

ALA (2004). Wastewater System Performance Assessment Guideline (Guideline). American Lifelines Alliance, USA. https://www.americanlifelinesalliance.com/pdf/aaaALAWWGuidelines070104DraftFinal.pdf

Alegre H., Covas D., Monteiro A. J. and Duarte P. (2006). Water Infrastructure Asset Management: A Methodology to Define Investment Prioritization. Water Distribution Systems Analysis Symposium 2006. Presented at the Eighth Annual Water Distribution Systems Analysis Symposium (WDSA), American Society of Civil Engineers, Cincinnati, Ohio, United States, pp. 1–22, https://doi.org/10.1061/40941(247)29

Alegre H., Cabrera E. and Merkel W. (2009). Performance assessment of urban utilities: the case of water supply, wastewater and solid waste. *Journal of Water Supply: Research and Technology–AQUA*, **58**, 305, https://doi.org/10.2166/aqua.2009.041

Alegre H., Coelho S. T., Covas D. I. C., Almeida M. do. C. and Cardoso A. (2013). A utility-tailored methodology for integrated asset management of urban water infrastructure. *Water Supply*, **13**, 1444–1451, https://doi.org/10.2166/ws.2013.108

Almeida M. C., Fernandes F., Charneca N. and David M. C. (2009). Rehabilitation of a large sewer: methodology for the Alcantara interceptor sewer. In: Strategic Asset Management of Water Supply and Wastewater Infrastructures, H. Alegre and M. C. Almeida (eds), IWA Publishing, London, UK, pp. 299–311.

Almeida M. D. C., Telhado M. J., Morais M., Barreiro J. and Lopes R. (2020). Urban resilience to flooding: triangulation of methods for hazard identification in urban areas. *Sustainability*, **12**, 2227, https://doi.org/10.3390/su12062227

Almeida M. do. C., Brito R. S., Jorge C. and Cardoso M. A. (2021). Performance assessment system to wastewater utilities strategic planning. *Water*, **13**, 2489, https://doi.org/10.3390/w13182489

Ana E. V. and Bauwens W. (2010). Modeling the structural deterioration of urban drainage pipes: the state-of-the-art in statistical methods. *Urban Water Journal*, **7**, 47–59, https://doi.org/10.1080/15730620903447597

Aven T. (2017). Improving risk characterisations in practical situations by highlighting knowledge aspects, with applications to risk matrices. *Reliability Engineering & System Safety*, **167**, 42–48, https://doi.org/10.1016/j.ress.2017.05.006

Aven T. (2018). The call for a shift from risk to resilience: what does it mean? *Risk Analysis*, **39**, 1196–1203, https://doi.org/10.1111/risa.13247

Aven T. and Thekdi S. (2018). The importance of resilience-based strategies in risk analysis, and vice versa. In: IRGC Resource Guide on Resilience (Vol. 2): Domains of Resilience for Complex Interconnected Systems, B. D. Trump, M.-V. Florin and I. Linkov (eds), EPFL International Risk Governance Center, Lausanne, Switzerland, pp. 33–38.

AWWA (2018). AWWA Asset Management Definitions Guidebook. American Water works Association, Asset Management Committee, USA. https://www.awwa.org/Portals/0/AWWA/ETS/Resources/AMGuidebook.pdf?ver=2018-12-13-100101-887

Bertrand-Krajewski J.-L. (2021). Integrated urban stormwater management: evolution and multidisciplinary perspective. *Journal of Hydro-Environment Research*, **38**, 72–83, https://doi.org/10.1016/j.jher.2020.11.003

Borgovini R., Pemberton S. and Rossi M. (1993). Failure Mode, Effects and Criticality Analysis (FMECA) (No. F30602-91-C-0002). Reliability Analysis Center, Rome, NY, USA.

Burns P., Hope D. and Roorda J. (1999). Managing infrastructure for the next generation. *Automation in Construction*, **8**, 689–703, https://doi.org/10.1016/S0926-5805(98)00115-0

Cardoso M. A., Almeida M. C. and Silva M. S. (2016). Sewer asset management planning – implementation of a structured approach in wastewater utilities. *Urban Water Journal*, **13**, 15–27, https://doi.org/10.1080/1573062X.2015.1076859

Carriço N., Covas D. and Almeida M. do C. (2021). Multi-criteria decision analysis in urban water asset management. *Urban Water Journal*, **18**(7), 558–569, https://doi.org/10.1080/1573062X.2021.1913613

CSA W211 (2021). Management Standard for Stormwater Systems (Canadian Standard). Canadian Standards Association, Toronto, Ontario, Canada.

Davies J. P., Clarke B. A., Whiter J. T. and Cunningham R. J. (2001). Factors influencing the structural deterioration and collapse of rigid sewer pipes. *Urban Water*, **3**, 73–89, https://doi.org/10.1016/S1462-0758(01)00017-6

Debères P., Ahmadi M., De Massiac J.-C., Le Gauffre P., Cherqui F., Werey C. and Nirsimloo K. (2012). Deploying a sewer asset management strategy using the Indigau decision support system. Proceedings of 4th Leading Edge Conference on Strategic Asset Management (LESAM), Mühlheim, Germany, p. **11**.

Doran G. T. (1981). There's a S.M.A.R.T. way to write management's goals and objectives. *Management Review*, **70**, 35–36.

EN 16323 (2014). Glossary of Wastewater Engineering Terms (European Standard). European Committee for Standardization, Brussels, Belgium.

EN 752 (2017). Drain and Sewer Systems Outside Buildings: Sewer System Management (European Standard). European Committee for Standardization, Brussels, Belgium.

ERSAR (2023). Guide for Assessment of the Quality of Water and Waste Services Provided to Users: 4th Generation Assessment System. ERSAR – The Water and Waste Services Regulation Authority, Lisbon, Portugal.

Feliciano J., Almeida R., Santos A., Ramalho P., Ganhão A., Covas D. and Alegre H. (2016). Assessing human resources renovation needs in water utilities. *Water Practice and Technology*, **11**, 728–735, https://doi.org/10.2166/wpt.2016.078

Fletcher T. D., Shuster W., Hunt W. F., Ashley R., Butler D., Arthur S., Trowsdale S., Barraud S., Semadeni-Davies A., Bertrand-Krajewski J.-L., Mikkelsen P. S., Rivard G., Uhl M., Dagenais D. and Viklander M. (2015). SUDS, LID, BMPs, WSUD and more – The evolution and application of terminology surrounding urban drainage. *Urban Water Journal*, **12**, 525–542, https://doi.org/10.1080/1573062X.2014.916314

GASB 34 (1999). Statement No. 34 of the Governmental Accounting Standards Board Basic Financial Statements—and Management's Discussion and Analysis—for State and Local Governments (Governmental Accounting Standard No. GS34), Governmental Accounting Standards Series. Governmental Accounting Standards Board, Norwalk, CT, USA.

Gavrikova E., Volkova I. and Burda Y. (2020). Strategic aspects of asset management: an overview of current research. *Sustainability*, **12**, 5955, https://doi.org/10.3390/su12155955

GIOC (2021). Green Stormwater Infrastructure Asset Management Resources Toolkit (Toolkit). Green Infrastructure Ontario Coalition, Ontario, Canada, https://giexchange.org/wp-content/uploads/2021/12/GSI-AM-Resources-Toolkit-Final-Dec-17.pdf

Haldane A. G. (2014). The age of asset management?

IAM (2014a). The IAM Competences Framework. Part 1: Asset Management Competence Requirements Framework. The Institute of Asset Management, Bristol, United Kingdom.

IAM (2014b). The IAM Competences Framework. Part 2: Guidance. The Institute of Asset Management, Bristol, United Kingdom.

IAM (2015). Asset Management: an Anatomy, Version 3. ed. The Institute of Asset Management, Bristol, United Kingdom.

IEC 31010 (2019). Risk Management – Risk Assessment Techniques (International Standard). International Electrotechnical Commission, Geneva, Switzerland.

ISO 55000 (2014). Asset Management – Overview, Principles and Terminology (International Standard). International Organization for Standardization, Geneva, Switzerland.

ISO 55001 (2014). Asset Management – Management Systems – Requirements (International Standard). International Organization for Standardization, Geneva, Switzerland.

ISO 24516-3 (2017). Guidelines for the Management of Assets of Water Supply and Wastewater Systems – Part 3: Wastewater Collection Networks (International Standard). International Organization for Standardization, Geneva, Switzerland.

ISO 31000 (2018). Risk Management – Guidelines (International Standard). International Organization for Standardization, Geneva, Switzerland.

ISO 8930 (2021). General Principles on Reliability for Structures – Vocabulary (International Standard). International Organization for Standardization, Geneva, Switzerland.

ISO 31073 (2022). Risk Management – Vocabulary (International Standard). International Organization for Standardization, Geneva, Switzerland.

ISO 8000-1 (2022). Data Quality – Part 1: Overview (International Standard). International Organization for Standardization, Geneva, Switzerland.

Jones M., Williams W. and Stillman J. (2014). The evolution of asset management in the water industry. *Journal AWWA*, **106**, 140–148, https://doi.org/10.5942/jawwa.2014.106.0114

Kapetas L. and Fenner R. (2020). Integrating blue-green and grey infrastructure through an adaptation pathway approach to surface water flooding. *Philosophical Transactions of the Royal Society A: Mathematical, Physical and Engineering Sciences*, **378**, 20190204, https://doi.org/10.1098/rsta.2019.0204

Laakso T., Ahopelto S., Lampola T., Kokkonen T. and Vahala R. (2017). Estimating water and wastewater pipe failure consequences and the most detrimental failure modes. *Water Supply*, **18**, 901–909, https://doi.org/10.2166/ws.2017.164

Langeveld J. G., Cherqui F., Tscheikner-Gratl F., Muthanna T. M., Juarez M. F.-D., Leitão J. P., Roghani B., Kerres K., do Céu Almeida M., Werey C. and Rulleau B. (2022). Asset management for blue-green infrastructures: a scoping review. *Blue-Green Systems*, **4**, 272–290, https://doi.org/10.2166/bgs.2022.019

Le Gauffre P., Cherqui F., de Mordant de Massiac J. C., Joannis C., Werey C. and Rozan A. (2010). Indicateurs de Performance Pour la Gestion Patrimoniale des Réseaux D'assainissement Urbains. Rapport final. Projet ANR RGC-U INDIGAU. Irstea, Lyon, France.

Mancuso A., Compare M., Salo A., Zio E. and Laakso T. (2016). Risk-based optimization of pipe inspections in large underground networks with imprecise information. *Reliability Engineering & System Safety*, **152**, 228–238, https://doi.org/10.1016/j.ress.2016.03.011

Marlow D., Heart S., Burn S., Urquhart A., Gould S., Anderson M., Cook S., Ambrose M., Madin B. and Fitzgerald A. (2007). Condition Assessment Strategies and Protocols for Water and Wastewater Utility Assets (No. 03-CTS-20CO). Water Environmental Research Foundation (WERF), American Water Works Association Research Foundation (AwwaRF), United States Environmental Protection Agency (U.S. EPA), London, UK.

Marlow D., Davis P., Trans D., Beale D. and Burn S. (2009). Remaining Asset Life: A State of the Art Review. WERF, Alexandria, VA, USA.

MASC (2009). Road Asset Management Plan. Mount Alexander Shire Council.

OFWAT, DEFRA (2006). The Development of the Water Industry in England and Wales. Office of Water Services & Department for Environment Food and Rural Affairs, The Crown, UK, https://www.ofwat.gov.uk/wp-content/uploads/2015/11/rpt_com_devwatindust270106.pdf

PAS 55-1 (2008). Asset Management. Part 1: Specification for the Optimized Management of Physical Assets. Institute of Asset Management (IAM), in collaboration with the British Standards Institution (BSI), London, UK.

Patt A. G. and Schrag D. P. (2003). Using specific language to describe risk and probability. *Climatic Change*, **61**, 17–30, https://doi.org/10.1023/A:1026314523443

Rausand M. (2011). Risk Assessment: Theory, Methods, and Applications, Statistics in Practice. John Wiley & Sons, Inc., Hoboken, NJ, USA.

Rausand M., Barros A. and Høyland A. (2020). System Reliability Theory: Models, Statistical Methods, and Applications. 3rd edn, Wiley Series in Probability and Statistics. John Wiley & Sons, Inc., Hoboken, NJ.

Schneider J., Gaul A. J., Neumann C., Hográfer J., Wellßow W., Schwan M. and Schnettler A. (2006). Asset management techniques. *International Journal of Electrical Power & Energy Systems*, **28**, 643–654, https://doi.org/10.1016/j.ijepes.2006.03.007

Smith D. J. (2017). Reliability, Maintainability and Risk: Practical Methods for Engineers. 9th edn, Butterworth-Heinemann, an imprint of Elsevier, Kidlington, Oxford, United Kingdom.

Srinivasan R. and Parlikad A. K. (2017). An approach to value-based infrastructure asset management. *Infrastructure Asset Management*, **4**, 87–95, https://doi.org/10.1680/jinam.17.00003

Stanić N., Langeveld J. G. and Clemens F. H. L. R. (2014). HAZard and OPerability (HAZOP) analysis for identification of information requirements for sewer asset management. *Structure and Infrastructure Engineering*, **10**, 1345–1356, https://doi.org/10.1080/15732479.2013.807845

SüwVO Abw (2013). Verordnung zur Selbstüberwachung von Abwasseranlagen – Selbstüberwachungsverordnung Abwasser (Landesverordnung). Der Minister für Umwelt, Raumordnung und Landwirtschaft des Landes Nordrhein-Westfalen.

Tafuri A. N. and Selvakumar A. (2002). Wastewater collection system infrastructure research needs in the USA. *Urban Water*, **4**, 21–29, https://doi.org/10.1016/S1462-0758(01)00070-X

Too E. and Tay L. (2008). Infrastructure asset management (IAM): evolution and evaluation. In: CIB International Conference on Building Education and Research: Building Resilience Conference Proceedings, R. Haigh and D. Amaratunga (eds), School of Built Environment, University of Salford, UK, pp. 950–958, https://www.irbnet.de/daten/iconda/CIB11298.pdf

Tscheikner-Gratl F., Caradot N., Cherqui F., Leitão J. P., Ahmadi M., Langeveld J. G., Gat Y. L., Scholten L., Roghani B., Rodríguez J. P., Lepot M., Stegeman B., Heinrichsen A., Kropp I., Kerres K., Almeida M. do C., Bach P. M., Vitry M. M. de, Marques A. S., Simões N. E., Rouault P., Hernandez N., Torres A., Werey C., Rulleau B. and Clemens F. (2019). Sewer asset management – state of the art and research needs. *Urban Water Journal*, **16**, 662–675, https://doi.org/10.1080/1573062X.2020.1713382

UCLA (2016). UCLA Sustainability Charter. UCLA Sustainability Committee, The University of California, Los Angeles, USA.

van Duin B., Zhu D. Z., Zhang W., Muir R. J., Johnston C., Kipkie C. and Rivard G. (2021). Toward more resilient urban stormwater management systems – bridging the gap from theory to implementation. *Frontiers in Water*, **3**, 1–21, https://doi.org/10.3389/frwa.2021.671059

Vanier D. J. (2001). Why industry needs asset management tools. *Journal of Computing in Civil Engineering*, **15**, 35–43, https://doi.org/10.1061/(ASCE)0887-3801(2001)15:1(35)

WCED (ed.) (1987). Report of the World Commission on Environment and Development: Our Common Future, Oxford Paperbacks. Oxford University Press, Oxford, New York.

WRC (2023). Sewerage Risk Management. Water Research Centre, Swindon, UK.

doi: 10.2166/9781789063059_0051

Chapter 3

Rules and regulations

Frédéric Cherqui[1], *Bert van Duin*[2,3,4], *Nathalie Hernandez*[5], *Karsten Kerres*[6], *Bardia Roghani*[7] *and Franz Tscheikner-Gratl*[7]

[1]*University of Lyon, INSA-LYON, Université Claude Bernard Lyon 1, DEEP, F-69621, F-69622, Villeurbanne, France*
[2]*City of Calgary, City & Regional Planning, Calgary, Canada*
[3]*University of Calgary, Calgary, Alberta*
[4]*University of Alberta, Edmonton, Alberta*
[5]*Stein Infrastructure Management GmbH, Bochum, Germany*
[6]*FH Aachen – University of Applied Sciences, Aachen, Germany*
[7]*NTNU – Norwegian University of Science and Technology, Trondheim, Norway*

ABSTRACT

This chapter covers the legal framework and technical regulations that must or should be observed for the strategic asset management (AM) of urban drainage systems over the entire life cycle (planning, construction, maintenance, and dismantling). A distinction is made between rules and regulations that deal with network management in general (strategic level) and rules and regulations that address the management of individual network components and/or certain activities (e.g., CCTV-inspection and condition assessment of reaches or stormwater basins). These activities are subsumed under 'operative level'.

It should be noted that both the legal framework and the applicable technical regulations vary widely from region to region. In some cases, different regulations apply even in different provinces or federal states of a country. Against this background, only case studies can (such as regulations that apply to Germany, France, Colombia or Canada) and will be presented in this chapter.

It is thus made clear that AM in the sense of ISO 55000 to 55002 (AM) enables a structured approach to a multi-layered field of tasks. In this way, goals and conflicting goals can be identified and prioritized at various levels and, in conjunction with the continuous improvement process in accordance with ISO 9000 and 9001 (Quality Management Systems), efficient ways can be found to achieve these goals.

Keywords: Asset management, quality management, PDCA-circle, sewer system management, law, ordinance, generally recognised rules of technology (standard), state of the art (advisory leaflet).

3.1 INTRODUCTION

While guidance pertaining to the design and construction of sanitary (and combined) conveyance systems has matured across the globe, there is, as explained in Chapter 1, considerable diversity as it pertains to stormwater management systems and nature-based solutions reflecting the differences in topography and climate and the immature nature of those fields. This is compounded by the diversity of the social aspects and make-up of our communities and the different sensitivity of the receiving water bodies and watersheds. However, there is growing consensus on asset management (AM) best practice approaches as explained in Chapter 2. Regardless, even though available information on AM best practices has increasingly been translated into rules and regulations, there are inherent challenges which are connected back to different value systems as explained below.

In the European Union (EU), the basis for sewer management, that is, the planning, construction, cleaning, and maintenance (for example, service, inspection, and rehabilitation) of drainage systems, is EN 752 (2017) 'Drain and sewer systems outside buildings – Sewer system management'. It is based on the principle of the continuous improvement process with the PDCA (Plan/Do/Check/Act) circle as the basic principle of quality management according to ISO 9000 (2015) and ISO 9001 (2015). It describes the process of both considering the requirements of a drainage system or its system components and the planning and implementation of measures to meet these requirements (see Table 3.1).

When the concept according to EN 752 (2017) is followed, the network operator must determine the following in detail for each of the activities listed above:

- What are the basic objectives of the maintenance activities?
- What costs or risks are associated with the 'avoidance' of maintenance activities?
- What costs and what benefits (e.g., risk reduction compared to 'avoidance') are associated with the maintenance activity?
- At what intervals should maintenance activities ideally be carried out, considering the above-mentioned aspects?

Particularly regarding costs and risks associated with 'carrying out' or 'avoiding' maintenance activities, there are very different assessments and conflicting goals, which can be illustrated by the nature of the questions shown in Figure 3.1.

This risk or cost trade-off can be discussed using the following example: for the purpose of groundwater and soil protection, discharge of wastewater into the soil (or groundwater) via the sewer system (exfiltration) must be avoided. If this goal is pursued consistently, all existing sewer

Table 3.1 PDCA circle according to ISO 9000 (2015), ISO 9001 (2015) and EN 752 (2017), respectively.

Process Step (PDCA Circle)	Content Description/Definition According to	
	ISO 9000 (2015) and ISO 9001 (2015)	**EN 752 (2017)**
Plan	Establish the objectives of the system and its processes, and the resources needed to deliver results in accordance with customers' requirements and the organization's policies and identify and address risks and opportunities.	Planning – developing the plan of measures to be taken.
Do	Implement what was planned.	Implementation of the plan.
Check	Monitor and (where applicable) measure processes and the resulting products and services against policies, objectives, requirements, and planned activities, and report the results.	Appropriate level of investigation of all aspects of the performance of the drain and sewers system.
Act	Take actions to improve performance, as necessary.	Assessment of the performance by comparison with the performance requirements including identification of the reasons for any observed performance failures.

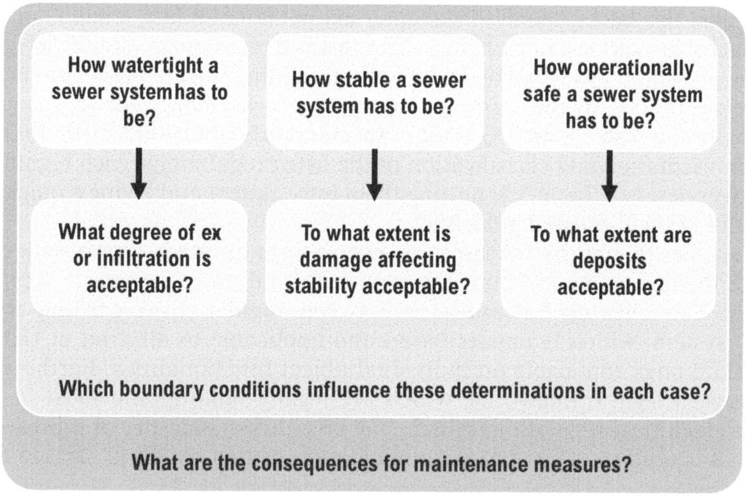

Figure 3.1 An example for a decision-making process regarding costs and risks associated with 'carrying out' or 'avoiding' maintenance activities

systems should be leak-tested and, as a rule, considerable parts of the network might have to be sealed afterwards. The costs associated with these measures would be very high and possibly even much higher than the costs and risks (both of which are difficult to determine in practice, especially since health or environmental costs are often not clearly quantifiable) associated with exfiltration.

Against this background, in Directive 91/271/EEC (1991) it was formulated that wastewater treatment requirements should be considered for collection systems and that the design, construction and maintenance of collection systems shall be undertaken in accordance with the best technical knowledge not entailing excessive costs.

Since it is impractical to provide a uniformly acceptable definition of what constitute reasonable costs, the above-mentioned determinations are incumbent on the local conditions and, by extension, the local network operators. The framework for planning, construction, and maintenance of drainage systems that applies to the operator is therefore derived from the applicable legal requirements and the state of the art or the recognized rules of technology. The nature, content and scope of this framework vary greatly from region to region. In the case of stormwater control measures (SCMs) this challenge is exacerbated by the diversity of the relevant approaches, and types and dimensions of relevant infrastructure, which are a direct function of topography and climate. Not only do different rules apply in different countries, but different legal requirements may even apply within a single country. For example, in Germany, sewer operations in different cities take place within different legal frameworks. In addition, a considerable number of regulations exist regarding the maintenance of sewer systems. Listing them all would go beyond the scope of this chapter, and even beyond the scope of the book as a whole. However, anyone involved with the design, construction, operation, and maintenance of drainage infrastructure should try to understand not only the locally relevant regulations but also the rationale for those regulations. Knowledge and understanding of the conditions in other locales may help in evaluating whether it would be appropriate to update the local regulations and approaches. For example, the Canadian Infrastructure Benchmarking Initiative (see https://nationalbenchmarking.com), which is the successor of the earlier National Water and Wastewater Benchmarking Initiative, provides a platform for discussions of such nature.

Regardless, against this background, this chapter cannot (and, in fact, should not even attempt to) present a comprehensive overview of globally valid rules and regulations. Rather, a legal and technical framework will be presented by means of examples highlighting selected regulations that the authors believe should be of interest to the UDAM community. To be more precise, ones that follow the notion of the globally applicable Plan/Do/Check/Act standards (ISO 9000, 2015; ISO 9001, 2015) are included here. For orientation and classification of the listed regulations, each regulation is preceded by the following overview (see Table 3.2) putting them into context and giving a quick overview of the main characteristics of the discussed guideline.

Given the differences in maturity of sanitary versus storm drainage systems, inevitably, a special focus is placed on the topics of condition gathering and condition assessment of piped UD systems outside buildings. The chapter differentiates between regulations regarding the functionality of the whole UD system, which is impact based and applicable to all kind of UD infrastructure, including SCMs, and ones applicable on individual object functionalities. Furthermore, it must be differentiated between legal frameworks, which are legally binding and must be complied with, and standards and technical regulations which may be either mandatory or represent the technical state of the art that is used to express a recommendation or that which is advised but not required or binding.

Table 3.2 Overview classification provided for each presented regulation.

Criteria	Localization
Validity	Global Europe National state Federal state or province
Nature	Law/Ordinance Generally recognized rules of technology (standard) State of the art (Advisory Leaflet)
Observation level	System (strategic level) Object (operative level): • Not specified • General • Sewers and manholes (gray infrastructure) • Special structures (gray infrastructure) • Nature-based solutions (blue-green infrastructure)
AM activity	Not specified General Investigation (condition gathering and condition assessment) Rehabilitation (replacement, renovation, and repair) Maintenance and cleaning
PDCA-circle	Not specified General Plan Do Check Act

3.2 SYSTEM FUNCTIONALITY

3.2.1 Legal framework

3.2.1.1 Water framework directive (Directive 2000/60/EC (2000))

Criteria	Localization
Validity	Europe (Members of European Union) (Members (2023): Austria, Belgium, Bulgaria, Croatia, Cyprus, Czech Republic, Denmark, Estonia, Finland, France, Germany, Greece, Hungary, Ireland, Italy, Latvia, Lithuania, Luxembourg, Malta, Netherlands, Poland, Portugal, Romania, Slovakia, Slovenia, Spain, Sweden)
Nature	Law/Ordinance
Observation level	System (strategic level)
AM Activity	Not specified
PDCA-Circle	Not specified

The Water Framework Directive (WFD) was issued on October 23, 2000, to standardize European water protection law and has been its foundation ever since. The WFD replaces many of the older water protection directives, while others remain alongside it, such as the Nitrates Directive (91/676/EEC, 1991), the Urban Wastewater Directive (91/271/EEC, 1991), and the Flood Risk Management Directive (2007/60/EC, 2007). It is itself supplemented by specific directives, such as the Groundwater Directive (2006/118/EC, 2006) or the Directive on Environmental Quality Standards (2008/105/EC, 2008) in the field of water policy.

Due to its holistic approach, the WFD has an impact on almost all regulations concerning water protection in the EU. Its requirements are reflected nationally not only in laws but also in numerous ordinances.

The objective of the WFD is establishing a framework for EU action in the field of water policy, with the aim of achieving good water status by 2015. Where this was not possible for natural or technical reasons or the measures would be disproportionately expensive, exceptions, that is, deadline extensions until 2027 or target reductions, could be claimed with appropriately substantiated justification.

Among others, the Directive states, that

- Water is not a commercial product like any other but, rather, a heritage which must be protected, defended, and treated as such.
- (26) Member States should aim to achieve the objective of at least good water status by defining and implementing the necessary measures within integrated programs of measures, considering existing community requirements. Where good water status already exists, it should be maintained. For groundwater, in addition to the requirements of good status, any significant and sustained upward trend in the concentration of any pollutant should be identified and reversed.
- (22) This Directive is to contribute to the progressive reduction of emissions of hazardous substances to water.

From the last point, it can be deduced that the planning, construction, and operation of drainage systems must minimize exfiltration (e.g., by ensuring tightness) and hydraulic performance for reasons of water protection. This aspect is specified in Directive 91/271/EEC (1991).

3.2.1.2 Urban Wastewater Treatment Directive (Council Directive 91/271/EEC, 1991)

Criteria	Localization
Validity	Europe (Members of European Union) (Members (2023): Austria, Belgium, Bulgaria, Croatia, Cyprus, Czech Republic, Denmark, Estonia, Finland, France, Germany, Greece, Hungary, Ireland, Italy, Latvia, Lithuania, Luxembourg, Malta, Netherlands, Poland, Portugal, Romania, Slovakia, Slovenia, Spain, Sweden)
Nature	Law/Ordinance
Observation level	System (strategic level)
AM activity	Not specified
PDCA-circle	Not specified

Council Directive 91/271/EEC of May 21, 1991 concerning urban wastewater treatment covers the collection, treatment, and discharge of urban wastewater and the treatment and discharge of wastewater from certain industrial sectors. The objective of the Directive is to protect the environment from the adverse effects of the abovementioned wastewater discharges.

For this reason, the member states, among others, shall ensure that that all agglomerations are provided with collection systems for urban wastewater

- at the latest by December 31, 2000 for those with a population equivalent (PE) of more than 15,000, and
- at the latest by December 31, 2005 for those with a PE of between 2000 and 15,000.

Where the establishment of a collection system is not justified either because it would produce no environmental benefit or because it would involve excessive cost, individual systems or other appropriate systems which achieve the same level of environmental protection shall be used.

Collection systems described above shall satisfy the requirements of Annex I (A) of Directive 91/271/EEC. There, it is required that collection systems should consider wastewater treatment requirements and that the design, construction, and maintenance of collection systems shall be undertaken in accordance with the best technical knowledge not entailing excessive costs, notably regarding:

- volume and characteristics of urban wastewater,
- prevention of leaks,
- limitation of pollution of receiving waters due to combined sewer overflows (CSO).

The implementation of the Wastewater Directive by the Member States and investment programs established in connection with it is regularly reviewed at two-year intervals. The tenth report (COM/2020/492, 2020) covers the period up to and including 2016 and involves more than 23,600 agglomerations.

According to the report, the collection and treatment of urban wastewater has improved in the EU over the past decade, with compliance rates of 95% for collection. Funding and planning remain the biggest challenges for the water services sector. To ensure compliance with the Wastewater Directive, all Member States (still including the UK) estimated in 2016 that nearly €229 billion of investment is required. According to a similar OECD estimate (OECD, 2020), EU countries and the UK will need to spend an additional €253 billion between 2020 and 2030 to achieve and maintain compliance with the Wastewater Directive. Current spending in many member states has been found to be too low to achieve and sustain compliance.

In response to the evaluation findings, the European Commission (EC) launched an impact assessment to evaluate policy options for adapting the Wastewater Directive in a sustainable way.

In October 2022, the EC presented a proposal for the revision of the Urban Wastewater Directive. Possible changes are known insofar as the EC – before submitting a Commission proposal – conducts a so-called impact assessment.

Policy options considered in the impact assessment for the revision of the EU Urban Wastewater Directive include the topics of combined sewage and surface runoff from urban areas as follows:

- For large settlements there are legally binding EU targets for the management of combined sewer and surface runoff.
- There are legally binding EU targets for combined water and surface water runoff management for all urban areas.
- Member States apply a risk-based approach to reducing CSOs. Settlement areas > 50,000 PE must have a strategic plan for the management of combined water and surface runoff based on prevention or implementation of green infrastructure.

3.2.1.3 Federal Water Act Germany (Wasserhaushaltsgesetz)

Criteria	Localization
Validity	National state (Germany)
Nature	Law/Ordinance
Observation level	System (strategic level)
AM activity	Not specified
PDCA-circle	Not specified

The Federal Republic of Germany has made the protection of the natural foundations of life a state objective in its constitution (Article 20a). This also includes the protection of water bodies. The Water Resources Act (Wasserhaushaltsgesetz (WHG)), which originally dates to 1957, forms the core of water protection law (WHG, 2023). Its purpose is to protect water bodies as a component of the natural balance, as the basis for human life, as habitat for animals and plants, and as a usable resource through sustainable water management (Section 1 of WHG).

The WHG aims to create the legal conditions for the orderly management of surface and groundwater in terms of quantity and quality, and to control human impacts on water bodies. The WHG stipulates that water bodies must be safeguarded as a component of the natural balance and as habitat for animals and plants and must be managed in such a way that they serve the public good and, in harmony with it, also the benefit of individuals. Avoidable impairments of their ecological functions should be avoided (that is, adoption of precautionary principle). Overall, a high level of protection for the environment shall be ensured.

The objectives and management rules of the WFD (2000/60/EC, 2000) form a central element of the WHG. The details of the specific directives (Directive on Environmental Quality Standards 2008/105/EC (2008), Groundwater Directive 2006/118/EC (2006)) have been implemented through statutory instruments (Surface Water Protection Ordinance (OGewV, 2020), Groundwater Protection Ordinance (GrwV, 2016)). A comprehensive authorization basis for the federal government to issue legal ordinances on water management in accordance with the management objectives is found in section 23 of the WHG. This authorization offers the possibility to elaborate the regulations, which are kept concise in the WHG, in more detail and serves the uniform implementation of EU law at the federal level.

The principles of wastewater disposal include:

- Wastewater shall be disposed of in such a way that the public welfare is not impaired. The public goods may also be served by the disposal of common wastewater through decentralized systems.

- Stormwater runoff shall be infiltrated close to the site or discharged into a body of water directly or via a sewerage system without being mixed with wastewater, insofar as this does not conflict with any regulations under water law or other public law or with water management concerns.

Accordingly, wastewater facilities shall be constructed, operated, and maintained in such a way that the requirements for wastewater disposal are met. In addition, construction, operation, and maintenance of wastewater treatment plants must be in accordance with the state of the art as well as (special) structures for sewage in accordance with the generally recognized rules of technology. If existing wastewater systems do not meet the above requirements, the necessary measures must be carried out within a reasonable period.

Anyone who discharges wastewater into a body of water or into a wastewater facility is further obliged:

- to have the wastewater examined by competent personnel or by a suitable body in accordance with a statutory order or the official decision authorizing the discharge of wastewater (self-monitoring), and
- to monitor its condition, functionality, maintenance, and operation as well as the type and quantity of wastewater and its constituents itself. In accordance with a statutory order pursuant to paragraph 3, the operator shall prepare and keep records and submit them to the competent authority upon request.

3.2.1.4 State Water Act North Rhine-Westphalia (Landeswassergesetz)

Criteria	Localization
Validity	Federal State (North Rhine-Westphalia, Germany)
Nature	Law/Ordinance
Observation level	System (strategic level)
AM activity	Not specified
PDCA-circle	General

The North Rhine-Westphalia (NRW) State Water Act (LWG NRW, 1995) applies to the water bodies and their parts listed in the Federal Water Act (WHG, 2023), see Section 3.2.1.3, as well as to actions and facilities that affect or may affect the water bodies and their uses. In detail, regulations of the Federal Water Act are specified. For drainage systems, the following is stipulated in detail, among other things:

- Operators are required to prepare a wastewater disposal concept (Abwasserbeseitigungskonzept) at regular intervals (see also administrative regulation on the preparation of wastewater disposal concepts (LANUV, 2008)).
- With the wastewater disposal concept, the local authorities shall submit to the competent authority an overview of the status of public wastewater disposal (wastewater and stormwater disposal) and of the measures to fulfil their obligations, in particular the chronological sequence and estimated costs of the measures required. The wastewater disposal concept shall be resubmitted every six years. It shall be drawn up by the operator. The Ministry responsible for the environment shall determine by administrative regulation which information must be included in the wastewater disposal concept and in which form it is to be presented.
- To achieve the objectives established in the management plan, as well as to ensure the fulfilment of the municipal obligations, the competent authority may object to the wastewater disposal concept. Measures and deadlines can be set if the municipality, without compelling reason, fails to implement necessary measures or delays their implementation.
- The wastewater disposal concept must also contain statements on how stormwater runoff can be disposed of in the drainage areas in compliance with the Federal Water Act and urban development and which measures are still required for stormwater runoff disposal.
- In this context, the effects on the existing drainage situation and on groundwater and surface waters are to be presented, considering measures for climate impact adaptation, among other things.

- Operators must ensure the proper construction and operation of wastewater facilities (see also Self-Monitoring Ordinance – Sewerage System (SüwVO Abw, 2013)).
- Construction and operation (the maintenance of wastewater facilities also includes the upkeep of the building fabric) of wastewater facilities shall be based on the respective valid rules of technology.

The Ministry responsible for the environment shall be empowered to make regulations on the methods and deadlines for carrying out the inspection of the condition and functionality, the recognition of inspections carried out, the necessity and deadlines for remediation, information and advice, the requirements for expertise and the conditions for the recognition or withdrawal of expertise.

3.2.1.5 French Climate and Resilience Law (Loi N° 2022-217 Du 21 Février 2022 relative À La différenciation, La décentralisation, La déconcentration Et portant diverses mesures De simplification De l'action publique locale)

Criteria	Localization
Validity	National state (France)
Nature	Law/Ordinance
Observation level	System (strategic level)
AM activity	Not specified
PDCA-circle	General

The Climate and Resilience Law (JORF, 2022) was officially promulgated and published in France in the Official Journal in August 2021 (and amended in 2022). The law is based on the assumptions that increasing land development is leading to

- acceleration of biodiversity loss,
- global warming as urbanized areas no longer absorb CO_2,
- increased risk of floods related to increased imperviousness,
- reduction in capacity of agricultural land,
- increased expenditures on networks required by land development, and
- desertification and the decline in value of rural land.

The objective of this law is to achieve 'zero net artificialization' or development (for example, from urban sprawl) by 2050, by halving land artificialization rate every prior decade, while at the same time encouraging the infiltration and evaporation of stormwater. As far as UD is concerned, this law will have a strong influence on urban renewal operations to favor the reuse of already urbanized sectors (i.e., vacant dwellings, industrial, or commercial wastelands, etc.) and to encourage the design and construction of slightly more compact operations that integrate green spaces. The law will also encourage the disconnection of rainwater and therefore, when possible, the replacement of end-of-life drainage pipes or networks by SCMs.

3.2.1.6 New territorial organisation of the French Republic (Loi portant nouvelle organisation territoriale De La République)

Criteria	Localization
Validity	National state (France)
Nature	Law/Ordinance
Observation level	System (strategic level)
AM activity	Not specified
PDCA-circle	General

The NOTRe law (nouvelle organisation territoriale de la République – new regional organization of the Republic) was officially promulgated and published in 2015 (JORF, 2015). It established a new specific competence, the GEPU – Gestion des Eaux Pluviales Urbaines – (Urban Stormwater Management) and helped evolving the distribution of drinking water and sanitation competences by moving them from single municipalities to associations of several municipalities, implying a mutualization of technical competences at this regional scale. The clustering of organizations is particularly relevant: in 2015, there were more than 33,000 organizations dedicated to drinking water or drainage, this number decreased to 26,000 in 2020. The clustering is intended to continue in the coming years: it enables organizations to reach a sufficient size for resources to be dedicated to the management of the system. More information regarding this law can be found in Torre and Bourdin (2023).

3.2.1.7 Observatory for data on public water and sanitation services (Loi N° 2006-1772 Du 30 Décembre 2006 Sur L'eau Et Les Milieux Aquatiques)

Criteria	Localization
Validity	National state (France)
Nature	Law/Ordinance
Observation level	System (strategic level)
AM activity	Not specified
PDCA-circle	Check

In 2006, the French government introduced a new law (JORF, 2006) whose primary objective is to transpose in French law the European framework directive (see Section 3.2.1.1). Four main objectives were established:

- good water status by 2015 (revised since),
- improve access to water for all,
- greater transparency in the operation of the public water service, and
- the reorganization of freshwater fishing.

The third objective (transparency) has resulted in the implementation of an annual benchmarking of all water and sanitation utilities in France, with the creation of an Information System on the public water and sanitation organizations (SISPEA). SISPEA collects and disseminates on a national scale data on the organization, management, pricing and performance of public water and sanitation organizations. A dedicated website (https://www.services.eaufrance.fr) allows each local authority to enter its performance indicators for the year. This site also allows public consultation of all the indicators by service or in aggregate. There are in total 41 performance indicators including 20 performance indicators related to sanitation.

Regarding sanitation, some performance indicators are dedicated to the quality of service (complaint rate) and to infrastructure management (asset knowledge and management index, average rate of network renewal, etc.). The last summary report (https://www.services.eaufrance.fr/mediatheque) provides more information on the benchmarking and on national performance indicators.

3.2.2 Technical regulations

3.2.2.1 ISO 9000 and 9001: quality management systems

Criteria	Localization
Validity	Global (ISO Members) (ISO is a global network of national standards bodies and includes (2023) 127 member bodies, 38 correspondent members and 3 subscriber members.)
Nature	Generally Recognised Rules of Technology (Standard)
Observation level	System (strategic level)
AM activity	Not specified
PDCA-circle	General

The quality management concepts and principles described in ISO 9000 (2015) and ISO 9001 (2015) give an organization (as a network operator) the capacity to meet challenges presented by an environment that is profoundly different from recent decades. The context in which an organization must operate nowadays is characterized by accelerated change, globalization of markets and the emergence of increasingly large amounts of data as a principal resource. The impact of quality extends beyond customer satisfaction: it can also have a direct impact on the organization's reputation.

ISO 9001 (2015) promotes the adoption of a process approach when developing, implementing, and improving the effectiveness of a quality management system, to enhance customer satisfaction by meeting customer requirements.

The concept described in ISO 9001 (2015) is therefore also valid for sewer system management (as reflected in EN 752 (2017)) and will be presented in the following. Understanding and managing interrelated processes as a system contributes to the organization's effectiveness and efficiency in achieving its intended results. This approach enables the organization to control the interrelationships and interdependencies among the processes of the system, so that the overall performance of the organization can be enhanced.

The process approach involves the systematic definition and management of processes, and their interactions, to achieve the intended results in accordance with the quality policies and strategic direction of the organization. Management of the processes and the system as a whole can be achieved using the PDCA cycle (Figure 3.2) with an overall focus on risk-based thinking aimed at taking advantage of opportunities and preventing undesirable results.

The PDCA cycle can be applied to all processes and to the quality management system as a whole and can be briefly described as follows:

- Plan: establish the objectives of the system and its processes, and the resources needed to deliver results in accordance with customers' requirements and the organization's policies and identify and address risks and opportunities.

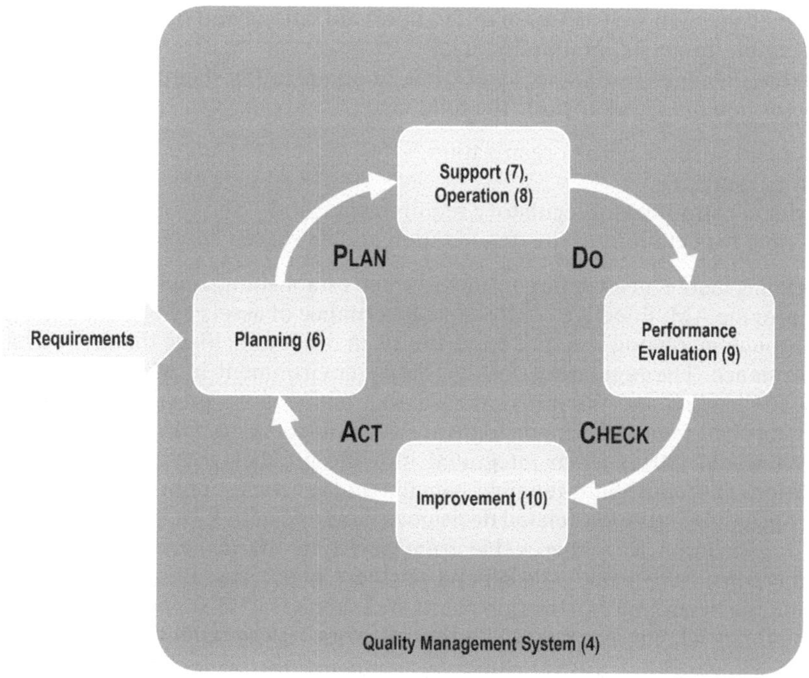

Figure 3.2 Representation of the PDCA cycle structure (numbers in brackets refer to the clauses in ISO 9001 (2015)).

- Do: implement what was planned.
- Check: monitor and (where applicable) measure processes and the resulting products and services against policies, objectives, requirements, and planned activities, and report the results.
- Act: take actions to improve performance, as necessary.

3.2.2.2 ISO 55000 to 55002: AM

Criteria	Localization
Validity	Global (ISO Members) (ISO is a global network of national standards bodies and includes (2023) 127 member bodies, 38 correspondent members and 3 subscriber members.)
Nature	Generally Recognised Rules of Technology (Standard)
Observation level	System (strategic level)
AM activity	Not specified
PDCA-circle	General

ISO 55000 (2014) provides an overview of AM, its principles and terminology, and the expected benefits from adopting AM. They can be applied to all types of assets and by all types and sizes of organizations. ISO 55001 (2014) specifies requirements for an AM system within the context of the organization. ISO 55002 (2018) provides guidelines for the application of an AM system, in accordance with the requirements of ISO 55001 (2014).

These International Standards can be applied to all types of assets and to all kinds and sizes of organizations. They are intended to be used for managing physical assets (like UD systems) in particular, but they can also be applied to other asset types.

These standards define an asset as an item, thing or entity that has potential or actual value to an organization. The value will vary between different organizations and their stakeholders, and can be tangible or intangible, financial, or non-financial.

The factors that influence the type of assets that an organization requires to achieve its objectives, and how the assets are managed, include the following:

- the nature and purpose of the organization,
- its operating context,
- its financial constraints and regulatory requirements, and
- the needs and expectations of the organization and its stakeholders.

These influencing factors need to be considered when establishing, implementing, maintaining, and continually improving AM. Effective control and governance of assets by organizations is essential to realize value through managing risk and opportunity, in order to achieve the desired balance of cost, risk, and performance. The regulatory and legislative environment in which organizations operate is increasingly challenging and the inherent risks that many assets present are constantly evolving. The fundamentals of AM and the supporting AM system introduced in ISO 55000 (2014), ISO 55001 (2014), and ISO 55002 (2018) when integrated into the broader governance and risk framework of an organization, can contribute tangible benefits and leverage opportunities. AM translates the organization's objectives into asset-related decisions, plans and activities, using a risk-based approach. It enables an organization to realize value from assets in the achievement of its organizational objectives. What constitutes value will depend on these objectives, the nature and purpose of the organization and the needs and expectations of its stakeholders. AM supports the realization of value while balancing financial, environmental, and social costs, risk, quality of service, and performance related to assets.

The benefits of AM can include, but are not limited to the following:

- Informed asset investment decisions: enabling the organization to improve its decision-making and effectively balance costs, risks, opportunities, and performance.
- Managed risk: reducing financial losses, improving health and safety, goodwill and reputation, and minimizing environmental and social impacts, can result in reduced liabilities such as insurance premiums, fines, and penalties.
- Improved services and outputs: assuring the performance of assets can lead to improved services or products that consistently meet or exceed the expectations of customers and stakeholders.
- Demonstrated social responsibility: improving the organization's ability to, for example, reduce emissions, conserve resources, and adapt to climate change, enables it to demonstrate socially responsible and ethical business practices and stewardship.
- Improved organizational sustainability: effectively managing short- and long-term effects, expenditures, and performance, can improve the sustainability of operations and the organization.
- Improved efficiency and effectiveness: reviewing and improving processes, procedures and asset performance can improve efficiency and effectiveness, and the achievement of organizational objectives.

To direct, coordinate, and control AM activities, an AM system can be used by the organization. An AM system is a set of interrelated and interacting elements of an organization, whose function is to establish the AM policies and AM objectives, and the processes, needed to achieve those objectives. In this context, the elements of the AM system should be viewed as a set of tools, including policies, plans, business processes, and information systems, which are integrated to give assurance that the AM activities will be delivered.

Of great importance here is, among other things, the elaboration of a strategic AM plan (SAMP). The SAMP includes:

- the plan for the asset portfolio, that is, the high-level activities to be undertaken on the assets to enable the delivery of organizational objectives, and
- the plan for the AM system, that is, the enhancements needed for the processes, resources, and other capabilities necessary for the effective implementation of the SAMP and the AM Plans (AMPs).

The structure and hierarchy of the AMPs, and their interrelationships should be defined by the organization in the SAMP after considering factors such as:

- its divisional/departmental structure, with objectives and accountabilities defined for the different units,
- the groupings of assets managed by these units,
- the geographical spread, size, complexity, and criticality of asset systems, and
- the number of assets within each asset type.

The strategic AM objectives should be developed for the entire asset portfolio after considering the AM policies, organizational objectives, stakeholder expectations, and external and internal factors influencing the organization and its AM system. The strategic AM objectives should be cascaded down to the level at which separate AM plans are developed and should relate to required performance targets for asset systems and individual assets. The SAMP also provides the direction and approach for the development of the AM plans.

When all the AM plans have been developed, using an iterative top-down and bottom-up process, the outputs from these plans should be:

- aggregated at the portfolio level,
- prioritized using the decision-making framework, while being balanced with the available funding and other resource constraints (see ISO 55000, 2014, Annex D).

In developing any time-bound and resource constrained AM plans for an individual asset, or a grouping of assets (for example, by asset type, asset system, facility, or unit in the organization), it is important that certain common methods/strategies are applied to ensure consistency and cohesiveness in achieving the overall strategic AM objectives. These strategies can be developed in advance in terms of asset life-cycle strategies for the different asset types or strategies for generic activity types (that is, asset creation/acquisition, utilization/maintenance, replacement/disposal, etc.) and applied in developing the AM plans.

3.2.2.3 EN 752: Sewer system management

Criteria	Localization
Validity	Europe (CEN Members) (CEN's National Members are (2023) the National Standardization Bodies (NSBs) of the 27 European Union countries, United Kingdom, the Republic of North Macedonia, Serbia and Türkyie plus three countries of the European Free Trade Association (Iceland, Norway, and Switzerland).)
Nature	Generally Recognized Rules of Technology (Standard)
Observation level	System (strategic level)
AM activity	General
PDCA-circle	General

EN 752 (2017) 'Sewer system management' provides the framework for planning, construction, maintenance, operation, and rehabilitation (servicing) of drainage systems outside buildings by referring to the methods of quality management and AM listed in Section 3.2.2.2. EN 752 is concretized and supplemented by detailed standards on 'investigation & assessment', 'planning & construction' or 'management & monitoring' of drainage systems (e.g., EN 14654-2 (2021) 'Drain and sewer systems outside buildings – Management and control of activities – Part 2: Rehabilitation').

According to EN 752 (2017), the requirements must be defined considering the overall system in such a way that the original planning objectives (see ISO 55000, 2014; ISO 55001, 2014; ISO 55002, 2018) continue to be met in the event of additions or changes to the system. The same applies to changes in the general conditions. These objectives are the basis for the integrated sewer system management, including AM of the drain and sewer systems and are the basis for determining the 'functional requirements' and the 'performance requirements'.

The objectives of drain and sewer systems are among others:

- Public health and safety:
  Drain and sewer systems are provided to manage rainwater and surface runoff while minimizing hazards to the public (including flooding hazards).
  As poorly designed, constructed or maintained systems can cause health or safety hazards to the public, the objective is to design, construct, operate, maintain, and rehabilitate the system to minimize the health and safety risks associated with the conveyance of wastewater and runoff.
- Environmental protection:
  The objective is to design, construct, operate and maintain the system to minimize the impact on the environment. These include impacts on water quality, water ecology or local water balance.
- Sustainable development:
  The objective is to design, construct, operate, maintain, and rehabilitate the system at the best environmental, social, and economic costs so that, in particular, it uses materials that minimize the depletion of finite resources, can be operated with the minimum practicable use of energy and can be constructed, operated and, at the end of their life, decommissioned with the minimum practicable impact on the environment.

Table 3.3 Examples of the relationship between objectives and functional requirements taken from EN 752 (2017).

Functional Requirements Regarding to the System	Subgoal			
	Public Health and Safety	Occupational Health and Safety	Environmental Protection	Sustainable Development
Protection from sewer flooding	XXX	XX	XXX	–
Protection of groundwater	XXX	–	XXX	XXX
Protection of surface receiving water bodies	XXX	X	XXX	XX
Structural integrity and design working life	XXX	XXX	XXX	XXX
Watertightness	XXX	X	XXX	XX

To ensure that the objectives listed above can be achieved, functional requirements shall be established. Essential functional requirements are among others (see also Table 3.3):

- Protection from (sewer) flooding:
  Sewer flooding shall be limited to nationally or locally prescribed frequencies considering the health and safety effects of the sewer flooding, costs arising from sewer flooding.
- Protection of groundwater and of surface receiving water bodies:
  Receiving water bodies shall be protected from pollution within nationally or locally prescribed limits. The measures to achieve the standards required under EU WFD 2000/60/EC (2000) are summarized in the water framework river basin plan (see also Section 3.2.1.1).
- Structural integrity and design working life:
  Drains, sewers, and other components shall be planned, designed, constructed, maintained, operated, and rehabilitated to ensure their structural integrity over the design working life.
- Watertightness:
  New drains, sewers, and ancillary structures, other than those specifically designed for infiltration drainage, shall be watertight in accordance with the testing requirements of EN 1610 (2015). Existing drains, sewers and auxiliary structures shall be watertight in accordance with national or local testing requirements.

The basis for the operation and rehabilitation of the drainage system is integral UD management (and also integral sewer management). It is based on the principle of the continuous improvement process with the PDCA circle as the basic principle of quality management according to ISO 9000 (2015) and ISO 9001 (2015) (see Figure 3.2) and describes the process of considering the requirements of a drainage system or its system components as well as the planning and implementation of measures to meet these requirements.

Accordingly, integral sewer management is divided into the following sections:

- Investigation of all performance aspects of the drainage system to an appropriate extent (CHECK):
  The investigation is the first stage in integrated sewer system management. The investigation and planning of rehabilitation works should be carried out on defined catchment areas in phases if appropriate, so that all problems and their causes can be considered together. The procedures described in EN 752 (2017) can be applied in any drain and sewer system, but detailed application should take account of the age, location, and type of system, the materials used in its construction, together with functional and climatic factors.

- Assessment of performance by comparison with performance requirements, including identification of causes for any observed performance deficiency (ACT):
  The performance of the system shall be assessed against the performance requirements. The performance assessment shall include the evaluation of risks of failure to achieve the performance requirements. The assessment shall be carried out in accordance with EN 13508-1 (2012).
- Planning the actions to be taken (PLAN):
  Integrated solutions shall be developed that fulfil the performance requirements, considering future conditions. Development of solutions for rehabilitation of existing drain and sewer systems should be carried out in accordance with EN 14654-2 (2021).
  One or a combination of the following possible approaches can be used to formulate the rehabilitation plan (with advantages and disadvantages of these approaches listed in EN 752 (2017); a temporal context is given for example by DWA-A 143-14 (2017):
  - Asset value approach:
    A financial-based approach that is used to ensure that rehabilitation is carried at a rate that ensures that the value of the system at the end of a specified period does not fall below a specified threshold.
  - Area-related approach:
    One part of the network is selected having common characteristics, requirements, and properties, for example: a pumping station catchment area, a groundwater protection zone, a residential or commercial area or the whole of a specific street.
  - Condition-based approach:
    All drains and sewers are inspected and those drains or sewers that do not meet some specified threshold conditions are rehabilitated.
  - Multi-utility approach:
    The rehabilitation of wastewater and storm drainage systems is carried out jointly with rehabilitation of other utility services in the same location in one construction activity.
  - Functional-related approach:
    This is built around the need for changes to improve the performance of the system (e.g., the need to reduce or eliminate discharges to receiving water bodies) and takes the opportunity to do other rehabilitation work where this can be done more efficiently at the same time.
  - Reactive approach:
    This involves responding to failures and problems as they are identified. Examples can include restoring the structural integrity following a collapse, removing a complete or partial blockage to restore the flow, or increasing hydraulic capacity following a sewer flooding incident.
- Implementation of the plan (DO):
  The integrated sewer system management plan does not generally contain the necessary details to proceed directly to the implementation of the work. To implement the work usually involves applying the stages of investigation, assessment and planning to produce a work programme which outlines a series of projects, in line with the plan, to ensure that the drain and sewer system meets the performance requirements. The outline for each project in the work programme describes:
  - the objectives of the proposed project, and
  - work specifications describing an individual project in sufficient detail to carry out the work.
- Work programmes and specifications can be developed for work in connection with new development, rehabilitation as well as operations or maintenance activities. Further guidance on the management and control of implementation works is given in EN 14654-1 (2021).

3.2.2.4 Iranian guide to the operation and maintenance of sewage collection networks

Criteria	Localization
Validity	National state (Iran)
Nature	State of the art (Advisory Leaflet)
Observation level	System (strategic level)
AM activity	General
PDCA-circle	General

Iran has two main set of rules for managing and maintaining sewage networks. The first is the 'Guide to the Operation and Maintenance of Sewage Collection Networks' (IRN 520, 2009), which is an Advisory Leaflet used by engineers, technicians, and other people involved in the operation of the network. It was created by the Executive Technical System Office of the Vice President for Planning and Strategic Supervision. This document has been put together in accordance with worldwide standards and guidelines to provide useful solutions that offer a thorough productivity viewpoint while adhering to the concepts, methods, and techniques of engineering management based on national facilities.

The paper includes crucial approaches and tried-and-true methods for organizing and carrying out critical duties in various wastewater collecting networks, including network element analysis and defect prevention. More specifically, this standard covers technical aspects of pumping station operation and maintenance, inspection of the sewer networks (including internal inspection of pipes, sealing test, sediments), types of potential network damages (including displacement of pipes, corrosion, cracks, and pipeline collapse), flushing and removing obstructions in sewers (flushing with pressure and mechanical methods), safety and health in urban sewage collection, as well as training of operation and maintenance staff. It is advised to routinely inspect the components of these networks to assess the status of sewage collection networks and determine the necessary actions to improve the current situation.

Another section of this guide suggests an analysis procedure in accordance with Figure 3.3 to help pick the appropriate strategy based on the current state of the drainage network elements. The primary factors for establishing the extent of essential corrective actions, according to this graphic, are their conformity to the seriousness of damages, the requirement for an increase in the volume of flow, technical constraints, and the economic effectiveness of the suggested measures.

3.2.2.5 DWA-A 143-14: Development of a rehabilitation strategy

Criteria	Localization
Validity	National state (Germany)
Nature	Generally recognized rules of technology (standard)
Observation level	System (strategic level)
AM activity	Rehabilitation
PDCA-circle	General

The DWA-A 143-14 (2017) worksheet of the German Association for Water, Wastewater and Waste (DWA) has adopted essential aspects of the EN 752 (2017), see Section 3.2.2.3, and specifies the process of the development and implementation of a (rehabilitation) plan on a strategic level. A rehabilitation strategy for sewer systems therefore defines the future rehabilitation action of a network operator to achieve the self-defined strategical, technical, economic, and legal sub-goals with continuous control of the implementation (i.e., monitoring). With the intent of long-term planning that accounts for the entire network, EN 752 (2017), which shows the design framework of holistic/integral sewer management, must also be observed when establishing the rehabilitation strategy. As a

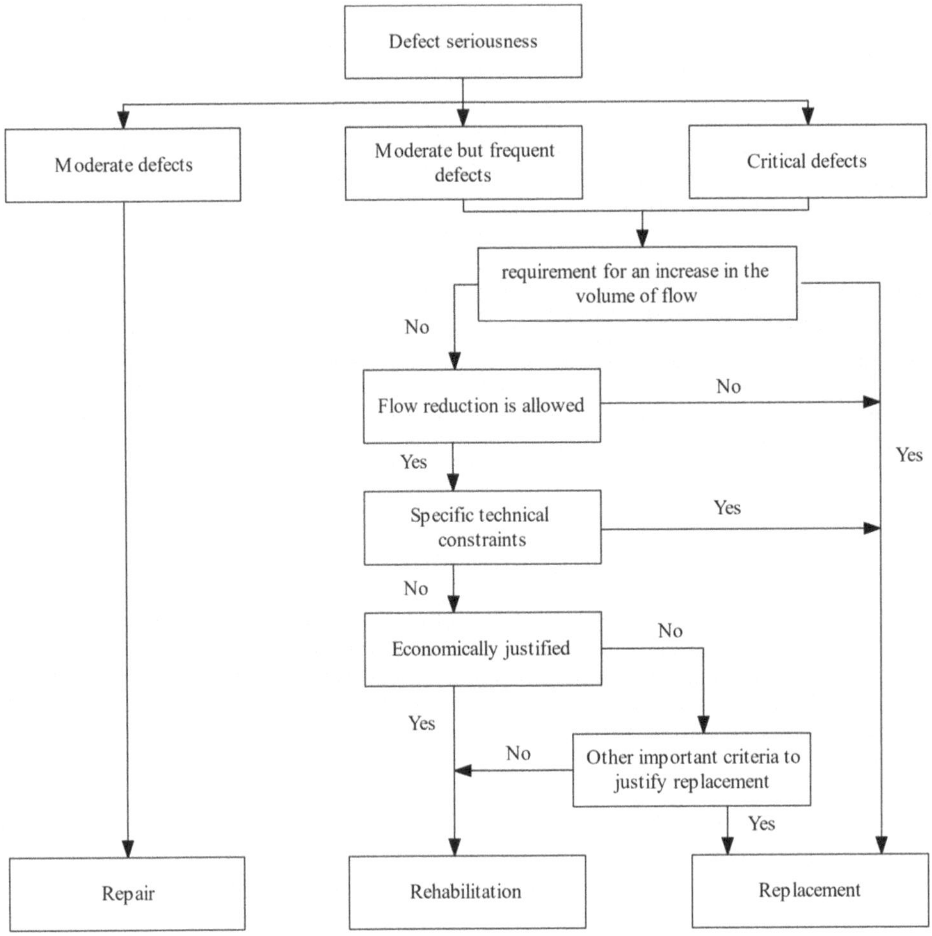

Figure 3.3 Proposed procedure to choose the best approach to fix the network faulty elements as described in IRN 520 (2009).

rule, the strategic approaches defined in EN 752 (2017) have not been adopted in their purest form. Rather, the question arises as to how the guiding ideas anchored in the approaches can be considered in the creation of an optimized renovation strategy individually tailored to a specific network.

In addition to the different, partially overlapping objectives, the approaches also have different implementation periods. For example, the event-based (reactive) approach, which primarily aims to remedy systemic failure that has occurred or is imminent, is a short-term approach, while the net asset value approach is a long-term task.

With the help of Figure 3.4, the approaches can be placed in their temporal context. In this way, it becomes clear which approach is decisive for the respective observation horizons in the development of the strategy with its guiding principles. The periods under consideration are defined as follows:

- long-term observation period (e.g., more than 12 years),
- medium-term observation period (e.g., 6–12 years), and
- short-term observation period (e.g., up to 6 years).

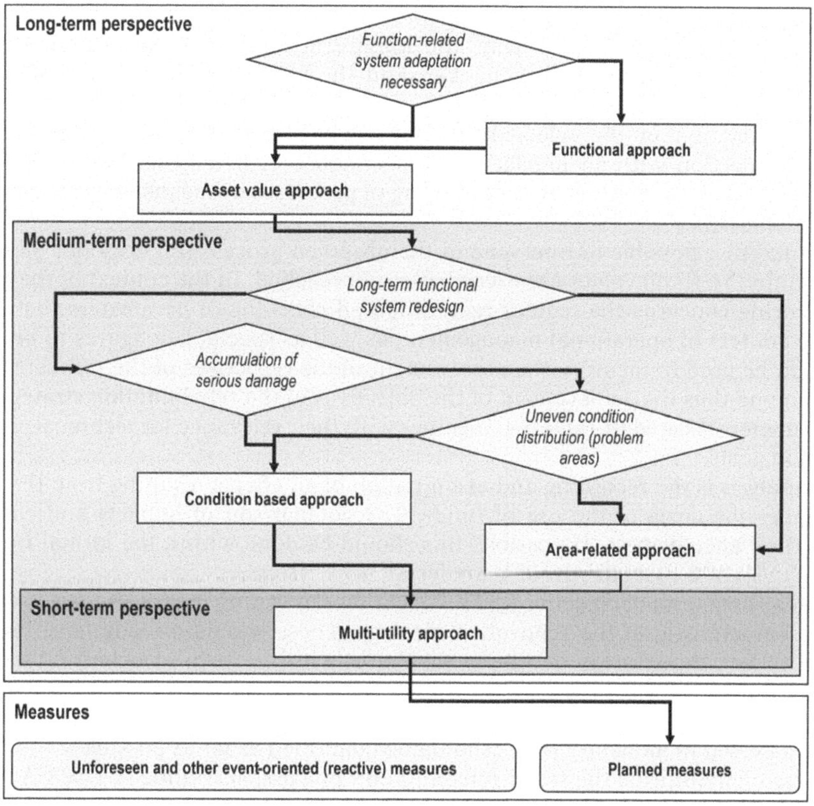

Figure 3.4 Strategic approaches in a temporal context as provided by DWA-A 143-14 (2017).

By adapting the approaches to the network needs (e.g., network modification, complete renovation of a sub-area, consideration of several sectors), it is easier to identify factors influencing the strategy for the respective periods under consideration when creating the individual strategy and to take them into account accordingly when creating the strategy. Thus, in the event-based approach, the idea of coordination recedes into the background, and the factors necessary for this become less important (e.g., data availability from planning in other sectors). In the same way, the preservation of the substance of the existing system is of secondary importance if it is changed drastically anyway within the framework of a function-based approach. At this point it becomes clear that strategic planning also requires continuous revision to account for changed conditions.

The development of a suitable rehabilitation strategy can be done with different tools. Simple estimates of the rehabilitation needs and the rehabilitation activities to be derived from them are possible based on empirical values. Benchmarking can also be used to transfer the surveys and experiences of other sewer network operators to one's own rehabilitation activities. For a more detailed consideration and planning of the required rehabilitation activities, the preparation of corresponding rehabilitation concepts is necessary. For smaller and younger sewer networks, a static analysis without detailed consideration of the progressive deterioration can be considered sufficient according to the regulations.

In many cases, such rehabilitation concepts are available for the entire sewer network or the respective district under investigation. In the case of larger and older sewer networks, it is advantageous to also include further deterioration of the condition in the analysis through appropriate aging modelling. In this way, the long-term effects of the rehabilitation strategy on the condition development can be checked and the selected strategy can be, if necessary, adapted.

For further evaluation of the selected strategy consisting of the approaches, accompanying monitoring in connection with an impact analysis should be carried out. Monitoring is a generic term for all types of direct recording and monitoring of processes by means of technical aids or other observation systems.

Monitoring makes it possible to intervene in the observed process if it does not meet the desired course or if certain threshold values are exceeded or not reached. In the context of the rehabilitation strategy, monitoring concerns the regular recording and checking of parameters that are generally required in the context of operational management, as well as special key figures to be derived from these, which can be used to measure the achievement of the objectives of the respective approaches to rehabilitation and thus the achievement of the objectives of the rehabilitation strategy. A selection of suitable parameters is listed in Table 3.4, together with their relevance for technical, economic, and legal remediation goals.

An impact analysis is the recording and examination of all effects resulting from the use of funds. This also includes the costs of the use of funds. The comparison of impacts and costs allows an assessment of their effectiveness. If possible, this should be done within the formal framework of a cost–benefit analysis or a cost-effectiveness analysis.

The medium-to-long-term perspective and the fact that renovation measures have long-term effects (benefits) are characteristic of the renovation strategy. These should already have been estimated in bandwidths during the strategy development phase (with the help of deterioration models and applying scenario techniques) and compared with the investment costs. If the effects can be expressed in monetary units, dynamic cost comparison methods should be used. Monetarily non-valuable effects that cannot be assessed in monetary terms should be quantified as far as possible.

In the process of monitoring the sewer rehabilitation process, according to DWA-A 143-14 (2017), the funds used should also be subjected to an impact analysis to retrospectively assess the extent to which they have contributed to the efficient achievement of the rehabilitation goals. This question cannot be answered by monitoring some few key figures, as the effects of different measures overlap over time. Possible approach independent effects of interest are the development of the number of

Table 3.4 Examples of possible parameters for monitoring in a given year (DWA-A 143-14, 2017).

Monitoring Parameters	Relevance for Sub-goal			Comments
	Technical	Economical	Legal	
Network length (km)	+	+	+	Generally required
Length of rehabilitation (km) (replacement and renovation)	+	+		
Investment in rehabilitation in (€)		+		
Number of repairs (dimensionless)	+	+		
Residual book value (€)		+		
Average remaining useful life (a)		+		
Technical substance value (dimensionless)		+		
Comparison of inspection findings with predicted condition (empiric data to adjust the deterioration model)	+			Additionally required for asset valuation

assets in critical and non-critical condition, its connection with the rehabilitation budget and the review and if necessary, adjustment of interventions. Other effects depend on the applied approach:

- Asset value approach:
  - Asset value gains (investments minus asset value of replaced or renovated old facilities).
  - Losses on residual book values of the replaced old assets.
- Area-related approach:
  - Notional savings (benefits) from comparison of actual costs with the costs determined by adding up notional individual measures (for multi-sector projects).
  - Coordination and coordination effort.
  - Losses on residual book values of replaced old assets.
  - Benefits of the measures for urban development.
- Condition-based approach:
  - No further specific assessment parameters beyond those mentioned above.
- Multi-utility approach:
  - Fictitious savings by comparing the actual construction costs with the costs determined by adding up fictitious individual measures.
  - Coordination and coordination effort.
  - Losses on remaining useful lives of replaced old assets.
  - Losses on residual book values of replaced old assets.
- Functional-related approach:
  - Effectiveness of the measures on the combined sewage inflow to the treatment plant.
  - Effectiveness of the measures on the number of flooding events.
  - Effectiveness of the measures on the number of discharge events in the combined system.
- Reactive approach:
  - Analysis of the direct and indirect costs of damage that could have been avoided by a foresighted remediation strategy.
  - Effectiveness of the measures regarding road safety obligations.

3.2.2.6 DWA-M 149 series: condition detection and assessment

Criteria	Localization
Validity	National state (Germany)
Nature	State of the art (Advisory Leaflet)
Observation level	System (strategic level)
AM activity	Investigation (condition gathering and condition assessment)
PDCA-circle	General

Condition recording and assessment is part of an overall process for integral sewer management as described in EN 752 (2017) and EN 14654-2 (2021), among others. General objectives and requirements for drainage systems outside buildings are formulated in European standard EN 752 (2017). These general requirements are also underpinned at the European level by further detailed standards.

The EN 13508 series (EN 13508-1, 2012; EN 13508-2, 2011) of standards is authoritative for the area of condition recording and assessment of drainage systems outside of buildings from a structural, operational, and environmental point of view. They are adopted by the Codes of Practice of the DWA-M 149 series.

DWA-M 149-1 (2018) represents the basic document for the DWA-M 149 series of Advisory Leaflets and is intended to support the application of this series by providing the following information, among other things:

- Overview of the topic of condition assessment and evaluation.
- Reference to relevant regulations.
- Supplementary information on aspects of condition monitoring and assessment that are not covered by additional regulations.
- Definition of terms that are used across the board for the topic of condition monitoring and assessment.

The DWA-M 149 series includes the following parts, among others:

- DWA-M 149-2 (2013): Coding system for optical inspection.
- DWA-M 149-3 (2015): Assessment after visual inspection (see also Section 3.3.2).
- DWA-M 149-4 (2008): Detection of bedding defects and cavities using geophysical methods.
- DWA-M 149-5 (2010): Optical inspection.
- DWA-M 149-6 (2016): Pressure testing of in-service drainage systems with water or air.
- DWA-M 149-7 (2016): Assessment of the environmental relevance of the structural/operational condition.
- DWA-M 149-10 (yet unreleased): Substance classification.

The purpose of condition recording and assessment in general depends on the components of a drainage system to be considered and their life cycle. It includes the areas of planning, construction, operation, and rehabilitation. The requirements to be considered can be of a structural, operational, environmental, and/or hydraulic nature, although hydraulic investigations are not the subject of DWA-M 149-1 (2018).

In accordance with the specifications in EN 1610 (2015), the sewers/pipes and manholes/inspection openings considered in DWA-M 149-1 (2018) are understood as structures in which the interaction of components and bedding/filling under the existing environmental conditions is the basis for stability and operational safety (engineering structure). Consequently, the outer boundary of the pipeline zone is defined here as the boundary to groundwater and soil or the environment in general. As a rule, the object of operation is the conditions within an object. An example of possible purposes of condition recording and assessment under the described boundary conditions is shown in Table 3.5. Depending on the purpose of the investigation, an investigation can concern individual or several requirements.

Table 3.5 Examples of possible parameters for monitoring in a given year (DWA-A 143-14, 2017).

Request	Purpose of Condition Detection and Condition Assessment		
	Structural	**Operational**	**Environmental**
Demand planning	• Monitoring by the operator (initial/repeat inspection) • Determination of condition development • Acceptance of work	• Self-monitoring by the operator (initial/repeat inspection) • Function control • Operation optimization (e.g., cleaning strategy)	• Hazard analysis
Special purpose	• Inventory • Acute damage (e.g., collapses, subsidence) • Preservation of evidence	• Emissions (e.g., odor) • Infiltration • Operational disturbances	• Case-by-case consideration

Demand planning includes different tasks depending on the purpose, such as demand planning for rehabilitation or operational optimization (e.g., cleaning/inspection strategy). The aim of demand planning is to forecast the necessary resource requirements based on the necessary measures according to the need for action. Self-monitoring by the operator of a sewer system basically covers the entire catchment area. It is prompted (in Germany) by the legal requirements (that is, water law requirements, self-monitoring ordinances of the federal states or similar), or self-requirements (sewer management, inventory). The condition of the sewer system is to be determined within the scope of self-monitoring.

Similarly, repeated inspections can cover the entire catchment area, sub-catchments, or individual objects. They are determined by legal requirements (that is, water law requirements, self-monitoring ordinances of the federal states, or similar), or defined by the needs (e.g., planned remediation, observation, determination of status development). Self-monitoring ordinances usually require repeat inspections at fixed time intervals, regardless of other boundary conditions. For a demand-oriented repeated inspection, a time-differentiated inspection concept must be drawn up.

3.2.2.7 CSA W211-21: Management standard for stormwater systems

Criteria	Localization
Validity	National state (Canada)
Nature	Generally recognized rules of technology (standard)
Observation level	System (strategic level)
AM activity	Not specified
PDCA-circle	Not specified

This Canadian Standard (CSA W211, 2021), which is intended for use by persons or organizations that own, manage, and/or operate a stormwater system, outlines minimum requirements and recommendations for the management of a stormwater system established by an operating authority. It also provides a framework to ensure that persons responsible for the stormwater system are provided with the information necessary to support sound decision making. The scope of this Standard covers both (i) publicly owned stormwater systems, and (ii) privately owned stormwater infrastructure (e.g., commercial and industrial) that are connected to a stormwater system, albeit in areas not subject to permafrost. It focuses on the management of the planning, design, construction, closure, or abandonment of a stormwater system rather than those activities themselves.

It provides a standardized guidance process to manage a stormwater system to meet regulatory requirements, reflect sound policy decision making, and demonstrate a high standard of care to minimize legal, environmental, social, and economic risks, especially given a changing climate and hydrologic conditions. It also provides a documentation protocol to ensure all changes and decisions made in policy and procedures are well documented and identifies steps for operating authorities to internally audit their management system. It is written with flexibility such that it could be applicable to any size of stormwater system.

This Standard provides:

- an outline of the key policies and procedures needed for effective management of a stormwater system,
- a framework for increasing public safety by identifying and mitigating environmental, social, legal, and economic risks to the system, and
- the ability to demonstrate enhanced environmental outcomes through the effective application and continual improvement of the stormwater management system.

3.2.2.8 CSA W210-21: Prioritization of flood risk in existing communities

Criteria	Localization
Validity	National state (Canada)
Nature	Generally recognized rules of technology (standard)
Observation level	System (strategic level)
AM activity	Not specified
PDCA-circle	Not specified

This Standard (CSA W210, 2021) for existing communities (as opposed to new or recently planned communities) is guided by a number of principles and criteria which are intended to support the development and application of this Standard in various jurisdictions across Canada. It is intended to support unbiased and transparent decision-making related to those locations within a municipality which are deemed to be most vulnerable to flooding. Once identified, these areas would be the priority for more advanced flood-risk analysis, development, and design of solutions, and/or flood mitigation investment. The principles include:

- the consideration of the full range of flood mechanisms present in a community,
- the consideration of interdependency between various flood mechanisms in establishing flood risks,
- the consideration of site-specific conditions, as well as existing community planning policies and principles, in the determination of flood risk, and
- the consideration of flood-exacerbating factors such as climate change (future frequency and severity of precipitation), urban intensification, and changes in upstream land uses that affect long-term resilience.

It provides the methodology to establish a framework to help the prioritization as a function of the technical maturity, data availability, and data adequacy, all expressed as a 'readiness index', and ultimately the options for conducting risk assessments, based on a municipality's maturity or readiness level. The approach advocates a systematic and consistent set of steps at three levels of assessment (i.e., foundational, intermediate, and advanced) to allow municipalities to establish their locations of greatest flood risk, and from this understanding, start to plan and budget for remediation activities.

3.2.2.9 CSA W218-23: Methodology for completing natural asset inventories

Criteria	Localization
Validity	National state (Canada)
Nature	Generally recognized rules of technology (standard)
Observation level	System (strategic level) and object (operative level)
AM activity	Investigation (condition gathering and condition assessment)
PDCA-circle	Not specified

This Canadian Standard (CSA W218, 2023), the first of a series of Natural Asset standards, provides a minimum requirement for development and reporting of a natural asset inventory. The Standard is designed to be sufficiently flexible that it can be applied in any jurisdictional context. It can also be used to include not only natural assets within a given jurisdictional boundary, but also those in adjacent jurisdictions that provide important services. As such, it provides:

- Definitions for terms that are routinely used in the development of natural asset inventories.
- Minimum requirements for natural asset inventories, including
  - Data gathering and processing requirements.
  - Inventory structure requirements.

  - Documentation requirements.
  - Requirements for maintaining and continually improving a natural asset inventory.

Although not mandated, minimum requirements for completing a condition assessment for natural assets are provided should the user(s) decide to complete one. The Standard is implemented in alignment with the AM approach for built infrastructure of the jurisdictions having authority and/or recognized regional, national or international practices, including ISO 55001 (2014).

3.3 OBJECT FUNCTIONALITY

3.3.1 Legal framework

Regarding the functionality of the individual objects of a drainage system, no legal regulations are known at a European or (at least for Germany) national level. In the individual German federal states, there are different specifications regarding the maintenance of drainage systems: in some federal states, reference is only made to the state of the art; in others, however, there are very strict or clearly defined legal requirements. As a case study for such clearly defined legal requirements, the legal framework applicable in the federal state of North Rhine-Westphalia is listed below.

3.3.1.1 Administrative regulation on the preparation of wastewater disposal concepts

Criteria	Localization
Validity	Federal State (North Rhine-Westphalia, Germany)
Nature	Law/Ordinance
Observation level	Object (gray infrastructure)
AM activity	General
PDCA-circle	Not specified

To fulfill the WFD as well as the duties formulated in the State Water Act, operators in North Rhine-Westphalia must plan, construct, expand or adapt wastewater facilities to the generally recognized rules of wastewater technology within reasonable periods of time (LANUV, 2008).

The status of public wastewater disposal as well as the chronological sequence and estimated costs of the measures necessary for the construction, rehabilitation, maintenance, or replacement of wastewater facilities to fulfil the wastewater disposal obligation must be presented.

The respective measures are, among others, to be assigned to the following headings according to type:

- A2: Sewer system – rehabilitation measure for hydraulic reasons.
- A3: Sewer system – rehabilitation measure for structural reasons.
- A4: Wastewater sewer system – measures for extraneous water remediation.
- A5: Combined sewer system – measures for extraneous water remediation.
- A11: Measures in the watercourse to compensate for the negative impacts of combined sewer and stormwater discharges, insofar as they are relevant to wastewater.
- A12: Infiltration system.

The prioritization of a measure must be based on the achievement of the management objectives resulting from the Federal Water Act and any existing programme of measures pursuant to the Federal Water Act, and thus in particular on the prevention of hazards and the protection of public welfare. In the case of measures accompanied by administrative orders or other decisions, it is necessary to indicate the associated deadlines.

3.3.1.2 Ordinance on the self-monitoring of wastewater facilities

Criteria	Localization
Validity	Federal State (North Rhine-Westphalia, Germany)
Nature	Law/Ordinance
Observation level	Object (gray infrastructure)
AM activity	General
PDCA-circle	General

The operator of a sewer network shall monitor for condition and functionality by himself and draw up instructions for self-monitoring for this purpose. Examples of the facilities to be monitored, the respective scope of inspection and the frequency of the respective inspection are shown in Table 3.6. Further regulations have been made for house and/or property connection pipes (if part of the public sewer system), sewage pumping stations, rainwater clarification basins, stormwater overflow basins, storage sewers, rainwater retention basins, and so on.

Table 3.6 Extend, type and frequency of monitoring of wastewater facilities according to SüwVO Abw (2013) for sewers and manholes.

Facility	Examination	Type of Examination	Frequency
Sewers (including tie-ins of connecting sewers)	Determination of filings	Visual inspection or visual inspection, on foot	According to the operating or rinsing schedule; otherwise, every two years
	First-time recording of the status (1996–2006)	CCTV or visual inspection	Review of the status after completion of the initial version (2006–2020 and every 15 years thereafter)
	Review of the status after completion of the initial version (1.1.2006 to 31.12.2020 and every 15 years thereafter)		Five percent of the sewers every year, but the entire network every 15 years
Manholes	Determination of the general condition, visible damage to manhole covers, dirt traps and bars as well as to the manhole body, leaks, inflow of extraneous water, discharges	Visual inspection	In connection with the self-monitoring of the sewers

3.3.2 Technical regulations

3.3.2.1 DWA-A 147: Operating efforts for drain and sewer systems

Criteria	Localization
Validity	National state (Germany)
Nature	Generally recognized rules of technology (standard)
Observation level	Object (gray infrastructure as well as nature-based solutions)
AM activity	Inspection (condition gathering and condition assessment), maintenance and cleaning
PDCA-Circle	General

Worksheet DWA-A 147 (2017) contains requirements for the operation and maintenance of drainage systems as defined in EN 752 (2017). The requirements formulated there for operation and maintenance basically result from two different sources:

- Legal requirements:
  Wastewater systems must be operated in accordance with the generally recognized rules of technology. These are specified in the DWA rules and regulations and in European and national standards. In addition, some federal states have issued self-monitoring ordinances, some of which contain very different requirements for the performance of individual operational tasks (compared to previous explanations in Sections 3.2.1 and 3.3.1).
- Self-selected quality requirements:
  In addition to the legal requirements, operators can also define quality objectives that they want to ensure within the scope of their task performance. The subject of such quality objectives can be, for example, special measures to reduce odors. Both the legal requirements and the self-selected quality requirements usually include work that can be planned, the frequency of which results from maintenance schedules or the execution of which is linked to the occurrence of certain events. Troubleshooting and damage repairs are important operational tasks. Concrete requirements for these tasks exist neither in legal requirements nor in technical regulations. However, operators must ensure through organizational precautions and suitable personnel that the elimination of faults and the implementation of repair measures can be carried out within reasonable time periods.

Operational work can be distinguished as per the following aspects:

- object-related work,
- non-object-related work,
- plannable work, or
- event-dependent work.

In the appendix of the worksheet, the object-related, plannable operational work for the various wastewater systems is listed in detail. Examples concerning inspections include:

- Inspection of sewers:
  - Inspection according to inspection plans to determine the structural condition with documentation according to Code of Practice (DWA-M 149-5, 2010) by inspection or camera inspection.
  - Frequency: every 10–20 years (10–5% per year)
    - Lower range (5% per year):
      Repeated inspections in sewers; network areas with good structural condition; newly constructed sewers; high-quality construction methods and pipe materials; low environmental hazards.
    - Upper range (10% per year):
      When timely monitoring of the development of the structural condition is required; in sensitive sewer network areas (e.g., poor structural condition, increased environmental hazard according to the wastewater composition); with high demands on availability and precautions against operational disturbances; high demands on the need for actual inspection results.
    - Special cases at higher frequency:
      Active subsidence areas; in the case of structurally endangered sewers in single cases; in the case of contractual obligations (e.g., crossings with traffic routes such as railway facilities).
    - Special cases at lower frequency:
      Outfall sewers and relief sewers operating under backwater conditions provided that the effort required to carry out the inspection is disproportionate to the risk of danger.
    - Further requirements in water catchment areas depending on the location in the various water protection zones, leakage tests and additional inspections may be required.

  - Optimization:
    Supplementary inspection activities in accordance with operator-specific specifications, for example, visual inspections in sewers of high operational significance to initiate preventative maintenance work.
- Inspection of retention soil filters:
  - Retention soil filter basin:
    Visual inspection, especially of the inlet and outlet, for deposits and transport of filter material; frequency according to experience (guideline: every 3 months).
  - Filter body:
    Visual inspection, checking the filter vegetation, checking for contamination, uniform feeding, waterlogging and displacement of filter material; frequency according to experience (guideline: every 3 months).
    Examination of the filter material including examination of the permeability and sorption capacity of the filter material (e.g., through laboratory tests every 5–10 years).
  - Drainage system:
    Camera inspection of the drainage system to check for disturbing root intrusion, precipitation, and deposits; frequency according to experience (guideline: every 5–10 years or in the case of significantly prolonged waterlogging times and waterlogging).
  - Optimization:
    Connection to a remote monitoring/effective system; automatic check of the operating conductivity of the filter body to detect incipient clogging.
- Inspection of systems for the infiltration of rainwater:
  - Above ground facilities:
    - Visual inspection:
      Check for deposits, contamination, condition of the turf, increased accumulation of leaves, foreign growth, damage to fixtures such as retaining walls, pipe inlets and outlets, waterlogging due to silting; frequency according to experience; guideline: every 3 months.
    - Soil survey:
      Sampling from the soil layer at various depths and examination for pollutants in the case of discharge of polluted stormwater runoff every 10–20 years.
  - Underground facilities:
    - Visual inspection:
      Checking the manholes regarding their proper infiltration function; frequency according to experience (guideline: once a year). Inspection of the infiltration pipes by camera inspection to check the infiltration function.
  - Optimization:
    - Integration of the facilities into green spaces to the extent possible, so that joint maintenance of the surfaces is possible; choose low-maintenance vegetation.

3.3.2.2 EN 13508 part 1–3: Investigation and assessment of drain and sewer systems outside buildings

Criteria	Localization
Validity	Europe (CEN Members) (CEN's National Members are (2023) the National Standardization Bodies (NSBs) of the 27 European Union countries, United Kingdom, the Republic of North Macedonia, Serbia and Türkyie plus three countries of the European Free Trade Association (Iceland, Norway, and Switzerland).)
Nature	Generally recognized rules of technology (standard)
Observation level	Object (gray infrastructure)
AM activity	Investigation (condition gathering)
PDCA-circle	Check

While EN 13508-1 (2012), with reference to EN 752 (2017), specifies the general requirements for drainage systems, which are mainly operated as gravity systems, and derives from them requirements for their maintenance. EN 13508-2 (2011) specifies a coding system for describing the observations made inside sewers and drains, manholes and inspection openings during a visual inspection (as described in Chapter 4). EN 13508-3 will be an extension of Part 2 and defines a coding system for sewer cleaning and sewer condition control.

The coding system introduced by Part 2 comprises a set of codes to describe damage and condition characteristics found in drains and sewers. A distinction is made between header information, which relates to a pipeline as a whole (all associated codes begin with the letter A), and information on individual observations within the sewers and drains (all associated codes begin with the letter B). Similarly, header information and information on individual observations for manholes and inspection chambers are coded with the first letters C and D.

Header information is to be entered at the start of the inspection. It shall include, but not be limited to

- pipe reference and/or the two node references,
- direction of inspection,
- the method of inspection,
- date of inspection, and
- whether the drain or sewer was pre-cleaned.

Each observation shall be recorded using a principal code or combination of principal codes broadly describing the condition characteristic and, where necessary, the following supplementary information:

- characterization (up to two codes which describes the feature in more detail),
- quantification (up to two values which quantify the feature),
- circumferential location (up to two clockface references which locate the position of the observation around the circumference),
- joint (identifies when the observation is associated with a joint),
- longitudinal location (the distance from the stated reference point including a method of re-cording observations which continue over a significant length),
- photograph and/or video reference, and
- remarks (text which describes aspects of the observation which cannot be described by any other way).

Examples for records of two different observations are given in Tables 3.7 and 3.8. They refer to details of codes relating to the fabric of the pipeline shown exemplary in Table 3.9.

Table 3.7 Example of the record for a longitudinal crack at the top of the pipe 10.5 m from the starting point taken from EN 13508-2 (2011).

Longitudinal Location	Continuous Defect Code	Main Code	Characterization		Quantification		Circumferential Location		Joint	Photo Ref.	Video Ref.	Remarks
			1	2	1	2	1	2				
10.5		BAB	B	A			12				00:10:30	

Table 3.8 Example of the record for a 100 mm diameter intruding lateral connection (intruding half of the diameter of the main pipe) 16.5 metres from the start point (two codes are necessary to describe the feature) taken from EN 13508-2 (2011).

Longitudinal Location	Continuous Defect Code	Main Code	Characterization		Quantification		Circumferential location		Joint	Photo Ref.	Video Ref.	Remarks
			1	2	1	2	1	2				
16.5		BCA	E	A	100		9				00:12:20	
16.5		BAG			50		9				00:12:20	

Table 3.9 Details of codes relating to the fabric of the pipeline – example fissures (BAB) taken from EN 13508-2 (2011).

Main Code	Additional Information	Description
BAB	Characterization 1	The nature of the fissure: surface crack (A) – a crack only on the surface; crack (B) – crack lines visible on the pipe wall, pieces still in place; fracture (C) – crack visibly open in the pipe wall, pieces still in place.
	Characterization 2	The orientation of the fissure: longitudinal (A) – a crack or fracture which is mainly parallel to the axis of the pipe; circumferential (B) – a crack or fracture which is mainly around the circumference of the pipe; complex (C) – a group of cracks or fractures which cannot be described as longitudinal or circumferential; helical (D); radiating from a point (star fissure) (E).
	Quantification	The width of the fissure in millimeters.
	Circumferential location	The position of the fissure.

3.3.2.3 DWA-M 149-3: Assessment after visual inspection

Criteria	Localization
Validity	National state (Germany)
Nature	State of the art (Advisory Leaflet)
Observation level	Object (gray infrastructure)
AM activity	Investigation (condition assessment Priority for rehabilitation)
PDCA-circle	Check

In Germany, the DWA-M 149-3 (2015) Code of Practice regulates the procedure for classifying and assessing the structural/operational condition of drains and sewers outside buildings as subtask of the overall rehabilitation process. An assessment of the hydraulic or environmentally relevant rehabilitation requirements is not covered under this Code of Practice.

The aim is to present the structural/operational rehabilitation requirements, producing rankings or priorities for rehabilitation. The result of the classification and assessment of the structural/operational condition is an important component for the development of refurbishment strategies and measures, for which, however, further information is necessary. Those additional details are regulated under EN 752 (2017) and worksheet DWA-A 143-14 (2017), see also Section 3.2.2.

The condition classification carried out according to DWA-M 149-3 (2015) links the condition characteristics with requirements in terms of tightness, structural condition, and operational safety and, if necessary, boundary conditions through

- Classification of the individual condition:
  The basis for a condition classification is the description of the structural/operational condition as per EN 13508-2 (2011). The condition classification as per the EN codes should be carried out

Table 3.10 Classification of sewers and drains – example fissures (BAB) taken from DWA-M 149-3 (2015).

Main Code	Characterization		Requirements			Unit	Condition Class (Determination of Condition Classes: Condition Class 0: Very Severe Defect (Imminent Danger) Condition Class 1: Severe Defect Condition Class 2: Medium Defect Condition Class 3: Slight Defect Condition Class 4: Minor Defect)					Scope
	1	2	T	S	O		0	1	2	3	4	
BAB	A	A,B,C,D,E		X		mm					All	
	B,C	A,B,C,D,E	X			mm		≥ 3	≥ 2 <3	<2		
		A		X		mm	≥ 8	≥ 5 <8	≥ 3 <5	≥ 1 <3	<1	General
				X		mm	≥ 3	≥ 2 <3	≥ 1 <2	≥ 0.5 <1		$\geq$DN 300
				X		mm	≥ 5	≥ 3 <5	≥ 2 <3	≥ 2 <3	<1	$>$ DN 300 $\leq$DN 500
				X		mm	≥ 8	≥ 4 <8	≥ 3 <4	≥ 2 <3	<2	$>$ DN 500 $\leq$DN 700
		B		X		mm					all	
		C,D,E		X		mm	Case-by-case consideration					

line by line for each individual condition. Tables are used for the classification, which are used for relevance for key influential factors (e.g., nominal size, material type). If these influential factors are not known from the master data of the inspection the worst case is to be assumed. A distinction is made between the protection goals of tightness (T), stability (S), and operational safety (O). An example is given in Table 3.10.

- Assessment of the individual condition of the object:
  The object is classified based on the condition classes of the individual findings. The decisive factor is the respective minimum value, which is determined separately according to the protection goal or aggregated as an overall value.

3.3.2.4 NS-058: Technical aspects for inspection of sewer networks and structures

Criteria	Localization
Validity	National state (Columbia)
Nature	State of the art (Advisory Leaflet)
Observation level	Object (gray infrastructure)
AM activity	Investigation (condition assessment Priority for rehabilitation Wear and tear, rehabilitation expense)
PDCA-circle	Check

In Colombia, Bogota's water and sewerage company has adopted the national sewer assessment standard NS-058 (2010) since 2001. The standard covers the circuit-camera television technology's (CCTV) inspection guidelines in which the found defects are classified as operational and structural defects. At the same time, each category is ranked as to the defect types. According to the defect's type, the severity and the defect's location, the standard yields a score from 0 to 165 points for structural defects and from 0 to 10 for operational defects (see Tables 3.11 and 3.12).

Once scores have been identified for each defect found in the sewer asset, the standard provides a categorization of the structural and operational conditions for each sewer asset. Both structural and operational states are ranked into five grades. However, the methodology to arrive at the score differs:

Table 3.11 Structural defects and their scores according to NS-058 (2010).

Code	Type of Defect	Description	Score
1.1.1.1	Deformation or deflection	Vertical (>3% and 7.5%; >7.5% and 12.5%; or >12.5%)	20, 80, or 165
		Horizontal (>12.5%)	165
1.1.1.2	Crack or facture	Superficial crack (circular, complex, helicoidal, or longitudinal)	2
		Crack (circular, longitudinal, complex, or helicoidal)	10 or 80
		Fracture (circular, longitudinal, complex, or helicoidal)	40 or 80
1.1.1.3	Breakage or collapse	Breakage (<1/4; or +1/4)	80 or 165
		Collapse (complete loss)	165
1.1.1.4	Seal material introduced in the sewer asset	The seal is a ring (visible displacement, less than a half, more than a half, broken)	1, 2, 5, or 8
		Other seal (<5%; 5–20%; or >20%)	1, 2 or 5
		Without seal	5
1.1.1.5	Shifted joint	Longitudinal (1 and 1.5 width; >1.5 width; or displacement joint over NS–073 of visible soil)	1, 2 or 80
		Radial (1 and 1.5 width; >1.5 width; or 10% diameter)	2 or 80
		Angular (1 and 1.5 width; >1.5 width; or 10% diameter)	2 or 80
1.1.1.6	Superficial damages	Roughness, pelling (mild, moderate, high, or very high)	5, 20, 120, or 165
		Aggregate (visible, leaving on the surface, absent) (mild, moderate, high, or very high)	5, 20, 120, or 165
		Reinforcement (visible, and leaving on the surface or corroded) (mild, moderate, high, or very high)	5, 20, 120, or 165
		Surface abrasion, corrosion waste or porous pipes (mild, moderate, high, or very high)	5, 20, 120, or 165
		Mechanical damage (mild, moderate, high, or very high)	5, 20, 120, or 165
		Chemical attack – damage above or below water level (mild, moderate, high, or very high)	5, 20, 120, or 165
		Non-obvious cause or other damages (mild, moderate, high, or very high)	5, 20, 120, or 165
1.2.1.1	Defects related to brick or masonry assets	Displacement	80
		Absent bricks (<1/4; or +1/4)	120 or 165
		Collapsing collector in brick	165
1.2.1.2	Lack of mortar	Lack of mortar (<15 mm; 15–50 mm; or >50 mm)	10, 20, or 40

Table 3.12 Operational defects and their scores according to NS-058 (2010).

Code	Type of Defect	Description	Score
1.1.2.1	Obstruction by connection	Obstruction by connection (<5%; 5–20%; 20–50%; 50–75%; or >75%)	1, 2, 5, 8, or 10
1.1.2.2	Trees' roots	Root blocking connection	5
		Independent fine roots	1
		Complex root mass (<5%; 2–20%; 20–50%; 50–75%; or >75%)	2, 4, or 10
1.1.2.3	Bonded or sedimentary deposits or soil income	Incrustation (<5%; 2–20%; or >20%)	2, 4, or 10
		Materials glued to the wall (<5%; 2–20%; 20–50%; 50–75%; or >75%)	1, 2, 5, 8, or 10
1.1.2.4	Other obstacles	Brick pieces, broken piece of pipe, other objects in the bottom of the pipe, protuberance through the wall, board wedge or construction inside the structure	10
1.1.2.5	Infiltration	Sweating, drip, jet, or pressure water jet	3, 5, or 10

the structural grade reflects the sum of the scores of all structural defects found in the sewer asset, see NS-058 (2010) Table 3 of the appendix – Part B.1 while the operational grade reflects the quotient of the sum of the scores of all operational defects found in the sewer asset and the sewer asset's length, see NS-058 (2010) Table 4 of the appendix – Part B.1. For both the structural and operational rankings, Grade 1 represents excellent structural or operational conditions while Grade 5 means critical structural or operational conditions. For more details about the diagnosis of each grade and suggested recommendation by NS-058 (2010), see Tables 3.13 and 3.14.

Table 3.13 Classification of structural grades, diagnosis, and recommendations according to NS-058 (2010).

$\sum$ Structural Scores	Grade	Diagnosis	Recommendations
> 10	1	With little to no structural defects which are not important for the structural stability of the system	New inspection within 4–5 years to verify the structural state of the sewer asset
10–39	2	The defects found are minor and do not compromise the stability of the network in the short term	Maintenance actions to correct the damage found and make a new inspection within 3–4 years to analyze the structural risk
40–79	3	Defects that could generate structural problems	Maintenance actions to correct defects by prioritizing them according to their severity or qualification. Make a new inspection within 2–3 years to verify with the result, the actions carried out and that the structural risk has not increased
80–164	4	Defects with high importance and could generate structural problems	Rehabilitation actions that prevent the damage propagation by prioritizing defects according to the severity or qualification. Make a new inspection within 1–2 years to analyze the performed actions
165+	5	Collapsed sewer asset	Replacement actions of an urgent nature

Table 3.14 Classification of operational grades, diagnosis, and recommendations according to NS-058 (2010).

$\frac{\sum \text{Operational Defects}}{\text{Sewer Asset's Length}}$	Grade	Diagnosis	Recommendations
<0.5	1	With little to no operational defects which are not important for the operation of the system	New inspection within 2–3 years to verify the operational state of the sewer asset
0.5–0.9	2	The defects found are minor, and do not compromise the operation of the network in the short term	Maintenance actions to correct the most important damage found and make new inspection with 1–2 years to analyse the operational risk
1–2.4	3	The defects found could affect the normal operation of the network	Maintenance actions that avoid the operational failure of the system. New inspection within 1 year to verify the actions' results
2.5–4.9	4	The defects found of high importance that could generate operational problems that limit the free water flow	Rehabilitation actions that prevent damage propagation. New inspection within 8–12 months to analyze the actions' results
>5	5	Sewer asset has a level of clogging greater than or equal to 60%, and/or flow of water is slow or zero	Emergency actions of cleaning and unblocked of urgent nature. Make inspections of surrounding networks to determine the clogging cause

3.3.2.5 DWA-M 149-10 (under preparation): classification of structural substance

Criteria	Localization
Validity	National state (Germany)
Nature	State of the art (Advisory Leaflet)
Observation level	Object (gray infrastructure)
AM activity	Investigation (condition assessment Wear and tear, rehabilitation expense)
PDCA-circle	Check

The structural condition of the pipe materials (i.e., reflecting the wear and tear) of the pipes within a sewer network is an important parameter for medium-term planning support and for the development of rehabilitation strategies as a part of integral sewer management according to EN 752 (2017): The pipe material condition classification (or the evaluation of the residual wear capacity) supplements the decision-making basis for infrastructure management and rehabilitation planning.

Wear and tear of drains and sewers may result in a restriction of the functionality (e.g., tightness, stability, and operation) of the entire unit under consideration but not necessarily urgency with which rehabilitation must take place. It results accordingly from the sum of all indications or damage patterns impairing the function. Analogous to DWA-M 149-1 (2018), the functional capability used here includes operational limitations, but not hydraulic functional limitations as in capacity problems.

Table 3.15 Examples for the allocation of damage codes in damage type and damage severity, with codes according to EN 13508-2 (2011).

Type of Damage	Characteristic of Damage		
	Penetrating	Surface	Unrelated to the Building Structure
Punctual damage	BAD A	BAK A	BAH A
Perimeter damage	BAI	BAB A B	BAL L
Longitudinal damage	BAAA A/B	BAF A	BAK B

The process for quantifying the wear and the classification of the structural condition of the pipe materials of drains and sewers comprises the following steps:

- Pre-processing of observations (damage codes and damage patterns):
  Individual observations are combined into damage patterns as all observations are assigned a 'type of damage' and a 'damage severity'. The approach distinguishes between the damage types 'punctual damage', 'circumferential damage' and 'longitudinal damage' as well as the damage characteristics 'penetrating damage' (wall-penetrating and thus corresponding damage), 'surface damage' and 'damage unrelated to the building structure' (pipe wall or seal). Examples of this classification are given in Table 3.15. Observations at the same physical location are further processed within the same type of damage (for example, penetrating damages, surface damages or damages unrelated to the structure) and the same damage characteristics (for example, punctual damages, perimeter damages or longitudinal damages) to 'damage patterns' as follows:
  - Penetrating (wall penetrating and thus corresponding) damages:
    - Punctual damage: All findings with identical station and position are combined into one damage within the damage type point damage.
    - Circumferential damage: All findings with identical stations are combined into one damage within the type of damage.
    - Longitudinal damage: All findings with overlapping stations are combined into one damage within the damage type line damage.
  - Surface damages: Observations describing surface damage are grouped into damage pictures adopting the same approach as used for the penetrating damage.
  - Damage unrelated to the structure: Each finding is considered as an individual damage pattern.
- Calculation of wear:
  The absolute wear (weighted total damage length [m]) results from the sum of the damages weighted with respect to damage type, damage length and condition class. The calculation of the damage weight is carried out using the length of the damage (e.g., point damage: 0.3 m), the starting weight (e.g., point damage: 8) and the class weight (e.g., condition class 2: 0.25). All parameters are multiplied. The wear corresponds to the ratio of the absolute wear and object extension (i.e., length).
- Substance classification:
  The value range of the wear or structural substance is subdivided into substance classes, to be able to interpret the determined substance of a sewer and in a practical context (see Table 3.16).

Table 3.16 Definition of the classification of structural substance.

Class of Structural Substance Class	Wear (%)	Verbal description
5	<5	Very good to complete substance/negligible to no wear and tear Usually monitoring (scheduled inspection), no measures for substance restoration required
4	≥5 <15	Good substance/low wear Hardly any investment measures required for complete substance restoration, high probability for repair procedures
3	≥15 <33	Sufficient substance/progressive wear and tear predominantly repair processes with increasing investment share for the restoration of substance
2	≥33 <67	Poor substance/high wear and tear predominantly investment measures with decreasing repair share for substance restoration
1	≥67 <95	Very poor structural substance/critical wear and tear measures to restore substance will mostly be investment-based, repair share very low
0	>95	Structural substance completely used up/complete wear measures to restore the substance in the rules only possible for investment

3.3.2.6 Regulation no. 677: Iranian instructions for performing sewer pipe video inspections

Criteria	Localization
Validity	National state (Iran)
Nature	Generally recognised rules of technology (standard)
Observation level	Object (gray infrastructure)
AM activity	Investigation (condition assessment Priority for rehabilitation)
PDCA-circle	Check

The IRN 677 (2015) is the second key standard utilized in the maintenance and operation of Iran's sewage networks. The requirements for video inspection, sewer classification based on condition, and the contractor's quality control system are the main issues covered by this standard. The basis for preparing this standard was two well-established international standards, as for example, EN 13508-2 (2011). It is important to note that, according to this standard, each sewer's condition grade is determined twice, independently, once based on structural defects and once based on service defects. After that, sewer pipes are graded based on Tables 3.17 and 3.18.

Table 3.17 Sewer pipe condition grade considering structural failures (IRN 677, 2015).

Maximum Score	Pipe Condition Grade
Less than 10	1
10–39	2
40–79	3
80–164	4
More than 165	5

Table 3.18 Sewer pipe condition grade considering service failures (IRN 677, 2015).

Average Score Per Meter of Pipe	Maximum Score	Pipe Condition Grade
Less than 0.5	Less than 1	1
0.5–0.9	1–1.9	2
1–2.4	2–4.9	3
2.5–4.9	5–9.9	4
More than 5	More than 10	5

3.3.2.7 DWA-A 143-14: Development of a rehabilitation strategy

Criteria	Localization
Validity	National state (Germany)
Nature	Generally recognized rules of technology (standard)
Observation level	Object (gray infrastructure) and network
AM activity	Investigation (condition assessment Asset value)
PDCA-circle	Check

The change of the magnitude of asset values should be monitored to recognize a gradual deterioration in a (sub)network or in an object at an early stage to allow for the development of preventive or corrective measures (DWA-A 143-14, 2017).

The net asset value (av) of an object (e.g., a sewer pipe) as a network component can at most correspond to its replacement value (rv); it is reduced by the aging or deterioration of the system. If the age-related change in condition is described by the relationship between the age and the technical service life of the network component in question, its net asset value can be approximately formulated as

$$\mathrm{av}_{\mathrm{Object}} = \mathrm{rv}_{\mathrm{Object}} \cdot \left(1 - \frac{\mathrm{age}_{\mathrm{Object}}}{\text{technical service life}_{\mathrm{Object}}}\right)$$

When considering a sewer network, the network-related net asset value results from the integration of the values of the network components.

The determination of the appropriate (residual) useful life of an asset or network is of considerable importance. The inferred residual book value of an asset can only be used as a measure of the asset value if the useful life of the asset on which it is based corresponds to the normal useful life. As a rule, however, the useful lives of assets cannot be expected to be uniform, but differ, for example, due to different construction qualities during certain phases of construction. To estimate the deviation of an expected object-specific useful life from the normal operating life, the individual object condition can be compared with the typical condition for the age. The latter can be determined by a statistical evaluation of the condition data at a network level. A possible perspective for the assessment of a network is the formation of the relative asset value (rav):

$$\mathrm{av}_{\mathrm{Network}} = \sum \mathrm{rv}_{\mathrm{Object}} \cdot \left(1 - \frac{\mathrm{age}_{\mathrm{Object}}}{\text{technical service life}_{\mathrm{Object}}}\right)$$

$$\mathrm{rav}_{\mathrm{Network}} = \frac{\mathrm{av}_{\mathrm{Network}}}{\mathrm{rv}_{\mathrm{Network}}}$$

An interpretation of the relative asset value is given in Table 3.19.

Table 3.19 Interpretation of the relative asset value according to DWA-A 143-14 (2017).

Relative Asset Value	Verbal Description
$\ll$0.5–0	If a relative asset value is found to be significantly below 0.5, it can be deduced that more remediation should be carried out. However, it cannot yet be deduced that the current level of rehabilitation is insufficient. The investments increase the current capital input for the benefit of future periods without carrying out unnecessary or uneconomical remediation measures.
About 0.5	An (age-)homogeneous network, in which reinvestments and disposals are in balance over a longer period of time, will ideally have a relative net asset value of approximately 0.5.
$\gg$0.5–1	A relative asset value of between 0.5 and 1 is typical for networks that are only a few decades old. In this 'young' phase, which is probably characterized by a low frequency of damage, high depreciation income is offset by a low physical renovation requirement. Due to the still high asset value, the network operator can make greater use of repair procedures in this phase, as long as he keeps an eye on the long-term development of the asset value.

3.3.2.8 EN 14654-2: Drain and sewer systems outside buildings – management and control of activities – Part 2: rehabilitation

Criteria	Localization
Validity	Europe (CEN Members) (CEN's National Members are (2023) the National Standardization Bodies (NSBs) of the 27 European Union countries, United Kingdom, the Republic of North Macedonia, Serbia and Türkyie plus three countries of the European Free Trade Association (Iceland, Norway, and Switzerland).)
Nature	Generally recognized rules of technology (standard)
Observation level	Object (gray infrastructure)
AM activity	Rehabilitation
PDCA-circle	General

EN 752 (2017) provides the framework for the design, construction, rehabilitation, maintenance, and operation of drainage systems outside buildings and thus addresses a higher system level. This standard is supported by more detailed standards on the investigation, planning, construction, organization, and monitoring of drainage systems. These include, among others, the standards of the EN 14654 series 'Management and monitoring of operational measures in drainage systems outside buildings'. Part 2 of this series of standards deals with the rehabilitation of sewer systems. Even though the introductory Chapters 5 and 6 of EN 14654-2 (2021) deal with the preparation of a rehabilitation plan and a rehabilitation concept (and thus mimic essential aspects of EN 752 (2017) which addresses the system level), it concentrates essentially on the planning of measures (Chapter 7) and the implementation of measures (Chapter 8) and therefore primarily focuses on the object level. According to EN 14654-2 (2021), the planning of the rehabilitation measures should contain all information and drawings required for the implementation of the measures. The starting point for the preparation of the measure planning is the measure description and the measure objectives from the renovation concept, considering the performance requirements (see EN 752, 2017).

The preparation of the measure design includes:

- Review of the measure description and the measure objectives:
  The measure description and objectives set out in the redevelopment concept should be reviewed to ensure that they still apply. New development or other utility-related works associated with the measure should be reviewed. If there have been changes in the nature or timing of the development or other works that could affect the measure or its phasing, the concept should be revised.
- Further investigation:
  Investigations required to prepare the action plan may include, but are not limited to, further investigation of the existing drainage system, other investigations to determine the feasibility of alternatives or to provide information for design calculations, or investigations to determine the impact of the proposed works.
- Further assessment (if required):
  Regarding the assessment of existing drainage systems, reference is made to EN 13508-1 (2012). The assessment should be carried out with a sufficient level of detail to allow solutions to be selected for the measure. Following the assessment, the measure objectives should be reviewed to ensure that they still apply.
- Preparation of the action plan:
  Considering the results from the investigation and assessment, several possible detailed variants for the measure should be developed in accordance with EN 14654-2 (2021) as part of the remediation concept described for the measure. The solutions identified as technically feasible should be compared with each other to determine the optimal solution using the approach described in EN 752 (2017). For the chosen variant, further detailed calculations should be carried out and a detailed description of the measures (including drawings) should be prepared to provide all the information necessary for the execution of the measures.

In this context, according to EN 14654-2 (2021), performance indicators should be chosen to determine whether the measure has been executed in accordance with the measure planning and whether the objectives set out in the remediation concept or measure have been met. The performance indicators should be based on easily measurable parameters. They should be used to assess the compliance of the completed measure. Examples of indicators may include:

- percentage of extraneous water collected in the wastewater pipe/sewer,
- percentage of exfiltration,
- number of floods per year and per kilometer,
- number of blockages per year and per 1000 km,
- cracks per kilometer,
- number of collapses per year and per 1000 km,
- and sedimentation rate.

3.4 CONCLUSION

The previous case studies made it clear that AM in general, and UDAM in particular, is a multidimensional field due to the large number of system elements, the maintenance tasks, and the different levels of consideration (see Figure 3.5). These levels are, for example:

- at a strategic level or operational level.
- at a system level or object level.
- consider the drainage system as a water management facility or as an economic asset.

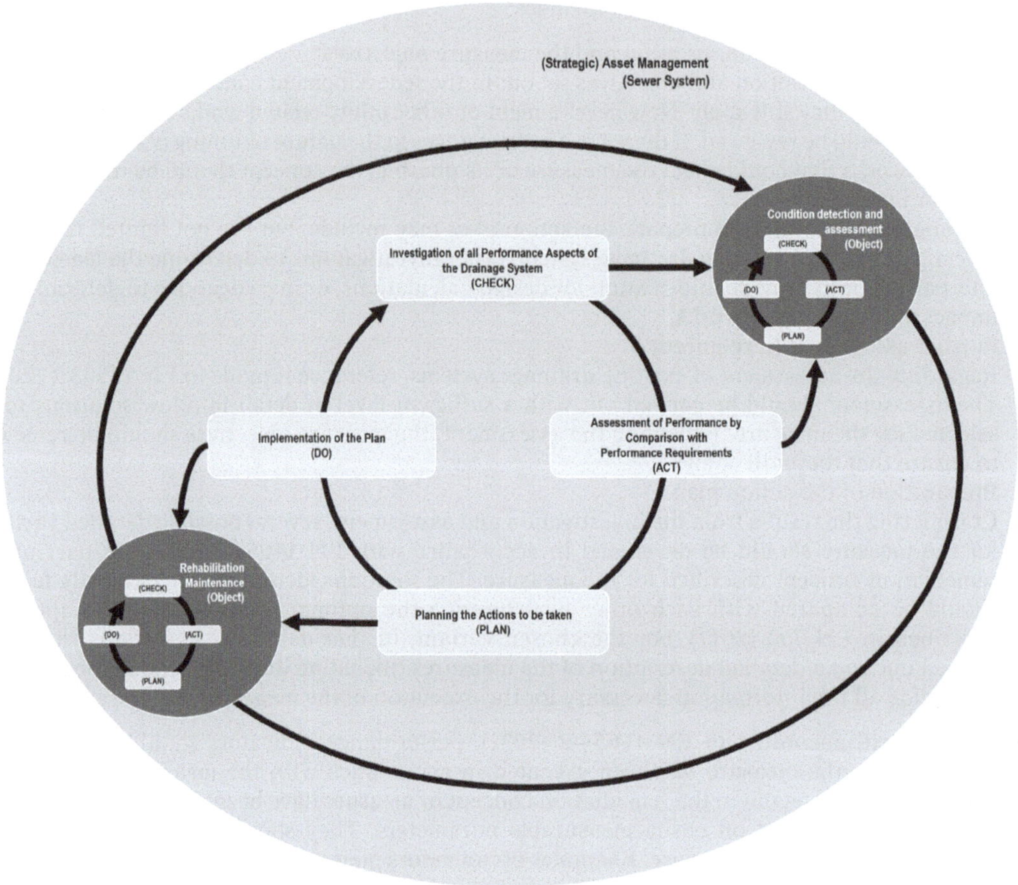

Figure 3.5 Strategic and operative UDAM and PDCA cycle.

Reflecting this, legal requirements and technical regulations vary widely. The ordinances, laws, and regulations presented accordingly represent only a small part of the framework conditions applicable. The selection was essentially limited to the maintenance tasks of condition assessment and rehabilitation.

The following is noted:

- Maintenance is not a one-time task but takes place continuously. In this respect, maintenance procedures as well as maintenance cycles must be defined for all maintenance tasks.
- Effective and quality-assured action is possible at all levels when the PDCA cycle is applied (see also Figure 3.2). Numerous ordinances, laws, and regulations address this in various ways.

REFERENCES

91/271/EEC (1991). Council Directive 91/271/EEC of 21 May 1991 concerning urban waste-water treatment. Council of the European Communities.

91/676/EEC (1991). Council Directive of 12 December 1991 concerning the protection of waters against pollution caused by nitrates from agricultural sources. Council of the European Communities.

2000/60/EC (2000). Directive 2000/60/EC of the European Parliament and of the Council of 23 October 2000 establishing a framework for Community action in the field of water policy. European Parliament and of the Council of the European Union.

2006/118/EC (2006). Directive 2006/118/EC of the European Parliament and of the Council of 12 December 2006 on the protection of groundwater against pollution and deterioration. European Parliament and of the Council of the European Union.

2007/60/EC (2007). Directive 2007/60/EC of the European Parliament and of the Council of 23 October 2007 on the assessment and management of flood risks. European Parliament and of the Council of the European Union.

2008/105/EC (2008). Directive 2008/105/EC of the European Parliament and of the Council of 16 December 2008 on environmental quality standards in the field of water policy, amending and subsequently repealing Council Directives 82/176/EEC, 83/513/EEC, 84/156/EEC, 84/491/EEC, 86/280/EEC and amending Directive 2000/60/EC of the European Parliament and of the Council. European Parliament and of the Council of the European Union.

COM/2020/492 (2020). Tenth report on the implementation status and programmes for implementation (as required by Article 17 of Council Directive 91/271/EEC, concerning urban waste water treatment). Council of the European Communities.

CSA W210 (2021). Prioritization of Flood Risk in Existing Communities (Canadian Standard). Canadian Standards Association, Toronto, Ontario, Canada.

CSA W211 (2021). Management Standard for Stormwater Systems (Canadian Standard). Canadian Standards Association, Toronto, Ontario, Canada.

CSA W218 (2023). Specifications of Natural Asset Inventories (Canadian Standard). Canadian Standards Association, Toronto, Ontario, Canada.

DWA-A 143-14 (2017). Sanierung von Entwässerungssystemen Außerhalb von Gebäuden – Teil 14: Entwicklung Einer Sanierungsstrategie (Arbeitsblatt). DWA Deutsche Vereinigung für Wasserwirtschaft, Abwasser und Abfall e. V., Hennef, Germany.

DWA-A 147 (2017). Betriebsaufwand für Kommunale Entwässerungssysteme – Betriebsaufgaben und Häufigkeiten (Arbeitsblatt). DWA Deutsche Vereinigung für Wasserwirtschaft, Abwasser und Abfall e. V., Hennef, Germany.

DWA-M 149-1 (2018). Zustandserfassung und -Beurteilung von Entwässerungssystemen Außerhalb von Gebäuden – Teil 1: Grundlagen (Merkblatt). DWA Deutsche Vereinigung für Wasserwirtschaft, Abwasser und Abfall e. V., Hennef, Germany.

DWA-M 149-2 (2013). Zustandserfassung und -Beurteilung von Entwässerungssystemen Außerhalb von Gebäuden Teil 2: Kodiersystem für die Optische Inspektion (Merkblatt). DWA Deutsche Vereinigung für Wasserwirtschaft, Abwasser und Abfall e. V., Hennef, Germany.

DWA-M 149-3 (2015). Zustandserfassung und -Beurteilung von Entwässerungssystemen Außerhalb von Gebäuden Teil 3: Beurteilung Nach Optischer Inspektion (Merkblatt). DWA Deutsche Vereinigung für Wasserwirtschaft, Abwasser und Abfall e. V., Hennef, Germany.

DWA-M 149-4 (2008). Zustandserfassung und -Beurteilung von Entwässerungssystemen Außerhalb von Gebäuden – Teil 4: Detektion von Lagerungsdefekten und Hohlräumen Mittels Geophysikalischer Verfahren (Merkblatt). DWA Deutsche Vereinigung für Wasserwirtschaft, Abwasser und Abfall e. V., Hennef, Germany.

DWA-M 149-5 (2010). Zustandserfassung und -Beurteilung von Entwässerungssystemen Außerhalb von Gebäuden – Teil 5: Optische Inspektion (Merkblatt). DWA Deutsche Vereinigung für Wasserwirtschaft, Abwasser und Abfall e. V., Hennef, Germany.

DWA-M 149-6 (2016). Zustandserfassung und -Beurteilung von Entwässerungssystemen Außerhalb von Gebäuden – Teil 6: Druckprüfungen in Betrieb Befindlicher Entwässerungssysteme mit Wasser Oder Luft (Merkblatt). DWA Deutsche Vereinigung für Wasserwirtschaft, Abwasser und Abfall e. V., Hennef, Germany.

DWA-M 149-7 (2016). Zustandserfassung und -Beurteilung von Entwässerungssystemen Außerhalb von Gebäuden – Teil 7: Beurteilung der Umweltrelevanz des Baulichen/Betrieblichen Zustands (Merkblatt). DWA Deutsche Vereinigung für Wasserwirtschaft, Abwasser und Abfall e. V., Hennef, Germany.

EN 13508-1 (2012). Investigation and Assessment of Drain and Sewer Systems Outside Buildings – Part 1: General Requirements (European Standard). European Committee for Standardization, Brussels, Belgium.

EN 13508-2 (2011). Investigation and Assessment of Drain and Sewer Systems Outside Buildings – Part 2: Visual Inspection Coding System (European Standard). European Committee for Standardization, Brussels, Belgium.

EN 14654-1 (2021). Drain and Sewer Systems Outside Buildings: Management and Control of Activities – Part 1: General Requirements (European Standard). European Committee for Standardization, Brussels, Belgium.

EN 14654-2 (2021). Drain and Sewer Systems Outside Buildings: Management and Control of Activities – Part 2: Rehabilitation (European Standard). European Committee for Standardization, Brussels, Belgium.

EN 1610 (2015). Construction and Testing of Drains (European Standard). European Committee for Standardization, Brussels, Belgium.

EN 752 (2017). Drain and Sewer Systems Outside Buildings: Sewer System Management (European Standard). European Committee for Standardization, Brussels, Belgium.

GrwV (2016). Grundwasserverordnung vom 9. November 2010 (BGBl. I S. 1513), die Durch Artikel 3 des Gesetzes vom 4. August 2016 (BGBl. I S. 1972) Geändert Worden ist (Verordnung). Bundestag, Berlin, Germany.

IRN 520 (2009). Guide to the Operation and Maintenance of Sewage Collection Networks (Advisory Leaflet). Executive Technical System Office of the Vice President for Planning and Strategic Supervision, Tehran, Iran.

IRN 677 (2015). Instructions for Performing Sewer Pipe Video Inspections (Advisory Leaflet). Executive Technical System Office of the Vice President for Planning and Strategic Supervision, Tehran, Iran.

ISO 9000 (2015). Quality Management Systems – Fundamentals and Vocabulary (International Standard). International Organization for Standardization, Geneva, Switzerland.

ISO 9001 (2015). Quality Management Systems – Requirements (International Standard). International Organization for Standardization, Geneva, Switzerland.

ISO 55000 (2014). Asset Management – Overview, Principles and Terminology (International Standard). International Organization for Standardization, Geneva, Switzerland.

ISO 55001 (2014). Asset Management – Management Systems – Requirements (International Standard). International Organization for Standardization, Geneva, Switzerland.

ISO 55002 (2018). Asset Management – Management Systems – Guidelines for the Application of ISO 55001 (International Standard). International Organization for Standardization, Geneva, Switzerland.

JORF (2006). Loi N° 2006-1772 du 30 Décembre 2006 sur L'eau et les Milieux Aquatiques, Légifrance, service public de la diffusion du droit par l'internet, Paris, France. https://www.legifrance.gouv.fr/dossierlegislatif/JORFDOLE000017758328/ (last accessed: 8/3/24)

JORF (2015). Loi no 2015-991 du 7 Août 2015 Portant Nouvelle Organisation Territoriale de la République, Légifrance, service public de la diffusion du droit par l'internet, Paris, France. https://www.legifrance.gouv.fr/loda/id/JORFTEXT000030985460/ (last accessed: 8/3/24)

JORF (2022). Loi no 2022-217 du 21 Février 2022 Relative à la Différenciation, la Décentralisation, la Déconcentration et Portant Diverses Mesures de Simplification de L'action Publique Locale, Légifrance, service public de la diffusion du droit par l'internet, Paris, France. https://www.legifrance.gouv.fr/jorf/id/JORFTEXT000045197395 (last accessed: 8/3/24)

LANUV (2008). Verwaltungsvorschrift über die Aufstellung von Abwasserbeseitigungskonzepten Wassergesetz für das Land Nordrhein-Westfalen. Der Minister für Umwelt, Raumordnung und Landwirtschaft des Landes Nordrhein-Westfalen.

LWG NRW (1995). Wassergesetz für das Land Nordrhein-Westfalen (Landesgesetz). Der Minister für Umwelt, Raumordnung und Landwirtschaft des Landes Nordrhein-Westfalen.

NS-058 (2010). Aspectos Técnicos Para Inspección de Redes y estructuras de Alcantarillado (Norma Técnica de Servicio). Empresa de Acueducto y Alcantarillado de Bogotá, Bogotá, Colombia.

OECD (2020). Financing Water Supply, Sanitation and Flood Protection: Challenges in EU Member States and Policy Options. OECD Publishing, Paris, France.

OGewV (2020). Oberflächengewässerverordnung vom 20. Juni 2016 (BGBl. I S. 1373), die Zuletzt Durch Artikel 2 Absatz 4 des Gesetzes vom 9. Dezember 2020 (BGBl. I S. 2873) Geändert Worden ist (Verordnung). Bundestag, Berlin, Germany.

SüwVO Abw (2013). Verordnung zur Selbstüberwachung von Abwasseranlagen – Selbstüberwachungsverordnung Abwasser (Landesverordnung). Der Minister für Umwelt, Raumordnung und Landwirtschaft des Landes Nordrhein-Westfalen.

Torre A. and Bourdin S. (2023). The French territorial reform of the regions: objectives, risks and challenges for some forgotten territories. *International Journal of Public Administration* **46**, 761–772, https://doi.org/10.1080/01900692.2021.2018456

WHG (2023). Wasserhaushaltsgesetz vom 31. Juli 2009 (BGBl. I S. 2585), das Zuletzt Durch Artikel 5 des Gesetzes vom 3. Juli 2023 (BGBl. 2023 I Nr. 176) Geändert Worden ist (Bundesgesetz). Bundestag, Berlin, Germany.

doi: 10.2166/9781789063059_0093

Chapter 4

Investigate the condition of an asset

F. Boogaard[1,2], F. Cherqui[3], F.H.L.R. Clemens-Meyer[4,5], M.L. Lepot[3,6], W. Shepherd[7] and M. van der Valk[8]

[1]Deltares, PO Box 177, 2600 MH Delft, The Netherlands
[2]Hanze University of Applied Sciences, Groningen, The Netherlands
[3]University of Lyon, INSA-LYON, Université Claude Bernard Lyon 1, DEEP, F-69621, F-69622, Villeurbanne, France
[4]SkillsInMotion BV, Esdoornlaan 11, 3454 HH, De Meern, The Netherlands
[5]NTNU – Norwegian University of Science and Technology, Trondheim, Norway
[6]Un Poids une mesure, Lyon, France
[7]University of Sheffield, Civil and Structural Engineering, Sheffield, UK
[8]van der Valk + de Groot, Poeldijk, The Netherlands

ABSTRACT

Monitoring the technical state of an urban drainage (UD) system is at the core of asset management, the deployment of visual inspection technology (either using direct visual access for inspection of applying photo and/or video cameras) was and has remained the main method of gathering information on the technical state. Despite some known fundamental shortcomings visual inspection is expected to remain the main source of information for inspection for the foreseeable future. This chapter discusses the virtues of visual inspection but also provides insight into other technologies that have been tried and/or deployed on a more limited scale but do offer access to more and more exact information when compared to the visual methods. Although not much experience is available, inspection techniques for nature-based solutions will be discussed as well.

Keywords: inspection techniques, condition assessment, visual inspection, outlook.

4.1 INTRODUCTION

4.1.1 Background

Urban drainage (UD) systems are capital intensive structures and, in most cases, installed along other types of networks (e.g., roads, potable water, power/energy supply). The order of magnitude of their overall length is millions of km (e.g., 392,000 km of sewer pipes in France alone in 2008, Enquête Eau, 2008). Such systems can be divided into several different asset types, as discussed in Section 3.3.1 of Chapter 8. They comprise both buried assets such as pipes and surface measures such as storm water controls.

UD infrastructure will, like any asset, deteriorate with time and use. These assets are often many decades old and remain in place beyond their expected life duration. Causes of their deterioration vary with the type of structure. However, there are common processes: ageing of the structures themselves, corrosion/erosion, loadings and impacts from the surrounding environment, lack of maintenance, errors induced by improper maintenance or upgrades in the sewer system and misuse of the facilities by citizens. Interested readers can refer to Chapter 5 which considers these processes in detail.

It is important to understand the condition of UD assets to ensure that their capacity remains sufficient and to minimize the risk of catastrophic failures. A UD network should convey wastewater and storm water runoff to a treatment facility to prevent pollution and health risks and should also reduce the likelihood of flooding. An asset in poor condition will often have a reduced capacity usually due to a reduction in cross-sectional area, increased roughness, and/or increased head loss. Poor understanding of asset condition can lead to underestimation of network repairs, renovation, and renewal costs. There will be a moment in time at which it will be more cost effective to replace an asset rather than to continue to use a deteriorating asset which is beyond its expected service life. This is especially the case where a utility faces regulatory fines for causing pollution and compensation pay-outs for flooding, but even increased infiltration will have a cost impact due to increased pumping costs.

Climate change is likely to increase the pressure on UD assets. In most design standards, sewer pipe diameters/shapes are often chosen to ensure a sufficient velocity during dry weather conditions (in order to reduce sedimentation and hence the likelihood of H_2S production) and provide capacity for the design flow. Climate change is expected to result in more extreme droughts, heat waves and rain events, thus leading to more favorable H_2S production conditions and higher velocities during rain events. Thus, corrosion and erosion processes will likely increase during the coming decades. Knowledge of asset condition and how it is changing will therefore become more important as processes affecting deterioration change and potentially even accelerate.

There is an increasing use of blue/green infrastructures which are often multi-use spaces, for example, the land is also used for sport or leisure activities, gardening, and so on. Such plural purposes can lead to degradation/soil compaction and therefore, decreasing hydraulic/infiltration capacities. Any infiltration-based solution can be affected by decreasing infiltration capacity due to clogging of pore spaces.

Engineered drainage structures suffer from a range of defects: the structure can deform or crack, the material can corrode and erode, joints can move and become non-water tight, allowing roots and soil ingress, debris can build up along the pipe invert as well as become attached to the walls, and the structure can be damaged due to work on surrounding infrastructure, or be subject to poor modifications – for example, adding new pipe connections. Inspections are vital to understand the current condition of assets and to plan future maintenance, rehabilitation, and replacement. The comparatively low value attributed to water and the complexities associated with buried pipes in a range of materials has resulted in UD inspection receiving less attention than pipes for higher value and higher risk activities such as the transport of oil and gas. As a consequence, apart from CCTV inspection, not many inspection technologies (Figure 4.1) are applied on a wide scale in UD.

The first part of this section gave a general overview (but not exhaustive) of the main causes of sewer system degradation. Overall, such processes lead to reduction of UD service through decreased capacities. Consequences are wide ranging: pollution of surface/bathing water with possible impacts on public health, urban flooding events, increased energy consumption of pumps, structural collapses that could lead to above ground road or building damage.

4.1.2 Which data to collect

It has been established that UD network assets deteriorate in different ways, are made of different materials and come in different shapes and sizes, which are not always properly recorded in asset databases. Furthermore, the date at which an asset had been originally constructed will have an impact

Figure 4.1 Typical CCTV inspection crawlers showing differing sizes (left) and tethered crawler entering pipe from a benched manhole (right). (Courtesy Will Shepherd).

on many parameters, for example, concrete technology, pipe jointing and manufacturing techniques have changed with time. Traditionally, inspections have been visual, and hence subjective and narrative based, however, sensor technology is developing rapidly allowing objective measurements, albeit with some degree of uncertainty.

Table 4.1 gives a very brief overview of some of the most frequent defect types. Defects can broadly be divided into those which are primarily structural, that is, their main impact is on the structural integrity of the sewer; and operational, that is, their main impact is on the functioning of the sewer, usually through reduced capacity. Some inspections record other information, such as the presence of vermin, or water colour, which have no clear impact on either the structure or serviceability. Clearly there are linkages between structural and operational defects and some defects fall into both categories, especially a well progressed structural defect which can reduce the pipe cross section and hence reduce capacity. This chapter cannot exhaustively list all the required data to assess the status of UD systems and Chapter 8 is focusing on this subject. However, general specifications of needed data are included.

Table 4.1 Example defects.

Category	Defect Type	Details
Structural	Deformation	The pipe cross-sectional shape has been deformed, quantified by a percentage change in dimension and orientation.
Structural	Crack	The pipe wall is displaying a crack. The magnitude varies from a surface crack to an open crack, the orientation can be longitudinal, circumferential, helical, or a combination.
Structural	Joint defects	The joint is displaced perpendicularly or longitudinally with respect to the pipe direction. The magnitude in millimetres or relative to the pipe wall thickness is recorded
Operational	Roots	Plant roots have entered the sewer through joints or defects, the size of the roots and magnitude of cross-sectional area reduction is recorded.
Operational	Debris	Foreign solid materials are present in the pipe, the nature of the material – fine, coarse, grease, and so on and the reduction in cross-sectional area is recorded.
Operational	Infiltration	Water is entering the pipe at a joint or defect, the rate of the infiltration is recorded.

To conduct investigations into the status of systems, high-quality data are required. 'High quality' implies that two aspects need to be respected: objectivity and accuracy of data. Objectivity is rather important, especially in the common case of visual inspections. Dirksen *et al.* (2013) demonstrated that the output of visual inspection could be quite different depending on the operator in charge of coding the defects. Subjective data interpretation might therefore lead to erroneous decisions. To ensure the quality of data, a validation step is vital. Clemens-Meyer *et al.* (2021) discuss this aspect in detail. Important considerations include the plausibility and consistency of the data. Clearly, the quality and maintenance status of any equipment used for the collection of data will impact on the accuracy of the data collected, as will the training and experience of the operators.

Additionally, to the clearly stated needs of objectivity and accuracy and despite the technological capabilities of new sensors, a few limitations of inspection techniques should be highlighted. At the time we write this chapter, some relevant data cannot currently be practically and non-destructively measured in the field at large scales: for example, wall thickness (i.e., the previous example assumes the initial thickness of the pipe) nor infiltration rates over a large basin or trench.

Depending on the technologies and the design of the inspection devices, some blind spots remain for example, CCTV cameras cannot take footage below the water level, or a fixed camera will not be able to see behind an intrusive connection. The size of blind spots can be reduced with the combination of different technologies, smarter design of inspection devices and/or improved inspection procedures (e.g., one survey from downstream to upstream and another one in the opposite direction). Data users must be aware of those limitations prior to using such data, for example, for planning maintenance operations.

4.1.3 Layout of the chapter

The present chapter gives a non-exhaustive overview of the existing (Section 4.2) and emerging (Section 4.3) technologies for inspection of key UD assets. The boundary between both categories (i.e., existing and emerging) is blurred, and strongly depends on the expertise of the reader. The authors applied the following rule to distinguish as to what is presented: while existing technologies seem mature enough to be deployed at a larger scale (having received prior effort in the development of industrial solutions), emerging method still require research effort to (un)validate those techniques for asset inspection. This chapter ends with conclusions about the technologies presented in this book.

4.2 EXISTING INSPECTION TECHNIQUES

4.2.1 Background and overview

Until the 1950s, the only method for sewer pipe inspection was person-entry, which not only has health and safety implications, but also limits the minimum pipe size that can be surveyed (which is likely to be 900–1800 mm, subject to local working practices and occupational health and safety requirements). Clearly, in the early years before electric lighting provided good illumination, visual inspections would be challenging to complete efficiently. Furthermore, these large diameter pipes generally only account for a small proportion of the network, for example, the WRc Sewerage Rehabilitation Manual estimated that only 5% of the UK sewer network was large enough for manual inspection. Early camera inspections used black and white cameras fitted to a sled dragged through the pipe and could survey pipes down to 200 mm in diameter. As with all video technologies, sewer CCTV has developed significantly, enabling nowadays high-quality colour images and recording, initially to VCRs and later digitally.

Whether by person-entry or using remote CCTV, visual inspections are subjective and limited to visible features, that is, nothing beyond the internal pipe surface, or below the water surface can be inspected.

This section considers inspection techniques that are commercially available, even though some might be rarely used. The techniques are divided into functionality-oriented and defect-oriented inspections. Functionality-oriented inspections are where the inspection is directly measuring a parameter that can represent the structural or operational performance, for example, internal size or infiltration, whereas defect-oriented inspections use techniques which identify a range of different defects in the pipe, such as cracks and blockages, but might not accurately measure them.

4.2.2 Functionality-oriented inspection

This section is divided into different parameters being measured and within each section a selection of the more commonly used techniques is described.

4.2.2.1 Geometry

The internal geometry of a pipe is vital to maintain its design capacity, but it can change due to deformation, corrosion, and erosion processes. Several techniques can be applied to measure the inner geometry of sewer pipes: choice among them is mainly driven by the expected accuracy, pipe materials, available budget and the time required to conduct the inspections. Several techniques or prototypes have been published during the last decades, with a wide range of applications; they mainly differ with respect to costs, detectable defects, and accuracies.

Acoustic technologies have been developed and tested since the late 1970s (Morgan & Crosse, 1978). They offer quick and cheap inspections but only a few defects are detectable (i.e., rough estimation of their location and barely any quantification is possible). While using a speaker and several microphones along a reach, Bin Ali *et al.* (2011) highlighted the feasibility of lateral connection and blockage mapping with a signal processing algorithm (time and frequency domain). Acoustic measurements could also be performed with sonar – such a technology allows accurate measurement of the inner geometry below the free surface and in pressurized pipe (for more details, see Section 4.2.2.8).

Optical measurements are feasible using Manhole Zoom Camera (MZC). Such cameras, accurately positioned in the middle of the cross section allow the recording of footage or video from which defects and geometry anomalies can be detected. However, the maximal inspection distance varies from 20 to 30 m and the accuracy is relatively low. Plihal *et al.* (2016) compared both acoustic and MZC measurements: acoustic measurements seemed less sensitive to inspection conditions (e.g., flow, displaced joints) and offered a maximal inspection distance two orders of magnitude larger than the optical ones. The combination of both types of measurement remained highly suggested by the authors.

Laser profiling of the internal parts of a pipe is commonly used for checking ovality and deformation. A laser beam or beams measure(s) the internal cross section of the pipe, thus any differences from the expected shape and size can be inferred to be a defect. As reported by Clemens *et al.* (2015), the alignment and orientation of the laser is however critical to ensure accuracy. Clemens *et al.* (2015) went on to suggest a method for correcting for alignment and orientation; this was further developed and described by Stanić *et al.* (2017), who showed an increase in measurement accuracy of 5–10 times. Testing by Lepot *et al.* (2017b) showed that the technology was robust to most types of challenges, except for reflective and transparent materials which are not common in sewers. Meijer *et al.* (2022) report on the application of stereovision to identify anomalies in sewer pipes realizing comparable accuracies as Lepot *et al.* (2017a, 2017b, 2017c).

Investigating the use of vision-based and laser-based techniques for 3D mapping of sewers, Bahnsen *et al.* (2021) carried out a review of the different technologies. They also assessed the impacts of available light and the in-pipe water level, concluding that the time-of-flight sensors provided superior output to camera-based systems.

4.2.2.2 Roughness

The hydraulic roughness of pipes is often considered as an important (design) parameter related to the hydraulic performance of UD systems (for both the hydraulic capacity and the minimum dry weather flow velocity). In the course of time the hydraulic capacity can decrease due to sedimentation, FOG deposits and for certain materials corrosion (see also Chapter 5). This notion, however, is only

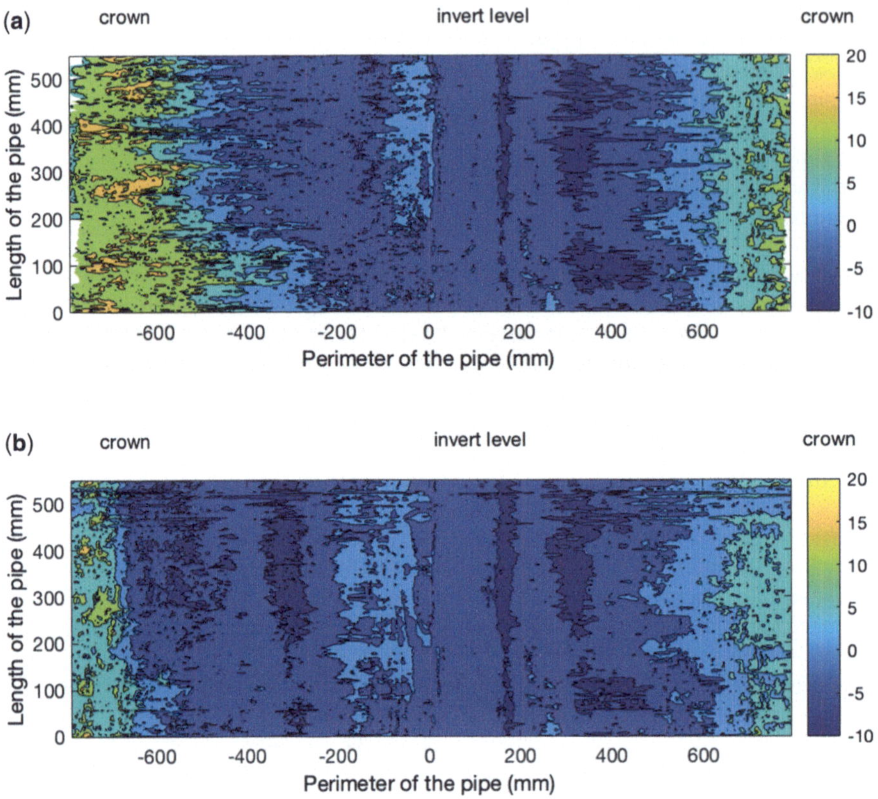

Figure 4.2 Mapping 3D inner geometry in order to derive K_n values from a first (top) and second (bottom) version of a laser scanning prototype. (*Source*: Stanić *et al.*, 2017).

seldom taken into account when making a new design (even though design manuals often prescribe a material dependent value for either manning's 'n', or Nikuradse's k_n equivalent roughness value). Determining the in-situ wall roughness as such of a sewer pipe seems for now a practical impossibility, determining the overall hydraulic capacity based on for example, calibration of a hydraulic model using measurements has been shown to be possible (see also Chapter 5). Stanić *et al.* (2017) reported on the application of laser scanning (Figure 4.2) of excavated concrete sewer pipes that had seen ~ 90 years of service in a combined sewer system that showed clear signs of crown-corrosion and found that the uncorroded parts of the pipe wall had k_n value of ~ 1.5 mm while the corroded parts (upper half of the pipes) showed k_n values of up to 12 mm. This implies that the generally accepted design values of 1.5 mm seem to hold true for sustained periods for the DWF situation; however, when the system fills up the hydraulic roughness increases significantly. For a case study Stanić ($k_n = 20$ mm) estimated a 100% increase of the head losses compared to the design situation (even without taking into consideration the presence of sediments and/or FOG deposits).

4.2.2.3 Inclination/invert levels

Measuring the invert elevations of individual pipe segments of a stretch of sewer may become important in situations in which subsidence may occur. Given that sewer inclinations can be as little as 0.2%, measurement resolution and accuracy is important. In the course of time uneven settling may

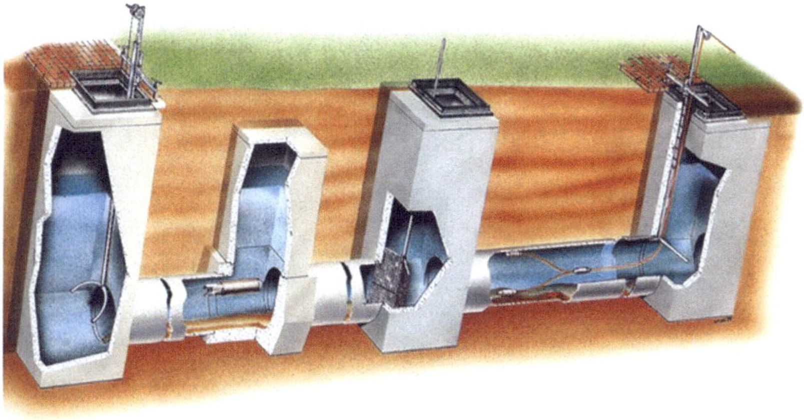

Figure 4.3 ATU system layout in a sewer. (Courtesy van der Valk + de Groot).

lead to the formation of stagnant pockets of (waste)water where sediments may accumulate giving rise to anaerobic conditions triggering biochemical induced corrosion (see also Chapter 5) and H_2S generation/emissions leading to potentially deadly working environment for practitioners. This can be done using the ATU method (Afzetting Traceer Unit (Dutch)/Deposition Tracking Unit (English)), see Figures 4.3 (system layout), 4.4 (surface setup) and 4.5 (measurements). An alternative method based on inclination measurement was described in some detail by Dirksen *et al.* (2014).

When making estimations of the amount of work involved in cleaning out a stretch of sewer from either sediments or FOG encrustations, a method based on pressure measurement is deployed in practice. The basic idea is simple: two pressure sensors are slowly dragged through a stretch of sewer (filled with water). One of these sensors is dragged along the bottom while the other, lighter than water, is dragged along the soffit of the pipe. When recording the location (i.e., its position along the length of the pipe), an indication of the amount of material to be cleaned out is obtained. As part of the procedure, a torpedo is sent through the pipeline with a pulling wire directed to the reception manhole (see Figure 4.3). After calibration of the equipment with a third sensor, the sensors can be pulled through the pipeline and the measurement can be started.

Figure 4.4 shows an example of the measuring set up. Figure 4.5 shows an example of the measuring results.

This asks, however, for prior detailed knowledge of the original profile of the sewer to be investigated. There are some limitations to the application:

- The stretch of sewer must be filled with water, so it is ideal for application in inverted siphons or pressure mains.
- As the variation in thickness of sedimentation and/or encrustation can be small (i.e., order of magnitude 1–2 cm) the pressure sensors must be accurately calibrated and have a high accuracy and resolution.
- The latter demand implies that accurate results can only be obtained when the sensors are dragged at a low pace to avoid disturbances on the pressure read out, caused by the movement of the sensors.

When interpreting the results, the following aspects have to be taken into account:

- Dragging a sensor over a sediment bed is an invasive method, this implies that the obtained results may show a systematic deviation (most likely an underestimation).

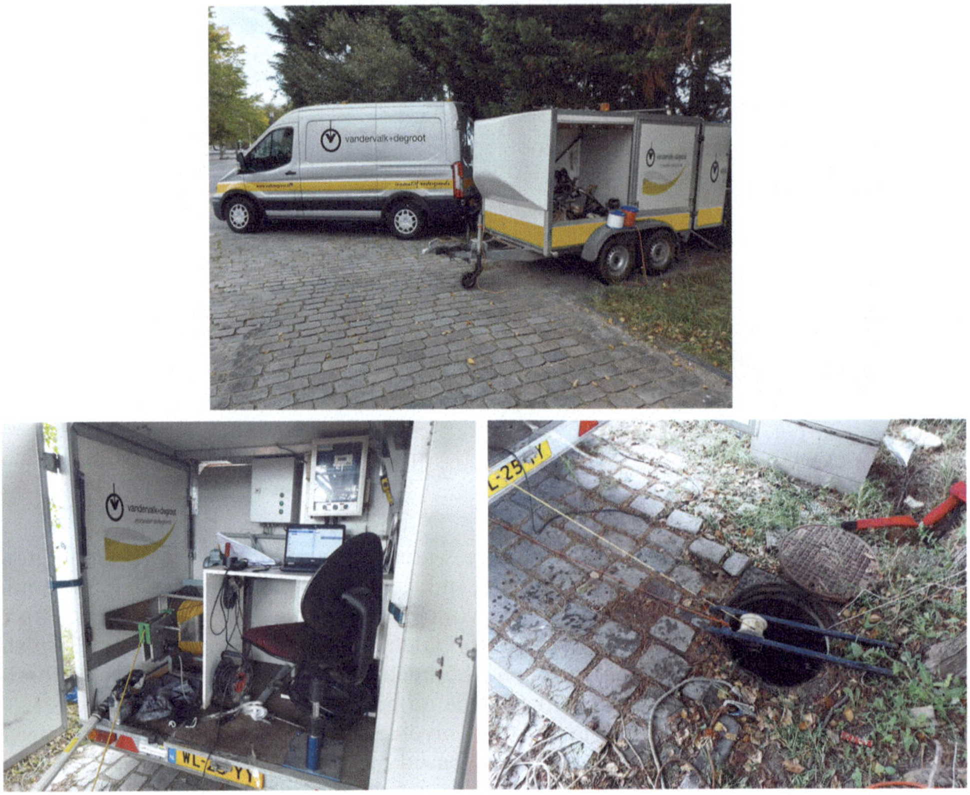

Figure 4.4 Set-up at the manhole. (Courtesy van der Valk + de Groot).

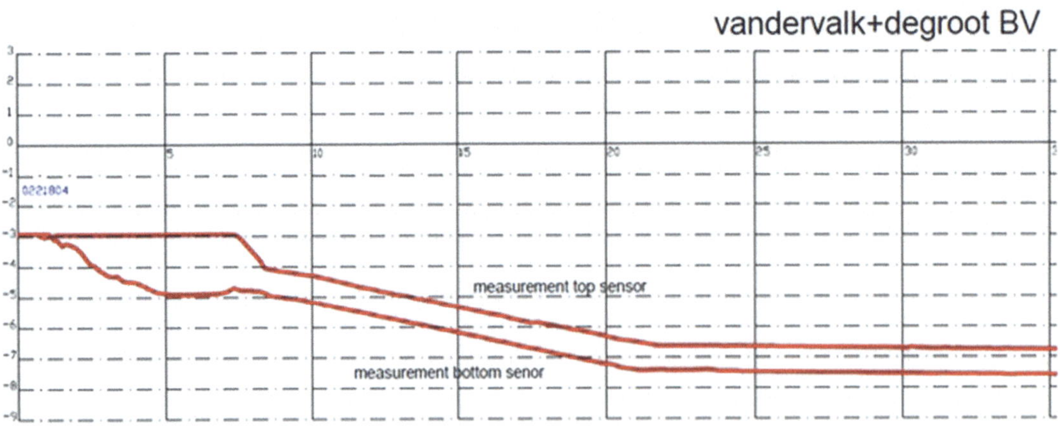

Figure 4.5 Start of the measurement of a culvert and a waterway. (Courtesy van der Valk + de Groot).

- The result is only valid at one point of the cross section, while the exact 3D position of this point is unknown.
- Given the previous point, when one wishes to estimate the volume of sediment the assumption of a homogeneous sediment level over the width has (implicitly) to be made.
- A further source of uncertainty is the position along the length of the pipe, although this seems to be relatively small compared to the other sources mentioned.

In spite of the limitations mentioned, the method described is applied with success in practice, has shown to be robust and produces results that at least allow indications of the amount of materials to be removed to be obtained. Apart from that, it also allows for a verification of the effectiveness of a cleaning operation and/or for performing an acceptance test after (re)construction.

The depth of pipelines and possible deformations of horizontal directional drilled pipelines can be reported. The ATU system shows the critical locations of subsidence and sludge deposits, so measures can be taken for preventive or corrective maintenance. In the context of this book, the most regular application of the ATU method is to determine the amount of FOG deposits in siphons, see for example, Chapter 5 (Section 5.8.2, Figure 5.44). However, other applications of the technology can be highlighted as well. For example, when raising a terrain in soft soil conditions, for example, at waste dump locations or at land reclamation sites the settlement of the area concerned can be measured by installing an HDPE pipeline at the location of the area to be filled. The monitoring of the settlements can be done with the ATU system by periodic measurement of the HDPE pipeline. No definite information on accuracy is available, but this is estimated to be in the order of magnitude of 1–2 cm in level. It is crucial to have the pressure sensors calibrated to a high level of accuracy in order to obtain useable results.

It has to be emphasized that there is no guarantee that the track followed by the sensors along the length of the pipe is always parallel to the length axis of the pipe under study; therefore, systematic deviations with an unknown magnitude may, and will, occur.

The method proposed by Dirksen *et al.* (2014) is only applicable in a cleaned-out sewer and utilizes a sewer tractor (IBAK KRA85) on which an inclinometer and a distance measuring device are mounted.

Figure 4.6 provides a sketch of the setup: the tractor is pulled, and the displacement is counted by the number of rotations of the pulley (with one revolution equal to 10 cm). The values for the distance and inclination measured are stored for further analysis.

Dirksen *et al.* (2014) showed that the tested equipment had an overall uncertainty of 0.05%, although when the tractor wheels are in different pipe sections, that is, straddling a joint (see Figure 4.7), the random error increased to 0.14%. A systematic error of 0.09% was determined; with this being close to the unit's 0.1% resolution, it was suggested that the resolution should be increased to 0.01%. Overall, it was suggested that as the random errors cancel out, these measurements of inclination can be useful and that systematic errors might be reduced by an improved calibration methodology.

From the measured inclinations and distances, an invert profile of the sewer can be derived. In a case study Dirksen *et al.* (2014) arrived at a confidence interval (95%) of 0.032 m (see also Figure 4.8).

4.2.2.4 Wall thickness

The remaining thickness of pipe walls is routinely monitored in industrial applications by techniques including guided waves and electro-magnetic acoustic resonance (EMAR); however, such technologies are rarely applicable to buried UD pipes because they had been developed for ferrous pipes and/or are applied from the outer surface of the pipe. The use of laser measurements, as discussed in Section 4.2.2.1 can be used to determine the difference between the expected and actual internal dimensions and thus infer remaining wall thickness; however, it cannot account for losses in the wall occurring outside the pipe. Pipe penetrating radar (PPR) is applied commercially and can provide accurate measurements of wall thickness, reinforcement location and developing voids outside the pipe (Ékes & Mahmood, 2019). PPR is however a contact method, and the accuracy is a function of environmental

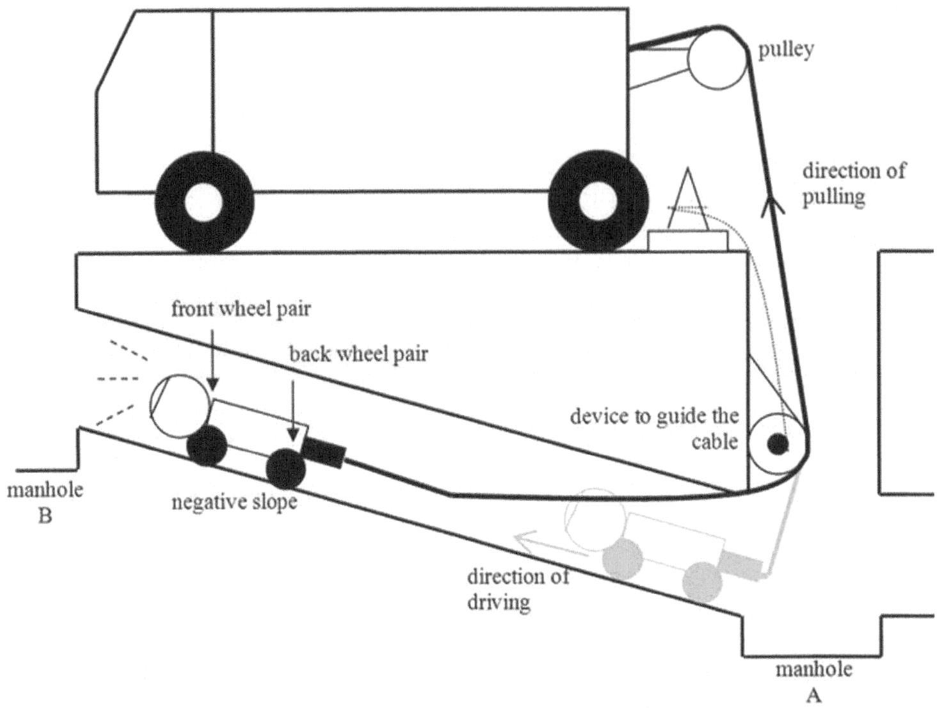

Figure 4.6 Set-up of the inclinometer. (Courtesy J. Dirksen).

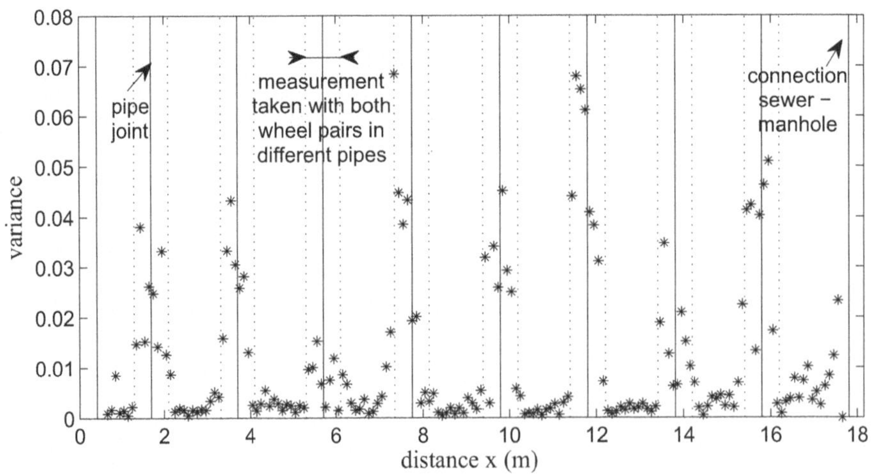

Figure 4.7 Variance in the slope measurement as a function of distance. The variance in the measurements increases around the individual pipe joints. This is due to very local differences in invert level around these joints.

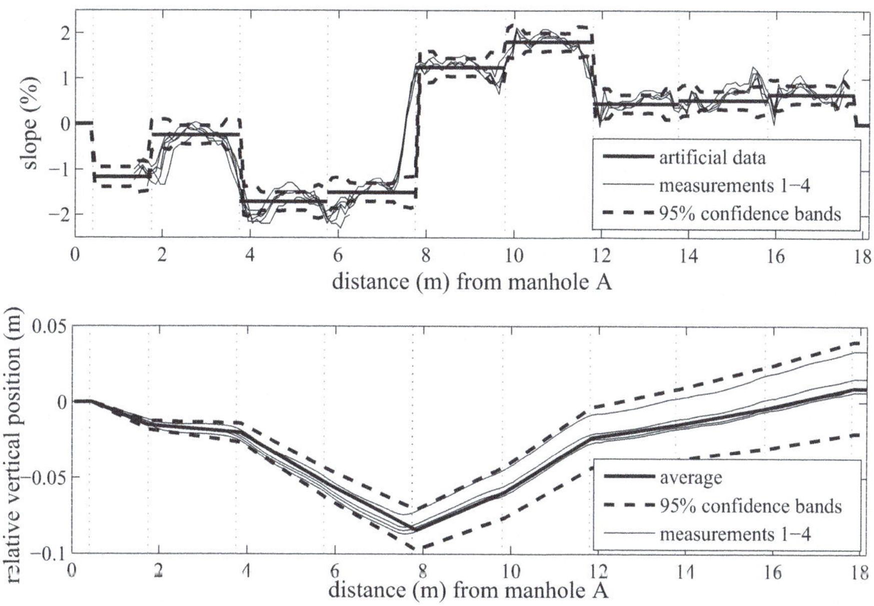

Figure 4.8 Confidence intervals for a case study over a sewer length of 17.35 m. (Courtesy J. Dirksen).

factors, calibration and survey quality. While good results have been obtained for concrete pipes, the presence of a layer of soft, deteriorated concrete reduces its accuracy. Such deteriorated concrete should therefore be scraped off to ensure an accurate measurement, further increasing the time and cost of such inspections. Wall thickness variations can be quite localized; thus, the survey technique needs to take this into account, otherwise uncertainties will be significant.

4.2.2.5 Material properties

When assessing material properties hitherto the most applied method is to take samples for lab testing, as in-situ methods are, to the authors' knowledge, lacking. There is some literature on the potential of applying in-situ acoustic methods to determine material properties of plastic pipes (see Makris *et al.* 2023); however, the uncertainties encountered are prohibitive for practical application so-far.

Even when taking core samples for determining material properties (i.e., tensile strength, porosity, density, etc.) in concrete, care has to be taken with the use of the data obtained. Stanić *et al.* (2016) reported on the effect of the heterogeneous nature of old concrete pipes with respect to the uncertainty of tensile strength results. They arrived at the following conclusions:

- The heterogeneity of concrete pipes increases with time; newly produced pipes show very homogeneous material properties (the latter may be contributed to the modern well-controlled production conditions, when compared to pipes produced on the construction site prior to ~1960).
- In old pipes up to 112 samples per meter may be needed to obtain a reliable estimate of tensile splitting strength; for new pipes this would only be seven samples.

This implies that, given the fact that taking core samples is a destructive method; one has the choice between either taking a few samples and work with uncertain values or taking so many samples that one knows exactly how strong the pipe was prior to testing while having caused significant structural damage by the testing.

Another observation made relating to core samples is that the correlation between material properties as obtained from core samples and the results of CCTV inspection reported along generally accepted standards, is virtually absent (see Stanić *et al.* 2013), which underlines the notion that visual inspection does not necessarily relate to physical properties. This is also illustrated by a very detailed study published by Luimes *et al.* (2023) in which the penetration depth of corrosion processes in the wall of several old (used) concrete sewer pipes was quantified and proved to be very inhomogeneous.

These findings illustrate the need for the development of non-destructive methods for producing reliable estimates of material properties.

4.2.2.6 Infiltration and (il)licit inflows

Several techniques are available to measure unexpected flows entering a sewer (e.g., infiltration, illicit inflows, or cross-connections). They are mainly based on discharge measurements, tracer experiments, and thermography.

For decades, discharge measurements and their correlations with rainfall data have supported sewer managers in the investigation of infiltration and/or cross-connections in separate sewers. For combined or domestic sewage, the statistical analysis of discharge data during dry weather conditions over long-term time series is likely the widest applied method to quantify infiltration. Seasonal tendencies of the minimal flowrate (i.e., typically, in the middle of the night) deliver information on existing infiltration (de Bénédittis & Bertrand-Krajewski, 2005). In separate sewer systems, a positive correlation between discharges and rainfall data indicates cross-connections contributing toward the sewage.

Tracer experiments are widely used either to quantify infiltration or pinpoint cross-connections in separate sewers (Ellis & Butler, 2015). With regard to the latter, dye experiments, smoke tests (Beheshti *et al.*, 2015) or coloured and numbered 3D-printed pills (Lepot *et al.*, 2016) can be used. If dye experiments or pills require access to private property, smoke tests can be applied from the public area (see Figure 4.9).

Changes in temperature are a strong indication of an incoming flow, which could be infiltrated via licit or illicit connections. Thermography is mainly used for detecting infiltration or illicit connections. Fiber optics are deployed commercially at a relatively small scale offering distributed temperature sensing (DTS). DTS can detect inflows and illicit connections along a reach (Figure 4.10), while using the difference in thermal footprint of different water sources (see e.g., Hoes *et al.* 2009; Vosse *et al.* 2013). This method can provide measurements for a long period, but it is spatially limited by the length of the cable.

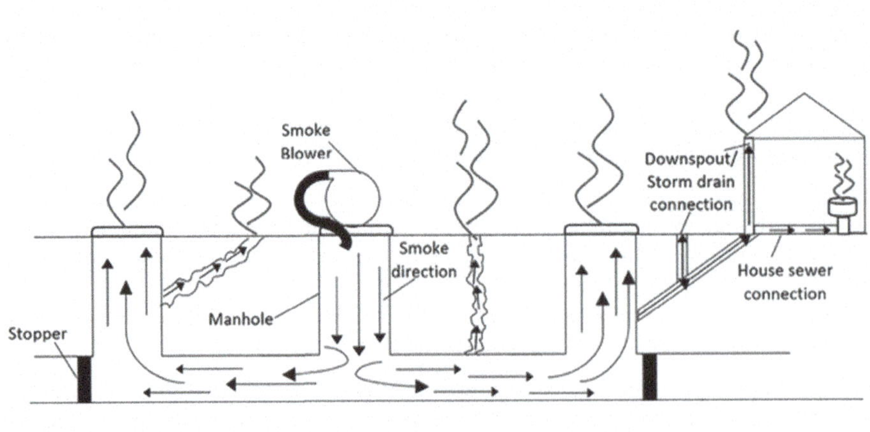

Figure 4.9 Principle of smoke test to pinpoint cross-connections in separate sewer system. (Reprinted with permission from Beheshti and Saegrov (2019).

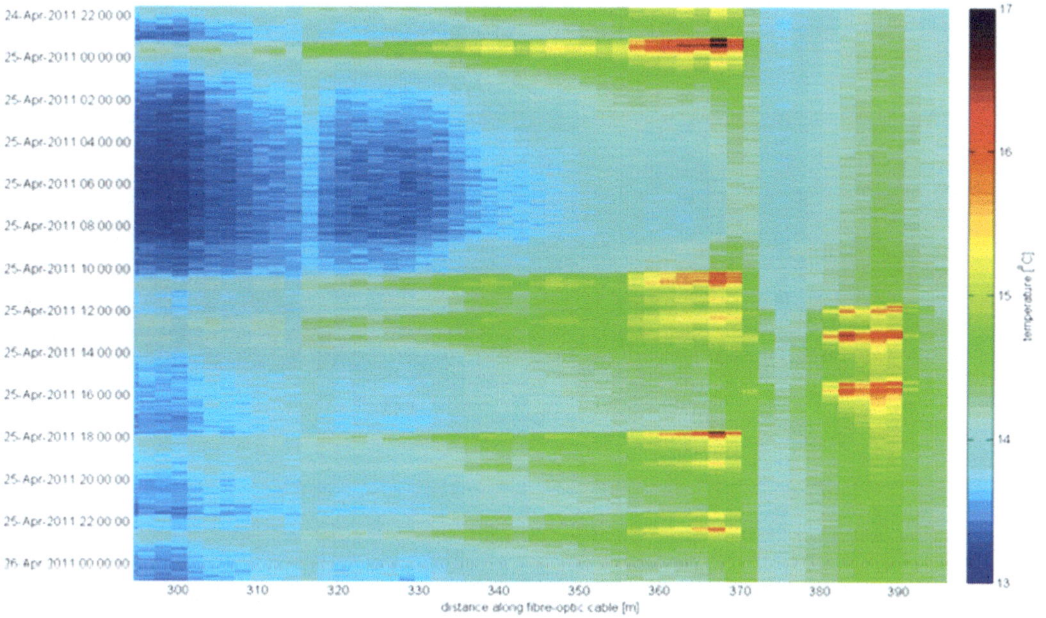

Figure 4.10 Example of connection locations with fibre optics (Courtesy R.P.S. Schilperoort).

Mounting an infra-red camera (Lepot *et al.*, 2017a) on a moving/flying platform inside the pipe could increase the inspectable length. However, the inflows being detected are not always continuous as shown in Figure 4.10; as a result, this inspection will only identify active inflows (e.g., during a rain event or when an illicit connection is in use). This technique offers a good sensitivity but is sensitive to the type of connection, the rate between lateral and main discharges and the temperature difference between both flows (Figure 4.11).

Both solutions (DTS and internal use of an IR camera) deliver accurate inflow locations since the thermal effect of the lateral connection can be monitored immediately downstream and can be applied without service disruption. In comparison to CCTV footage, the thermal fingerprint allows the type of connection to be identified (i.e., warm or cold) giving an additional key to pinpoint the source of infiltration.

While only applicable to stormwater pipes, Chandler and Lerner (2015) developed a simple, low-cost method to identify some illicit connections. The method relies on the presence of optical brighteners, which do not occur naturally, but which are present in laundry detergents, toilet paper, and cleaning products. Optical brighteners have a strong affinity for fabrics such as cotton and can be seen under ultra-violet light. In the study, cotton tampons, free from optical brighteners were placed in stormwater pipes and shown to successfully identify illicit connections.

Panasiuk *et al.* (2022) directly compared a range of the aforementioned techniques, concluding that the methods studied showed variable performance and that a combination of two or three methods would improve the efficiency of investigations.

At a catchment scale, various methods to estimate the amount of groundwater infiltrating into a UD system over a long period are available (see e.g., Weiß *et al.* 2002). These methods use the measured inflow to a waste-water treatment plant (WWTP) and discriminate between days with and without rainfall. Further, it is assumed than on dry days the discharge equals the theoretical amount based on the number of inhabitants and known industrial contributions (see Figure 4.12). These methods

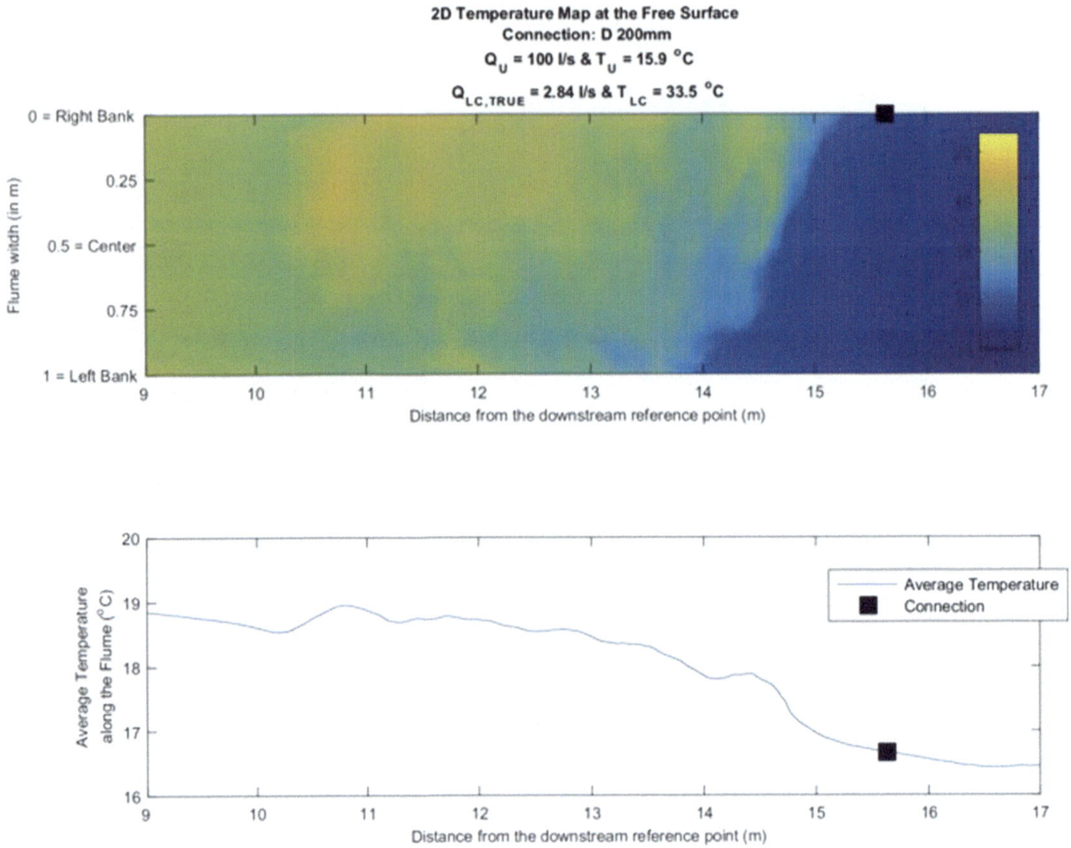

Figure 4.11 Temperature map reconstructed from IR camera footage for a warm lateral connection. (Source: Lepot *et al.*, 2017a).

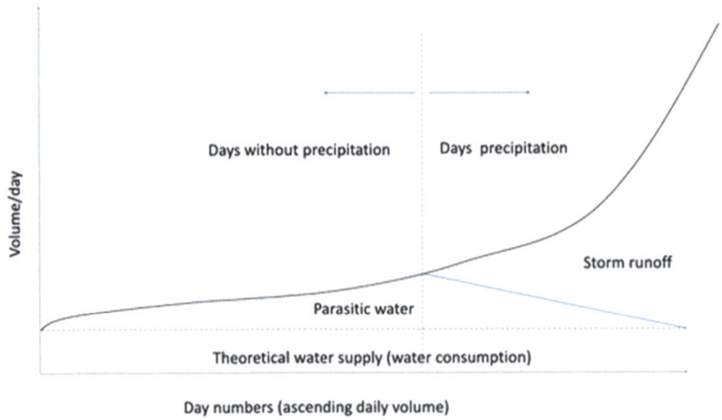

Figure 4.12 Illustration of the 'triangle method' to quantify inflow/infiltration based on very basic measurements. (Courtesy Deltares).

cannot be considered accurate, but they have the advantage of being simple to implement and use data that are generally readily available.

Kracht *et al.* (2003) presented an elegant method to quantify the infiltration at a catchment scale using small time windows. Their method relies on a difference in isotopic composition of the water in a UD system and the infiltrating groundwater. Basically, each waterbody has a specific ratio R between the isotopes of oxygen, defined as

$$^{18}R(O) = \frac{[^{18}O]}{[^{16}O]} \tag{4.1}$$

The 'Vienna Standard Mean Ocean Water' or VSMOW, which is a world-wide reference water, defines the so-called δ value:

$$^{18}\delta = \frac{^{18}_{\ s}R}{^{18}_{\mathrm{VSMOW}}R} - 1 \tag{4.2}$$

Hence, a negative value implies a lighter isotope composition; a positive value indicates a heavier isotope composition compared to the reference water composition. This can be seen as a 'fingerprint' of a water body. Provided that the values of the water in the UD system and the groundwater differ, this can be used to quantify the relative amount of groundwater infiltrating:

$$b = \frac{^{18}_{\mathrm{DWF}}\delta - {}^{18}_{\mathrm{DW}}\delta}{^{18}_{\mathrm{INF}}\delta - {}^{18}_{\mathrm{DW}}\delta} \tag{4.3}$$

In which, DWF indicates dry weather flow, INF indicates infiltrating water (i.e., groundwater, surface water) and DW indicates drinking water. It is assumed that in a system without parasitic water the isotope composition of wastewater would be equal to that of drinking water and that the source of infiltration is known and stable in terms of its δ value.

Given the known uncertainty in the determination of the values, the uncertainty in the relative amount of infiltration can be determined. Of course, this method falls short when the composition of the groundwater is not homogeneous, or when the drinking water is produced from different sources. The advantage of this method, however, is that it is relatively cheap to perform and allows the evolution over time of the infiltration to be evaluated and relationships to be determined with, for example, the variation in groundwater levels. Schilperoort (2004) reports on the application of the isotope method in a number of case studies,

4.2.2.7 Exfiltration

Exfiltration phenomena can be detected with other technologies. Aerial thermography seems to be a rather effective way to pinpoint locations of leakage at a large scale (Park *et al.*, 2020): areas where exfiltration occurs present a different thermal pattern (i.e., two images are required: one in the morning, one in the afternoon). Even though this method appears promising, weather and local environment (e.g., shadows) could induce serious bias in the measurements.

Ground penetrating radar (GPR) presents capabilities to measure exfiltration and their consequences. The back-scattered electromagnetic signal allows the positioning of the pipe itself, voids induced by leakage and high soil moisture (Ghozzi *et al.*, 2018; Sonkamble & Chandra, 2021).

Sewer exfiltration can be quantified using tracers; this generally involves injection of tracers at two different locations and comparing the mass recovered at a downstream location. Through careful selection of tracers and analysis of the results, quantification of exfiltration volumes can be estimated with good accuracy, as discussed by Rieckermann *et al.* (2007). However, the methodology is time-consuming to apply on anything but a small scale and requires careful application to minimize uncertainties.

Mobile geo-electrical measurement involves moving an electrode through a (partially) filled pipe and inserting a second electrode (grounding pin) into the earth. Stegeman *et al.* (2022) have shown this method to accurately identify the location of a leak; however, there is no clear dependence between the recorded current and the size of or flow from the leak.

In addition to the application of isotopes as natural tracers, literature (e.g., Rieckermann *et al.*, 2007) also describes the use of injected tracers (typically lithium salts or heavy water) to determine in- and exfiltration. These methods, however, prove to be hard to apply in day-to-day practice, requiring specialized personnel and are therefore limited to special cases. For details the reader is referred to Ellis and Bertrand-Krajewski (2010), who provide a comprehensive text on methods to determine in- and exfiltration. Schilperoort (2004) provides a comparison between the isotope method and more conventional methods (like the method proposed by Weiß *et al.*, 2002) using a range of case studies.

Determining the location(s) of in- or exfiltration is less straightforward. The presence of cracks or defective joints indicates the possibility of in- or exfiltration; however, this is not a given fact (unless of course groundwater is observed to enter on CCTV footage, see e.g., Figure 4.13).

In literature various methods are described to identify leakages. Distributed temperature sensing (Hoes *et al.*, 2009) may be applied. Using a focussed electro leak location (FELL) (Eiswirth & Heske, 2000; Wilmut *et al.*, 2005; Wolf, 2003; Stegeman *et al.*, 2022) as shown in Figure 4.14, the location of leakages can be determined with some accuracy, the main assumption here is that electrical

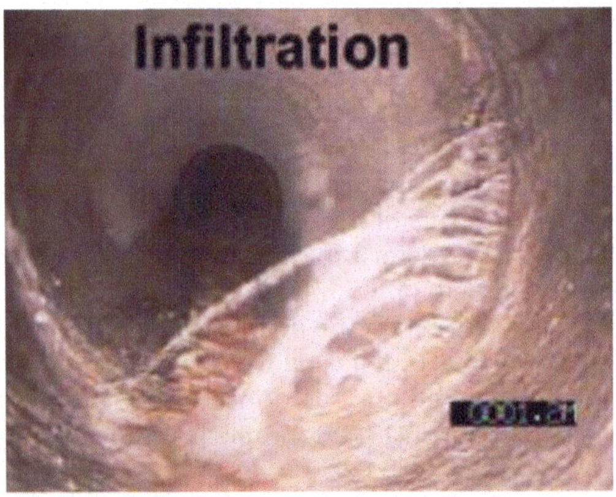

Figure 4.13 Example of clearly observable infiltration.

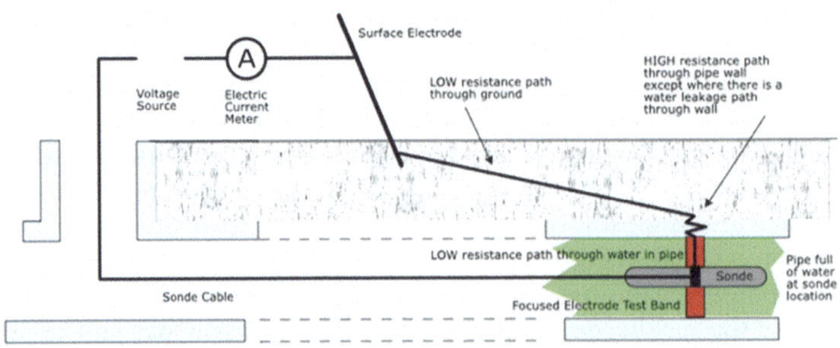

Figure 4.14 Simplified schema for FELL. (After Wilmut *et al.*, 2005).

conductivity is directly proportional to the in- or exfiltration of water, which has, to the authors' knowledge, not been proven. A detailed evaluation of the FELL is provided by Vermeulen (2022). Quantifying exfiltration is a subject on which not much literature exists. Vollertsen and Hvitved-Jacobsen (2003) report that the exfiltration rate at pilot scale experiments reached a stable value after some days and stayed constant over longer periods of time. They further report the formation of a clogging zone that strongly determines the exfiltration process.

Leakage in pressure mains can be detected at an early stage using a range of methods:

- Pressure testing (either with air or with water) allows a rough quantification.
- FELL.
- The use of tracer experiments as mentioned in Section 4.2.2.6 (these methods can, to a certain extent, quantify leakage; localization is possible but at a relatively low resolution determined by the accessibility of the system and assuming the leakage is near the injection point and the tracer is well mixed).
- Acoustic methods: leak noise correlation (Almeida *et al.*, 2014; Brennan *et al.*, 2016), allows localization, albeit very small leaks are still hard to pinpoint. The Smartball measuring system (Pure Americas Inc., 2011) passes through the mains recording the location of leaks to within 1.8 m; operators have however reported challenges in retrieving the un-tethered Smartball.
- Time domain reflectometry (see e.g., Fatemi Aghda, *et al.*, 2018).
- Distributed acoustic sensing (DAS) (see e.g., Stajanca *et al.*, 2019).
- Distributed strain sensing (DSS) (see e.g., Zhang *et al.*, 2019).

A disadvantage of the latter two methods is that wires need to be installed, preferably during construction which is mostly not an option for existing systems. Managing authorities are well advised to consider doing so in newly built situations, as TDR and DTS allow for relatively easy and almost continuous monitoring for leakages.

A recent and comprehensive overview on leakage detection is found in Sadeghikhah *et al.* (2022).

4.2.2.8 Sedimentation

Sediment deposits are detectable and, sometimes, quantifiable. Several technologies can be used for this purpose depending on the measuring conditions. If the pipe is dry (often implying a service disruption in the case of sanitary or combined sewer systems), standard CCTV or lately tested laser profiling/LIDAR detection offer capabilities to detect and, for the latter, measure sediment deposit volumes. Stanić *et al.* (2017) and Lepot *et al.* (2017b) demonstrated the capability of a laser profiler (Figure 4.15) to map in 3D and quantify sand deposits in a dry pipe with good accuracy (i.e., with a standard deviation between the real and estimated sand volumes less than 3.5%).

If the flow cannot be bypassed, acoustic sonar seems to offer a robust and reliable method to assess sediment volumes. Lepot *et al.* (2017c) tested an acoustic sonar placed on a floating platform in various sewer structures and pipes. Figure 4.16 depicts the output of a sonar mapping in a settling tank.

4.2.2.9 Infiltration capacity for permeable pavements

When testing the hydraulic capacity of pavement in terms of clogging, the question arises: 'what permeability should pavements have?' Referring to the guidelines in place when the system under investigation was originally designed (e.g., DIBt (2012), SLG (2020)) is a good starting point in the absence of better information. Many guidelines recommend using the ASTM D 3385-09 double-ring infiltration test to determine the infiltration rate of permeable pavement surfaces (Schmitt *et al.*, 2007; Schönberger *et al.*, 2005). It suggests undertaking a minimum of three different tests at three locations on the pavement. All three tests need to demonstrate an average infiltration rate of equal to or greater than 194 mm/h (540 L/s/ha) to be deemed to comply. Also, full-scale tests are used to visualize if the permeable pavement still performs satisfactorily over time (see Figure 4.17).

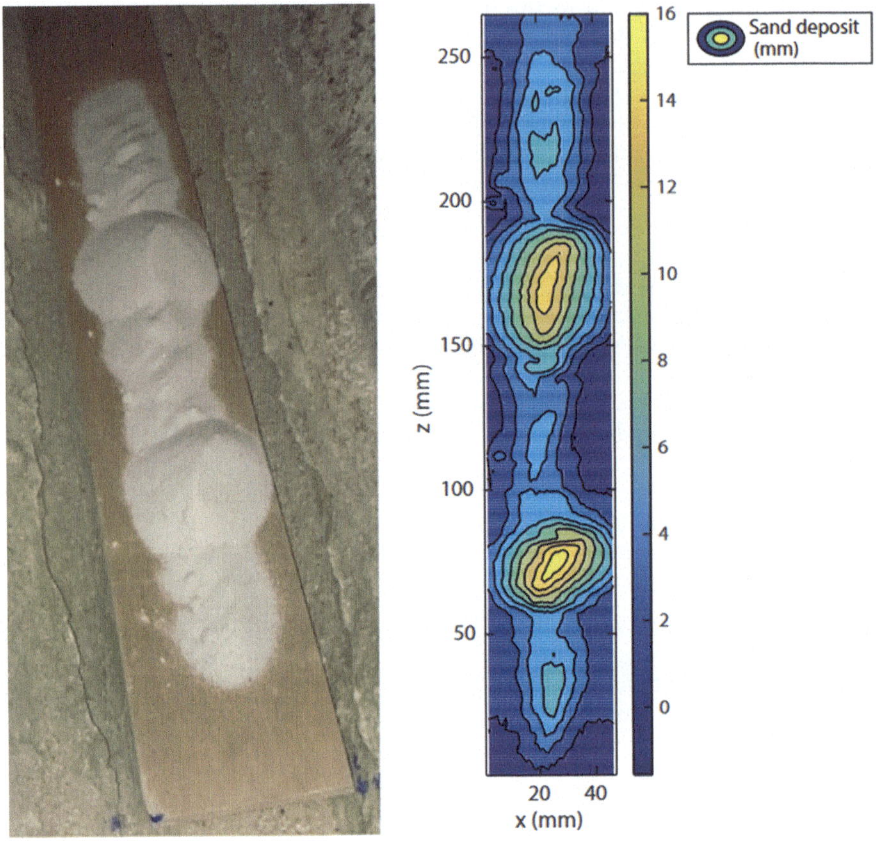

Figure 4.15 Quantification of the sand deposit volume using laser scanning: photo of the experiment (left) and scanned sand relief (right). (Source: Lepot *et al.*, 2017b).

The requirements will differ by jurisdiction. For example, newly installed permeable pavements in the Netherlands should demonstrate a minimum infiltration capacity of 194 mm/h. Over time the infiltration capacity will become lower due to clogging. In Germany, according to the DWA-A 138 (2005), the dimensioning of a surface infiltration system without above-ground storage facilities uses a rainfall intensity which corresponds to a rain event of 10-minute duration ($T = 10$ min), which is reached or exceeded once every 5 years ($n = 0.2$).

As the number of permeable pavement installations increases, the need for a proper tool to measure their surface infiltration functionality, especially with respect to clogging, is also increasing (Veldkamp *et al.*, 2022; Winston *et al.*, 2012). Measuring infiltration rates accurately in the field is not easy to do and a variety of infiltration test procedures have been utilized in the past. However, the results have generally been inconsistent, and have shown a large variation in the range of infiltration rates measured. Currently, there is no single standard agreed method for measuring the surface infiltration through permeable pavements even though numerous studies have tried to measure the surface infiltration rate of permeable pavement systems (e.g., Bean *et al.*, 2007; Fassman & Blackbourn, 2010). This has generally been done by measuring the infiltration rate of water through a particular section of the pavement surface. While a variety of infiltration test procedures have been used, most are based on some type of modified single- or double-ring infiltrometer test.

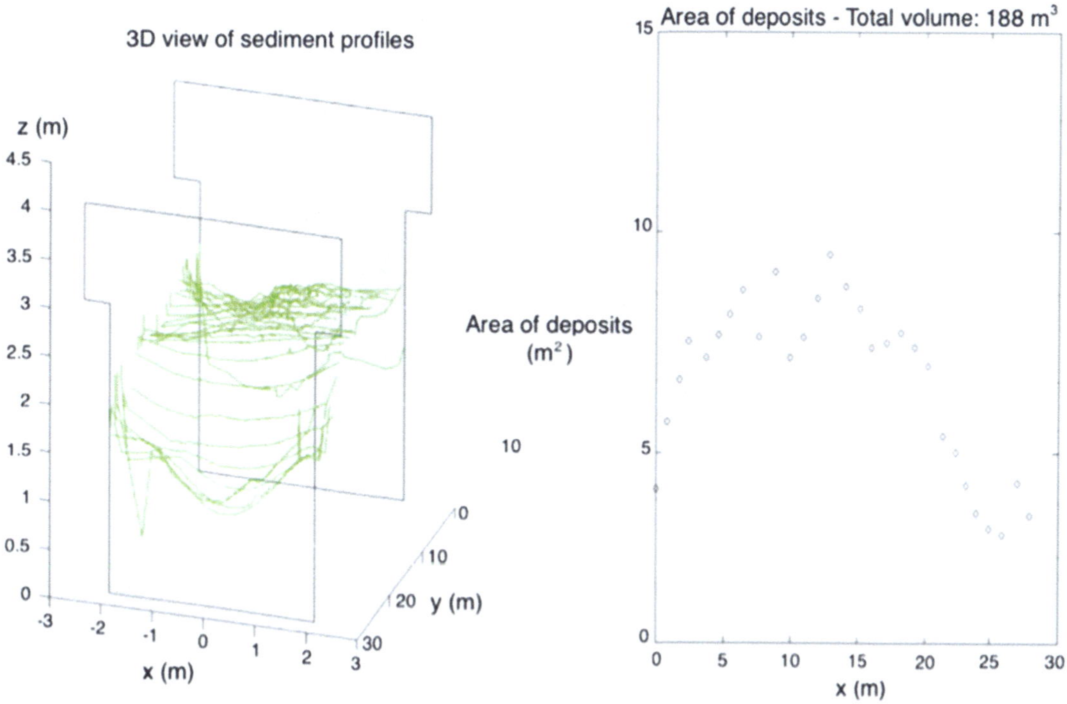

Figure 4.16 Sediment volume in a settling tank: 3D left on the left, area of deposits along different cross sections on the right. (Source: Lepot *et al.*, 2017c).

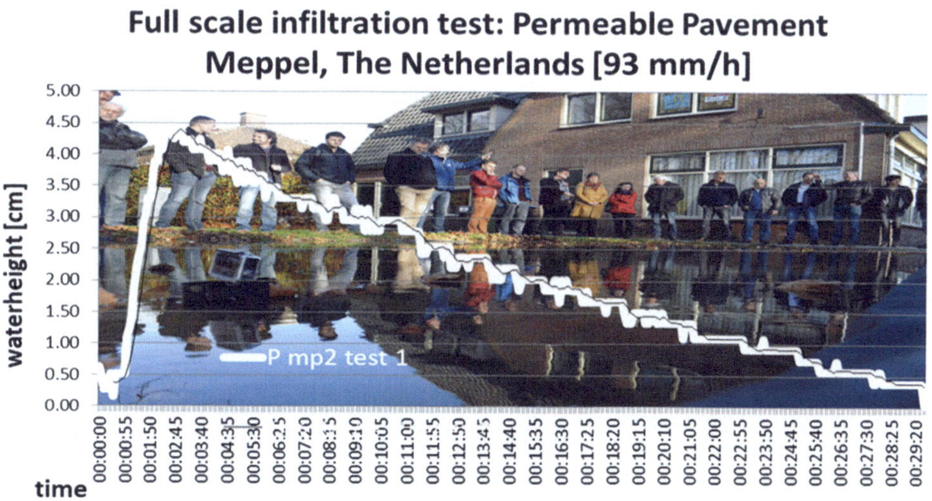

Figure 4.17 Full-scale test with stakeholders from municipality and water authority waiting for the water to infiltrate through permeable pavement in Meppel, The Netherlands. (Courtesy F. Boogaard).

Figure 4.18 In-situ test for infiltration of a permeable pavement in Germany. (Courtesy F. Boogaard).

Ring infiltrometer tests (e.g., ASTM C 1701/C 1701M, 2009) and variations to this (Li *et al.*, 2013) were originally developed to determine the hydraulic conductivity of in-situ field soils for evaluation of their irrigation properties. Water is generally supplied to the rings using either a constant head or a falling head method (see Figure 4.18 illustrating the practice in Germany). The flow rate of the water is then divided by the cross-sectional area of the ring to calculate the infiltration rate (usually reported in mm/h). Several studies have demonstrated a high degree of spatial variability between different infiltration measurements performed on the same pavement location (e.g., Boogaard, 2015; Borgwardt, 2006; García-Serrana *et al.*, 2017; Pezzaniti *et al.*, 2009). The single- and double-ring infiltrometer tests are based on the infiltration rate through a small area of the pavement that is used to represent the infiltration rate of the total pavement area. For example, the area of the inner ring of the test (ASTM C 1701/C 1701M, 2009) is 0.0707 m^2. The minimum area recommended by Dutch guidelines is even smaller, at only 0.01 m^2. Using such small areas for testing could potentially lead to erroneous results, as several studies have demonstrated a high degree of spatial variability between different infiltration measurements undertaken on the same pavement installation.

Previous research demonstrated that more accurate infiltration results may be produced by significantly increasing the area of the pavement surface being tested (Boogaard & Lucke, 2019). By inundating a much larger area of pavement during testing, it was shown that any spatial variations in infiltration capacity were effectively averaged out, and this produced more reliable infiltration data. A full-scale infiltration testing (FSIT) method (see Figure 4.19) was applied by most of the studies to determine the surface infiltration rate of over 100 existing permeable pavement installations in the Netherlands (Veldkamp *et al.*, 2022).

4.2.2.10 Infiltration capacity for bioswales

The infiltration capacity of swales is usually estimated by measuring the rate at which water soaks away at small test pits or boreholes (Palhegyi, 2010) or ring infiltrometer tests. Several studies have demonstrated a high degree of spatial variability between different infiltration measurements since the results were based on the infiltration rate through a very small area that is used to represent the total infiltration area (Boogaard, 2015). Studies showed large spatial variation in infiltration rates with individually measured infiltration values varying by a factor of 100, concluding that about 20 measurements at each swale are needed to reduce the uncertainty (Ahmed *et al.*, 2015;

Figure 4.19 Full-scale infiltration test (FSIT) where a large part of permeable pavement is flooded is applied in the Netherlands. (Courtesy F. Boogaard).

Boogaard, 2022). Previous research demonstrated that more accurate research results on the infiltration capacity of sustainable UD systems can be collected by significantly increasing the test area as discussed in last paragraph with permeable pavement (Boogaard & Lucke, 2019). By inundating the whole swale or a large area of the swale during testing, it was shown that any spatial variations in infiltration capacity were effectively averaged out, and this produced more reliable infiltration data. The full-scale infiltration testing (FSIT) method (see Figure 4.20) was applied by most of the studies to determine the surface infiltration rate of over 100 existing swale installations in the Netherlands (Boogaard, 2022).

Figure 4.20 Example of hydraulic test on a swale with the full-scale test in Groningen (tank truck on the right), measurements by loggers in the swale and hand measurement of students from Hanze Applied University of Groningen. (Courtesy F. Boogaard).

4.2.3 Defect-oriented inspection

The inspection techniques discussed in the previous section focused on the measurement of specific parameters related to UD system functionality. This section focuses on techniques which provide a broader, but often less specific and/or more subjective description of pipe deterioration. In the majority of cases these methods are based on the interpretation of images and/or footage. These so-called vision-based inspection techniques applied in practice (most notably CCTV) suffer from a spectrum of disadvantages:

- Only the internal surface of buried assets is observed, so damage or corrosion at the outer wall of a pipe is not observed; similarly, only the top surface of an infiltration basin can be inspected.
- No physical quantification can be accurately made (e.g., the loss of wall thickness due to corrosion cannot be quantified, nor can the hydraulic roughness be quantified, nor the clogging level of an infiltration system).
- What is observed is the result of an unknown combination of deterioration processes, giving no clues on time scales or rates of deterioration.

In addition, the normally applied standards (e.g., EN 13508-2, 2011) are formulated in such a manner that they are prone to errors made by inspectors, as these norms rely heavily on human observation, which is known to be unreliable (see Dirksen *et al.*, 2013). In the past few years, the development of computer vision combined with deep learning technology has shown the potential to outperform human observers in terms of consistency, thus showing potential to reduce error rates. Nevertheless, the main disadvantages of vision-based inspection remain. It has to be acknowledged however that in the course of the past 40 years a lot of CCTV data has been collected and has been shown to be valuable where it comes to planning of replacements and/or rehabilitation for the mid- and long term (i.e., >5 years ahead) and for larger areas.

4.2.3.1 Vision-based inspection

As discussed earlier in this section, inevitably visual inspection was the original form of inspection, and this remains the most used technique. This technique can be broadly divided into static and mobile techniques.

4.2.3.1.1 Static vision-based inspection

A static inspection is usually carried out at a manhole from the ground surface. The simplest basic technology solution is a mirror-based solution which involves a 45-degree mirror and a light source. This gives a basic view of the pipe closest to the manhole but is very limited for inspecting long lengths of pipe due to the lack of adequate lighting, potential deviations in the line of the pipe and the detail that can be seen. Where resources are constrained, mirror-based inspections can still provide valuable information.

Zoom cameras, or stationary CCTV are not new, but are not widely used. The approach consists of inserting a camera with a controllable focal length lens into a manhole to take images. Two key advantages are that there is no need to clean the pipe and the inspection is quick, both meaning that the inspection is significantly cheaper than a full CCTV survey. There are however clearly limitations in the distance that can be seen from the manhole, the proper identification of defects and the need for the pipe to be straight – without any horizontal or vertical deviations. Zoom cameras could be, however, a useful tool to prioritize defects for inspection. They have also been used in planning sewer cleansing strategies (Plihal *et al.*, 2014).

4.2.3.1.2 Mobile vision-based inspection

The advent of cameras which could relay an electronic image to a screen, termed closed circuit television (CCTV), led to the first applications in sewers in the 1950s. While the earliest CCTV cameras were dragged on a sled using winches at the upstream and downstream manhole, the CCTV

Figure 4.21 Picture of a mobile vision-based inspection system showing the lights (in front), the camera (between the lights), the swivel head (camera + lights), and the motorized crawlers (left part). (Courtesy F. Cherqui).

camera soon became mounted on motorized crawlers, thus reducing deployment time and allowing the inspection to take place from a single manhole (Figure 4.21). The crawlers themselves have developed with time, allowing the camera to be centered within the pipe and also to rotate to focus on defects. CCTV crawlers can commonly travel up to 500 m along a pipe, they can pass through manholes, but due to the tether cannot negotiate anything other than shallow radius bends. The smallest CCTV cameras are 'push cameras' which are deployed without a crawler just being pushed along the pipe by the operator; however, the orientation of the camera can change, and the jerky motion can be an issue. Some crawlers that allow deployment of lateral surveying cameras are now available.

CCTV cameras are operated from above ground, usually in a specially fitted vehicle, from which the cable is deployed, and which includes a display showing the camera image and controls for the crawler. The recording systems allow the video to be annotated with locations and highlighting defects during the survey, although the survey analysis can also take place afterwards based on pre-recorded footage. The former approach can allow manipulation of the camera to better assess any defects, while the latter allows a faster survey, minimizing time spent blocking traffic and maximizing the survey distance. The cable drum incorporates a counter, which is converted to a distance travelled by the crawler which is also recorded on the video images. In recent years, alternative methods to deploy CCTV have been used, including floating platforms and drones (or unmanned autonomous vehicles), particularly in larger diameter pipes. This has been reported to increase the daily inspection length by up to 50% to 900 m (Flyability, 2023).

Another alternative visual inspection technique, known as digital scanning or side scanning uses high-definition wide-angle cameras to record the entire pipe wall as the crawler moves through the pipe without stopping. These images can be viewed live but are best suited for recording later analysis where the images can be unwrapped to view the entire pipe wall in a way not possible with conventional CCTV recordings (Feeney *et al.*, 2009). The images produced with digital scanning do need careful processing to avoid distortion, but this is automated. As with other vision techniques, only the pipe above the waterline is inspected. While the operator cannot pan and zoom the camera, the resulting processed images will generally be more comprehensive.

Sewer inspection using CCTV was a logical step from person-entry visual inspection and has developed a strong place in the market, with defect codes being developed around this dominant inspection method. However, beyond the limitations of subjective analysis and subsequent aggregation, visual inspection – especially via CCTV – cannot provide a full description of pipe condition. A key limitation is what cannot be seen; first, to enable the crawler to pass through the pipe and to remove elements that would foul the lens or mask defects; the sewer is usually pre-cleaned. While Thornhill and Wildbore (2005) noted that higher reporting accuracies were obtained after cleaning, that cleaning will inevitably wash away evidence such as ingress of soil and sedimentation which can indicate network defects and performance issues. Second, the entire internal surface of the pipe is rarely visible, particularly in combined and wastewater sewers due to a continuous level of flow (unless flow is temporarily bypassed or blocked). Third, the inside of the pipe is not all of the pipe; the external walls of the pipe can be affected by corrosion; the bedding of the pipe can deteriorate, and voids can form in the surrounding ground which impact on both the structural stability of the pipe and that of the ground itself. Furthermore, the pipe material can appear to be visually intact, but has in fact lost structural integrity. It should also be considered that the image recorded by a camera lens can be subject to varying degrees of distortion, which may affect the perception of a defect. This is discussed further by Martins *et al.* (2020), along with suggestions as to how this can be overcome.

4.2.3.1.3 Direct visual inspection

When possible, the operator can perform the visual inspection directly by walking through the sewer pipes, without the need for a closed-circuit television. Regarding pipes, this method is commonly used for large-diameter sewer pipes (i.e., >1.8 m although the definition of 'large' remains subjective) where robotic crawlers might have limitations due to the size of the pipes or the complexity of the sewer network. During their inspections, operators can capture detailed visual information about the condition of the sewer (including nuances that automated methods might miss). Such activity remains however challenging due to the hazardous environment (i.e., it being a confined space entry environment subject to potential flooding, toxic gases, limited oxygen, bacteria, and other contaminants), limited access entry or exit points, limited reach with some areas of the pipes difficult to see because they are located too high or are not accessible, and time and costs as they require more time and resources than CCTV inspection. Man-made visual inspections are therefore limited to large diameter pipes (which are often the oldest ones) for which no mobile vision-based inspection is possible or pertinent. Defect inventory is often based on the same system as that used for televised pipe inspections, for example, most man-made pipe inspections in Europe are based on the EN 13508-2 coding system (although this coding system seems to be mainly dedicated to CCTV inspections).

When considering other drainage solutions ranging from swales to underground detention basins, direct visual inspection remains the main investigation method. To the knowledge of the authors, no norm or national guideline provides recommendations on visual inspection of stormwater control measures (SCMs). Inspection reports range from free text (describing the observations) to detailed checkbox forms (requiring the operator to give a condition to each component of the solution). For example, AECOM (2015) provides for each type of SCM a maintenance form divided into six sections (i.e., general inspection, inlet structure, pre-treatment, main treatment, outlet structure, and emergency overflow) containing each a list of items for which the condition should be chosen (i.e., either 'good,' 'marginal,' 'poor,' or 'N/A') with the possibility to add a comment. Seattle Public Utilities (2009) guidelines provide a very detailed description (with illustrative pictures) to assess the level of service (from A – excellent effort to D – poor effort) for the landscape and vegetation (i.e., aesthetics, vegetation, mulch, weeds, erosion, and bare spots), system functionality (e.g., bioretention, biofiltration, bioretention, conveyance, etc.), the hardscape and infrastructure (e.g., sedimentation structures, grates and debris screens, outlet structures, flow control structures, etc.), porous pavement, and other elements (e.g., culverts, irrigation systems, etc.). The Melbourne Water (Browne *et al.*, 2017)

WSUD Audit Guidelines and associated toolkit (Excel spreadsheet including macros) provide a very exhaustive audit form dedicated to assessing each SCM using a scale that awards each item (e.g., erosion, blockage, damage or removal of structures, sediment accumulation, erosion, standing water, rubbish, plant health, plant cover, etc.) from one point (good condition) to three points (poor condition). For each item, a detailed description is provided to help the operator choose the most pertinent condition. Each audit is uploaded in the database: individual assessment of items is used to identify maintenance (for 'moderate' condition) and rectification actions (for 'poor' condition). The overall condition of the asset is used for prioritizing maintenance and corrective works; however, other factors (such as significance of the asset, catchment sensitivity, and visibility) are also considered. Meijer *et al.* (2022) reported on the application of stereo vision method on sewer inspection; they concluded that stereo vision can be a good source of information in terms of recognizing defects. A precondition for stereo vision-based methods is that enough texture on the surface of the pipe wall is present, which normally is no problem for elements that have seen a few years of service. For new elements this may not be the case; hence, applications in determining the quality of newly built systems are not recommended.

4.2.3.1.4 Defect coding and uncertainty

The physical inspection of assets is only part of the answer. As CCTV surveying has traditionally been a purely subjective visual approach to assessing the condition of sewer pipes, coding systems have been developed to enhance uniformity in the reporting of asset conditions. Such coding also results in a level of compatibility between surveys that can allow some understanding of overall network condition and allows prioritization of maintenance. The earliest defect coding system was developed in the UK in the late 1970s following an identified need to significantly enhance information about sewer system conditions to allow cost-effective rehabilitation. The resulting Manual of Sewer Condition Classification (MSCC), currently in its fifth edition (WRc, 2013), introduced a standardized recording system incorporating condition codes (e.g., cracked; fractured; collapsed) and service codes (e.g., root intrusion; deposits (attached or settled); infiltration) and a clock system for specifying the location of the defect around the circumference of the pipe. MSCC also suggests a comprehensive set of survey headers to give a full understanding of the survey and traceability. The MSCC was complemented by the Sewerage Rehabilitation Manual (SRM) in 1983 (now an online handbook, retitled Sewerage Risk Management (WRc, 2022)). SRM introduced the concept of critical sewers and devised a scoring system for each defect, which forms the basis for the Internal Condition Grading (ICG), which ranks sewers into 5 grades where 1 is 'Acceptable Structural Condition' and 5 is 'Collapsed or Collapse Imminent'. This ICG can be extended into a Structural Performance Grade, considering additional factors, such as frequency of surcharge conditions and the surrounding soil type.

The MSCC and SRM have been a strong inspiration for different codes internationally. Thornhill and Wildbore (2005) provided a comprehensive summary, but it suffices to say that Australia, Canada, the United States, and many countries in Asia and the Pacific Rim follow similar principles, with appropriate modifications for local conditions.

The European Standard, EN 13508-2 (2011), developed from various standards from different countries in Europe, has also been influential in more recent developments around the world. Within Europe, EN13508-2 allows national annexes which tend to fuse each country's pre-existing codes with the standard, thus creating a very large number of possible codes, many of which are not directly relevant to the condition and performance of the sewer (e.g., presence of vermin, colour of water). The EN 13508-2 coding system uses a combination of letters and numbers to identify defects. The letters indicate the type of defect, while the numbers indicate the severity of the defect. The main code starts with the letter B when the observation concerns a drain or sewer. The second letter indicates a category of observation: BA for codes related to the fabric, BB related to operation, BC are inventory codes and BD are other codes. The last letter of the main code corresponds to a specific observation (e.g., BAA for a deformation, BBA for roots, BCA for a connection, or BDC for a finishing node). Figure 4.22 provides two examples of observations and associated codes according to the EN 13508-2 (2011) norm.

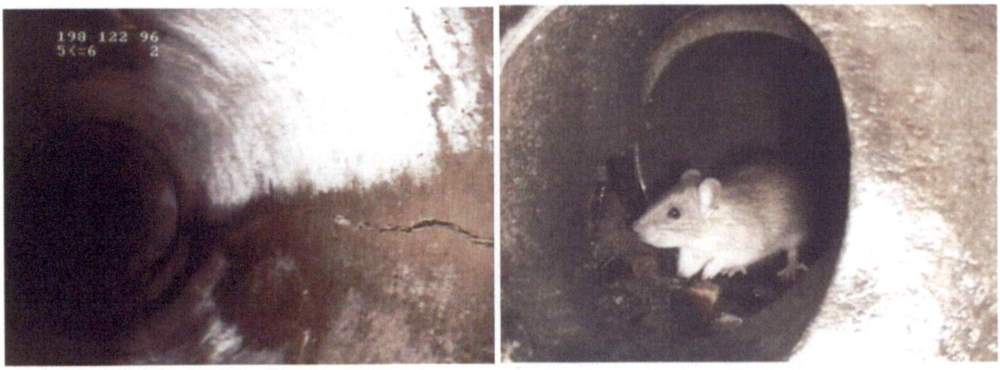

Figure 4.22 Example of observations coded according to the EN 13508-2 (2011) norm. On the left, BAB C A 03 & BAB B A 04: one fracture (BAB C) longitudinal (A) located at 3 h, and one crack (BAB B) longitudinal (A) located at 4 h. On the right, BBH A B 1: vermin (BBH), rat (A) in a connection (B) and one corresponds to the number of vermin observed.

In contrast, the Japanese standard was developed independently and focuses on identifying and quantifying defects which relate directly to hydraulic performance. As a result, the Japanese standard has a much smaller and hence simpler set of defect codes (Tait & Kazemi, 2022; van der Steen *et al.*, 2014). The Japan defect classification system for sewer pipe inspection (Japanese Sewerage Works Association, 1998) classifies the defects into 10 categories: crack, delamination (i.e., separation of the pipe wall into layers), scour (i.e., erosion of the pipe bottom or sides), corrosion (i.e., deterioration of the pipe wall by chemical or electrochemical reactions), erosion (i.e., gradual wearing away of the pipe wall by flowing water), collapse (i.e., complete or partial failure of the pipe wall), deformation (i.e., change in shape of the pipe wall), foreign object, tree root intrusion, and other. The severity of each defect is rated on a scale of 1–5, with 1 being the least severe and 5 being the most severe. The severity rating is based on the size, location, and type of defect. The classification system also provides recommendations on the speed of inspection, based on the severity of the defect. Severe defects (with rating of 4 or 5) should be inspected at a slow speed, so that the inspector can get a good look at the defect and identify its cause. Less severe defects (with rating of 1 or 2) can be inspected at a faster speed.

Table 4.2 (Tscheikner-Gratl *et al.*, 2019) presents an inventory of existing protocols to classify the pipe condition into several possible states depending on the level of complexity and the states considered. The condition(s) are usually assessed by aggregating all defects (with their respective importance). The scoring method is very often a weighted sum of the defects which is then divided by the length of the pipe (to enable the comparison of pipes of different lengths). Such a method is simple and easy to understand and can be improved by specifically considering major defects that lead to the pipe being considered to be in a poor condition.

While the coding systems bring some uniformity into the collection of CCTV inspection data, as with any use of data and models, the underlying quality and related uncertainties must be taken into account. Dirksen *et al.* (2013) studied the uncertainty of subjective human operation and interpretation by separating error sources into three categories: first, the collection of images whereby the collected images do not show defects which are present due to lens fouling, insufficient lighting or an inspection carried out too quickly. Second, the accuracy of identifying the defect or lack of defect, an error being represented as either: a false positive (FP) where a defect is recorded that does not exist in reality; or a false negative (FN) where a defect is present in the pipe but not recorded. FPs might be due to markings on the pipe, inadequate lighting/shadows, and so on. FNs could arise

Table 4.2 Existing protocols to assess the condition of a sewer pipe.

Type	Description	References
Comprehensive	Overall assessment of pipe's need for rehabilitation based directly on observed defects or combining the conditions below	Chughtai and Zayed (2011); EN 752 (2017); Kley *et al.* (2013); WRc (2013); Zhao *et al.* (2001)
Structural	Assessment of pipe physical condition by considering defects leading to deterioration and ultimately the collapse of the pipe	Ahmadi *et al.* (2014); Chughtai and Zayed (2011); EN 752 (2017); Khazraeializadeh *et al.* (2014); Kley *et al.* (2013); WRc (2013); Zhao *et al.* (2001)
Integrity (structural)	Assessment of structural condition with reference to strategic rehabilitation planning to determine remaining service life and structural integrity values of sewers (currently not standardized)	DWA-Themen T4 (2012); Kley *et al.* (2013)
Operational	Assessment of defects leading to an increase of operational interventions necessary on the pipe	Ahmadi *et al.* (2014); ATV-M 143–2 (1999); Chughtai and Zayed (2011); EN 13508–2 (2011); EN 752 (2017); NASSCO (2016); WRc (2013); Zhao *et al.* (2001)
Environmental	Assessment of defects leading to pollution of water (groundwater or surface water)	DWA-M 149–7 (2016); EN 752 (2017)
Hydraulic or serviceability	Assessment of defects that will perturbate the flow	Ahmadi *et al.* (2014); Arbeitshilfen Abwasser (2018); EN 752 (2017); Micevski *et al.* (2002); ÖWAV-RB 22 (2015)
Malfunctions	Consequences of defects on facility operations, for example ongoing corrosion, blockage, excessive spillage, sand silting, and so on.	Ahmadi *et al.* (2014); Kley *et al.* (2013); Le Gauffre *et al.* (2007)

Source: Tscheikner-Gratl *et al.* (2019).

from operators being distracted, there being multiple defects meaning the operator misses some or the operator fails to spot a less common defect. Third, errors in the description of the defect can be due to human error, ambiguity in the defect classification codes or improper understanding of the codes. Dirksen *et al.* (2013) presented a study of the effectiveness of condition surveys, showing that in identifying the presence of a defect FPs are rare; however, FNs have a probability of around 25%. When combined with the defect description, incorrect observations were over 50%. These data should therefore be treated as having a significant degree of uncertainty. Dirksen *et al.* (2013) listed a series of recommendations including the use of simpler coding systems and the use of other sources of data to assess sewer condition and performance. In order to reduce or eliminate operator errors, many utilities are now requiring CCTV data to be viewed by two operators in order to provide verification.

Caradot *et al.* (2018) investigated inspection uncertainties through a statistical approach. They found that the condition of the pipe has an impact on the probability of an incorrect condition grade but added a note of caution that their results were from a single city. Roghani *et al.* (2019) reproduced a comprehensive table listing 14 factors which can affect visual inspection results, many of which overlap with observations by Dirksen *et al.* (2013) but it is perhaps useful to stress the potential importance of time constraints and fatigue.

Beyond the fallibility of operators and coding systems themselves, further uncertainty can be incorporated in the inspection results through the aggregation process. Rahman and Vanier (2004) presented a useful comparison of how different condition assessment protocols can result in significantly different assessments of the same sewer network, even if the basis of the protocols is the same (i.e., MSCC). For example, in Canada, the City of Edmonton's modifications to MSCC showed 18% of sewers to be in poor condition, compared to only 3% if WRc's standard methodology was applied.

It should be noted that most if not all defect classification codes have been developed based on the assumption of visual, usually CCTV, inspection data. While many techniques discussed in this chapter provide measurement of a defect, the amalgamation and classification of data collected from one or more of these techniques continue to present some challenges.

4.2.3.1.5 AI-based assessment of CCTV data

The preceding section clearly identified limitations of CCTV analysis due to the human assessment of the captured images. As the power and storage of computers increased rapidly in the 2000s, interest in automating the analysis of CCTV gained traction.

Due to the time consuming and therefore costly nature of visual inspection of CCTV data, combined with acknowledged uncertainty in the subjective analysis, AI algorithms for analysing CCTV data have been investigated for several years with varying degrees of success. One of the earliest examples was the work of Sinha and Fieguth (2006) who used a neuro-fuzzy classifier with promising results. This area has received significant attention over the intervening years with Moradi *et al.* (2019) and Li *et al.* (2022) providing significant reviews of automated analysis of CCTV images. Many of these automated analysis techniques used image processing to identify potential defects and subsequently classified these using machine learning. More recent studies have used artificial neural networks and deep learning for feature extraction and classification. An example of such classifications is shown in Figure 4.23. Comparing the performance of different algorithms is challenging as they are not usually tested on the same dataset and different metrics are used; however, to give some indication, Li *et al.* (2022) report accuracies between 64.8% and 98.2%.

A key challenge in developing AI approaches to analyse CCTV images is the procurement of suitable datasets; furthermore, in order to provide confidence in the models, it is important for utilities to be able to compare outputs on a common dataset. Two open datasets are currently available, Sewer-ML from Denmark (https://vap.aau.dk/sewer-ml/), released in 2021 and containing 1.3 million images, and a UK database released in 2023 containing 27,000 images, but with the intention of extending the library to at least 1000 images per defect code

While clearly an area of ongoing research and development, the technologies are now sufficiently mature to have been commercialized by several companies, for example, the Swiss founded Pallon (https://www.pallon.com/) and the Australian founded Vapar (https://vapar.co/), both of whom cite significant customer bases. The apparent popularity of this technology no doubt hinges on strong water company knowledge around CCTV inspections and the existing capabilities to undertake such surveys. However good the automated analysis is, it is always limited by the quality of the CCTV data (e.g., lighting) and the inability to see below the water level

4.2.3.2 Acoustic inspection

Acoustic sensing has shown promise and some commercial applications. The method is based on the analysis of reflected sound intensities. Romanova *et al.* (2013) described a study comparing CCTV surveys with acoustic survey data and reported accuracies of 75–85% when considering four types of defects – connections, joints, cracks, and junctions. The surveys were conducted from a manhole at one end of the pipe; some defects can be missed due to a blind zone close to the sensor, overlapping reflections if there are many defects, weak reflections or shadowing from larger obstructions. Horoshenkov (2012) further presented results from the application of acoustic inspection for measuring water level, sediment levels and hydraulic roughness. Sewerbatt™ applied this technology commercially with some favorable trial results and worldwide sales (https://www.isleutilities.com/news/case-study-collaborative-trial-sewerbatt-acoustic-sensing-technology). Use of acoustic sensing for blockage detection has also been proven, with some analyses able to predict the cross-sectional

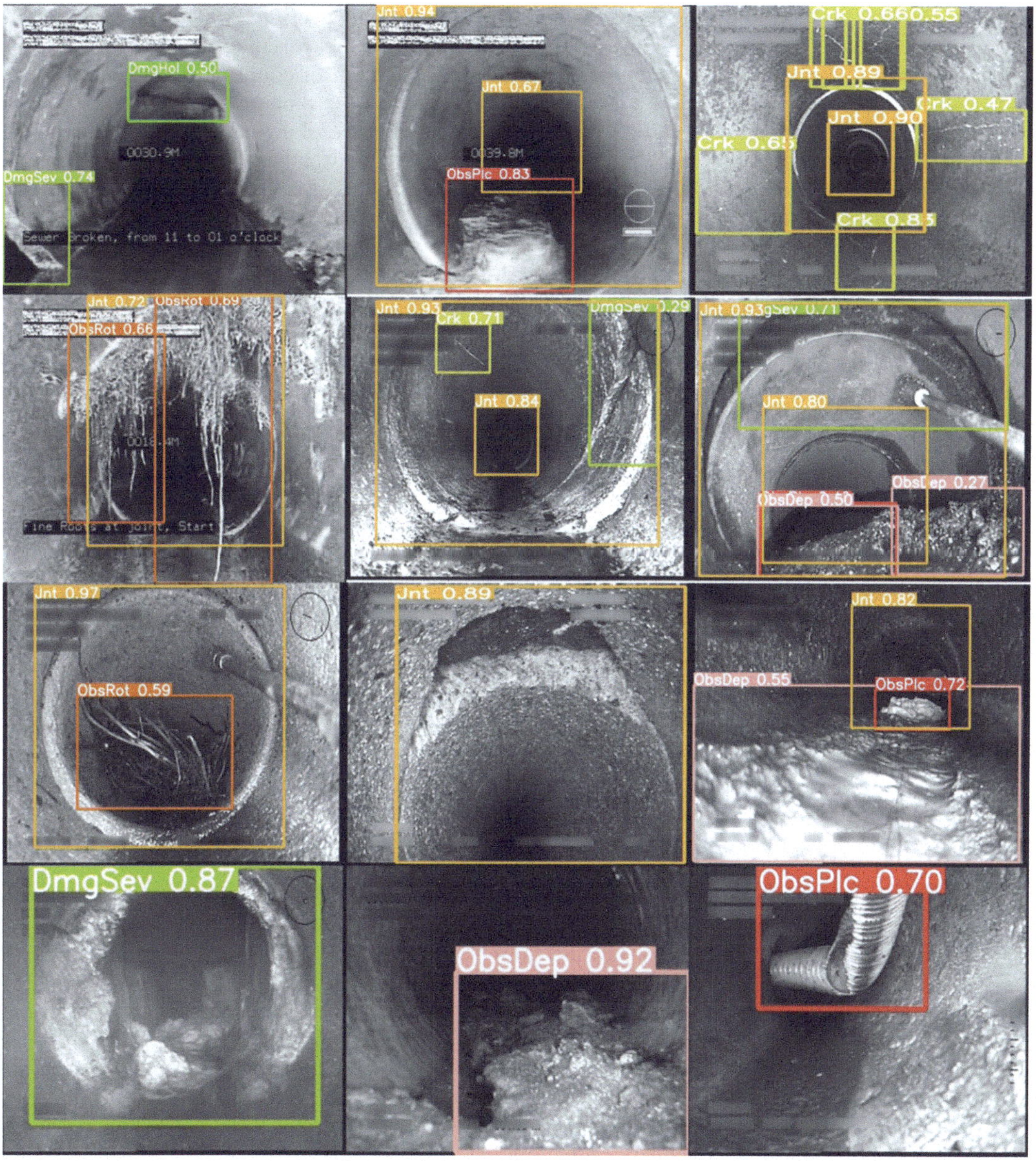

Figure 4.23 Examples of automated defect detection and classification. (After Tait and Kazemi, 2023).

area and length of the blockage with errors of less than 20% and 30%, respectively (Duan *et al.*, 2015). Development of acoustic sensing in sewer pipes is ongoing, with Yu *et al.* (2023) exploring the potential of microphone arrays to enhance the collected data and allow defect classification through data analytics techniques

Ultrasonic techniques, which use high-frequency inaudible acoustic waves, are widely used in non-destructive testing, but the materials and buried nature of UD pipes present challenges due to wave attenuation. Iyer *et al.* (2012) presented promising results from the inspection of 60 mm thick concrete: they identified cracks, fractures and holes using C-Scan (multiple measurement) imaging. Inspection of materials with ultrasonic techniques however requires some form of coupling between the sensor and the object. The technique used by Iyer *et al.* (2012) involves the pipe being immersed in water, limiting applicability. Towlson *et al.* (2022) have investigated the use of air-coupled ultrasonic arrays for characterization of blockages and pipe wall defects. The research provided promising results, particularly for pipe wall defects where a 2 mm through hole was identified from a distance of 200 mm. For identification of blockages an area between the pipe wall and the centre of the pipe suffers from superposition of reflections cancelling the signal. Blockages were however successfully identified and useful information about the number of transducers and the aperture over which the transducers should be spread was reported. This is a developing area of research, the relatively short distance (up to 1 m) over which the ultrasonic waves transmit requires the use of some form of crawler to position the array.

Sonar is another technology that has been applied for condition assessment of sewer pipes (Selvakumar *et al.*, 2014). The principle is similar to acoustic sensing except that the sound waves pass through the fluid; thus, it is used in filled pipes (e.g., siphons and rising mains) or below the water line. Analysis of the reflections allows detection of defects such as cracks, corrosion and deflection, as well as sediment build up, although the latter of course obscures a view of the pipe wall, as can solids in the water.

A relatively new development is the application of sound to detect leaks and gas pockets in pressure mains. This allows for identifying the presence of gas/air pockets and leakages in pressure mains due to the sounds they produce (see e.g., Pure Americas Inc, 2011).

Apart from the application of passive acoustic methods, detection of leakage can, reportedly, also be achieved by active acoustic methods (see e.g., Lee *et al.*, 2023).

4.2.3.3 Sensor combinations

It has been acknowledged that single sensors each have their own limitations; thus, inspection systems incorporating multiple sensors have been developed. In the late 1990s, several multi-sensor platforms were developed, as reviewed by Wirahadikusumah *et al.* (1998) and Duran *et al.* (2002). The German prototype, KARO was a tethered device capable of operating for up to 400 m in 200 mm diameter pipes, incorporating a camera, 3D optical sensor, ultrasonic sensors, and microwave radar. Data from the multiple sensors was fused through mathematical techniques. KARO was claimed to be capable of measuring obstacles, cracks, and wall thickness as well as conditions beyond the pipe wall as well as 3D positioning. It is believed that the system was not developed beyond the prototype stage. In Australia, the PIRAT system was developed adding internal geometry measurements via laser above the waterline and sonar below the water surface, to CCTV. Automated recognition, rating and classification of pipe defects were carried out by AI software. Early tests were reported to give superior results to CCTV for concrete and vitreous clay pipes, but as with KARO it did not move beyond prototype stage. Sewer scanner and evaluation technology (SSET) was developed in Japan and incorporates an optical scanner and gyroscope alongside CCTV. An unwrapped image of the whole pipe is output; defects can be colour coded and annotated. A non-crawler-based technique is the use of a manhole zoom camera with acoustic reflectometry (Plihal *et al.*, 2016). This was also shown to provide better information than the individual techniques. Based on the available literature, SSET seems to be the combined sensor technology that has seen the most uptake. While combinations of sensors do provide better information than the individual techniques, there are increases in costs and complexity both in the hardware and data analysis.

4.3 EMERGING TECHNOLOGIES

The technologies described up to this point all focused on the interior of the pipe; however, that is clearly only a part of the pipe. The material within and on the outside of the walls can deteriorate and voids can form outside the pipe. Hao *et al.* (2012), described techniques such as ground penetrating radar, which can identify voids in the ground. They also described ultrasonic-guided waves, which are used in other fields and can identify pipe corrosion; however, this technique faces significant difficulties when applied to granular and non-homogenous materials such as concrete.

There is growing interest in developing autonomous, untethered robots for UD inspection. Autonomous robotics have the potential to inspect pipes more cost effectively due to longer deployment times, less human labour, less highway disruption. Lower unit inspection costs could therefore allow a greater proportion of the networks to be surveyed more frequently. Such robots have been predominantly developed for larger (e.g., >600 mm diameter) pipes (Kolvenbach *et al.*, 2020; Spectar, 2023) and are still mainly prototypes. Ongoing and recent research has investigated robots for smaller pipes (e.g., 200–300 mm diameter), including the Danish ASIR (2018) project and Pipebots in the UK (Shepherd *et al.*, 2021). Autonomous robotics have been developed to use a range of different sensors; key considerations for the deployment of sensors on autonomous robots are the power, data volume, and data processing requirements. Autonomous robots also have significant challenges for localization and navigation in sparse piped networks with no GPS coverage, as well as environmental challenges to overcome due to variable depths and velocities of water, sewage matter such as rags and grease and the potential for explosive gases. Regarding localization, there has been a significant amount of work which has investigated different sensing and processing algorithms, as summarized by Aitken *et al.* (2021); however, the problem is not solved. As with condition assessment, sensor fusion seems to be a promising way forward.

While fibre optics were previously discussed for distributed temperature sensing, there has been potential shown for permanently installed fibre optic sensors to be used as distributed strain sensors which are able to carry out some structural assessment and monitor hydraulic parameters (Ainger *et al.*, 2021). This area is receiving further attention, as reviewed by Prisutova *et al.* (2022), with applications for both condition and flow measurement. There is a range of fibre optic measurement methods and installation techniques which affect cost and applicability for retrofitting. Overall, Prisutova *et al.* (2022) suggested that further research is needed on the application to partially full pipes. There is some commercial application of the technique in sewer pipes (e.g., nuron, https://www.nuron.tech/water/)

Another emerging technology entering the field of UDAM is machine learning: apart from the obvious application in processing inspection footage or defects, the quantification of some processes using cameras has been subject to the application of machine learning techniques. For example, Moreno-Rodenas *et al.* (2021) reported on the successful application of a semantic segmentation algorithm on video footages obtained in wastewater pumping stations to estimate the accumulation of FOG deposits. Overall, the use of images (i.e., photo, video) as source of information is an emerging field. From video footage, using particle image velocimetry (PIV), particle tracking velocimetry (PTV) or optical flow (OF) algorithms, footage information on velocity and vorticity fields can be obtained (see e.g., Duinmeijer, 2020) using PIV. Another emerging application is the use of spectral cameras for extracting potential on the composition of the wastewater (see e.g., Lechevallier and Rieckermann, 2020). In addition, the use of infrared cameras for detecting inflow has been reported (see e.g., Lepot *et al.*, 2017a).

4.3.1 Emerging technologies and lack of standards

Despite the fact CCTV is the most popular technology for piped network inspection, it has some serious drawbacks. During the last decades, some new monitoring techniques have been designed and tested in the field. While many of these solutions look promising since they partially or totally compensate for the disadvantages of CCTV, to the authors knowledge, there is no standard to deploy such inspections in the field and process the data. Additionally, several techniques are specific for one type or one group of defects and, therefore, could not identify all standard defects listed in the broadly accepted standards.

Section 4.3 described a plethora of emerging technologies, some of which are available commercially, while others are still in or never progressed beyond the prototype stage for various reasons. It is however clear that moving beyond the subjective data provided by human interpreted visual imagery to objective sensor measurements has many potential benefits to enhance the data collected from inspection of UD networks. Existing defect coding and classification frameworks have been developed based on the data available from the ubiquitously used CCTV surveys. Such codes and classifications are likely neither sufficient to record the detail available from alternative inspection technologies, nor to record the changes between the potentially more frequent surveys. The existing coding and classification schemes also focus on the structural attributes of the pipes; however, the hydraulic performance and remaining lifespan is generally of more concern to the responsible utilities.

An interesting development is the application of stereo imaging, LIDAR and 3D laser scanning (see Wang *et al.*, 2022). The use of stereo allows for a 3D reconstruction based on which at least some geometrical information on the inside of the construction can be obtained. A broad application of these technologies seems to be hampered by existing standards and existing workflows with managing organizations and enterprises providing inspection services.

4.3.2 Automated mapping of underground infrastructures

Despite much effort being put into obtaining accurate maps of the underground infrastructure (basically answering the fundamental AM question: 'what am I managing and where is it?') there is still a need for a system able to more or less autonomously map a network of underground pipes. In practice, one needs to open each individual manhole in order to obtain observations on invert elevation, geometrical dimensions, and so on.

In theory, one could construct a sensor platform that moves, either self-propelled or floating on the (waste)water stream and collects data on profile dimensions and keeps track of its position. The latter is the main challenge, as underground autonomous navigations is not a simple hurdle to overcome as GPS systems typically do not function when deployed subsurface. An alternative method can be the application of accelerometers, from which, after integrating twice, theoretically the relative 3D position of the device as a function of time can be reconstructed. However, due to the accumulation of the measuring uncertainties combined with the numerical integration errors such an approach can only work when a regular update on the position can be obtained from an independent source. One might, for instance, consider supplying the device with a database with the position of landmarks (e.g., manholes); such information can be used in conjunction with a distance sensor of the device that detects the presence of landmarks (e.g., manholes, see e.g., Thielemann *et al.*, 2008). This is a field under development in which combinations of methods (e.g., accelerometers, stereo vision, acoustic distance metering, etc.) are deployed to keep the uncertainty in the position within an acceptable envelope.

4.4 CONCLUSIONS AND OUTLOOK

This chapter has focused on the plethora of inspection techniques which have been and continue to be developed for the inspection of UD assets. It is clear that CCTV remains the most popular method of inspection, despite limitations as to which defects can be identified with costs often affecting the frequency of inspection and subjectivity of the footage interpretation by the operator.

Several other techniques are available (at various maturity levels) to provide objective measurements for specific defects. In order to fully understand asset condition, a range of techniques needs to be applied which requires more research and development effort for them to be widely used. The potential for robotic inspection has attracted significant interest over the years and is currently an area of research. While fully autonomous robots continuously inspecting sewer pipes seem to remain some way off, the benefits of some degree of autonomy and lack of reliance on human operators and tethers could start to bring benefits in terms of coverage, cost, and measurement techniques in the nearer future. The inspection of nature-based solution assets remains the least mature part of this field, likely due to a lack of comprehension of the ageing processes of such solutions and status criteria (e.g., infiltration).

Even if the lack of process comprehension or industrial development could be blamed for the limited use of other inspection techniques (i.e., besides CCTV), the absence of standards definitely counteracts a wider use of other techniques.

Researchers, industrialists, and legislators need to work together to deploy the presented (and coming) technologies at a larger scale. This collaboration appears mandatory to conduct numerous and accurate inspections to hold sewage and stormwater service under unavoidable and growing constraints: climate change and limited resources. The authors hope this chapter will contribute to solving this challenge we all face.

REFERENCES

AECOM (2015). Georgia Stormwater Management Manual Appendix E, September, 138 p. https://cdn.atlantaregional.org/wp-content/uploads/operation-and-maintenance-guidance-document.pdf (accessed 23 August 2023).

Ahmadi M., Cherqui F., de Massiac J. C., Werey C., Lagoutte S. and Le Gauffre P. (2014). Condition grading for dysfunction indicators in sewer asset management. *Structure and Infrastructure Engineering*, **10**, 346–358, https://doi.org/10.1080/15732479.2012.756916

Ahmed F., Gulliver J. S. and Nieber J. L. (2015). Field infiltration measurements in grassed roadside drainage ditches: spatial and temporal variability. *Journal of Hydrology*, **530**, 604–611, https://doi.org/10.1016/j.jhydrol.2015.10.012

Ainger M., Kingsley S., Horoshenkov K. and Tait S. (2021). Optical fibre pressure sensing apparatus employing longitudinal diaphragm. GB2567610B.

Aitken J. M., Evans M. H., Worley R., Edwards S., Zhang R., Dodd T., Mihaylova L. and Anderson S. R. (2021). Simultaneous localization and mapping for inspection robots in water and sewer pipe networks: a review. *IEEE Access*, **9**, 140173–140198, https://doi.org/10.1109/ACCESS.2021.3115981

Almeida F., Brennan M. J., Joseph P., Withfield S., Dray S. and Paschalini A. (2014). On the acoustic filtering of the pipe and sensor in a buried plastic water pipe and its effect on leak detection: an experimental investigation. *Sensors*, **14**, 5595–5610, https://doi.org/10.3390/s140305595

Arbeitshilfen A. (2018). Arbeitshilfen Abwasser – Planung, Bau und Betrieb von abwassertechnischen Anlagen in Liegenschaften des Bundes. Bundesministerium des Innern, für Bau und Heimat, Berlin, Germany.

ASIR (2018). Autonomous Sewer Inspection Robot, collaborative project funded by Innovation Fund Denmark. https://inlocrobotics.com/en/engineering/autonomous-robots/ (accessed 19 September 2022).

ASTM (2009). Standard Test Method for Infiltration Rate of In Place Pervious Concrete 1, https://cdn.standards.iteh.ai/samples/71163/3c358f91d5144ad4b082effaf2b02d27/ASTM-C1701-C1701M-09.pdf

ATV-M 143-2 (1999). Optische Inspektion -Inspektion, Instandsetzung, Sanierung und Erneuerung von Abwasserkanälen und -leitungen (Optical/visual Inspection – inspection, renovation, replacement and rehabilitation of sewers). Deutsche Vereinigung für Wasserwirtschaft, Abwasser und Abfall e. V., Hennef, Germany.

Bahnsen C. H., Johansen A. S., Philipsen M. P., Henriksen J. W., Nasrollahi K. and Moeslund T. B. (2021). 3D sensors for sewer inspection: a quantitative review and analysis. *Sensors*, **21**, 2553, https://doi.org/10.3390/s21072553

Bean E. Z., Hunt W. F. and Bidelspach D. A. (2007). Evaluation of four permeable pavement sites in eastern North Carolina for runoff reduction and water quality impacts. *Journal of Irrigation and Drainage Engineering*, **133**, 583–592, https://doi.org/10.1061/(ASCE)0733-9437(2007)133:6(583)

Beheshti M., Sægrov S. and Ugarelli R. (2015). Infiltration/inflow assessment and detection in urban sewer system. *Vann*, **01**, 24–34.

Beheshti M. and Saegrov S. (2019). Detection of extraneous water ingress into the sewer system using tandem methods – a case study in Trondheim city. *Water Science & Technology*, **79**(2), 231–239, https://doi.org/10.2166/wst.2019.057

Bin Ali M. T., Horoshenkov K. V. and Tait S. J. (2011). Rapid detection of sewer defects and blockages using acoustic-based instrumentation. *Water Science and Technology*, **64**(8), 1700–1707, https://doi.org/10.2166/wst.2011.183

Boogaard F. C. (2015). Stormwater characteristics and new testing methods for certain sustainable urban drainage systems in The Netherlands. PhD, Delft University of Technology, https://doi.org/10.4233/UUID:D4CD80A8-41E2-49A5-8F41-F1EFC1A0EF5D

Boogaard F. C. (2022). Spatial and time Variable long term infiltration rates of green infrastructure under extreme climate conditions, drought and highly intensive rainfall. *Water*, **14**, 840, https://doi.org/10.3390/w14060840

Boogaard F. and Lucke T. (2019). Long-term infiltration performance evaluation of Dutch permeable pavements using the full-scale infiltration method. *Water*, **11**, 320, https://doi.org/10.3390/w11020320

Borgwardt S. (2006). Long-term in-situ infiltration performance of permeable concrete block pavement. Proceedings of the 8th International Conference on Concrete Block Paving, 6–8 November, San Francisco, CA, USA.

Brennan M. J., Kroll de Lima F., Almeida F., Joseph P. and Paschoalini A. T. (2016). A virtual pipe rig for testing acoustic leak detection correlators: proof of concept. *Applied Acoustics*, **102**, 137–145, https://doi.org/10.1016/j.apacoust.2015.09.015

Browne D., Godfrey M., Markwell K. and Boer S. (2017). WSUD Audit Guidelines (Version 1.0), Stormwater Victoria and E2Designlab, June, 131 p. https://musicauditor.com.au/HostedFiles/WSUD%20Audit%20Guidelines%20Draft.pdf (Access 23 August 2023).

Caradot N., Rouault P., Clemens F. and Cherqui F. (2018). Evaluation of uncertainties in sewer condition assessment. *Structure and Infrastructure Engineering*, **14**, 264–273, https://doi.org/10.1080/15732479.2017.1356858

Chandler D. M. and Lerner D. N. (2015). A low cost method to detect polluted surface water outfalls and misconnected drainage. *Water and Environment Journal*, **29**, 202–206, https://doi.org/10.1111/wej.12112

Chughtai F. and Zayed T. (2011). Integrating WRc and CERIU condition assessment models and classification protocols for sewer pipelines. *Journal of Infrastructure Systems*, **17**, 129–136, https://doi.org/10.1061/(ASCE)IS.1943-555X.0000052

Clemens F., Stanić N., Van der Schoot W., Langeveld J. and Lepot M. (2015). Uncertainties associated with laser profiling of concrete sewer pipes for the quantification of the interior geometry. *Structure and Infrastructure Engineering*, **11**, 1218–1239, https://doi.org/10.1080/15732479.2014.945466

Clemens-Meyer F. H. L. R., Lepot M., Blumensaat F., Leutnant D. and Gruber G. (2021). Data validation and data quality assessment. In: Metrology in Urban Drainage and Stormwater Management: Plug and Pray, J.-L. Bertrand-Krajewski, F. Clemens-Meyer and M. Lepot (eds), IWA Publishing, London, UK, 327–390, https://doi.org/10.2166/9781789060119_0327

de Bénédittis J. and Bertrand-Krajewski J. L. (2005). Infiltration in sewer systems: comparison of measurement methods. *Water Science and Technology*, **52**(3), 219–227, https://doi.org/10.2166/wst.2005.0079

DiBT (2012). Zulassungsgrundsätze Niederschlagswasserbehandlungsanlagen Teil 2: Wasserdurchlässige Beläge für Kfz-Verkehrsflächen für die Behandlung des Abwassers zur anschließenden Versickerung in Boden und Grundwasser (Abwasserbehandelnde Flächenbeläge), Fassung April 2012.

Dirksen J., Clemens F. H. L. R., Korving H., Cherqui F., Le Gauffre P., Ertl T., Plihal H., Muller K. and Snaterse C. T. M. (2013). The consistency of visual sewer inspection data. *Structure and Infrastructure Engineering*, **9**(3), 214–228, https://doi.org/10.1080/15732479.2010.541265

Dirksen J., Pothof I. W. M., Langeveld J. G. and Clemens F. H. L. R. (2014). Slope profile measurement of sewer inverts. *Automation in Construction*, **37**, 122–130, https://doi.org/10.1016/j.autcon.2013.10.014

Duan W., Kirby R., Prisutova J. and Horoshenkov K. V. (2015). On the use of power reflection ratio and phase change to determine the geometry of a blockage in a pipe. *Applied Acoustics*, **87**, 190–197, https://doi.org/10.1016/j.apacoust.2014.07.002

Duinmeijer S. P. A. (2020). On the free-surface driven motion of buoyant particles. PhD thesis, Delft University of Technology, ISBN 978-94-6366-271-0. https://repository.tudelft.nl/islandora/object/uuid%3Aa33fa2a9-f347-40a3-96be-51e880018974

Duran O., Althoefer K. and Seneviratne L. D. (2002). State of the art in sensor technologies for sewer inspection. *IEEE Sensors Journal*, **2**, 73–81, https://doi.org/10.1109/JSEN.2002.1000245

DWA-A 138 (2005). Arbeitsblatt DWA-A 138, Planung, Bau und Betrieb von Anlagen zur Versickerung von Niederschlagswasser. Planning, construction and operation of facilities for the infiltration of precipitation water. ISBN Print: 978-3-937758-66-4, ISBN E-Book: 978-3-96862-062-6.

DWA-M 149-7 (2016). Zustandserfassung und -beurteilung von Entwässerungssystemen außerhalb von Gebäuden – Teil 7: Beurteilung der Umweltrelevanz des baulichen/betrieblichen Zustands (Condition assessment and classification of urban drainage systems outside of buildings). Deutsche Vereinigung für Wasserwirtschaft, Abwasser und Abfall e. V., Hennef, Germany.

DWA-Themen T4 (2012). Leitfaden zur strategischen Sanierungsplanung von Entwässerungssystemen außerhalb von Gebäuden (Guideline on strategical rehabilitation planning of urban drainage systems outside of buildings). Deutsche Vereinigung für Wasserwirtschaft, Abwasser und Abfall e. V., Hennef, Germany.

Eiswirth M. and Heske C. (2000). Pipe Defect Characterisation by Multi-Sensor Systems (pp.1–12). No-dig 2000.

Ékes C. and Mahmood S. (2019). Looking into the Void: Condition Assessment with Pipe Penetrating Radar 198–204, https://doi.org/10.1061/9780784482490.021

Ellis J. B. and Bertrand-Krajewski J. L. (eds) (2010). Assessing Infiltration and Exfiltration on the Performance of Urban Sewer Systems (APUSS). IWA Publishing, London, UK.

Ellis J. B. and Butler D. (2015). Surface water sewer misconnections in England and Wales: pollution sources and impacts. *Science of the Total Environment*, **526**, 98–109, https://doi.org/10.1016/j.scitotenv.2015.04.042

EN 13508-2 (2011). Condition of Drain and Sewer Systems Outside Buildings – Part 2: Visual Inspection Coding System. European Commitee for Standardization, Brussels, Belgium.

EN 752 (2017). Drain and Sewer Systems Outside Buildings. European Commitee for Standardization, Brussels, Belgium.

Enquête Eau (2008). CGDD, SOeS/SSP – Ministère du Développement Durable. Enquête C.I.EAU, 2006/Source: L'eau du robinet dans notre quotidien, C.I.EAU, 2006.

Fassman E. A. and Blackbourn S. (2010). Urban runoff mitigation by a permeable pavement system over impermeable soils. *Journal of Hydrologic Engineering*, **15**(6), 475–485, https://doi.org/10.1061/(ASCE)HE.1943-5584.0000238

Fatemi Aghda S. M., Ganjalipour K. and Nabiollahi (2018). Assessing the accuracy of TDR-based water leak detection system. *Results in Physics*, **8**, 939–948, https://doi.org/10.1016/j.rinp.2018.01.026

Feeney C. S., Thayer S., Bonomo M. and Martel K. (2009). Condition Assessment of Wastewater Collection Systems, US Environment Protection Agency Report EPA/600/R-09/049, nepis.epa.gov/Exe/ZyPURL.cgi?Dockey=P1003ZQY.TXT (Accessed 6 April 2023).

Flyability (2023). Sewer Inspections by Drone: How Suez RV Osis Uses the Elios 2 to Ensure 100% Coverage When Inspecting Wastewater Infrastructure [WWW Document]. URL https://www.flyability.com/casestudies/suez-rv-osis-drone-sewer-inspections (Accessed 13 April 2023).

García-Serrana M., Gulliver J. S. and Nieber J. L. (2017). Infiltration capacity of roadside filter strips with non-uniform over-land flow. *Journal of Hydrology*, **545**, 451–462, https://doi.org/10.1016/j.jhydrol.2016.12.031

Ghozzi R., Lahouar S., Besbes K. and Souani C. (2018). Mapping of sewer lines using GPR: a case study in Tunisia. *Data*, **3**(4), 40, https://doi.org/10.3390/data3040040

Hao T., Rogers C. D. F., Metje N., Chapman D. N., Muggleton J. M., Foo K. Y., Wang P., Pennock S. R., Atkins P. R., Swingler S. G., Parker J., Costello S. B., Burrow M. P. N., Anspach J. H., Armitage R. J., Cohn A. G., Goddard K., Lewin P. L., Orlando G., Redfern M. A., Royal A. C. D. and Saul A. J. (2012). Condition assessment of the buried utility service infrastructure. *Tunnelling and Underground Space Technology*, **28**, 331–344, https://doi.org/10.1016/j.tust.2011.10.011

Hoes O., Schilperoort R. P. S., Luxemburg W. M. J., Clemens F. H. L. R. and van de Giessen N. C. (2009). Locating illicit connections in storm water sewers using fibre optic distributed temperature sensing. *Water Research*, **43**, 5187–5197, https://doi.org/10.1016/j.watres.2009.08.020

Horoshenkov K. (2012). Acoustical monitoring of water infrastructure. Société Française d'Acoustique. Acoustics 2012, Apr 2012, Nantes, France. https://hal.science/hal-00811121

Iyer S., Sinha S. K., Pedrick M. K. and Tittmann B. R. (2012). Evaluation of ultrasonic inspection and imaging systems for concrete pipes. *Automation in Construction*, **22**, 149–164, https://doi.org/10.1016/j.autcon.2011.06.013

Japanese Sewerage Works Association (1998). Japan Defect Classification System for Sewer Pipe Inspection. Japanese Sewerage Works Association, Tokyo.

Khazraeializadeh S., Gay L. F. and Bayat A. (2014). Comparative analysis of sewer physical condition grading protocols for the City of Edmonton. *Canadian Journal of Civil Engineering*, **41**, 811–818, https://doi.org/10.1139/cjce-2014-0077

Kley G., Kropp I., Schmidt T. and Caradot N. (2013). SEMA D1.1 – Review of Available Technologies and Methodologies for Sewer Condition Evaluation. Kompetenzzentrum Wasser Berlin, Berlin, Germany.

Kolvenbach H., Wisth D., Buchanan R., Valsecchi G., Grandia R., Fallon M. and Hutter M. (2020). Towards autonomous inspection of concrete deterioration in sewers with legged robots. *Journal of Field Robotics*, **37**, 1314–1327, https://doi.org/10.1002/rob.21964

Kracht O., Gresch M., de Benedittis J., Progiobbe V. and Gujer W. (2003). Stable isotopes of water as a natural tracer for infiltration into urban sewer systems. *Geophysical Research Abstracts*, **5**, 07852. http://www.cosis.net/abstracts/EAE03/07852/EAE03-J-07852-1.pdf

Lechevallier P. and Rieckermann J. (2020). Data and codes: towards non-contact pollution 624 monitoring in sewers with hyperspectral imaging, https://doi.org/10.25678/0007WY

Lee J. C., Choi Y. R. and Cho J. W. (2023). Pipe leakage detection using ultrasonic acoustic signals. *Sensors and Actuators A: Physical*, **349**, 114061, https://doi.org/10.1016/j.sna.2022.114061

Le Gauffre P., Joannis C., Vasconcelos E., Breysse D., Gibello C. and Desmulliez J. J. (2007). Performance indicators and multicriteria decision support for sewer asset management. *Journal of Infrastructure Systems*, **13**, 105–114, https://doi.org/10.1061/(ASCE)1076-0342(2007)13:2(105)

Lepot M., Hernandez B., Cedillo S., Goes B. H., Schellart A., Verlinden J. C. and Clemens F. H. (2016). 3D printed pills to investigate mis-connections. 8th International Conference on Sewer Processes and Networks.

Lepot M., Makris K. F. and Clemens F. H. (2017a). Detection and quantification of lateral, illicit connections and infiltration in sewers with infra-red camera: conclusions after a wide experimental plan. *Water Research*, **122**, 678–691, https://doi.org/10.1016/j.watres.2017.06.030

Lepot M., Stanić N. and Clemens F. H. (2017b). A technology for sewer pipe inspection (part 2): experimental assessment of a new laser profiler for sewer defect detection and quantification. *Automation in Construction*, **73**, 1–11, https://doi.org/10.1016/j.autcon.2016.10.010

Lepot M., Pouzol T., Aldea Borruel X., Suner D. and Bertrand-Krajewski J. L. (2017c). Measurement of sewer sediments with acoustic technology: from laboratory to field experiments. *Urban Water Journal*, **14**(4), 369–377, https://doi.org/10.1080/1573062X.2016.1148181

Li H., Kayhanian M. and Harvey J. T. (2013). Comparative field permeability measurement of permeable pavements using ASTM C 1701 and NCAT permeameter methods. *Journal of Environmental Management*, **118**, 144–152, https://doi.org/10.1016/j.jenvman.2013.01.016

Li Y., Wang H., Dang L. M., Song H.-K. and Moon H. (2022). Vision-based defect inspection and condition assessment for sewer pipes: a comprehensive survey. *Sensors*, **22**, 2722, https://doi.org/10.3390/s22072722

Luimes R. A., Scheperboer I. C., Suiker A. S. J., Bosco E. and Clemens F. H. L. R. (2023). Assessment of chemo-mechanical degradation of concrete sewer pipes through an integrated experimental approach. *Engineering Structures*, **293**(2023), 116590, https://doi.org/10.1016/j.engstruct.2023.116590

Makris K., Langeveld J., Clemens-Meyer F., Watts J., Begun H. and Horoshenkov K. (2023). Sonic assessment of physical ageing of plastic pipes. *Journal of Sound and Vibration*, **544**, https://doi.org/10.106/j.jjsv2022.1177393

Martins L. L., do Almeida M. C. and Ribeiro ÁS (2020). Optical metrology applied in CCTV inspection in drain and sewer systems. *ACTA IMEKO*, **9**, 18–24, https://doi.org/10.21014/acta_imeko.v9i1.744

Meijer D. W., Luimes R. A., Knobbe A. J. and Back T. H. W. (2022). RADIUS: robust anomaly detection in urban drainage with stereovision. *Automation in Construction*, **139**, 104285, https://doi.org/10.1016/j.autcon.2022.104285

Micevski T., Kuczera G. and Coombes P. (2002). Markov model for storm water pipe deterioration. *Journal of Infrastructure Systems*, **8**, 49–56, https://doi.org/10.1061/(ASCE)1076-0342(2002)8:2(49)

Moradi S., Zayed T. and Golkhoo F. (2019). Review on computer aided sewer pipeline defect detection and condition assessment. *Infrastructures*, **4**, 10, https://doi.org/10.3390/infrastructures4010010

Moreno-Rodenas A. M., Duinmeijer S. P. A. and Clemens-Meyer F. H. L. R. (2021). Deep learning monitoring of FOG layers in wastewater pumping stations. *Water Research*, **202**, 117482, https://doi.org/10.1016/j.watres.2021.117482

Morgan E. S. and Crosse P. A. E. (1978). The acoustic ranger, a new instrument for tube and pipe inspection. *NDT International*, **11**(4), 179–183, https://doi.org/10.1016/0308-9126(78)90052-4

NASSCO (2016). Pipeline Assessment Certification Program (PACP). National Association of Sewer Service Companies.

ÖWAV-RB 22 (2015). Betrieb von Kanalisationsanlagen (Operation of Urban Drainage Systems). Österreichischer Wasser- und Abfallwirtschaftsverband, Vienna, Austria, Vienna, Austria.

Palhegyi G. E. (2010). Designing storm-water controls to promote sustainable ecosystems: science and application. *Journal of Hydrologic Engineering*, **15**, 504–511, https://doi.org/10.1061/(ASCE)HE.1943-5584.0000130

Panasiuk O., Hedström A., Langeveld J. and Viklander M. (2022). Identifying sources of infiltration and inflow in sanitary sewers in a northern community: comparative assessment of selected methods. *Water Science and Technology*, **86**, 1–16, https://doi.org/10.2166/wst.2022.151

Park S., Lim H., Tamang B., Jin J., Lee S., Park S. and Kim Y. (2020). A preliminary study on leakage detection of deteriorated underground sewer pipes using aerial thermal imaging. *International Journal of Civil Engineering*, **18**(10), 1167–1178, https://doi.org/10.1007/s40999-020-00521-8

Pezzaniti D., Beecham S. and Kandasamy J. (2009). Influence of clogging on the effective life of permeable pavements. *Proceedings of the Institution of Civil Engineers – Water Management*, **162**(3), 211–220, https://doi.org/10.1680/wama.2009.00034

Plihal H., Kretschmer F., Schwarz D. and Ertl T. (2014). Innovative sewer inspection as a basis for an optimised condition-based maintenance strategy. *Water Practice and Technology*, **9**, 88–94, https://doi.org/10.2166/wpt.2014.010

Plihal H., Kretschmer F., Bin Ali M. T., See C. H., Romanova A., Horoshenkov K. V. and Ertl T. (2016). A novel method for rapid inspection of sewer networks: combining acoustic and optical means. *Urban Water Journal*, **13**, 3–14, https://doi.org/10.1080/1573062X.2015.1076857

Prisutova J., Krynkin A., Tait S. and Horoshenkov K. (2022). Use of fibre-optic sensors for pipe condition and hydraulics measurements: a review. *CivilEng*, **3**, 85–113, https://doi.org/10.3390/civileng3010006

Pure Americas Inc. (2011). SmartBall Inspection Report of North Beach Force Main, August 2011. https://your.kingcounty.gov/dnrp/library/wastewater/wtd/construction/NorthBeachFM/1108_NorthBeachForceMainInspectionReport.pdf (visited on 25 August 2023).

Rahman S. and Vanier D. J. (2004). An evaluation of condition assessment protocols for sewer management. Report B-5123.6. Ottawa (Canada): CNRC-NRC. Available online at: http://irc.nrc-cnrc.gc.ca/pubs/fulltext/b5123, 6, b5123.

Rieckermann J., Bareš V., Kracht O., Braun D. and Gujer W. (2007). Estimating sewer leakage from continuous tracer experiments. *Water Research*, **41**, 1960–1972. 10.1016/j.watres.2007.01.024

Roghani B., Cherqui F., Ahmadi M., Le Gauffre P. and Tabesh M. (2019). Dealing with uncertainty in sewer condition assessment: impact on inspection programs. *Automation in Construction*, **103**, 117–126, https://doi.org/10.1016/j.autcon.2019.03.012

Romanova A., Horoshenkov K. V., Tait S. J. and Ertl T. (2013). Sewer inspection and comparison of acoustic and CCTV methods. *Proceedings of the Institution of Civil Engineers – Water Management*, **166**, 70–80, https://doi.org/10.1680/wama.11.00039

Sadeghikhah A., Ahmed E. and Krebs P. (2022). Towards a decentralized solution for leakage detection-a review. *Water Science and Technology*, **86**(5), 1034–1054, https://doi.org/10.2166/wst.2022.263

Schilperoort R. P. S. (2004). Natural water isotopes for the quantification of infiltration and inflow in sewer systems. M.Sc. thesis, Delft University of Technology.

Schmitt T. G., Welker A., Illgen M., Bosseler B., Harting K. and Brüggemann T. (2007). Untersuchung des Abfluss und Versickerungsverhaltens. wasserdurchlässiger Flächenbeläge (Abschlussbericht No. IV-9-042 236 0020). Ministerium für Umwelt und Naturschutz, Landwirtschaft und Verbraucherschutz des Landes Nordrhein-Westfalen (MUNLV), Düsseldorf, Germany.

Schönberger O., Harting K., Gabryl P. and Nolting B. (2005). Prüfung wasserdurchlässiger Flächenbeläge nach mehrjähriger Betriebsdauer. IKT – Institut für Unterirdische Infrastruktur, Düsseldorf, Germany.

Seattle Public Utilities (2009) Green Stormwater Operations and Maintenance Manual, August, 25 p. https://www.seattle.gov/util/cs/groups/public/@spu/@usm/documents/webcontent/spu02_020023.pdf (last access 23.08.2023).

Selvakumar A., Tuccillo M. E., Martel K. D., Matthews J. C. and Feeney C. (2014). Demonstration and evaluation of state-of-the-art wastewater collection systems condition assessment technologies. *Journal of Pipeline Systems Engineering and Practice*, **5**, 04013018, https://doi.org/10.1061/(ASCE)PS.1949-1204.0000161

Shepherd W., Schellart A., Boxall J., Evans M. H., Yu Y. and Tait S. (2021). Passive sensing for buried wastewater collection and drainage networks. Proc 15th Int. Conf. on Urban Drainage, Melbourne, Australia.

Sinha S. and Fieguth P. (2006). Neuro-fuzzy network for the classification of buried pipe defects. *Automation in Construction*, **15**(1), 73–83, https://doi.org/10.1016/j.autcon.2005.02.005

SLG (2020). Merkblatt für versickerungsfähige Pflasterbefestigungen aus Beton: Grundlagen, Planung, Ausführung, Erhaltung. Betonverband Straße, Landschaft, Garten e. V., Bonn, Germany.

Sonkamble S. and Chandra S. (2021). GPR for earth and environmental applications: case studies from India. *Journal of Applied Geophysics*, **193**, 104422, https://doi.org/10.1016/j.jappgeo.2021.104422

Spectar (2023). Spectar Technologies – a Belgium-based startup with a clear mission to half the cost of sewer management in Europe [WWW Document]. https://spectar-technologies.com/ (accessed 14 April 2023)

Stajanca P., Chruscicki S., Homann T., Seifert S., Schmidt D. and Habib A. (2019). Detection of leak-induced pipeline vibrations using fiber optic distributed acoustic sensing. *Sensors*, **18**, 2842, https://doi.org/10.3390/s18092842

Stanić N., de Haan C., Tirion M., Langeveld J. and Clemens F. (2013). Comparison of core sampling and visual inspection for assessment of concrete sewer pipe condition. *Water Science and Technology*, **67**, 2458–2466, https://doi.org/10.2166/wst.2013.138

Stanić N., Langeveld J., Salet T. and Clemens F. (2016). Relating the structural strength of concrete sewer pipes and material properties retrieved from core samples. *Structure and Infrastructure Engineering*, **13**(5), 637–651, https://doi.org/10.1080/15732479.2016.1187631

Stanić N., Lepot M., Catieau M., Langeveld J. and Clemens F. H. (2017). A technology for sewer pipe inspection (part 1): design, calibration, corrections and potential application of a laser profiler. *Automation in Construction*, **75**, 91–107, https://doi.org/10.1016/j.autcon.2016.12.005

Stegeman B., Hoffmann R., Hopman V., Langeveld J. and Clemens-Meyer F. (2022). The assessment of a mobile geo-electrical measurement system: a study on the validity and contributing factors to quantify leakage in sewer systems. *Urban Water Journal*, **19**, 374–387, https://doi.org/10.1080/1573062X.2021.2020298

Tait S. and Kazemi E. (2022). D7.1. Report on methodologies used to identify in-pipe defects (1.0). Zenodo, https://doi.org/10.5281/zenodo.7266706

Tait S. and Kazemi E. (2023). Open Access Coding/Images for Automated detection of in-pipe defects in CCTV sewer surveys (1.0). Zenodo, https://doi.org/10.5281/zenodo.8193032

Thielemann J. T., Breivik G. M. and Berge A. (2010). Robot navigation and obstacle detection in pipelines using time-of-flght imagery. *Proc of The International Society for Optical Engineering*, https://doi.org/10111/12.838979

Thornhill R. and Wildbore P. (2005). Sewer defect codes: origin and destination. Underground Construction.

Towlson A. R. K., Croxford A. J. and Drinkwater B. W. (2022). Ultrasonic non-destructive characterization of blockages and defects in underground pipes. *IEEE Transactions on Ultrasonics, Ferroelectrics, and Frequency Control*, **69**, 2540–2554, https://doi.org/10.1109/TUFFC.2022.3183339

Tscheikner-Gratl F., Caradot N., Cherqui F., Leitão J. P., Ahmadi M., Langeveld J. G., Gat Y. L., Scholten L., Roghani B., Rodríguez J. P., Lepot M., Stegeman B., Heinrichsen A., Kropp I., Kerres K., do Almeida M. C., Bach P. M., de Vitry M. M., Marques A. S., Simões N. E., Rouault P., Hernandez N., Torres A., Werey C., Rulleau B. and Clemens F. (2019). Sewer asset management – state of the art and research needs. *Urban Water Journal*, **16**, 662–675, https://doi.org/10.1080/1573062X.2020.1713382

Van der Steen A., Dirksen J. and Clemens F. (2014). Visual sewer inspection: detail of coding system versus data quality? *Structure and Infrastructure Engineering*, **10**(11), 1385–1393, https://doi.org/10.1080/15732479.2013.816974

Veldkamp T. I. E., Boogaard F. C. and Kluck J. (2022). Unlocking the potential of permeable pavements in practice: a large-scale field study of performance factors of permeable pavements in The Netherlands. *Water*, **14**, 2080, https://doi.org/10.3390/w14132080

Vermeulen D. (2022). Detection of leaks: next-generation pipe inspection: Testing the FELL system under controlled field conditions. MSc Thesis, Available at http://resolver.tudelft.nl/uuid:b062ca56-d5aa-4bf2-9fa2-e20503545dd5 (accessed 14 September 2023).

Vollertsen J. and Hvitved-Jacobsen T. (2003). Exfiltration from gravity sewers: a pilot scale study. *Water Science and Technology*, **47**(4), 69–76, https://doi.org/10.2166/wst.2003.0223

Vosse M., Schilperoort R., de Haan C., Nienhuis J., Tirion M. and Langeveld J. (2013). Processing of DTS monitoring results: automated detection of illicit connections. *Water Practice and Technology*, **8**(3–4), 375–381, https://doi.org/10.2166/wpt.2013.037

Wang Y., Li P. and Li J. (2022). The monitoring approaches and non-destructive testing technologies for sewer pipelines. *Water Science and Technology*, **85**(10), 3107–3121, https://doi.org/10.2166/wst.2022.120

Weiß G., Brombach H. and Haller B. (2002). Infiltration and inflow in combined sewer systems: long-term analysis. *Water Science and Technology*, **45**(7), 11–19, https://doi.org/10.2166/wst.2002.0112

Wilmut C. G., Dsouza B. J. and Guidry B. A. (2005). Electro-scan technology – is it the "silver bullet" for rainfall dependent infiltration (Rdi) and exfiltration? *Proceedings of the Water Environment Federation*, **2005**(4), 697–706, https://doi.org/10.2175/193864705784291529

Winston R. J., Hunt W. F., Kennedy S. G., Wright J. D. and Lauffer M. S. (2012). Field evaluation of storm-water control measures for highway runoff treatment. *Journal of Environmental Engineering*, **138**, 101–111, https://doi.org/10.1061/(ASCE)EE.1943-7870.0000454

Wirahadikusumah R., Abraham D. M., Iseley T. and Prasanth R. K. (1998). Assessment technologies for sewer system rehabilitation. *Automation in Construction*, **7**, 259–270, https://doi.org/10.1016/S0926-5805(97)00071-X

Wolf L. (2003). Anwendung einer segmentierten geoelektrischen Sonde zur Leckagendetektion in Abwasserkanälen. (Application of a segmented geo electrical sensor for leakage detection in sewers). *Technisches Messen*, **70**(7–8), 346–351, https://doi.org/10.1524/teme.70.7.346.22647

WRc (2013). Manual of Sewer Condition Classification. WRc plc, Wiltshire, U.K.

WRc (2022). Sewerage Risk Management manual. Available at: https://srm.wrcplc.co.uk/

Yu Y., Worley R., Anderson S. and Horoshenkov K. (2023). Microphone array analysis for simultaneous condition detection, localization, and classification in a pipe. *The Journal of the Acoustical Society of America*, **153**, 367–383, https://doi.org/10.1121/10.0016856

Zhang S., Liu B. and He J. (2019). Pipeline deformation monitoring using distributed fiber optical sensor. *Measurement*, **133**, 208–213, https://doi.org/10.1016/j.measurement.2018.10.021

Zhao J. Q., McDonald S. E. and Kleiner Y. (2001). Guidelines for Condition Assessment and Rehabilitation of Large Sewers. Institute for Research in Construction, National Research Council Canada, Ottawa, Canada.

doi: 10.2166/9781789063059_0131

Chapter 5

Deterioration processes and modelling in urban drainage systems

S. Barraud[1], E. Bosco[2], F.H.L.R. Clemens-Meyer[3,4], G. Fernandes[5], Y. Le Gat[6], C. de Haan[7], R.A. Luimes[2], K. Makris[8], F.A.M. Rooyackers[2], I.C. Scheperboer[9], J. Skjelde[10] and A.S.J. Suiker[2]

[1]University of Lyon, Insa Lyon, DEEP, 1 1 Rue de la Physique, F-69621 Villeurbanne Cedex, France
[2]Department of the Built Environment, Eindhoven University of Technology, PO Box 513, 5600 MB Eindhoven, The Netherlands
[3]Department of Civil and Environmental Engineering, Norwegian University of Science and Technology, PO Box 8900, NO-7491 Trondheim, Norway
[4]SkillsInMotion BV, Esdoornlaan 11, 43454 HH De Meern, The Netherlands
[5]Univ Lyon, Université Claude Bernard Lyon 1, CNRS, ENTPE, UMR 5023 LEHNA, F-69518 Vaulx-en-Velin, France
[6]INREA-ETTIS, 50 avenue de Verdun, 33612 Cestas Cedex, France
[7]De Haan Watersolutions, Javastraat 104A, 6524 MJ Nijmegen, The Netherlands
[8]Partners4UrbanWater, Graafseweg 274, 6532 ZV Nijmegen, The Netherlands
[9]IKT Nederland, Tivolilaan 205, 6824 BV Arnhem The Netherlands
[10]COWI AS, PO Box 4220 Torgarden, N-7436 Trondheim, Norway

ABSTRACT

This chapter discusses and assesses processes underlying the loss of functionality of (elements) of urban drainage systems. The chapter does not pretend to cover all literature nor do the authors claim to cover all processes involved. As the body of literature experiences a rapid growth, the authors want to provide the reader a comprehensive overview of the literature to-date to be used as an entry for further exploration. Several processes are described and the possibility to model them in a deterministic manner is discussed. The last sections provide a brief introduction to the use of statistical models based on condition classes (as is applied in practice) along with a brief overview of advancements made in machine learning applications.

Keywords: processes, timescales, deterioration, modelling.

Symbol*	Description	Units
E_c	Young's modulus (concrete)	N/mm^2
G	Gravitation acceleration	m/s^2
$G_{I,c}$	Mode I toughness	N/mm^2
H	Capillar pressure head, height	m
H	Hydraulic head	m
$i(t)$	Infiltration as function of time	m/s

(Continued)

Symbol*	Description	Units
I	Cumulated infiltration	m
K_s	Hydraulic conductivity at saturation	m/s
M	Shape parameter	–
M_w	Water mass	kg
M_s	Solid mass	kg
N	Shape parameter	–
R	Radius	m
R	Isotope ration	–
T	Time	s
Q	Flux	m/s
$Q(t)$	Flow rate as function of time	m^3/s
S	Surface area	m^2
S_e	Effective saturation	
V_a	Air volume	m^3
V_f	Fluid volume	m^3
V_t	Total soil volume	m^3
V_w	Water volume	m^3
Z	Elevation	m
α	Empirical parameter, surface tension	m^{-1}
ϕ	Porosity	–
θ	Volumetric water content	–
θ_r	Water content at saturation	–
θ_s	Saturation	–
ρ	Mass density	kg/m^3
ρ_b	Dry bulk density	kg/m^3
ρ_p	Particle density	kg/m^3

*Other symbols are explained in the text where appropriate.

5.1 INTRODUCTION AND LAYOUT OF THE CHAPTER

Often the term 'ageing processes' is used in practice when it comes to describe changes in systems' functionality and connected asset management (AM) activities. The use of ageing has an implicit meaning attached, as it suggests these processes are all time driven, which to a certain extent is correct. However, it does not imply that all processes involved can be modelled as if they are *continuous* in time or are inherently time dependent. To avoid this bias, we will use the term *deterioration process* in this chapter.

Deterioration refers to a spectrum of processes, or sequences of events, that affect the functionality of either elements in a system, part of a system or a system as a whole. Such a definition encompasses a very wide range of topics. The present chapter does not pretend to cover all possible processes, or combination of processes that may be relevant in the context of urban drainage (UM). The reason for this is simple: we cannot be sure we are aware of all processes that contribute to deterioration, but we are aware that, when it comes to the processes we are familiar with, knowledge is limited at best (i.e., there is a variety of known or suspected unknowns, and most likely a lot of unknown unknowns as well). A further complication is the diversity of processes encountered: physical, chemical, and biological processes and combinations or sequences of these, sometimes in concert with a dose of human ignorance, are all in play. In addition, our systems consist of very diverse parts ranging from pipes to nature-based solutions (NBS) and, therefore, the diversity of failure mechanisms would overwhelm the scope of this book.

Consequently, this chapter merely tries to provide the reader with a selective overview of processes involved in deterioration in terms of characteristics (like time scales involved, or how to recognize a certain process or chain of events leading to a loss of functionality), introducing the very wide field of deterioration processes and associated models, touching upon statistical and deterministic models highlighting their strong and weak points. Sections 5.2 and 5.3 discuss the infiltration and clogging processes that are of importance in many NBS that have been implemented over the past few decades. Sections 5.4, 5.5, 5.6 and 5.7 discuss a range of ageing processes linked to PVC and concrete pipes (i.e., the most popular materials applied in UD) along with a range of processes influencing the properties of materials. Further processes like sedimentation, formation of FOG (i.e., fats, oils and grease) deposits is discussed, illustrated with case studies where available or appropriate. For some aspects comprehensive textbooks are available; for example, in the case of sewer sediments the reader is referred to Ashley *et al.* (2004), where it comes to micro-biological processes the reader is referred to Hvitved-Jacobsen *et al.* (2013).

Section 5.8 presents processes involved in loss of hydraulic capacity, while Section 5.9 goes into separation of flows and why they tend to keep mixing. Section 5.10 presents an introduction to statistical modelling of (piped) UD systems as there is not yet sufficient data available for NBS to merit such an approach. This is the most applied approach in practice although some understanding or knowledge of the underlying processes can be beneficial. A good understanding of these processes is critical when making decisions about AM activities (e.g., when the reasons for failure of an element are clear, this knowledge tends to provide important information on how to remedy the failure and at what time scale action should be taken; statistical models do not provide that sort of information). In Section 5.11 a brief overview of the application of machine learning techniques for deterioration modelling in Urban Drainage Asset Management (UDAM) is provided.

Statistical models are typically based on a set of reference images that allow for the identification and scoring of defect types, and subsequently the identification of a condition class (often from 0 to 5) indicating their 'hazard' level, with the lowest (or highest depending on the standard) condition class indicating an (almost) complete loss of functionality. This approach is basically descriptive, solely looking at the end result of an unknown number of deterioration processes. No information is obtained that allows for the quantification of a loss of functionality, for example, Class 4, which is caused by corrosion, does not denote the hydraulic roughness of a pipe wall nor the remaining 'healthy' material, nor the actual mechanical properties of the construction material.

This may be seen as the main reason that, for all practical purposes in UDAM, deterioration is modelled as a time-dependent process using statistical models that use a pool of (visual) observations to make prediction on the development of the condition class of elements for planning purposes (see e.g., Caradot *et al.*, 2020). Nevertheless, detailed deterministic descriptions of process and subsequent modelling of some of the processes involved in deterioration are available and can be used in AM to a certain extent. Their added value is found in:

- Knowledge of time and space scales of processes can be beneficial when judging a given situation.
- Understanding which processes may be responsible for the evolution of the technical state in a given situation, allowing for identifying and possibly removing the cause of the deterioration.

Given the complexity and the apparent lack of detailed knowledge, using very detailed deterioration models for practical applications seems unrealistic. Instead, a considerable body of literature is available on statistical models describing 'ageing' based on inspection data. For instance, applying condition classes discussed in Chapters 4 and 8. Apart from the complexity and limitations with respect to practical applicability, there are strong indications that processes occur in UD which are chaotic by nature. This implies that there is a fundamental limit regarding the prediction horizon of models representing these processes (e.g., rainfall, rainfall-runoff processes or the transport of solid objects in complex flow fields that occur in UD systems [see e.g., Rodrigues-Iturbe *et al.* 1989; Sivakumar *et al.* 2009, or Duinmeijer and Clemens, 2021]). To complicate matters even more, deterioration of UD systems may well be the result of failure of other urban infrastructures (see e.g., van Riel, 2016). This aspect, however, will not be addressed in this chapter to limit the scope as to what reasonably can be covered.

The interested reader who wants to keep up with the developments in the field is referred to the proceedings of the IWA/IAHR conference series *Sewer Processes and Networks* that is organized on a regular basis by the Working Group on Sewer Processes and Networks (https://www.sspwg.org/) and the activities of the IWA/IAHR working group on Urban Drainage Asset Management (https://udam.home.blog/)

5.2 INFILTRATION

5.2.1 Introduction

The process of infiltration is one of the key processes in UD systems, even in the classical (piped) systems where infiltration plays a role in the rainfall-runoff process. It is a major component of recently developed concepts such as 'sponge cities' (see e.g., Jiang *et al.*, 2018). In fact, in many NBS, infiltration is the essential process that is relied upon. It must be realized however that the quantification of the infiltration capacity is not straight forward and is inhomogeneous in time and space. A further aspect of infiltration is clogging which can be seen as the main factor that reduces the infiltration capacity over time.

Infiltration systems such as infiltration basins, raingardens, bioretention systems, swales, trenches, filter strips, soil cells with trees or shrubs, permeable pavements and so on have become widely implemented in urban strategies to manage stormwater over the past few decades.

They are recognized to provide significant benefits. For instance, they contribute to flood reduction when they are part of integrated stormwater management strategies (Davis *et al.*, 2012; Petrucci, 2012). They reduce the degradation of receiving water bodies (i.e., through smaller volumes of water and associated pollutant loads that are directly discharged), help to restore baseflows, which in some cases can be depleted due to urbanization (Hamel & Fletcher, 2014; Kauffman *et al.*, 2009; Smakhtin 2001) and can contribute to groundwater recharge (Bonneau *et al.*, 2018). When designed as NBS using vegetation, they also provide multiple co-benefits like enhanced landscape aesthetics, biodiversity preservation or creation and mitigation of the urban heat island (Oral *et al.*, 2020).

While they have become increasingly implemented, a recurrent issue is their durability over time, particularly due to the potential decrease in their permeability, which can jeopardize:

(1) Their hydraulic functioning. If clogging occurs, less water can be infiltrated, possibly resulting in permanent ponding and associated urban flooding.
(2) Their performance in terms of water quality. As already mentioned, an infiltration device is effective in stormwater treatment as a result of both hydrological processes (i.e., reduction of stormwater volume discharged to receiving water bodies leading to a reduced discharge of pollutants transported by the runoff) as well as physical and chemical processes (such as filtration, adsorption or precipitation) that occur during the percolation of water through a substratum or a soil. When clogging occurs, less water can be infiltrated, and the excess water is diverted or remains at the surface without treatment.
(3) Their aesthetic aspect and other uses of the space. Standing water is not only unaesthetic but it can also compromise the other uses of the space taken up by these assets when they are multifunctional. It can negatively impact vegetation and, if the water remains too long on the surface, it could contribute to the proliferation of mosquitoes (Al-Rubaei *et al.*, 2015; Valdelfener *et al.*, 2018). However, this potential proliferation only occurs when standing water remains for at least 5 days (Valdelfener *et al.*, 2018).

5.2.2 Soils

The Food and Agriculture Organization of the United States defines soils as '*a natural body consisting of layers (soil horizons) that are composed of weathered mineral materials, organic material, air and water. Soil is the product of the combined influence of climate, topography, organisms (flora, fauna and human) on parent materials (original rocks and minerals) over time. As a result, soil differs from its parent material in texture, structure, consistency, colour, chemical, biological and physical characteristics.*' Soil thus is a living medium, at the interface between the atmosphere and the bedrock.

Soil provides numerous ecosystem services such as nutrient cycling, provision of food, fibre, fuel and construction materials. It also has hydrological functions such as storing water for plants and fauna, regulating the water cycle including reducing floods, regulating the climate (mainly by evapotranspiration) and adsorbing and/or facilitating the breakdown of pollutants which limits groundwater contamination. However, human activities tend to negatively impact its quantity and quality as both freshwater and soil are increasingly consumed by the increasing number of humans. The Intergovernmental Panel on Climate Change report (Olsson *et al.*, 2019) on climate change and land shows that human activity has caused soil degradation on about 25% of the Earth's surface not covered by ice. For example, with monoculture, biodiversity and fertility of soil disappear and the soil surface is less strong and stable (Kutílek & Nielsen, 2015).

Soil is composed of gas, liquid and solid. The liquid and gas phases represent 30–60% of the soil volume. Some air is maintained in the soil, with a slightly different composition from the atmosphere. When water enters the unsaturated part of soil (i.e., the so-called vadose zone), the air within the pores is pushed away and replaced by water. Water can be found in soils in its liquid phase incorporating dissolved compounds. Its quantity and quality vary in the soil on both temporal and spatial scales as the result of climate, hydrological and paedological processes.

As a result of its granular nature, a soil can be seen as an assembly of particles (Figure 5.1). Basically, soils may be differentiated by two key characteristics: their texture and structure. The texture depends on both the type and shape of particles, differentiated according to their mean diameter size. According to the United States Department of Agriculture soil texture classification, clays have a diameter smaller than 0.002 mm, silt between 0.002 and 0.05 mm, and sand between 0.05 and 2 mm. In addition to the differences in size and shape of the particles, their packing is often irregular, creating the voids and channels that make up the pores (Figure 5.1), which is measured by the porosity (Table 5.1). The arrangement of the particles in the soil creates its structure. These two characteristics greatly influence the movement of water into and through the soil.

Soils can be analysed by measuring different parameters such as porosity, water content and effective saturation (Table 5.1).

The total porosity can be determined knowing the dry bulk density of the soil: in the field, a known volume of soil is excavated and then weighed, dried 24 h at 105°C and weighed again (Blake & Hartge, 1986). Using a gas pycnometer requires a sample chamber and a reservoir where a change of pressure is made to determine the volume (Danielson & Sutherland, 1986), thus allowing the determination of the particle density (after measuring the mass of the sample).

The water content can be determined most of the time by removing the water from the sample, as detailed by Gardner (1986), with direct and indirect methods. In the field, it can be measured

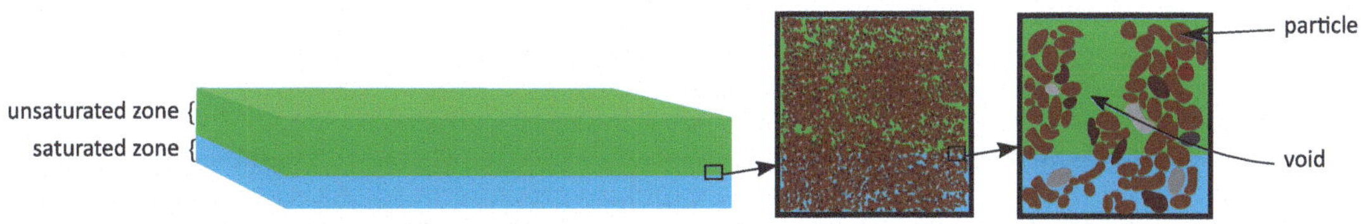

Figure 5.1 Scheme of soil texture and structure.

Table 5.1 Key soil parameters with the fluid volume V_f, the total volume V_t, the water volume V_w, the air volume V_a and the solid volume V_s, the dry bulk density ρ_b, the particle density ρ_p, the water mass M_w, the solid mass M_s, the water density ρ_w, the relative water content θ_r and the water content at saturation θ_s.

Parameter	Porosity (L^3/L^3)	Mass Water Content (M^3/M^3)	Volumetric Water Content (L^3/L^3)	Effective Saturation (No Dimension)
Calculation	$\varphi = \frac{V_f}{V_t} = \frac{V_w + V_a}{V_w + V_a + V_s}$ $= 1 - \frac{\rho_b}{\rho_p}$	$w = \frac{M_w}{M_s}$	$\theta = \frac{V_w}{V_s}$ $= \frac{\rho_b}{\rho_w} w$	$S_e = \frac{\theta - \theta_r}{\theta_s - \theta_r}$
Determination method	Total porosity with particle and bulk densities or gas pycnometer	Direct method: gravimetry, drying Indirect method: electrical conductivity, capacitance, neutrons probes…	Linked to the mass water content	Linked to the volumetric water content

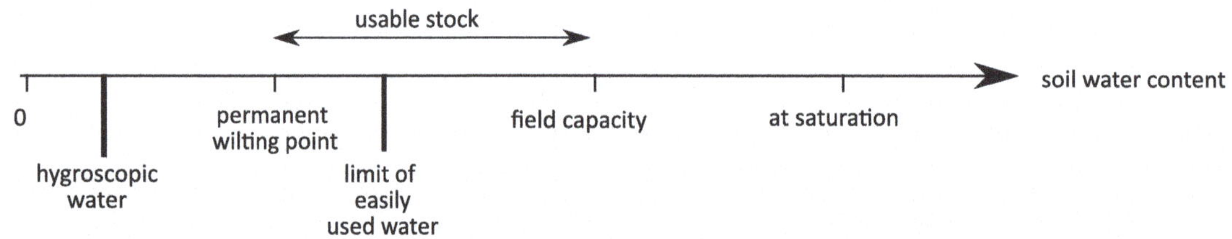

Figure 5.2 Key values of water content in soils linked to the capacity for vegetation to access it.

by a neutron probe (Bell, 1976) and other capacitance probe techniques (Robinson & Dean, 1993) like time-domain reflectometry (Whalley, 1993). The soil-water content is, for example, used to characterize the soil's ability to support vegetation growth (Cassel & Nielsen, 1986): between two values of water content (i.e., corresponding to the field capacity and the permanent wilting point of the soil), the vegetation can access the water they need (Figure 5.2). In hydrology studies, when soil is naturally saturated, the volumetric soil water content is called 'at saturation' and is written as θ_s. Thus, θ_s is the maximum water content value reached for a given soil under natural conditions.

The soil composition tends to be highly heterogeneous with different lithofacies (material). Sedimentological studies aim to characterize the geological and sedimentological features of soils, that is, the lithofacies and their arrangement in soil (Goutaland *et al.*, 2007). The groundwater table is defined by the elevation of the saturated zone where the volumetric water content is equal to the maximum value θ_s. If this value is not reached, the water content is lower than θ_s and the soil is unsaturated. The unsaturated zone of soil composing the first metres is also called the vadose zone (Figure 5.1). With respect to the role of soils in UD the prime focus is on the unsaturated parts of the soil.

5.2.3 Hydraulic properties of soils

Another important soil parameter, the water pressure head, h [L], is used to characterize the energy status of water into the soil and represents the capillary pressure head. As water moves slowly in soils, the kinetic energy is assumed negligible and the hydraulic head H [L] is then the sum of the pressure head (i.e., capillary pressure head), the elevation and the osmotic force. For a soil in osmotic equilibrium (i.e., with the same concentrations or equivalently similar quality of interstitial water), the total hydraulic head is given by:

$$H = h + z \quad (5.1)$$

where z [L] is the elevation and h [L] is the capillary pressure head. In the unsaturated zone, the pressure head h is negative, and its absolute value is called the water suction. Water flow is due to the gradient of the total hydraulic head H and the movement of water goes from high to low hydraulic heads (Angulo-Jaramillo *et al.*, 2016; Bruand & Coquet, 2005: *Les sols et le cycle de l'eau*; Hillel, 1998).

The capillary pressure head at standard temperature (20°C) and pressure conditions can be estimated for a vertical cylinder of radius r with the Jurin–Laplace law at standard conditions for temperature and pressure as

$$h = \frac{2\sigma\cos(\theta)}{r\rho g} \sim \frac{1.49}{r} \quad (5.2)$$

where both h and r are expressed in cm, with h corresponding to the height of liquid in the cylinder above a free water reference level, σ is the surface tension, θ is the contact angle between the water and the cylinder wall, ρ is the mass density and g is the gravitational acceleration. Note that the pressure head is inversely proportional to the cylinder radius (Equation (5.2)). Thus, if we consider soil porosity as a set of cylinders of different diameters, the finer the pore diameter, the lower the pressure head needed to activate (i.e., saturate) the pore. In other words, when the soil dries, the larger pores empty before the smaller ones.

A water retention curve defines, at soil–capillary water equilibrium, the relationship between the volumetric water content θ and the absolute value of the pressure head $|h|$. The S-shape of the curve depends strongly on both the texture and the structure of the soil (Figure 5.3a, b). Clays retain more water than sand and desaturate less easily. A less compacted soil for the same soil texture can have a higher water content at saturation, because of a higher porosity. Moreover, a phenomenon of hysteresis exists and changes the shape of the retention curve, as water is better retained by drying than by wetting soil (Figure 5.3c). Historically, Buckingham (1907) was the first to establish retention curves based on laboratory experiments on different textures of soils.

Van Genuchten (1980) developed the following analytical model for water retention curves:

$$\frac{\theta - \theta_r}{\theta_s - \theta_r} = \left(1 + |\alpha h|^n\right)^{-m} \quad (5.3)$$

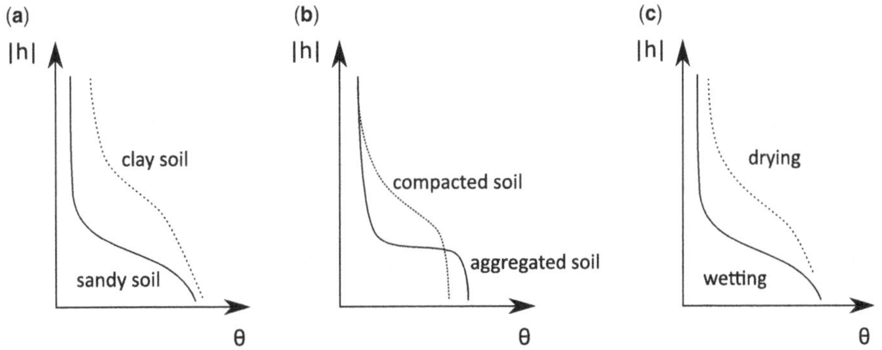

Figure 5.3 Influence of (a) texture and (b) structure on the water retention curve. (After Angulo-Jaramillo *et al.*, 2016).

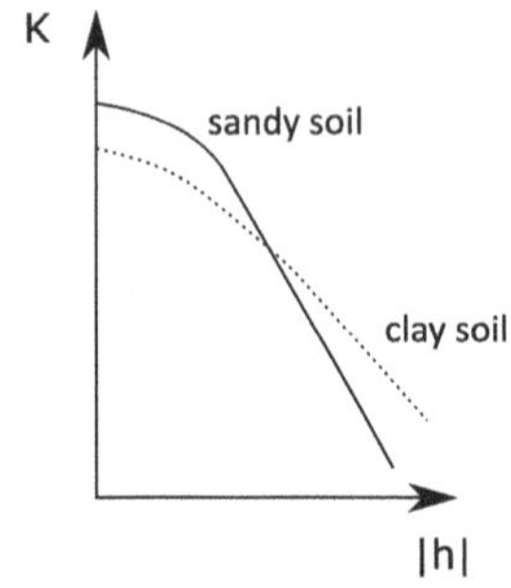

Figure 5.4 Water hydraulic conductivity curve.

with θ_r is the residual volumetric water content (at negative infinite pressure head, smaller or equal to the wilting point), θ_s is the saturated volumetric water content, α is an empiric parameter (per L) also called scale parameter for water pressure head and n and m is the shape parameters of the curve.

The hydraulic conductivity, K (L/T) describes the ability of a fluid to move through pores. The hydraulic conductivity is calculated as the ratio of water flow to hydraulic gradient. It is a function of the pressure head and has the same dimension as a velocity (L/T). This parameter highly depends on the water content in the soil, and the hydraulic conductivity at saturation K_s as an intrinsic parameter of soils. The water hydraulic conductivity curve represents the link between the hydraulic conductivity and the pressure head or water content (Figure 5.4). Indeed, when the soil is unsaturated, only part of the pores is activated and conducts water, resulting in a lower hydraulic conductivity. Conversely, at saturation, all pores are activated and the hydraulic conductivity increases.

The Darcy equation (see e.g., Hillel, 1998) represents steady water flow conditions in a vertical soil column, with z in an upward direction, under uniform and saturated soil conditions:

$$q = -K_s \frac{\partial H}{\partial z} \tag{5.4}$$

with q is the flux [L/T], $K_s$is the hydraulic conductivity at saturation and $\partial H/\partial z$ is the vertical hydraulic gradient. The minus sign means that water always moves from high to low values of total hydraulic head.

Under unsaturated conditions, the hydraulic conductivity depends on pressure head (or the water content), and Darcy's law can be generalized as

$$q = -K(h) * \frac{\partial H}{\partial z} = -K(h) * \left(\frac{\partial h}{\partial z} + 1\right) \tag{5.5}$$

Alternatively, $K(h)$ may be written as a function of the water content, $K(\theta)$. This last relationship is preferred since hysteresis affects the function $K(\theta)$ less than $K(h)$. For one-dimensional vertical flow, the combination of the continuity equation (Eq. 5.6) with the generalized Darcy's law (Equation (5.5)) gives the Richards equation that governs the flow of water in soils (Equation (5.7)).

$$\frac{\partial \theta}{\partial t} = -\frac{\partial q}{\partial z} \tag{5.6}$$

$$\frac{\partial \theta}{\partial t} = -\frac{\partial}{\partial z}\left(-K(h) * \left(\frac{dh}{dz} + 1\right)\right) \tag{5.7}$$

The Richards equation is valid for non-deformable soils (e.g., non-swelling and shrinking soils), under isothermal conditions, and ignores the interaction with the air flow in the soil pores, when water pushes the air away. For more details on the Richards equation refer to Angulo-Jaramillo *et al.* (2016). The solution of the Richards equation consists of calculating the changes in h and θ in both space and time using the hydraulic characteristic curves $h(\theta)$ and $K(\theta)$ of the soil and the initial and boundary conditions.

The pair of characteristic curves $h(\theta)$ and $K(\theta)$ (or $h(\theta)$ and $K(h)$) fixes the hydrodynamic identity of an unsaturated soil. Their estimation is one of the challenges in hydrology and hydrogeology. Different in situ experimental tests are used to estimate the parameters of the relationships (Equation (5.7)) from the analysis of the measured infiltration flux (see paragraph 5.2.5).

5.2.4 Infiltration component of the water cycle

Infiltration concerns the physical process of water entry into the soil from the soil surface. The infiltration flux or Darcy's velocity corresponds to the flow rate entering the soil divided by the infiltration surface:

$$i(t) = \frac{Q(t)}{S} \tag{5.8}$$

Cumulated infiltration I (L) is the integral over time of the infiltration flux (Equation (5.9)), which implicitly corresponds to the total volume of infiltrated water divided by the soil surface. It has the same unit as rainfall amount, that is, the unit of length.

$$I(t) = \int_{t=t_0}^{t} i(t)\mathrm{d}t \tag{5.9}$$

Water infiltrates first through the surface pores and moves vertically or horizontally because of gravitational or suction forces, thus replacing the air in the pores. When an amount of water is applied at the surface, a wetting front develops in the soil profile which moves downward. The soil between the surface and the wetting front is wetted at a water content that is in equilibrium with the water pressure head applied at the surface (assuming this remains constant over time).

The three variables detailed above, θ, h and K_s, characterize infiltration of water into the soil and the redistribution of this water after an infiltration (or rainfall or irrigation event). Infiltration thus depends on the capacity of the soil to let water move (i.e., as a function of its texture and structure) and on the changes in the energetic state between the surface, the groundwater and the unsaturated parts (Bruand & Coquet, 2005). The speed of infiltration depends also on the initial soil water. Other parameters not considered in the above equations and models include physical soil properties, particle dispersion, air retention, temperature and the presence of dissolved gases or salts in the water, and so on (Chossat, 2005).

For regular cases, where the infiltration of water is governed by capillary forces and gravity, allowing for the application of Richards' equations, the cumulative infiltration curve, $I(t)$, increases with an initial convex shape followed by an asymptote with a constant slope, illustrating the attainment of steady state conditions when water infiltrates mostly by gravity. Different equations exist to represent cumulative infiltration at the soil surface. Some are empirical, directly related to the soil in question studied (e.g. Horton, 1941), while others reflect the physics of the infiltration and movement of water in soils (e.g. Green and Ampt, 1911). However, many of these equations consider the soil as homogeneous, with a single porosity and hydraulic conductivity, and overlook the complexity and heterogeneity of the soil structure and texture. As water infiltrates mainly by capillary forces and gravity (Angulo-Jaramillo *et al.*, 2016), Philip (1957) was first to account for both the vertical and the horizontal infiltration components, introducing sorptivity as a '*measure of the capillary uptake or removal of water*'. Sorptivity is a crucial parameter with regards to the modelling of water infiltration into soils (Angulo-Jaramillo *et al.*, 2016).

5.2.5 *In situ* measurements and soil hydraulic characterization

5.2.5.1 *Field methods*

To determine the cumulative infiltration of water into soils in the field, the main concept consists of applying a known volume of water and recording the time needed for its complete infiltration. Two types of experimental methods exist. The first one, the unsaturated method, applies a negative pressure head on the soil surface, activating only the thinnest pores of the soil matrix. The wetting bulb created under the surface during the experiment is unsaturated and involves mostly the soil matrix. The second one, the saturated method, applies a null or positive pressure head on the soil, activating all pores, even the macropores. For both methods, the infiltrated volume is recorded as a function of time. The pressure boundary condition applied at the surface can be constant or variable over time.

The infiltration can take from a few minutes to several hours, depending on the volume injected, the initial soil condition or the intrinsic hydraulic properties of the soil. The spatial scale is also an important parameter to consider in determining the hydraulic parameters of the soil (Glaser *et al.*, 2019). Presented in this section are some field methods for both saturated and unsaturated soil conditions. In the field, most devices allow one to determine the infiltration rate $i(t)$ (L/T) and the cumulative infiltration $I(t)$ (L) [Equations (5.8) and (5.9), respectively]. The hydraulic properties of the soil are then estimated by the analysis of the measured $i(t)$ or $I(t)$ curves and fitting them to infiltration models.

One of the most used methods in civil engineering is called the Porchet essay (or 'percolation rate'). It consists of a hole filled with water until steady state conditions are supposed to have been reached. Then, a known volume of water is poured into the hole, and the infiltration over time is recorded. With Darcy's equation and the hypothesis of a uniform vertical hydraulic gradient, the hydraulic conductivity can be calculated. This method is easy and cheap but requires huge amounts of water and time. Moreover, it assumes both steady-state and one-dimensional flow conditions. Nevertheless, it is widespread and has been used for a long period of time for measuring infiltration fluxes between 0 and 10 m/day (Chossat, 2005).

Table 5.2 shows a brief overview of the main methods used in the field for the determination of the hydraulic parameters of unsaturated soils. In his book Chossat (2005) presented a table comparing the various devices found in the literature. Deb and Shukla (2012) also wrote a review of some devices.

5.2.5.2 *Modelling water infiltration to estimate soil hydraulic parameters*

The BEST (Beerkan Estimation of Soil Transfer parameters) protocol is one example of the various methods developed to determine water retention and hydraulic conductivity curves (Angulo-Jaramillo *et al.*, 2016). The BEST method combines the van Genuchten (1980) equation with the conditions of Burdine (1953) ($m = 1\text{–}2/n$ in Eq. 5–3) in describing the water retention curve, $K(\theta)$, and with the Brooks and Corey (1964) equation in describing the unsaturated hydraulic conductivity, $K(\theta)$. The method uses a single ring infiltration experiment (i.e., a Beerkan test) to measure the infiltration curve and combines the data with additional field measurements including the particle size distribution, soil dry bulk density, and the initial and final water contents. The particle size distribution is modelled analytically to estimate the shape parameters of the water retention and hydraulic conductivity curves $h(\theta)$. The cumulative infiltration curve is then fitted to analytical models to derive the hydraulic conductivity and the sorptivity. The BEST method is one example of an analytical inversion method for water infiltration. Three BEST methods can be considered, BEST-Slope, -Intercept and -Steady, which are all based on the same principle. They only differ on how the analytical infiltration model is adjusted to the experimental cumulative infiltration data. The three BEST methods have been coded in Scilab (https://www.scilab.org/) with an easy-to-use graphic interface. All codes are open-source and can be found on the BEST website (https://bestsoilhydro.net/)

Šimůnek *et al.* (2008, 2016) developed a commercially available software package called HYDRUS that models the transport of heat, water, and solutes in a 1-D, 2-D or 3-D variably saturated medium. The Richards' equation for variably saturated water flow and advection-dispersion type

Table 5.2 Main field devices with their characterization.

	Name	Principle	Tools/Device	Hydraulic Conductivity Range, Soil Type and Experimental Period	Advantages	Disadvantages	References
Surface infiltration	Tension infiltrometer	Infiltration at constant negative pressure head	Mariotte tower and water reservoir above a porous plate, from a few to a dozen cm Commercialized and possibly automatized	10^{-4}–10^{-7} m/s (minidisc infiltrometer) sand and silt approx. 30 min	Easy to use in the field more relevant results for near-saturated conditions one experimenter is enough tested and validated for different soil types and climates	Not representative of macropores can be perturbated by the sand layer which is added to improve conductivity	Angulo-Jaramillo *et al.* (2016); Ankeny *et al.* (1988); Castiglione *et al.* (2005); Chossat (2005)
	Single ring	Infiltration at null or low pressure head water decreases in the ring (variable charge) or in a reservoir (constant charge)	A ring from 5 to 70 cm of diameter and 10–30 cm of height in general in inox with or without reservoirs Commercialized and possibly automatized	All soils 30 min to a few hours	Easy to use in the field precise result for the hydraulic conductivity at saturation one experimenter is enough tested and validated for different soil types and climates for all types of pores	Tends to overestimate hydraulic conductivity at saturation can compact the soil can create preferential flows along the edges of the ring water can also put particles in suspension, which can clog the soil	Bouarafa *et al.* (2019); Chossat (2005); Jačka *et al.* (2014); Zhao *et al.* (2020a, 2020b)

(Continued)

Table 5.2 Main field devices with their characterization (*Continued*).

	Name	Principle	Tools/Device	Hydraulic Conductivity Range, Soil Type and Experimental Period	Advantages	Disadvantages	References
	Double ring	Infiltration at null or low pressure head considers the flow in the inner ring as 1D	Two concentric rings, one small in the centre (10–30 cm) and a biggest around (30–60 cm) Commercialized	10^{-4}–10^{-8} m/s 30 min to a few hours	Is supposed to be more precise and surface evaporation is limited tested and validated for different soil types and climates	Can compact the soil can create preferential flows along the edges of the ring water can also put particles in suspension, which can clog the soil limited representativeness of soil pores heterogeneities	Chossat (2005); Gregory *et al.* (2005)
Borehole infiltration	Guelph permeameter	Infiltration in a borehole constant pressure head	Tube of a few cm of diameter and high height (until 2 m) Commercialized	10^{-4}–10^{-7} m/s all soils approx. 1 hour	Tested and validated for different soil types and climates for all types of pores	Can destruct and compact the soil overestimates horizontal flows compared with vertical ones water can also put particles in suspension, which can clog the soil	Bosch and West (1998); Chossat (2005); Kodešová *et al.* (2010); Reynolds and Lewis (2012)
	Philip–Dunne permeameter	Infiltration in a borehole at falling pressure head	Flat hole with a few cm diameter	10^{-4}–10^{-6} m/s Silt loam, Sandy loam and sand	Simple measure of the hydraulic conductivity at saturation for soils with a small permeability tested and validated for different soil types and climates	Tube's installation can be complicated and allow for leaks to occur, similar to the Guelph permeameter water can also put particles in suspension, which can clog the soil	Alakayleh *et al.* (2019); Chossat (2005)

equations for heat and solute transport are numerically solved by HYDRUS. The flow equation can also account for dual-permeability or dual-porosity type flow. This software has been intensively used for the context of infiltration and solutes transfer modelling, with various add-on modules that expand its capabilities (Šimůnek *et al.*, 2016; Yilmaz *et al.*, 2013).

5.2.6 Addressing the complexity of processes

5.2.6.1 Pore characterization

The models and equations presented above are almost all based on Richards' equation and do not consider the heterogeneity of soils found in the field, according to Beven and Germann (1982). Thus, in the years after the publication of their paper, different equations were developed to represent macropores and preferential flows more precisely. As soils in the field are *never* homogeneous, they typically present pores of different sizes. The variety of pores may change the ability of the soils to move water as a function of associated water repellence, cracks and clogging.

Different types of pores are often distinguished: (I) the micropores, which arise from the inhomogeneous packing of the particles and which are responsible for governing water and solutes retention; (ii) the macropores, which are larger and are mainly created by roots, animal galleries and fractures; and finally, (iii) the mesopores, which are an intermediate class between the two previous classes. The Soil Science Society of America (https://www.soils.org/publications/soils-glossary) classifies the pore sizes as: micropores between 5 and 30 μm, mesopores between 30 and 75 μm, very fine to fine macropores between 75 and 2000 μm, and medium and coarse macropores above 2000 μm. The differences between the pore sizes are important for the representativeness of infiltration models. The presence of macropores tends to create preferential dual-porosity flows characterized by physical non-equilibrium (Beven & Germann, 2013).

Some soils are called water-repellent when infiltration is diminished or even prevented by a barrier that restricts the rate and amount of water absorption under dry conditions. The capillary rise in a very water-repellent soil will be zero, preventing soil particles from moistening, resulting in water droplets forming on the soil's surface which will frequently evaporate before being able to penetrate into the soil (Hallett, 2008). Water repellence can stem from an organic layer at the surface, which becomes hydrophobic when the soil is dry. The most used technique to determine water repellence is the water drop penetration time (WDPT) test, which measures how long it takes for a water drop to infiltrate. The presence of water repellent soils completely changes infiltration curves (Wang *et al.*, 2000).

Some clay soils can crack while drying, because of soil suction and soil mechanical properties, creating large voids (Morris *et al.*, 1992), which modify water infiltration (Cheng *et al.*, 2021). In arid and semi-arid areas, some clay soils can also shrink and swell which has an impact on the soil structure. As soil deformation is not reversible in all cases, it can introduce cracks and changes in porosity (Medjnoun & Bahar, 2016). Finally, clogging, which is characterized by a decrease in the soil's accessible pore volume, may also modify infiltration in soils (Jeong *et al.*, 2018). The processes mentioned above, and many others that reflect the complexity of the soil systems, favour the establishment of preferential flows.

5.2.6.2 Spatio-temporal variability

Even though the soil variability has the biggest impact on infiltration (Childs *et al.*, 1993), the spatial and temporal (i.e., reflecting both seasonal and weather impacts) scales should be identified and considered as they can influence the hydraulic parameters (Green *et al.*, 2003; Nielsen *et al.*, 1973). Infiltration depends on the spatial variability observed at all scales, from the micro-scale (i.e., mm) to the field scale (Gerke *et al.*, 2010; Hallett *et al.*, 2004; Haws *et al.*, 2004). The methods mentioned earlier give an estimation of the infiltration at a point scale, which could be different at other scales because of the slope of the land or soil heterogeneities. The conversion from small scale to larger scale parameters is still a subject of studies (Koppe *et al.*, 2022). Nevertheless, some methods have been developed to study infiltration directly at larger scales by multiplying the number of samples or tests

(Concialdi *et al.*, 2020; Zhao *et al.*, 2020b), by using non-invasive methods (Di Prima *et al.*, 2015; Gerke *et al.*, 2010), or by increasing the size of the device or the footprint of the experiment.

The parameters can also vary on a temporal scale between different seasons and even during one season. The infiltration pattern, the soil porosity and the main hydraulic parameters can change, mainly because of the evolving state of the soil (e.g., reflecting compaction, different activated zones over time, wetting and drying and so on, Childs *et al.*, 1993; Mubarak *et al.*, 2009). This seasonality may modify the runoff coefficient (Cerdà, 1996) and hydraulic parameters such as the infiltration rate (Cerdà, 1997). Moreover, the occurrence of several freeze-thaw cycles during the winter months can severely reduce the infiltration rate (Zaqout *et al.*, 2022).

5.2.6.3 Preferential flow quantification

To quantify preferential flow, Allaire *et al.* (2009) reviewed the techniques used in laboratories and in the field to investigate the soil–structure dynamics. The authors suggested that complementary measurements to infiltration should be performed to obtain additional information on both soil heterogeneity and water flow: imagery scanning, resin impregnation, skeletization, photos and excavation, smoke injection, geophysical devices (ground penetrating radar (GPR), electrical resistivity tomography (ERT) and electromagnetic induction (EMI), see below), ring and tension infiltrometers under varying pressure heads, tracers and dyes, through breakthrough curves of solutes and particles.

Tracers such as salts and dyes like Brilliant Blue FCF or rhodamine are widely used to investigate macropore flow and interactions between macropores and the soil–matrix (Anderson *et al.*, 2009a, 2009b; Kodešová *et al.*, 2010; Nimmo *et al.*, 2009; van Schaik *et al.*, 2010). The tracers and dyes are usually used with one of the infiltration devices presented above. After an infiltration measurement with dyes, soils need to be excavated. For monitoring the tracers' pathways in the soil under the infiltration device, it is necessary to couple the infiltrometer with a non-invasive device capable of detecting the tracers. A difference of electrical conductivity with salts can be determined, for example.

Non or slightly invasive studies of infiltration at various scales of time and space have found value in geophysical approaches to introduce new information about water infiltration processes. Some geophysical methods using surface electrical resistivity measurements (i.e., ERT), changes in the apparent electrical conductivity (like EMI) or pulses of radar (like GPR) can image subsurface structures and thus can be used to monitor water infiltration into soils. They allow a broad view of flow pathways, at different and large scales, are non-destructive, are repeatable and can connect the dynamics of infiltration to underground structures (Fan *et al.*, 2020). A comparison of the three devices may be found in the review by Fan *et al.* (2020).

ERT is made up of four electrodes to reduce the impact of contact resistance at the interface between the electrode and the soil pore water (Daily *et al.*, 2005). Two electrodes are subjected to a known current, and the potential difference between the other two electrodes is measured. Daily *et al.* (1992) were the first to use ERT in the field to follow water movement in the unsaturated zone. They made two kinds of infiltration experiments and used the finite element method (FEM) with a 7×14 mesh to model the reconstruction plane. The resistivity differences give an appreciation of the water mobility. More recently, ERT was used in a karst vadose zone (Watlet *et al.*, 2018). The methods for data gathering, storage, filtering, inversion and visualization were semi-automated, and data were time-lapse inverted to visualize resistivity variations.

EMI uses the fact that the physical and chemical characteristics as well as the morphology of soils are directly impacted by the depth and movement of water through the subsurface (Doolittle & Brevik, 2014). The EMI device is made of two sensors: one transmitter and one receiver. The fundamental electromagnetic field that they transmit causes electrical currents induced in the soil, and the sensor's receiver detects the secondary electromagnetic field produced by these currents. The latter is used to determine the 'apparent' or 'bulk' electrical conductivity (ECa) for the volume of soil profiled in circumstances referred to as 'operating under low induction numbers'. The EC changes as a function of the soil parameters, temperature and concentrations of ions in solution.

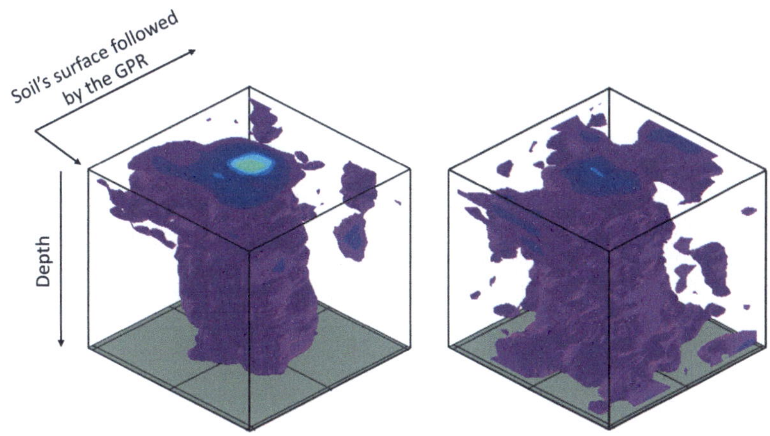

Figure 5.5 Two examples of GPR visualization in 3D of water infiltration and preferential flow pathways.

GPR consists of two antennas, one emitter and one receptor, and is linked to an acquisition system for data storage and visualization (Utsi, 2017). The emitter generates electromagnetic waves through the soil. If the pulse meets a change in the nature of the materials, (textures and structure) or in the hydric conditions, based on electromagnetic properties, the signal is returned to the receptor antenna. GPR is also used to characterize preferential flow paths (Guo *et al.*, 2014). Different methods for data treatment exist, specifically developed to track water in soils (Allroggen & Tronicke, 2015; Di Prima *et al.*, 2020). For example, Figure 5.5 illustrates, in three dimensional, the preferential flow pathways obtained from a single ring infiltration experiment.

5.2.7 Modelling of preferential flow

Šimůnek *et al.* (2003) presented a review of models depicting preferential and non-equilibrium flows in the vadose zone. The two main model types are based on double porosity or permeability approaches. Dual-permeability and dual-porosity models both presuppose that the porous medium is composed of two zones that interact, one involving the inter-aggregate porosity, macropores or fractures, referred to as the fast-flow matrix, and the other made up of micropores inside the soil aggregates or the rock matrix, referred to as the soil matrix. Dual-permeability approaches consider water flows in both the fast-flow and the matrix regions while water stays stagnant in the matrix region for the dual-porosity approach. Similar in principle, multi-porosity and/or multi-permeability models also contain additional interacting pore zones.

The BEST method is not always valid, as shown by Angulo-Jaramillo *et al.* (2019): water repellent soils can show a negative intercept and are not well represented; the infiltration rate of some very fine soils can be over-estimated; the intercept of very coarse soils are not consistently estimated; soils with macropores are only represented with the BEST-steady method while the estimated hydrodynamics parameters of sealed soils are not representative.

A BEST method was developed to tackle the challenges associated with the presence of macropores (Lassabatere *et al.*, 2019). The principle is the same, but the equations are repeated, differentiating between the parameters for two regions, the matrix and macropores. Thus, the bulk and local volumetric water content, the bulk and local hydraulic conductivities and the volume occupied by both regions have different values. Additional research is being conducted to adapt the BEST methods to water-repellent soils (Prima *et al.*, 2021; Yilmaz *et al.*, 2022).

5.2.8 Concluding remarks on infiltration processes

UD systems (Fletcher *et al.*, 2015) include heterogeneous and living soils that evolve over time. Cracks, roots and animal galleries can appear (Bedell *et al.*, 2021, 2013; Fernandes *et al.*, 2022; Saulais *et al.*, 2011), facilitating water infiltration. However, infiltration systems have shown a general decrease of their infiltration capacities over time (Al-Rubaei *et al.*, 2015), mainly caused by clogging (Le Coustumer *et al.*, 2009). The complexity of the soil heterogeneity and its evolution over time, coupled with the impact of environmental and physical processes, is still a challenging research topic.

To better design and manage infiltration facilities, the determination of the hydrodynamic parameters and the water movement models should be more precise and monitored. Regular monitoring and maintenance could help retain high infiltration capacities for these systems (Al-Rubaei *et al.*, 2015; Bouarafa *et al.*, 2019).

For field measurements, the combination of experimental devices, with the use of geophysical measurements and/or automated devices appears promising. For example, some automated single ring infiltrometers were developed, facilitating the data acquisition and allowing several experiments at the same time (Di Prima *et al.*, 2015) which can be used with the BEST methods to evaluate the risk of disfunctioning UD systems (Bouarafa *et al.*, 2019). The use of GPR proved very efficient for the characterization of infiltration bulbs and the understanding of processes in the vadose zone, allowing for the determination of the main preferential flow pathways in a non-destructive way (Di Prima *et al.*, 2023, 2020).

5.3 CLOGGING

5.3.1 Introduction

As the decrease of the permeability/infiltration rate (in natural soils and in NBS) over time is inevitable and has been observed at both centralized systems (e.g., Dechesne *et al.*, 2005; Gonzalez-Merchan *et al.*, 2012; Hunt *et al.*, 2011; Le Coustumer & Barraud, 2007; Marsalek *et al.*, 2006) and source control facilities (e.g. Barraud *et al.*, 2014; Bergman *et al.*, 2011; Blecken *et al.*, 2017; Emerson *et al.*, 2010; Le Coustumer *et al.*, 2009; Winston *et al.*, 2016; Yang *et al.*, 2022 and so many others) such as bioretention facilities, swales, trenches or porous pavements, it is important to understand

- the way it develops,
- its spatial distribution (i.e., where does clogging occur first or most?),
- the temporal dynamics (i.e., how long does it take for a device to get clogged?), or
- when should maintenance procedure reasonably be triggered?

and the predominant factors to limit or prevent it.

The large diversity of devices, configurations, substrates and environments constitutes a real challenge in addressing the issue of clogging. The scales at which infiltration occurs and the pressure exerted on facilities directly influence the spatial and temporal distribution of clogging. For example, there could be a big difference between centralized structures (e.g., infiltration basins collecting stormwater from a large catchment with large sediment inputs) and source control facilities such as swales, bioretention systems, trenches and permeable pavements draining much smaller areas. But, even for the same type of facility there will be significant variability, for example, hydraulic loading ratios, composition, the presence or absence of pre-treatment provisions, splash pads, overflows, implementation in different contexts, and so on.

For all kinds of systems, the decrease in permeability can be classified into three types. It can be:

- physical/mechanical (which tends to be the most important one) due to:
  - the input of suspended solids brought by the stormwater from the tributary catchments,
  - natural compaction or compaction resulting from human activities when the facilities are accessible to the public, or
  - internal erosion (e.g., erosion of the sides of a swale clogging its bottom).

- biological due to the development of micro-organisms (e.g., algae, fungi, bacteria and protozoa) that form biofilms on the surface. This has been observed at large infiltration basins with a bare bottom at the onset of the clogging subjected to hot weather conditions (Gonzalez-Merchan *et al.*, 2012). The use of vegetation or coarse gravels in the top layer generally eliminates this type of clogging.
- chemical: involving the precipitation and dissolution of minerals within the infiltrating substratum which can be accompanied by the development of bacteria reducing the porosity of the medium. As this clogging is closely linked to the previous ones, it is difficult to study it in isolation. However, it is minor compared to physical and biological clogging.

Clogging tends therefore to be the result of excessive deposition of debris and sediment (especially fine particles transported by runoff) or excessive inputs of organic matter (Gonzalez-Merchan *et al.*, 2012). This accumulation of material can be due to a variety of site-specific factors, often occurring in tandem, including:

- poorly specified infiltration media (e.g., undersized infiltration systems, low initial hydraulic conductivity, media composed of too fine materials (Beryani *et al.*, 2021; Le Coustumer *et al.*, 2009),
- a high hydraulic loading ratio (i.e., volumetric loading) for the size of the system, here (as done often) roughly estimated by the area of infiltration system divided by the size of the contributing catchment area,
- significant erosion of material or wash-off of deposits from the catchment, especially in the presence of construction sites with inadequate erosion and sediment control provisions which lead to high inputs of construction sediments (Azzout *et al.*, 1994; Barraud *et al.*, 2014; Silva *et al.*, 2010),
- compaction (i.e., the re-arrangement and settling of the infiltration medium due to the weight of water or to the action of users (Beryani, *et al.*, 2021; Blecken *et al.*, 2017),
- lack of maintenance.
- normal ageing process.

5.3.2 Approaches to determine the evolution of permeability

To assess the spatial distribution and temporal evolution of clogging, the general approach consists of defining one (or several) indicator(s), to evaluate or measure the performance and then compare the progression in the recorded values over time or against different parts of the facility of interest. Various approaches can be found in the literature as a function of the original intent, type and scale of the infiltration systems.

The most widely used indicator is the hydraulic conductivity (most often conducted for full saturation conditions to eliminate the variability of the media water content and preferably standardized using a specific temperature [e.g., 20°C] to get comparable results from one date to another). As explained in Chapter 4, the method consists of measuring the drop in water level in rings sunk in the media (e.g., calibrated simple ring infiltrometers or home-made devices [e.g., used by Beryani *et al.*, 2021; Angulo-Jaramillo *et al.*, 2016; Le Coustumer *et al.*, 2009; Gonzalez-Merchan *et al.*, 2014], or double ring infiltrometers [e.g., Al-Rubaei *et al.*, 2015; Kluge *et al.*, 2018]). This approach of measuring the hydraulic conductivity yields localized results and must therefore be repeated across the entire surface of the facility to provide an overview of the spatial distribution.

The method suits well and is easy to perform for small source control facilities with a rather flat bottom (e.g., swales, bioretention systems and rain gardens). Moreover, it requires few testing points because of the size of the facility. Beryani *et al.* (2021) noticed in a Swedish study on 36 biofilters less than 6 years old (and with footprints ranging from 8 to 700 m^2 accommodating runoff from catchments between 230 and 15,000 m^2) that there was no significant difference between the three tested locations used (i.e., near the inlet, outlet and middle point along the flow path). The experience of others was different in that order of magnitude differences were observed which is particularly problematic when such tests are conducted in conjunction with inspections at the time of project acceptance before it is turned over to the future owner. A large number of tests would then need to be conducted for the findings to be statistically significant (Erickson *et al.*, 2013). For that reason, flood tests with water supplied by either a hydrant or water truck have been conducted as illustrated in Section 9.5.2.

For systems with a large footprint (e.g., infiltration basin) approaches using the above hydraulic conductivity testing methods are not well suited to generate a good appreciation of the global hydraulic performance of the entire unit given its expected spatial heterogeneity being much higher. Thus, an alternate procedure should be applied which, incidentally, can also be used for smaller source control facilities. The aim is to evaluate the global hydraulic resistance by measuring the inflows, outflows (when an overflow is implemented) and water depths in the system during different events of a similar nature (e.g., events generating a water depth of more than 50 cm in an infiltration basin, or a few cm in an infiltration trench). The hydraulic resistance may be obtained by its calibration in, for instance, Bouwer's model with the data acquired (Bouwer, 2002). This method was applied many times including Gautier, 1998; Dechesne *et al.*, 2005; Le Coustumer *et al.*, 2007; Gonzalez-Merchan *et al.*, 2012 for infiltration basins and Proton, 2008; Barraud *et al.*, 2014; Emerson *et al.*, 2010 for swales or trenches. This method not only generates an appreciation of the global extent of clogging but also allows one to differentiate between the bottom and the sides of a facility. To be applied, the clogging process should have commenced with water visibly ponding. While this method has proven to be quite useful for research purposes and an overall understanding of the dynamic nature of clogging, it is rather difficult to carry out it for practical purposes due to the extensive data and relatively sophisticated monitoring needs.

Other global indicators may be used to get an appreciation of the drawdown or emptying time (e.g., the time taken for the maximum height of water in a facility to be completely drained out after a rainfall event) which has expectations or limits expressed in many design guidance documents or standards. For instance, Beryani *et al.* (2021) reported that most municipalities have a limit of 48 hours for biofilters, with some using shorter durations of 12–24 hours when there are concerns about the capacity of the facility to be restored for the next storm or in the presence of high pedestrian traffic or other governing factors.

Finally, the presence of standing water observed during inspections long after the end of an event is a simple visual indication of clogging. Such inspection is easy to perform and may trigger further investigations or preventive maintenance, see also Section 9.5 In the context of AM, the emergence of new methods to monitor the water depth with low-cost sensors or the use of still photos or video footage bear promise as to simplifying the tasks of drawdown assessment or the presence of standing water (Cherqui *et al.*, 2020; Yang *et al.*, 2022).

5.3.3 Where does clogging occur?

For most infiltration systems, including both centralized and source control facilities, clogging forms at the top of the infiltration media, only penetrating the first few cm. As the clogging results from sediment deposition and/or compaction, the clogging layer tends to be visible, easy to detect and clean or remove, if necessary. To a lesser extent, clogging may also occur at transition layers, specifically between the media of the infiltration facility and the substratum when the media is much more porous and permeable than the underlying soils. In that case it has been suggested that a geotextile between these layers can fix the clogging (e.g., Siriwardene *et al.*, 2007 on biofilters and Proton, 2008 on infiltration trenches), however in practise these geotextiles have shown to suffer from clogging as well, implementation of appropriate transitions in gradations of materials has been adapted instead (van Duin, 2023).

When a ponding zone exists within an infiltration facility, clogging mainly occurs at the bottom as shown in Figure 5.6 for a large infiltration basin (Gonzalez-Merchan *et al.*, 2012). The same process has been observed for smaller facilities like infiltration trenches (e.g., Proton, 2008) or biofilters (e.g., Blecken *et al.*, 2021). This means that the decline of the permeability of the sides tends to be negligible

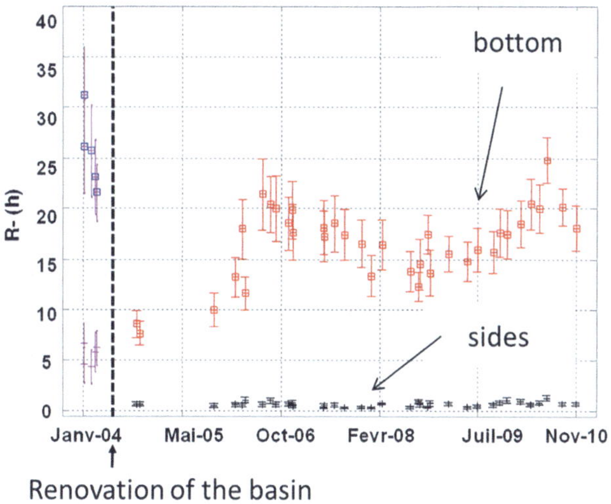

Figure 5.6 Example of evolution of the hydraulic resistance (*R*) of the sides and the bottom in the basin Django Reinhardt in Chassieu, France. The basin has an area of about 8000 m² accommodating a 185 ha catchment with a mean runoff coefficient equal to 0.4. In this case, the bottom got clogged but the basin continued to infiltrate properly along its sides. A basin is considered to have become clogged when *R* is higher than 24 hours. This occurred before it was renovated in 2004. The hydraulic resistance of the bottom decreased after spontaneous vegetation growth between 2006 and 2008. (*Source:* Gonzalez-Merchan *et al.*, 2012).

over time. However, in some cases, the sides can contribute to clogging of the bottom when their slopes are steep, non-vegetated and thus highly erodible (Blecken *et al.*, 2021).

For infiltration basins, the clogging process of the bottom shows a high degree of heterogeneity. Unsurprisingly, clogging develops primarily in the areas that are most exposed to the incoming water (i.e., primarily near the inlets). Over time, the footprint of the area influenced by the clogging extends until it occupies the entire surface. When the bottom is completely clogged, the sides provide the main avenue for infiltration. The provision of a forebay is generally a good practice to slow down this process. It also allows for easier maintenance.

Similarly, for linear infrastructure (e.g., swales, vegetated or non-vegetated permeable pavements along conventional roads or car parks), clogging first develops at the surface near the interface between the impermeable upstream areas and the infiltration facility (e.g., Al-Rubaei *et al.*, 2015).

For the case of permeable pavement structures such as porous asphalt, clogging is also superficial. Clogging tends to occur in areas subject to high tangential stresses (e.g., turning or manoeuvring areas) that leads to the stripping of the top layer material (Azzout *et al.*, 1994). It also occurs at the interface between the impermeable and permeable pavement areas, in the vicinity of construction activities or in areas fronting green spaces with overhanging vegetation (Azzout *et al.*, 1994; Blecken *et al.*, 2017). To restore the permeability of such systems, many experiments on maintenance techniques were carried out (e.g., Balades *et al.*, 1995; Chopra *et al.*, 2010; Drake and Bradford, 2013; Faure and Lemaire, 1990; Henderson and Tighe, 2011; Van Duin *et al.*, 2008; Winston *et al.*, 2016). Water pressure cleaning systems coupled with immediate vacuuming are among the most efficient methods to restore an acceptable permeability. For permeable asphalt, a pressure of 4 MPa for preventive maintenance and from 10 MPa to 30 MPa for preventive maintenance were suggested by Faure and Lemaire (1990), Balades *et al.* (1995). For full restoration, surface milling turned out to be inevitable (Azzout *et al.*, 1994; Winston *et al.*, 2016). However, standardized maintenance procedures still need to be developed. The major role that vegetation plays is addressed separately in Section 5.3.5.

5.3.4 The dynamics of clogging

The temporal dynamics of clogging are very difficult to assess with the question of the time it takes for a specific facility to get clogged even more so. The longevity of facilities is highly site and condition specific as a function of the original design configuration, uses, environment, climate, maintenance and operation routines, and so on). Virtually no long-term studies exist that monitor an infiltration facility after it had originally been installed to complete clogging. In addition, the procedures employed in assessing the evolution of clogging have proven to be highly variable and often questionable. For instance, do low hydraulic conductivities equate to the occurrence of clogging? Experience has shown it to be not so simple. Gonzalez-Merchan *et al.* (2012) showed an example of an infiltration basin that had been suspected to be clogged, based on point measurements showing low permeability of the bottom. However, this 'apparent' clogging did not compromise the overall hydraulic performance during heavy storms after 8 years of operation. Similarly, Le Coustumer *et al.* (2009), in their review of 37 small Australian biofiltration systems, with age ranging from a few months to three years, concluded that more than 40% of them were below the recommended conductivity of 180 mm/hour but that they (fortunately) continued to perform well because they had effectively been oversized.

Without standardized long-term studies that would facilitate true comparisons (i.e., by using the same methodology, the same indicators and the same considerations with respect to clogging), answering the question of up to what point a structure can be considered clogged or in need of maintenance is difficult, for the time being here one has to rely on exerts' judgement.

Despite these considerations, studies have been carried out all over the world. They illustrate the great diversity of the temporal evolution of clogging as a function of the local context. For large structures, the clogging dynamics reported in the literature are extremely diverse, ranging from a few years (Lindsey *et al.*, 1992) to decades (Dechesne *et al.*, 2002). The case of decades was observed on basins in the eastern part of the Lyon region in France implemented on fluvio-glacial soils and equipped with upstream settling tanks. The clogging was found to be progressive and take a long time, except in one particular case that corresponded to deficient rehabilitation (i.e., poorly removed sediment deposits).

With respect to smaller assets (e.g., bioretention systems, grassed swales, trenches), there is also a large variety of results (Yang *et al.*, 2022). They range from completely clogged assets to assets that, despite a decrease in permeability, still do the job in terms of infiltration after the same number of years in operation. For example, a German study (Kluge *et al.*, 2018), showed that the hydraulic conductivity of most of the 22 grassed swales studied with a substrate composed of sand and loam/silt still met the technical guideline requirement (i.e., between 10^{-3} and 10^{-6} m/s or 3.6 and 3600 mm/h) after 11–22 years of operation. The same trend was observed for biofilters reported by Austrian researchers (Haile *et al.*, 2016). Those facilities consisted of a small upstream sedimentation tank followed by a biofilter with a typical media composition. The results indicated that the hydraulic conductivity decreased 1–2 orders of magnitude after 5–7 years of operation but still complied with the Austrian design standard value which ranged from 10^{-3} to 10^{-5} m/s (36–3600 mm/h). This finding resulted in the recommendation of a preventative maintenance frequency of 7 years.

Other studies on bioretention cells (Paus *et al.*, 2014; Willard *et al.*, 2017) suggest that the assets studied can provide many years of effective infiltration (i.e., more than 6 and 7 years, respectively). The experience gained in Lyon, France with respect to a campus car park with a porous structure and permeable asphalt pavement showed that clogging had been quite slow even considering it had not had any maintenance for over 10–20 years. In contrast, a Swedish study (Al-Rubaei *et al.*, 2015) carried out on two grassed swales made of sand with a crushed stone layer underneath were found to have been completely clogged after 9 and 14 years, respectively. The explanation included the use of too fine infiltration media, compaction caused by using the swale as temporary roadside parking, and the possible accumulation of additional fines due to winter maintenance of the roads themselves.

A study conducted in Brazil (Barraud *et al.*, 2014) showed completion clogging of an infiltration trench composed of crushed stone (i.e., without vegetated topsoil) on a silty soil after only one year. The global hydraulic resistance had increased by more than 9 times over this period. The clogging was explained by the very high concentrations of suspended solids (reaching 2000 mg/L) brought to the system because of intense rain fall events.

Even if it is impossible to predict the lifespan of an infiltration system as far as clogging is concerned, some qualitative trends can be distinguished. Their design (e.g., sizing and composition integrating the consideration of environmental and social conditions) is crucial. The presence of vegetation as part of the assets, as discussed in the next section, may be able to reduce the advent of clogging.

5.3.5 Role of vegetation

Many infiltration assets are nowadays vegetated and designed as NBS. Dagenais *et al.* (2018) explored the scientific-based literature with respect to the global performance of bioretention systems and, in particular, the potential role of the vegetation in reducing clogging through four questions:

(1) Are vegetated systems more effective than unvegetated ones?
(2) Do plant species differ as to their effectiveness?
(3) Are native species more effective than exotic ones?
(4) Are diverse systems more efficient than monocultures?

With respect to the first question, the research studies confirm the beneficial effects of vegetation as to the evolution of permeability over time. The vegetated systems turned out to be more effective than the unvegetated ones. This has been demonstrated in both several column studies (e.g., Le Coustumer *et al.*, 2012 or Fowdar *et al.*, 2022) and in the field (Virahsawmy *et al.*, 2014; Gonzalez-Merchan *et al.*, 2012, 2014; Hatt *et al.*, 2007, 2009). The growth, senescence, death and degradation of plant roots create pores and macropores that can maintain soil porosity, permeability and thus retard clogging. The wind-induced movement of plant stems which loosens the surface of the filtering medium also helps to prevent surface clogging. This is true for both centralized and source control facilities as demonstrated by Gonzalez-Merchan *et al.* (2014): a study of an infiltration basin near Lyon, France compared the role of different zones covered by spontaneous types of plants (*Phalaris arundinacea*, *Polygonum mite*, *Rumex crispus*) found *in situ* versus similar non-vegetated zones during eight field campaigns in one year. The results clearly showed a positive impact of vegetation on the hydraulic performance, in particular during the growth of the plants over the summer months. The hydraulic conductivity over this period was two to four times higher than that for bare areas or in vegetated zones during the plant rest periods. The substratum of the basin was the original soil made of quaternary deposits.

Virahsawmy *et al.* (2014) carried out similar measurements albeit for smaller facilities, that is, seven raingardens less than ten years old in Melbourne, Australia. These sites had both vegetated and unvegetated areas. The vegetation species at the sites were the same consisting of grasses, sedges and rushes with an extensive and fibrous root system and a high proportion of fine roots. While the sites varied with respect to age and configuration. They had been all constructed using a similar fine sandy filter medium. The results confirmed that the infiltration rates around the plants were statistically much higher even when the sediment deposition was high.

With respect to the second question, scientific studies indicated that certain plant species are more suitable than others. For example, for bioretention systems, those with greater root mass density and root diameter were found to be most appropriate (Goh *et al.*, 2017; Hart, 2017; Le Coustumer *et al.*, 2012), in particular. In the presence of tree species (Fowdar *et al.*, 2022), Archer *et al.* (2002) found a positive correlation between the volume of roots and the saturated conductivity of soil, especially for older roots that had undergone shrinkage and expansion over time. This was also demonstrated on larger systems (i.e., an infiltration basin) where spontaneous vegetation had developed (Gonzalez-Merchan *et al.*, 2014).

With respect to the question pertaining to the use of native vegetation, no evidence was found to prove that native plants may be more efficient than exotic ones as far as retarding clogging is concerned. Larger centralized facilities are more prone to show the spontaneous introduction of various plant species (most often native but unfortunately often invasive species too) even if appropriate plant species had been selected and planted when the basin was first constructed (Saulais, 2011). Maintenance is thus needed whatever the nature of the vegetation originally planned.

The last question addresses the appropriateness of using a diverse mix of plants rather than a monoculture in infiltration systems. The question is mainly relevant for source control facilities. After all, large, vegetated infiltration basins are usually planted with different species in the first place. In the long run, the plant assemblage generally evolves with a simplification of the original mix due to the dominance of certain species and the spontaneous emergence of exogenous, possibly invasive plants. Here again, maintenance is necessary as for other vegetated areas (Saulais, 2011). For bioretention systems, Dagenais *et al.* (2018) showed a lack of evidence to confirm the superiority of a diverse set of plants and a strong need for experiments comparing systems of different plant richness. This will help in determining not only if there is a measurable benefit for biodiversity but also to determine the best combination of plants based for example on their functional traits even though plant diversity can provide many benefits such as aesthetic value (and thus social acceptance), contribution to local biodiversity, and possibly greater resistance to disturbance. However, the choice of mixed or unmixed vegetation appears to be a secondary issue as far as clogging is concerned, the only issue being the presence and nature of the plants used.

5.3.6 Concluding remarks on clogging

NBS relying on infiltration offer many benefits but are prone to clogging. Managing them in space and time is, or should be, part of AM strategies. This is even more necessary given the increasing number of infiltration facilities installed. However, their AM depends on several aspects:

The first aspect (although not addressed so far because it does not explicitly concern clogging) pertains to the complexity of the global management of these infiltration facilities:

- Where are the facilities located? It is essential to identify the location of the devices implemented in our urban context. This is particularly the case for highly dispersed source control measures whose multifunctional nature sometimes masks their primary hydraulic function.
- Who manages them? As these facilities tend to be multifunctional, and can be private or public, they often involve many actors in their management which introduces complexity in terms of both maintenance and intervention (Cossais *et al.*, 2017; Werey *et al.*, 2016). For example, in Bordeaux, France, the regional water department looks after a facility when it is 'grassed and not planted'. However, if the facility is planted, its management is to be provided by the Parks Department of the local municipalities (Werey *et al.*, 2016).

A second aspect concerns the factors to prevent or limit the evolution of clogging. According to the surveys carried out on these systems, it is recognized that their durability (i.e., their proper hydraulic functioning) depends largely on the initial design and construction of the systems. The most sensitive points are:

- Properly characterize the soils in place and carefully formulate the added substrates/growing media, avoiding fines where possible, and ensuring good initial hydraulic conductivity. Typical compositions of added substrates from different guidelines are summarised in Beryani *et al.* (2021), In semi-arid regions, some fines and wicking systems may be needed to provide adequate moisture to sustain the vegetation. The hydraulic conductivity may be less in the early days of the facility until the structure of the infiltration bed is established, requiring temporary gravel-based chimneys to prevent standing water. Skorobogatov (2014) provides an introduction to the relation between vegetation and hydrological properties of soils.
- Use vegetation to the greatest extent possible, with vegetation having greater root mass density and diameter being most appropriate.
- Analyse the upstream catchment carefully to ensure that it does not produce too many fines. Pre-treatment (e.g., forebays or sumps is also a good practice.
- Anticipate potential uses or abuses, especially those that may accelerate compaction.

Many relevant recommendations exist in published guidelines (e.g., County of San Diego, 2014; Fassman, *et al.*, 2013; Melbourne Water, 2023; NCDEQ, 2018).

A third aspect concerns the methods for identifying and monitoring the proper operation of the infiltration facilities whatever their original design composition. Different techniques and methods exist to aid in detecting or measuring the evolution of clogging, as a function of the type of the system, the objectives and the means that the facility managers can implement. Unfortunately, at this stage, none of them is standardized. The method can consist of a simple visual inspection, for instance for the presence of standing water one or two days after a (heavy) storm event, significant sediment deposits, eroded areas, deterioration of the vegetation or simply unaesthetic aspects. This can be accompanied by hydraulic conductivity measurements, utilizing one of the various methods available. It can also be based on continuous measurements of water depths or even of inflows and/or outflows.

Given that the context of the implementation of infiltration facilities is increasingly uncertain (e.g., with climate change resulting in longer dry periods coupled with more intense wet spells) while their use is increasing as part of overall urban stormwater management strategies, knowledge pertaining to their hydraulic efficiency is becoming vital. Therefore, the development of rigorous monitoring methodologies may become imperative. They need to be nimble and inexpensive given the number of facilities. The use of low-cost sensors and photo or video footage seems promising. In any case, as far as the potential for clogging is concerned, preventive maintenance must be encouraged. It is simpler, cheaper and more effective. Recommended maintenance actions are well detailed in the scientific and technical literature (e.g., Beryani *et al.* 2021; Blecken *et al.*, 2017; CIRIA, 2007; NCDEQ, 2018, TRCA, 2016) and discussed in more detail in Chapter 10.s

5.4 DETERIORATION OF PLASTIC PIPES

5.4.1 Introduction

Plastics are used for a wide range of commercial and industrial piping applications. Polyvinyl chloride (PVC), polyethylene (PE) and polypropylene (PP) are the most popular polymer materials for sewer systems. Fibre reinforced plastic (FRP) is an alternative for lighter pipes with higher stiffness requirements. Irrespectively of the construction material, numerous factors can potentially affect the physical, chemical and mechanical properties of plastic pipes. From the production process until the end of their operational lifetime, an overview of the governing factors and failure mechanisms is presented in this section.

5.4.2 Production phase

The most common production process for polymer particles is suspension polymerization. Additives and fillers are added to the polymer matrix during the production phase to improve the chemical and physical properties, respectively. The main applied additives include plasticizers and stabilizers which affect the behaviour and degradation rate of the product during its lifetime. In particular, the objective of plasticizers is the replacement of some of the monomers of the polymer chain, which leads to a higher degree of mobility and therefore higher flexibility. Stabilizers are used to enhance the material resistance against chemical attack (e.g., UV radiation) and other external factors (Cardarelli, 2008).

Each process within the production phase of plastic pipes and fittings can influence the durability of the end-product. For instance, the levels of water and oxygen during the polymerization stage affect the formation and quality of the produced PVC particles (Butters, 1982). Another factor is the level of gelation, which relates to the breaking of the boundaries of primary particles and the formation of a new continuous polymer chain (Visser, 2009). The gelation level is affected by the degree of polymerization (Fujiyama & Kondou, 2004) and plays a crucial role in the mechanical response of the polymer material (Moghri *et al.*, 2003). The mechanical properties are established based on the polymer's morphology (Benjamin, 1980; Kuriyama *et al.*, 1998), as well as its orientation and molecular mobility (Fillot *et al.*, 2006). Furthermore, impurities and voids in the polymer composition, usually referred to in the literature as inherent defects, are introduced during production forming crack initiators, while their occurrence seems to be inescapable (Johansson & Tornell, 1987). The wear observed at the surface of pipe extruders (Gladchenko *et al.*, 1997) is a possible source of inherent defects.

Additionally, imposing residual stresses during production is inevitable due to the different cooling rates between the inner and outer pipe surface (Siegmann *et al.*, 1981), influencing the mechanical performance of the produced pipe (Siegmann *et al.*, 1982). A faster cooling rate or a larger pipe wall thickness will most likely result in higher levels of residual stresses (Janson, 2003; Scholten *et al.*, 2016). Irrespective of their magnitude, residual stresses determine the crack propagation mechanisms as they modulate the stress profile through the pipe thickness (Chaoui *et al.*, 1987), increase the brittle-ductile temperature (Scholten *et al.*, 2016), and, accordingly, have a dreadful effect on the operational lifetime. The impact of the residual stresses is more profound in the case of pressurized pipes as the combination of internal pressure and tensile residual stresses is applied to the inner surface of the pipe (Hutař *et al.*, 2013; Poduška *et al.*, 2016).

Furthermore, nowadays non-pressurized PVC pipes are constructed with a structured wall composed of three layers, in which the middle layer consists of recycled or foamed material. This type can bear flexural deformations almost as efficiently as single-layer PVC pipes. Nonetheless, the performance is comparatively lower under conditions of tensile stresses (Makris *et al.*, 2021), which denotes a crucial potential failure factor in case of low-quality installation and/or differential soil settling.

5.4.3 Installation phase

The installation phase includes the excavation of an open trench, preparation of the soil bedding, laying of the pipe, and soil covering and compaction. Nonetheless, during transport and installation, scratches and dents can occur on the pipe surface. Such permanent deformations are potential stress risers, which under certain operational conditions can eventually result in failure. Additionally, poor soil compaction leads to pipe ovalisation, imposing high tensile stresses at the 12 and 6 o' clock positions (top and bottom) of the inner pipe surface and at the 3 and 9 o' clock positions(right and lift side) of the outer pipe surface. In pressurized pipes, the internal pressure is counteracted by a uniform external pressure on the pipe exerted by homogeneous soil embedding, lowering the probability of crack formation (Hutař *et al.*, 2011). An additional factor that can accelerate material degradation concerns the storage conditions prior to installation, especially concerning PVC pipes. Exposure to UV rays leads to photochemical degradation, affecting the mechanical properties of PVC pipes (Anton-Prinet *et al.*, 1999; Hussain *et al.*, 1995). Figure 5.7 presents a case of poor-quality installation in The Netherlands. The defect was classified as a 'complicated crack' according to visual inspection standard NEN 3399 and concerned a crack which had been initiated at the pipe crown and progressed to the side. The pipe was found during excavation to be in direct contact with the concrete cover of a pipe for district heating. In this case, the contractor was found to be responsible as an external piece of pipe was found to cover the damaged pipe section.

5.4.4 Operational phase

Four main ageing mechanisms can be observed during operation: physical ageing, mechanical degradation, chemical degradation and environmental stress cracking.

Physical ageing is a known mechanism in polymers which imposes changes on certain properties of a material as a function of time, even at a constant temperature and independent of external factors (Hutchinson, 1995). Amorphous (or glassy) polymers, such as PVC, experience physical ageing because they were cooled to a temperature below their glass transition temperature, and, hence, are not in a thermodynamic equilibrium state. In this non-equilibrium state, the glassy polymer has excessive thermodynamic properties and there is a continuous effort to reach the equilibrium state (Hutchinson, 1995). Physical ageing can be traced by reduction in volume and enthalpy, but also by changes in the mechanical properties (Rabinovitch & Summers, 1992). The polymer becomes stiffer and more

Figure 5.7 A case of poor-quality installation. An external piece of pipe was found to cover a PVC sewer pipe at the point of failure. (Courtesy K. Makris).

brittle (Figure 5.8), while its creep and stress relaxation rates decrease (Struik, 1977). In principle, physical ageing is an inevitable, although reversible process in polymers (Hutchinson, 1995), which is accelerated at higher temperatures (Visser *et al.*, 2011). The process of physical ageing occurs also in semi-crystalline materials, like PE and PP. In this case, physical ageing occurs in the amorphous part of the semi-crystalline polymer (Struik, 1977).

Mechanical degradation is the result of loadings that are applied on the pipe with their level surpassing the polymer's fracture threshold. It surfaces as fissures (e.g., crazes and cracks) or breaks.

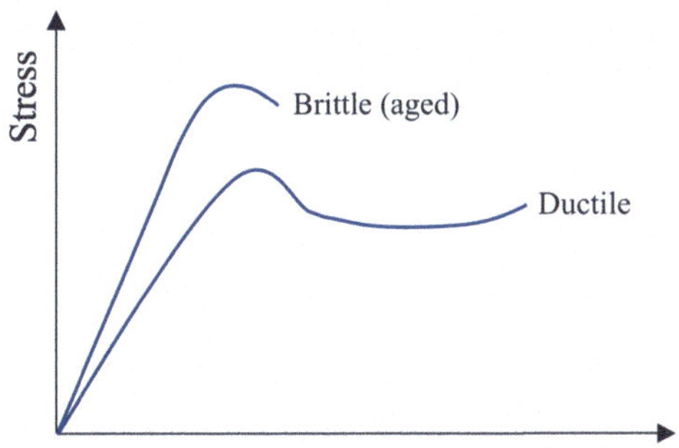

Figure 5.8 Effect of physical ageing on the response of amorphous polymers to deformation. (Courtesy K. Makris).

The loadings include stresses from internal pressure (if any), deflections due to the vertical loading of soil and the production phase (i.e., residual stresses). Additional stresses can be exerted by axial bending due to improper installation and soil settling. The quality of the pipe production can be a critical factor for the durability of polymer pipes, since cracks initiate from built-in voids and impurities. Subsequently, the propagation (rate and direction) of the cracks is determined by the magnitude and direction of the applied stresses. This failure mechanism is known as slow crack growth (SCG). A usual cause of failure for plastic pipes concerns external impacts (e.g., when a pipe is hit by an excavator). In this case, mechanical degradation surfaces as rapid crack propagation (RCP). Figure 5.9 illustrates the case of a breakage caused by external impact. The imprint of an excavator's bucket is obvious while the large number of other installations in the direct vicinity of the damaged pipe suggests extensive past excavation activities in the area.

Chemical degradation concerns the effect of chemical reactions that occur between the pipe material and the environment, inside and/or outside of the pipe. Chemical degradation can result in breakage of the polymer covalent bonds, which constitute the main back bone of a polymer chain. The breakage of covalent bonds leads to chain scission and molecular weight reduction. A known indication of chemical degradation is the discolouration that is often observed on PVC pipes (Figure 5.10). This effect is the result of dehydrochlorination, that is, HCl abstraction (Breen, 2006), which is caused by the formation of sequential conjugated polyenes (Arnold, 2003). The impact on the mechanical properties is usually expressed with the term 'stress corrosion cracking', which includes four individual stages (Choi *et al.*, 2005): initiation of microcracks, slow crack growth, clustering of cracks and clusters growth. Research on PVC sewer pipes that had served for 40 years before excavation revealed that discolouration is limited to the pipe surface (Makris *et al.*, 2021).

Last, environmental stress cracking (ESC) shares a similar failure pattern in terms of shape with slow crack growth. It is a purely physical process propagated by the applied stresses and accelerated by the effect of an active environment (Bishop *et al.*, 2000), since diffusion is the mechanism that increases the sensitivity to fractures due to formation of plasticized layers and surface energy reduction (Arnold, 2003). Research on the ESC mechanism under gas and liquid environments indicate that the combined effect of environment and stress intensity results in a lower load bearing capacity (Breen, 1993, 1994, 1995). In general, a wide range of amorphous and semi-crystalline polymers was studied to explore their ESC resistance (Robeson, 2013), concluding that signs of ESC may initiate under

Figure 5.9 A breakage of a PVC sewer pipe due to an external impact (hit by excavator). (Courtesy K. Makris).

Figure 5.10 PVC pipe with obvious signs of discolouration on the external side. (Courtesy K. Makris).

certain combinations of 'material-environment'. Finally, additional failure mechanisms to those of the polymer matrix can be found in FRP pipes. Fracture of one or multiple fibres could locally result in a lower stiffness (Alabtah *et al.*, 2021). Nonetheless, the addition of reinforcing fibres certainly adds to the durability, while their usual placement at two different directions within the pipe wall compensates for fibre failure, keeping the pipe operational (Fragoudakis, 2017). Further, detaching of the fibres from the polymer matrix, commonly known as debonding, can also be observed (Alabtah *et al.*, 2021).

5.5 BIOGENIC CORROSION OF CONCRETE

5.5.1 Introduction to biogenic sulphide corrosion of concrete sewer pipes

The assessment of the durability of concrete sewer pipes requires a thorough understanding of the underlying chemo-mechanical degradation processes. In aggressive chemical environments, concrete degradation may be promoted by the reaction of sulphates with the hydrated cement paste (Allahverdi & Skvara, 2000; Attiogbe & Rizkalla, 1988; Neville, 2004; Parker, 1945) or by alkali-silica reactions between reactive siliceous aggregates and the alkaline pore solution in hardened concrete (Farny & Kerkhoff, 1977; Rajabipour *et al.*, 2015). In both cases, the reaction products induce expansive strains in the concrete material, which may trigger micro-crack development. Another relevant chemo-mechanical damage process in concrete is carbonation, which is associated with the formation of calcium carbonates due to the reaction of carbon dioxide with the hydrated cement paste (Ismail *et al.*, 1993; Savija & Lukovic, 2016). This mechanism not only contributes to concrete degradation but also promotes corrosion of the steel reinforcement. This section particularly focuses on biogenic sulphide corrosion (Bielefeldt *et al.*, 2010; Cho & Mori, 1995; Davis *et al.*, 1998; Grengg *et al.*, 2015; Ismail *et al.*, 1993; Parker, 1945), which is a specific type of sulphate attack occurring in concrete sewer systems (Davies *et al.*, 2001). The content of this section is a summary of the work presented in Rooyackers *et al.* (2022).

5.5.2 Modelling of biogenic sulphide corrosion

The process of biogenic sulphide corrosion is schematically illustrated in Figure 5.11 and can be described as follows. The sewerage environment may be rich in sulphur compounds. Under anaerobic conditions, sulphate-reducing bacteria can reduce the sulphur compounds present in the wastewater into hydrogen sulphide gas (H_2S). This gas is released from the wastewater into the sewer atmosphere and then makes contact with the wet inner surface along the upper parts of the pipe, where it is

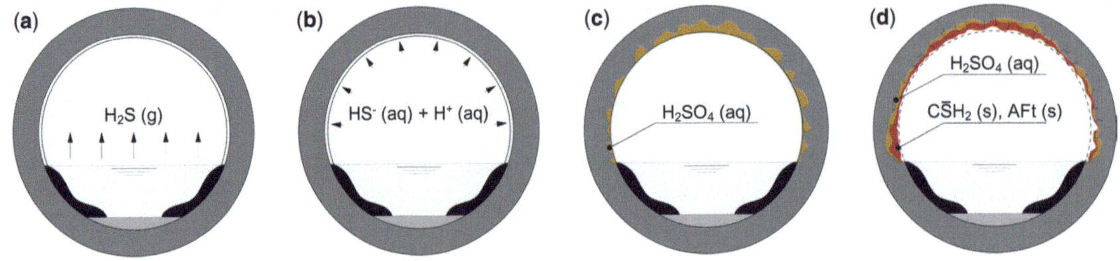

Figure 5.11 Schematic representation of the process of biogenic sulphide corrosion in concrete sewer systems. (a) Hydrogen sulphide gas (H_2S) is formed from sulphur compounds present in the wastewater which is subsequently released into the upper parts of the sewer pipe. (b) Hydrogen sulphide gas reaches the thin moisture layer present along the inner surface of the sewer pipe whereby it dissolves into ionic species HS^- and H^+. (c) The dissolved hydrogen sulphide species are converted into sulphuric acid (H_2SO_4) by bacteria, as indicated by the yellow colour, which may diffuse into the porous structure of cement. (d) sulphuric acid reacts with cement to form gypsum ($C\bar{S}H_2$) and ettringite (AFt), as indicated by the red colour, which may lead to the formation of micro-cracks and damage along the concrete pipe. The figure has been reprinted from Rooyackers *et al.* (2022).

dissolved into ionic species, that is, bi-sulphide (HS^-) and hydrogen ions (H^+). In the presence of sulphur-oxidizing bacteria, the dissolved hydrogen sulphide species are converted into sulphuric acid (H_2SO_4) (Bielefeldt *et al.*, 2010; Cho & Mori, 1995; Marquez-Penaranda *et al.*, 2015; Parker, 1945). Sulphuric acid may diffuse into the porous structure of the cement matrix, where it reacts with dissolved constituents of the hydrated cement paste, resulting in the formation of gypsum ($C\bar{S}H_2$) and ettringite (AFt) (Attiogbe & Rizkalla, 1988). This chemical process affects the mechanical resistance of concrete for two reasons. First, gypsum and ettringite form a porous corrosion layer with inferior mechanical properties compared to the original cement paste (Marquez-Penaranda *et al.*, 2015). Second, the expansive nature of these reaction products induces tensile stresses in the remaining, intact cement paste (Attiogbe & Rizkalla, 1988; Eriksen, 2003; Mittermayr *et al.*, 2015). If these stresses exceed the tensile strength of the cement paste, they may result in the initiation and propagation of micro-cracks (Monteny *et al.*, 2000). The net effect of these two aspects is that the amount of undamaged concrete material across the wall thickness of a sewer pipe is reduced over time, which can have serious consequences for the structural load bearing capacity of the pipe. As demonstrated in a recent experimental-numerical study (Scheperboer *et al.*, 2021), a reduction of 20% of the concrete wall thickness may already reduce the load bearing capacity of a sewer pipe by approximately 40%. Hence, it is important to obtain a thorough understanding of the process of biogenic sulphide corrosion and the consequent degradation of concrete.

This paragraph summarizes the computational modelling framework recently developed in Rooyackers *et al.* (2022), which can be used for simulating the coupled chemo-mechanical behaviour of concrete sewer pipes under biogenic sulphide corrosion. The process of biogenic sulphide corrosion was modelled by considering a source of sulphuric acid that is produced by sulphur-oxidizing bacteria, which gets into contact with the inner surface of a concrete sewer pipe. The consequent chemical processes were described in terms of a set of diffusion-reaction equations, which include the reactions of calcium hydroxide dissolution, calcium silicate hydrate dissolution and gypsum formation, as well as the corresponding diffusion-reaction processes of the associated ionic species. The diffusion-reaction equations were coupled with a continuum damage model that describes the process of material degradation up to complete failure. One coupling between the chemical and mechanical fields was realized through the definition of a chemically induced growth strain, which determines the expansion effect by gypsum formation on the stresses generated in the concrete. Another coupling followed from the introduction of a chemical damage variable that quantifies the degradation of concrete as a result of calcium silicate hydrate dissolution. Since the presence of damage promotes the diffusion of chemical species (i.e., damage-enhanced diffusion), the diffusion coefficients of the chemical processes were prescribed to be a function of the generated material damage, by which the model formulation became two-way coupled.

The chemo-mechanical model was implemented within a FEM framework, whereby the numerical update procedure of the coupled chemical and mechanical problems was performed in a so-called segregated fashion. The model predictions were validated by performing dedicated laboratory experiments on cuboidal concrete specimens exposed to a sulphuric acid solution over different time periods, which mimics the temporal development of biogenic sulphide corrosion in concrete sewer pipes. The finite element analyses thus represented the chemo-mechanical response of a cuboidal concrete sample that is exposed at its external boundaries to a specific sulphuric acid solution. The cube had edges of length $2L = 150$ mm. Due to symmetry of the problem in terms of the geometry and the initial and boundary conditions applied, the main features of the corrosion process in the three-dimensional configuration could be captured accurately by a two-dimensional analysis of the left upper quarter part of the concrete sample, in accordance with the $L \times L$ computational domain illustrated in Figure 5.12. The computational domain was discretized by 7310 quadrilateral finite elements, with a local refinement close to the left and upper external boundaries to accurately describe the relatively strong concentration gradients that initially may appear at these boundaries. The specific chemical and mechanical material parameters and the initial and boundary conditions applied on the system, respectively, can be found in Rooyackers *et al.* (2022).

5.5.3 Corrosion experiment

For the validation of the proposed chemo-mechanical model presented in Section 5.5.2, a corrosion experiment was performed in which a set of concrete specimens was exposed to a sulphuric acid solution for different time durations. At two opposite sides of the cube the upper part was equipped with a 50 mm cantilever extension, see Figure 5.13a, which enables hanging the cubes using a PVC support structure in a container with the sulphuric acid solution, as described in more detail below. The composition of the casted concrete specimens included coarse granite aggregates, sea sand and ordinary Portland cement – CEM I 42.5N. The composition is representative of the dry-cast concrete used since the 1950s for sewer pipes in the Netherlands. The water–cement ratio of the composition

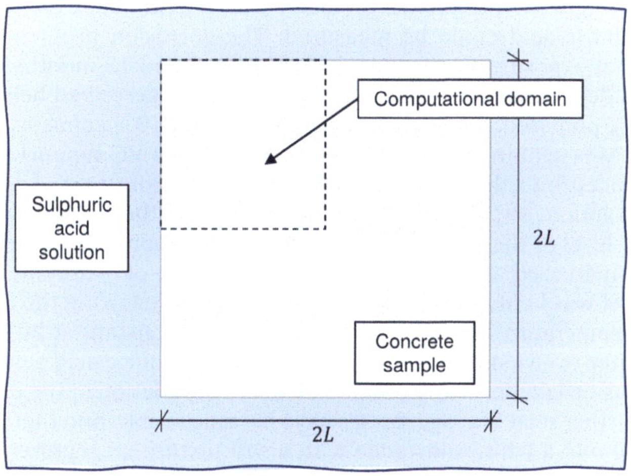

Figure 5.12 Cuboidal concrete sample exposed to a fixed, sulphuric acid solution. Due to symmetry, the computational domain considers one quarter of the sample geometry. The figure has been reprinted from Rooyackers *et al.* (2022).

Figure 5.13 Corrosion experiment on concrete specimens. (a) Cuboidal concrete specimen. b) Container and PVC support structure used for simultaneously exposing six concrete specimens to a sulphuric acid solution. (c) Corroded specimen after 12 months of exposure to a sulphuric acid solution with a pH of 1. (d) Cross-section of a corroded specimen after 12 months of exposure to a sulphuric acid solution with a pH of 1. From a phenolphthalein test the inner concrete region (pink) and the unspalled part of the outer gypsum layer (beige) can be clearly identified. The figure has been reprinted from Rooyackers *et al.* (2022).

was equal to 0.35. The formwork of the casted specimens was removed after 24 hours, after which the specimens were kept in water for a hydration period of 28 days. Subsequently, the dimensions of the hardened specimens were measured from 27 different spatial points that were uniformly distributed across the cube. Accordingly, their reference geometry was determined, from which corrosion depths after exposure to sulphuric acid could be measured. The corrosion profiles of the specimen were analysed for six different exposure times, namely 1, 2, 4, 6, 8 and 12 months. Since the assessment of corrosion requires a destructive analysis of the specimens (as described below), for each exposure time six specimens were prepared, resulting in a total number of 36 specimens. Per exposure time the sulphuric acid solution was kept in a separate container. Using a PVC support structure, the concrete specimens were suspended into the container, whereby the bottom part of the cubes was exposed to the sulphuric acid solution, see Figure 5.13b. The pH level of the solution was equal to 1. During the experiment the pH level of the sulphuric acid solution was monitored frequently, whereby it was observed that it slowly increased with time because of the release of hydroxide ions from the cement. The increase in pH level was kept within acceptable bounds by renewing the sulphuric acid solution once per month. The temperature in the experiment was kept constant at 20°C. Figure 5.13c shows a concrete specimen after twelve months of exposure to the sulphuric acid solution. The degradation effect of chemical corrosion can be clearly observed from the spallation of the outer part of the cube at the bottom. It can be further seen that aggregates have become visible, and that the cement in between the aggregates changed into a white substance with a soft texture, as representative of gypsum. The sediment found in the container underneath the specimen contained the aggregates and gypsum spalled from the specimen. In the experiment, the evolution of the spallation profile over time was not monitored in detail, because of the complexity following from the high sensitivity of spallation to the local morphology of the concrete material, in combination with the long duration of the experiment. For each of the six exposure times, the average value and standard deviation of the thickness of the corroded layer were computed from the dimensions of the corroded cube in question in relation to its original dimensions. For this purpose, the corroded cubes were halved by means of a tensile splitting test, after which the average thickness and standard deviation of the corroded layer were calculated from measurements taken across the specimen width at five different locations along the cross-sectional height of the corroded (bottom) part of the sample, as performed for six samples per exposure time. Hence, for each exposure time the two statistical values were calculated from 30 measurements in total. An additional refinement of the estimated corrosion layer thickness was necessary to account for the degree of corrosion generated inside the cube. This was done by performing a phenolphthalein test, whereby one of the fracture surfaces was sprayed with a phenolphthalein solution. In the presence of a relatively high pH value, the phenolphthalein took a pink colour, thereby indicating the region in which the sulphuric acid did not penetrate, and the concrete was uncorroded. Conversely, for a low pH value, the phenolphthalein remained transparent, and thereby identified the corrosion depth corresponding to the gypsum formed inside the specimen. For a concrete specimen exposed to the sulphuric acid concentration for a period of twelve months, Figure 5.13d depicts the fracture surface sprayed with phenolphthalein. In this figure the inner concrete region (pink) and the unspalled part of the outer gypsum layer (beige) are clearly recognizable.

5.5.4 Comparison between the experimental and numerical results

Figure 5.14a depicts the experimental (left) and numerical (right) quarter cross-sections of the concrete specimen after exposure to the sulphuric acid solution for a period of twelve months. It can be observed that the unspalled, experimental corrosion layer of gypsum (depicted in beige) has developed rather uniformly along the specimen boundary, with some small undulations caused by inhomogeneities (e.g., aggregates) present in the inner concrete region (depicted in pink). The maximum thickness of the gypsum layer is in good correspondence with that computed by the numerical simulation (depicted in red). Clearly, the undulations are absent in the numerical profile, since the concrete in the inner

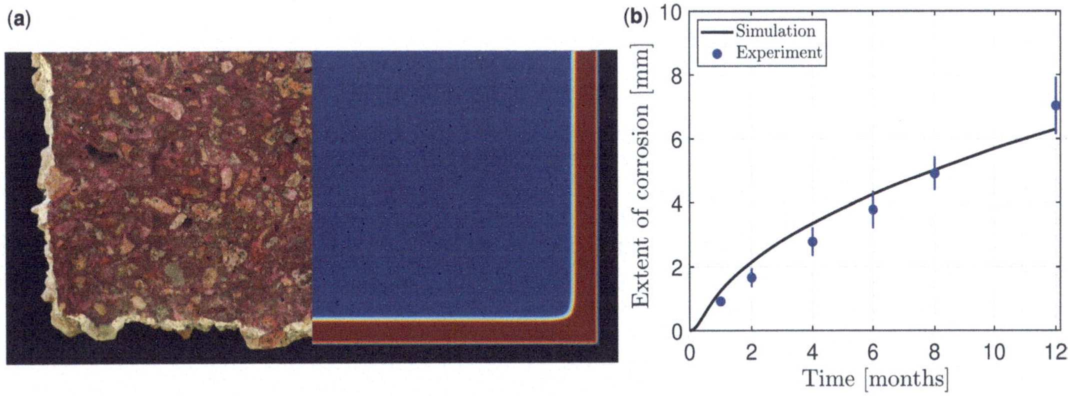

Figure 5.14 Comparison between experimental results and the numerical simulation results computed with the chemo-mechanical model of Rooyackers *et al.* (2022). (a) Quarter cross-section of one specific concrete specimen after exposure to a sulphuric acid solution for a period of 12 months. In the experimental profile (left) the unspalled gypsum layer is indicated in beige, and the inner concrete region in pink. In the numerical profile (right) the gypsum layer is indicated in red, and the inner concrete region in blue. (b) Time evolution of the corrosion depth. The mean values of the experimental data are indicated by blue solid circles; the corresponding error bar represents the standard deviation. The results of the numerical simulation are designated by the black solid line. The figure has been reprinted from Rooyackers *et al.* (2022).

concrete region (depicted in blue) is modelled as homogeneous. The specific effect of the aggregates on the chemo-mechanical response can be accounted for in the present model by explicitly simulating the heterogeneous concrete micro-structure, see for example, Bosco *et al.* (2020), which is a topic for future study. The evolution of the experimental corrosion depth with time is presented in Figure 5.14b for the six exposure times applied in the test procedure, with the average value and standard deviation indicated by the blue solid circles and the blue error bars, respectively. The trend computed by the numerical model is denoted by the black solid line. It can be observed that the rate of corrosion initially is relatively high, but that it gradually decreases with increasing time. The numerical result adequately corresponds to the experimental result for all six exposure times; at larger times it consistently falls within the standard deviation of the average experimental response, whereby the final corrosion depth corresponding to the gypsum layer formed after twelve months 6.3 mm. It was confirmed that the evolution of the corrosion depth with time as computed by the numerical model can be closely approximated by a power law of which the exponent equals 0.53, indicating that the development of the corrosion depth scales approximately with the square root of time. In summary, it may be concluded that the coupled chemo-mechanical model is able to adequately describe the important features of chemical corrosion observed in the experiments. This demonstrates that the model is suitable for being applied in the numerical analysis of biogenic sulphide corrosion processes taking place in in-situ sewer pipe systems to predict their long-term chemo-mechanical degradation behaviour under practical, relatively complex environmental conditions.

Finally, as systematically demonstrated in Rooyackers *et al.* (2022), the process of biogenic sulphide corrosion occurs as follows. Driven by the relatively low ionic concentrations of calcium and hydroxide ions at the external boundary of the cuboidal concrete sample, the calcium hydroxide and calcium silicate hydrates species present in concrete gradually dissolve from the boundary in an inward direction. The calcium ions generated by these dissolution processes subsequently react with the excess of sulphate ions to form gypsum. The excess of sulphate ions in the outer boundary region of the sample results from their inward diffusion, as caused by a relatively high ionic concentration at the boundary under biogenic reactions. The formation of gypsum ends at a specific material point once the calcium hydroxide and calcium silicate hydrates species have fully depleted. The thickness of the outer region increases with time, whereby the associated chemical growth strain is constrained by the inner concrete region. This induces a tensile stress in the inner core region that typically leads to mechanical damage, whereby the amount of damage increases from the core of the sample towards the interface with the outer gypsum layer. The speed of the growth of the gypsum layer and the amount of damage generated in the inner concrete region are significantly increased by the process of damage-enhanced diffusion. Furthermore, the diffusion coefficient has a strong impact on the degradation behaviour of the system, while the chemical growth strain influences the amount of damage but has almost no effect on the thickness of the gypsum corrosion layer.

In Luimes *et al.* (2023) the chemo-mechanical degradation of unreinforced concrete sewer pipes applied in domestic service locations is assessed through a systematic, integrated experimental approach, considering 18 new and 35 used sewer pipes. The characteristics and environmental conditions of the sewer pipes are reported, and data obtained from surface condition classification, residual alkalinity tests and XRD analyses are combined to identify the type and degree of chemical attack of the used sewer pipes. Concrete material properties are determined by material tests on sewer pipe samples, providing quantitative insight into the age dependency of the Young's modulus, compressive strength, tensile strength and mode I toughness. All relatively old pipes (installed in the 1920s and 1950s) show substantial chemical attack by biogenic sulphide corrosion (at the inner side of the pipe) and carbonation (at the outer side of the pipe). The time development of the corresponding corrosion depths on average follows a linear trend, whereby the corrosion rate for biogenic sulphide corrosion is about a factor of 1.3 larger than for carbonation. Due to these chemical processes, the mechanical properties of concrete may significantly depend on the age of the sewer pipe. In particular, the average compressive strength and average tensile strength decrease approximately linearly with the age of the pipe, in correspondence with relative reductions of, respectively, a factor of 1.7 and 1.5 over a period of almost 100 years. The values and time-dependent trends found for the concrete properties and corrosion depths of sewer pipes can serve as input for practical analyses and advanced numerical simulations on their bearing capacity and time-dependent degradation. The experimental results also emphasize the importance of regularly assessing the amount of chemical degradation of (especially older) *in situ* sewer pipes, by determining the (decrease in) effective wall thickness via core sampling and laser profiling.

5.6 SUBSURFACE SOIL EROSION AROUND A SEWER PIPE

5.6.1 Introduction

Soil erosion by water is a degradation process in which particles are detached from the soil structure and carried away by flowing water. This process can be activated both by overland and subsurface flow (Bernatek-Jakiel & Poesen, 2018). When soil erosion takes place by the way of subsurface flow, its temporal and spatial evolution of the erosion profiles are not visible on the surface, so that the process may remain unnoticed for a long period of time. This aspect makes it generally difficult to monitor and analyse subsurface erosion, which is likely the reason that the number of research studies on subsurface erosion – also referred to as internal erosion – are disproportional compared to those on surface erosion (Bernatek-Jakiel & Poesen, 2018; Poesen, 2018). However, subsurface erosion has been reported as a highly relevant and widespread process (Poesen, 2018), and is the cause of various catastrophic failure phenomena with large societal impact, such as sewer collapse, dam breaks, the formation of sinkholes and the instability of earth dams (Bonelli *et al.*, 2018; Davies *et al.*, 2001; Foster *et al.*, 2000; Walthan *et al.*, 2005; Yang *et al.*, 2019).

One important example of subsurface erosion processes is *soil piping*, whereby relatively large porous networks are created under preferential flow paths (Bernatek-Jakiel & Poesen, 2018; Bernatek-Jakiel *et al.*, 2017; Jones, 2010). Soil piping has been observed in various types of soils, such as loess-derived soils, organic soils, clayey soils and sandy soils (Bernatek-Jakiel *et al.*, 2017), and may be initiated when the soil contains a high percentage of fine particles that are smaller than the average pore space in the soil. When the process starts and small particles are eroded, the pore space increases and individual pores become connected and form channels, which increases the hydraulic permeability and may trigger the release of coarser particles (Bernatek-Jakiel & Poesen, 2018; Jones, 2010). The pipe network that develops may lead to subsidence of the soil structure, and eventually even to severe structural failure, such as an abrupt collapse of the pipe network and the overlying soil structure – termed a sinkhole – or landslides (Bernatek-Jakiel & Poesen, 2018; Walthan *et al.*, 2005). Whether or not a pipe network collapses depends on the loading conditions and on the level of migration of soil particles. If, under the specific hydraulic conditions applied, only the finer particles erode and the coarser particles to some extent maintain a particle contact force fabric, the soil structure preserves its load bearing capacity and does not collapse. This process is denoted as *suffusion* (Fannin & Slangen, 2014; Hunter & Bowman, 2018; Moffat *et al.*, 2011). Conversely, the erosion process is characterized as *suffosion* if, under a strong flow of water, the transport of fine particles is accompanied by the collapse of the contact force fabric of the coarser particles, whereby some form of catastrophic structural failure is induced (Fannin & Slangen, 2014; Hunter & Bowman, 2018; Moffat *et al.*, 2011).

Another important type of subsurface erosion is *void formation*, which may occur around buried concrete pipe systems, such as sewer pipes, industrial discharge lines, tunnels, culverts and storm drains (Meguid & Kamel, 2014; Peter *et al.*, 2018; Wang *et al.*, 2014). Erosion voids can emerge due to ageing or improper construction procedures of pipe systems, with defects (i.e., cracks or gaps) in the pipe system initiating a water flow that locally washes away the soil support. Under continuous growth the erosion void may eventually evolve into a suffosion sinkhole, whereby the bearing resistance of the supporting soil vanishes and the soil structure above the pipe collapses (Indiketiya *et al.*, 2019; Walthan *et al.*, 2005). Sinkhole failures have been reported to happen within the lifetime of the pipe system, as well as during the period of construction (Walthan *et al.*, 2005), indicating that the rate of the underlying erosion processes is strongly variable and determined by local hydro-mechanical conditions. Various experimental and numerical studies have addressed the effect of erosion voids on the stresses generated in pipe systems and in the surrounding soil, whereby the stress redistribution caused by the appearance of an erosion void is found to be characterized by its location, size and contact angle (Meguid & Dang, 2009; Peter *et al.*, 2018; Tan, 2007; Wang *et al.*, 2014; Yasuda *et al.*, 2017). During void erosion the load transfer from the surrounding soil to the pipe generally takes place more localized, whereby the load magnitude increases. Accordingly, the susceptibility of the pipe system to cracking and damage, and thus to catastrophic failure, also increases (Meguid & Kamel, 2014; Scheperboer *et al.*, 2021; Talesnick & Baker, 1999; Wang *et al.*, 2014).

The following section presents a coupled hydro-mechanical bulk erosion model that may be used for studying soil piping and erosion void formation near and around a sewer pipe system. The content of this section is a summary of the work presented in Scheperboer *et al.* (2022).

5.6.2 Modelling of subsurface soil erosion

The erosion model developed by Scheperboer *et al.* (2022) treats the soil as a two-phase porous medium composed of a solid phase and a liquid phase, and accounts for its elasto-plastic deformation behaviour caused by frictional sliding and granular compaction. The kinetic law characterizing the erosion process is assumed to have a similar form as the type of threshold law used in interfacial erosion models (Horton, 1945; Knapen *et al.*, 2007). The degradation of the particle structure under erosion is considered to reduce the effective elastic stiffness of the porous medium and increase its permeability. The reduction in elastic stiffness is accounted for via an *erosion degradation parameter* D_{er}, which depends on the actual porosity of the soil and quantifies how effectively the mechanical degradation under erosion takes place. Its value ranges as $0 \leq D_{er} \leq 1$, whereby the limit cases $D_{er} = 0$ and $D_{er} = 1$ represent uneroded and fully eroded soil materials, respectively. The coupled hydro-mechanical model was implemented in the commercial FEM program ABAQUS, whereby the solutions of specific problems were computed in an incremental-iterative fashion using a staggered update scheme. The details of the model formulation and the numerical implementation can be found in Scheperboer *et al.* (2022).

Two practical case studies are considered that relate to a sewer system embedded in a sandy soil structure. The first case study treats *soil piping* caused by *suffusion* near a sewer system subjected to natural ground water flow, and the second case study considers the formation of a *suffosion erosion void* under strong ground water flow near a defect sewer pipe. The defect is located at the bottom of the pipe and has a width of 20 mm, representing a gap created by a sudden failure of a pipe joint. The effects on the erosion profile and the soil deformation behaviour by plasticity phenomena are elucidated by comparing the computational results to those whereby the constitutive behaviour of the granular material is modelled as elastic. The results of this comparison study illustrate the importance of including an advanced elasto-plastic soil model in the numerical simulation of erosion-driven ground surface deformations and the consequent failure behaviour.

The specific geometry considered in the hydro-mechanical analyses is depicted in Figure 5.15. The soil structure consists of a top layer of dry sand with a thickness of 2.27 m, which is supported by a fully saturated sand layer with a thickness of 1.50 m, followed by a relatively thick, fully saturated clay layer. The location of the centre of the sewer pipe corresponds with the ground water table at 2.27 m depth. The inner diameter and wall thickness of the round sewer pipe are 400 and 65 mm, respectively. The sewer system is covered by 2 m of dry sand, which is representative of the conditions in The Netherlands (Bolderman *et al.*, 2003). The specific part of the geometry simulated with the FEM model is indicated in Figure 5.15 by the light shaded rectangular section of 6×3 m^2. The mechanical and hydrological initial and boundary conditions used in the FEM analyses and the material parameters of the sandy soil and concrete sewer pipe can be found in Scheperboer *et al.* (2022).

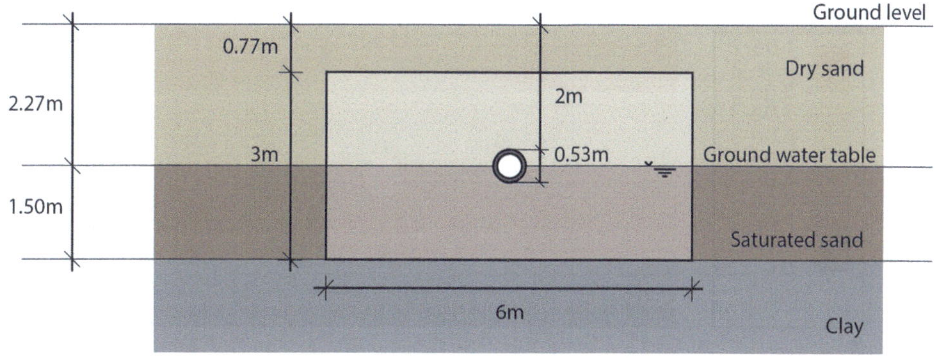

Figure 5.15 A round sewer pipe embedded in a stratified soil structure consisting of a 2.27 m thick, dry, sandy upper layer, followed by a 1.50 m thick saturated sandy layer, and finally a virtually impermeable, thick clay layer; the location of the centre of the sewer pipe corresponds to the ground water level at 2.27 m depth, and the light shaded area of $6 \times 3\ m^2$ indicates the computational domain used in the FEM simulations. The figure has been reprinted from Scheperboer *et al.* (2022).

The computational domain was modelled as two-dimensional, assuming a plane-strain condition in the out-of-plane direction. The soil structure with the elasto-plastic sand behaviour was discretized with 47,046 plane-strain 6-node iso-parametric coupled temperature-displacement elements, which were equipped with a 3-point Gauss quadrature. The sewer pipe was simulated as elastic and was discretized by 2089 plane-strain 6-node iso-parametric elements with a 3-point Gauss quadrature. As indicated in Figure 5.16, the mesh of the soil structure was refined towards the (coherent) interface with the sewer pipe to accurately simulate the local erosion profile. A preliminary mesh refinement study had indicated that the present discretization was sufficiently fine for obtaining converged numerical results for the two case studies.

5.6.3 Soil piping due to natural groundwater flow near a sewer system

The FEM simulation started with the application of the initial loading generated by the dead weight of the soil structure and the concrete pipe. The vertical settlement profile resulting from the dead weight loading is depicted in Figure 5.17a, and the corresponding deviatoric plastic deformations are illustrated in Figure 5.17b. The plastic deformations are reflected in the contour plot by the deviatoric invariant $\bar{\gamma}^p = \sqrt{(2/3)\gamma^p : \gamma^p}$ with γ^p the deviatoric plastic strain. It can be observed from

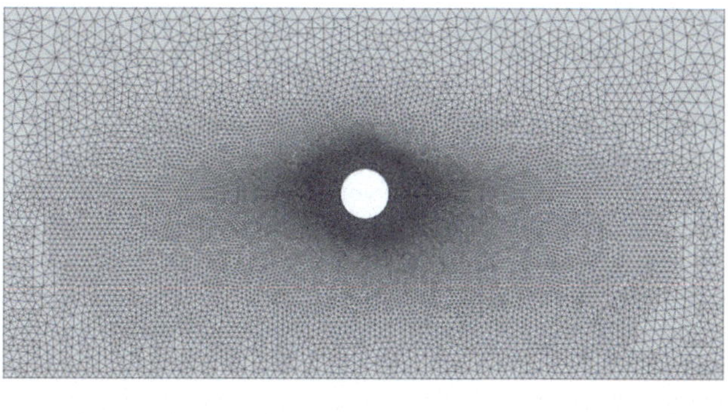

Figure 5.16 Finite element discretization used in the two case studies. The figure has been reprinted from Scheperboer *et al.* (2022).

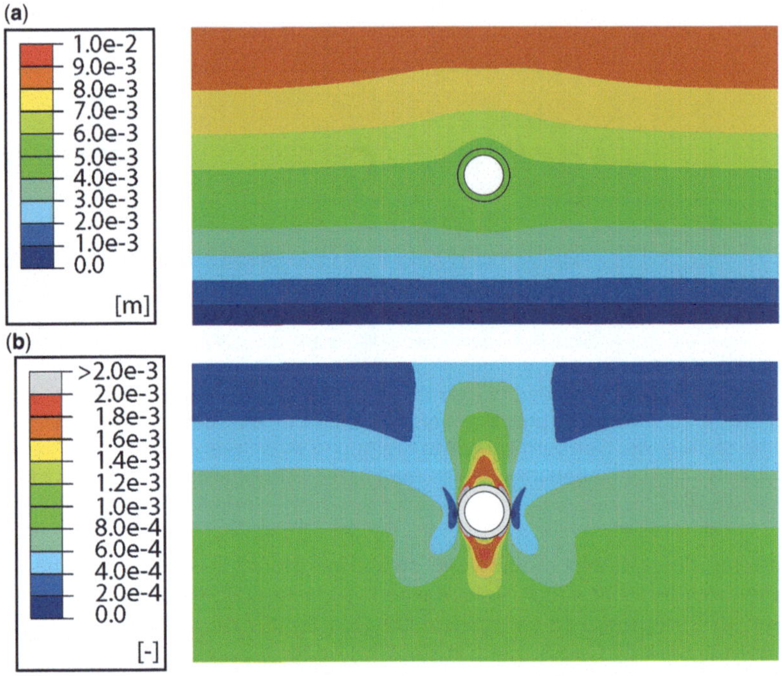

Figure 5.17 FEM solution resulting from the application of the initial, dead weight loading; the figure shows the spatial variation of (a) the vertical displacement w (in m) and (b) the deviatoric plastic strain invariant $\bar{\gamma}^p$. The figure has been reprinted from Scheperboer *et al.* (2022).

Figure 5.17a that the initial settlement profile was fairly uniform along the horizontal direction, with the settlements above the sewer pipe being slightly smaller than those at the left and right domain boundaries. Obviously, this difference may be ascribed to the relatively low effective density of the hollow sewer pipe. As indicated in Figure 5.17b by the grey-hatched area, the deviatoric plastic deformation was maximal at the top-left and top-right of the pipe circumference, reaching a value of $\bar{\gamma}^p = 0.45\%$. It can be further observed that the whole domain underwent plastic deformations after the application of the dead weight loading. The initial coefficient of lateral earth pressure for the non-cohesive sand could be computed from the ratio between the horizontal and vertical normal stresses, $K = \sigma_{xx}/\sigma_{zz}$, and varied between $K = 0.25$ and $K = 0.36$ going from the bottom to the top along the left (or right) boundary of the computational domain. In accordance with the range defined by the initial and maximum friction angles of the sand material, $29^\circ \leq \phi \leq 40^\circ$, this means that the sand along the left (or right) domain boundary experienced a horizontal stress state that approached the condition of active lateral earth pressure, as characterized by $K = (1 - \sin\phi)/(1 + \sin\phi)$ (Verruijt & van Baars, 2007). In contrast, right next to the sewer pipe the vertical stress generated by the dead weight loading was considerably smaller than the horizontal stress, in correspondence with $K = 3.29$, which was caused by a local arching of the sand material located along the upper half of the sewer pipe.

The application of a groundwater flux of 10^{-6} m/s (typical of *in-situ* conditions) at $t = 0$ at the left boundary of the computational domain initiated a groundwater flow from the left to the right domain boundary. Due to the relatively large permeability of the sand material, at sufficient distance below the sewer pipe the groundwater flow within a few minutes of time reached an almost uniform steady-state profile in the horizontal direction, whereby the flux value was equal, or close to, the applied boundary flux $\|\boldsymbol{q}_{fm}\| = 10^{-6}$ m/s. and the spatial pattern of $\boldsymbol{q}_{fm}$ had become virtually symmetric with respect to the vertical centreline of the FEM model. The time evolution of the erosion profile developing under the groundwater flow is sketched in figure 5.18, by plotting the erosion degradation parameter D_{er}, with $0 \leq D_{er} \leq 1$, at three different time instants, namely 7 years (Figure 5.18a), 13 years (Figure 5.18b) and 20 years (Figure 5.18c). The erosion profile started to grow at the bottom of the sewer pipe, and subsequently spread in a more or less horizontal direction towards the left and right domain boundaries, thereby creating a soil piping profile. Note that the green colour in the contour plot designates the area in which the erosion degradation parameter has reached the prescribed, maximum value for *suffusion* erosion, $D_{er} = D_{er}^{max} = 0.5$, and that the characteristic time scale associated with the development of the erosion profile is much larger than that of the process of ground water flow, that is, in the order of tens of years.

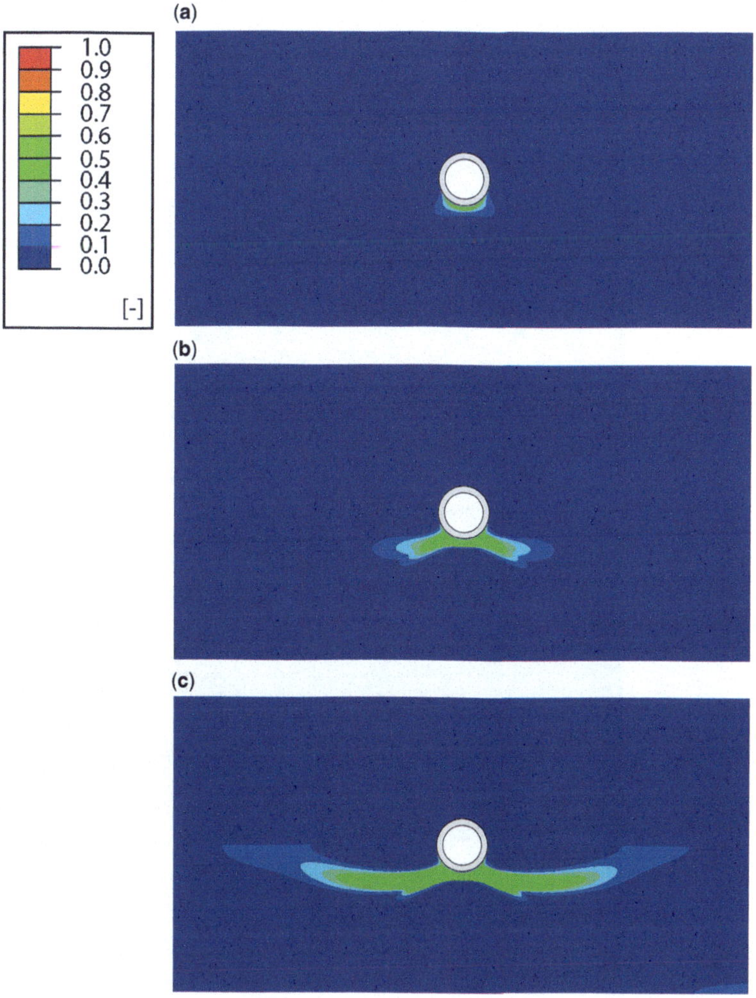

Figure 5.18 Time evolution of the erosion profile (soil piping) under natural groundwater flow (first case study), as characterized by the spatial development of the erosion degradation parameter D_{er} after (a) 7 years, (b) 13 years and (c) 20 years, whereby $D_{er}^{max} = 0.5$ (as representative of suffusion). The figure has been reprinted from Scheperboer *et al.* (2022).

The groundwater flux profiles after 7, 13 and 20 years are depicted in Figure 5.19 a, b and c, respectively, using the Euclidian norm of the flux, $\|\boldsymbol{q}_{fm}\|$, as the contour plot variable. It can be confirmed that the area in which the groundwater flux $\|\boldsymbol{q}_{fm}\|$ is larger than the critical threshold value of $q_{fm,cr} = 10^{-6}$ m/s (below which erosion does not occur) indeed corresponded to the geometry of the erosion profile shown in Figure 5.18. Further, the groundwater flux appeared to be maximal directly below the pipe, whereby the flux value increased when erosion developed, as caused by an increase in soil permeability.

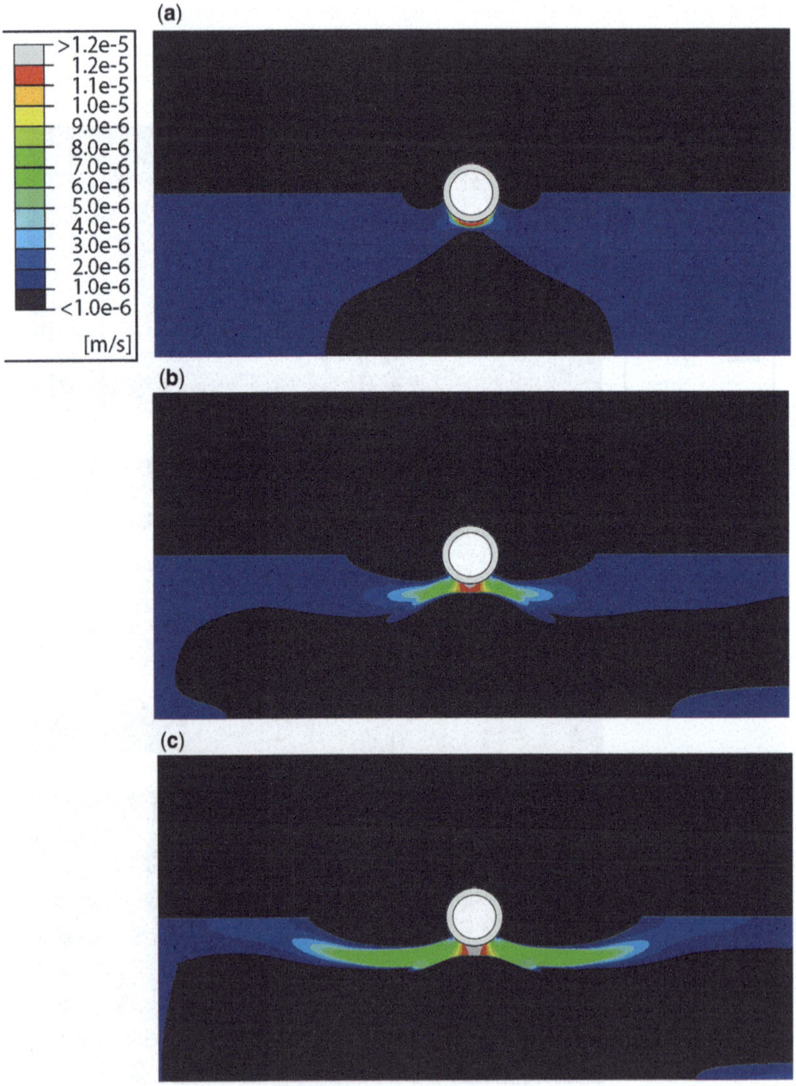

Figure 5.19 Time evolution of the porewater velocity field under natural groundwater flow (first case study), as characterized by the spatial evolution of the norm of the flux $\|\boldsymbol{q}_{fm}\|$ (in m/s) after (a) 7 years, (b) 13 years and (c) 20 years; the black colour indicates the area within which the value of $\|\boldsymbol{q}_{fm}\|$ is lower than the critical threshold value $q_{fm,cr} = 10^{-6}$ m/s for erosion. The figure has been reprinted from Scheperboer *et al.* (2022).

Figure 5.20 depicts the contour plot with the deviatoric plastic deformations, as evaluated after the application of the dead weight loading and an erosion process of 20 years. For clarity, in the contour plot the same scale division is used as for the deviatoric plastic deformations generated after the application of only the dead weight loading, see Figure 5.17b. In comparison with the deviatoric plastic deformations generated after the dead weight loading, left and right below the pipe the erosion process has led to a maximal increase in plastic strain of about a factor of two, in correspondence with a value of $\bar{\gamma}^p = 0.74\%$.

The evolution of the surface displacement, measured above the centre of the sewer pipe, is depicted in Figure 5.21. In order to clearly identify the displacement contribution caused by plastic deformations, the surface response is compared to that from a simulation in which erosion takes place on a fully elastic soil material. The elastic soil model was obtained from the elasto-plastic formulation by setting the yield strength artificially high. Figure 5.21 illustrates that for an elasto-plastic sand material the surface displacement monotonically increases with the development of erosion, whereby the value reached after 20 years equals 0.60 mm. The response for an elastic soil material shows a similar trend, whereby the deformation reached after 20 years is 25% less, that is, 0.45 mm. Since the surface displacements remained rather small, the practical suitability of using these as a monitoring parameter for the detection of a suffusion type of erosion under natural groundwater flow appears to be limited. This conclusion aligns with other scientific studies that report that soil pipes are only observable at the soil surface when a pipe roof collapses; accordingly, they are considered 'apparently inactive' over a long period of time, until clear surface evidence appears (Bernatek, 2015; Bernatek-Jakiel & Poesen, 2018; Verachtert *et al.*, 2011).

Figure 5.20 Spatial variation of the deviatoric plastic strain invariant $\bar{\gamma}^p$ after the application of the dead weight loading and an erosion process of 20 years under natural groundwater flow (first case study). The figure has been reprinted from Scheperboer *et al.* (2022).

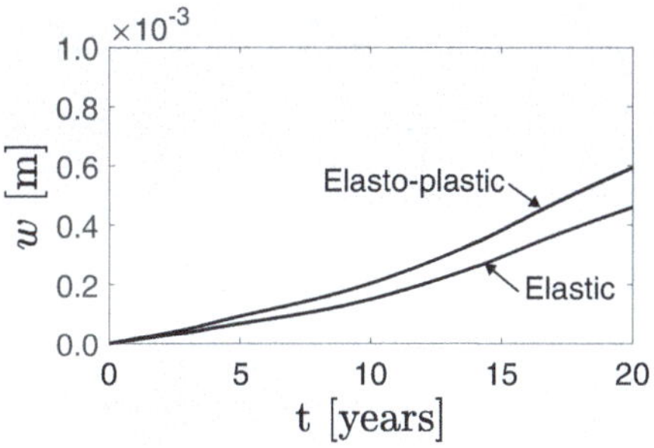

Figure 5.21 Time evolution of the vertical displacement *w* (in mm) of the ground surface – measured above the centre of the sewer pipe – under natural groundwater flow (first case study) for (i) an elasto-plastic sand material and (ii) an elastic sand material. The figure has been reprinted from Scheperboer *et al.* (2022).

5.6.4 Void formation due to strong groundwater flow near a defect sewer system

Like the process discussed in the previous section, the FEM analysis of soil erosion near the defect sewer pipe started with the application of the dead weight loading of the soil structure and the sewer pipe. The atmospheric pressure applied at $t = 0$ as a boundary condition across the gap width at the bottom of the sewer pipe initiated a relatively strong groundwater flow into the sewer pipe. The erosion profile caused by this flow profile is depicted in Figure 5.22 at three different moments in time, namely 47 days (Figure 5.22a), 93 days (Figure 5.22b) and 140 days (Figure 5.22c). It can be seen from the contour plot variable D_{er} that the erosion indeed started near the location of the gap, and subsequently extended along the bottom part of the pipe in a downward direction. At 140 days the erosion below the pipe had become quite severe and had induced a void with a depth of approximately half of the pipe diameter

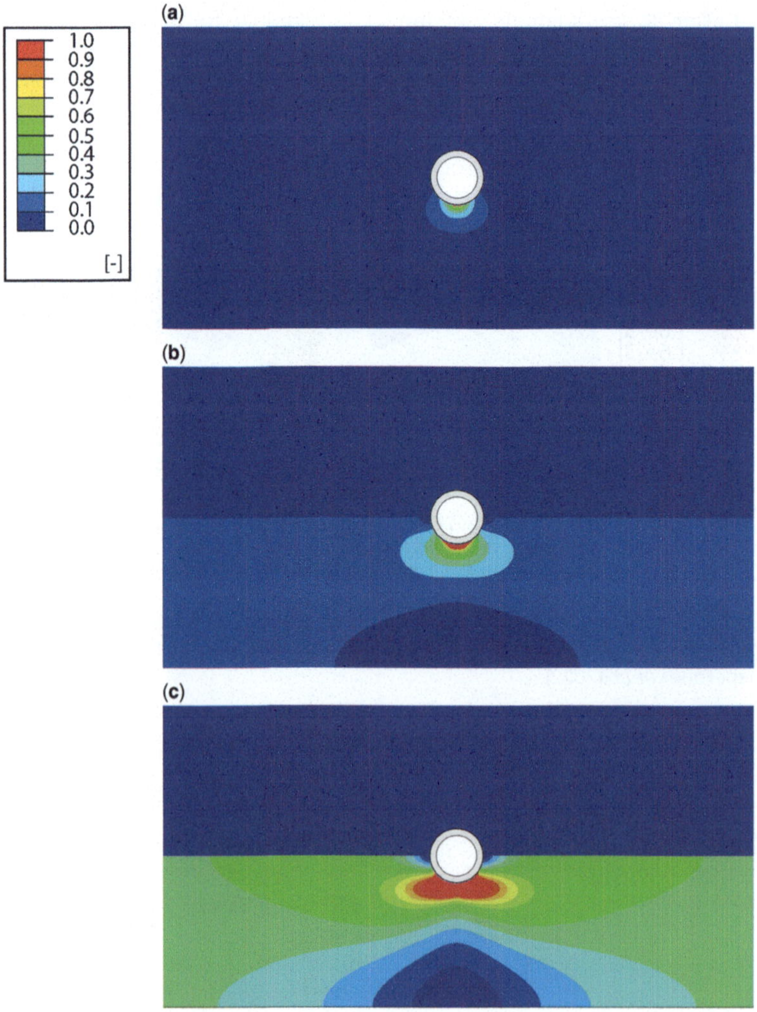

Figure 5.22 Time evolution of the erosion profile (void formation) under strong groundwater flow near a defect sewer pipe (second case study), as characterized by the spatial development of the erosion degradation parameter D_{er} after (a) 47 days, (b) 93 days and (c) 140 days, whereby $D_{er}^{max} = 0.93$ (as representative of suffosion). The figure has been reprinted from Scheperboer *et al.* (2022).

(red colour) within which the erosion degradation parameter was maximal, $D_{er} = D_{er}^{max} = 0.93$. Hence, the erosion void is of the *suffosion* type, with its volume for 95% filled by ground water, and only for 5% by remaining soil particles. It can be further observed that almost the complete saturated sand layer at this stage had undergone some degree of erosion, in agreement with an erosion degradation parameter larger than zero and below the maximum value, $0 < D_{er} < 0.93$. It is further interesting to observe that the erosion profile developed much faster than in the first case study on soil piping, see Figure 5.18, which clearly demonstrates that the characteristic time scale of an erosion process very much depends on the type of problem and the corresponding hygro-mechanical conditions.

The groundwater flow patterns after 47 days, 93 days and 140 days are illustrated in Figure 5.23a, b and c, respectively, and show to be in agreement with the erosion profiles in the area in which the

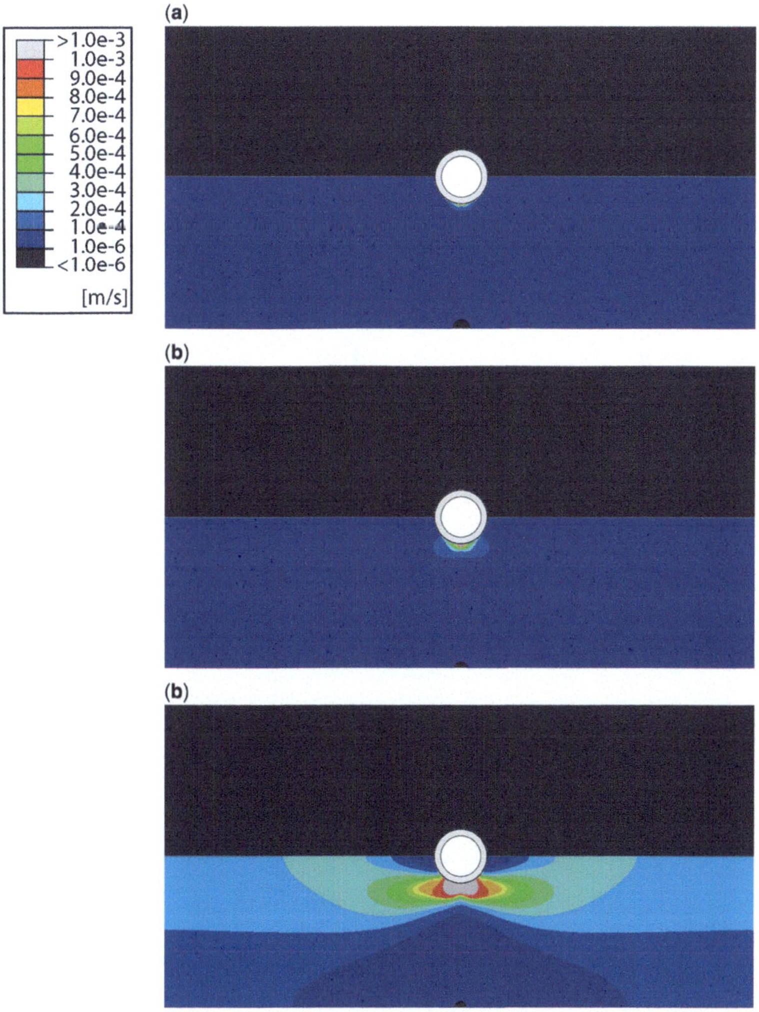

Figure 5.23 Time evolution of the porewater velocity field under a strong groundwater flow near a defect sewer pipe (second case study), as characterized by the spatial evolution of the norm of the flux $\|\boldsymbol{q}_{fm}\|$ (in m/s) after (a) 47 days, (b) 93 days and (c) 140 days; the black colour indicates the area within which the value of $\|\boldsymbol{q}_{fm}\|$ is lower than the critical threshold value $q_{fm,cr} = 10^{-6}$ m/s for erosion. The figure has been reprinted from Scheperboer *et al.* (2022).

flux threshold value of $q_{\mathrm{fm,cr}} = 10^{-6}$ m/s was exceeded. Within the erosion void the permeability of the remaining 'soil structure' was maximal and corresponded to a value of $k = 2.0 \times 10^{-2}$ m/s. This value is 200 times larger than the initial permeability $k_0 = 10^{-4}$ m/s of the sand material.

The deviatoric plastic deformations generated under the dead weight loading and the subsequent erosion process are illustrated in Figure 5.24 representing the passage of 140 days. In comparison with the deviatoric plastic deformations generated under only the dead weight loading, see Figure 5.23b, the deviatoric plastic deformation in the soil material directly left and right of the sewer pipe had substantially increased by more than a factor of 10 to a value of $\bar{\gamma}^{\mathrm{p}} = 5.4\%$. Furthermore, at the bottom-left and bottom-right of the pipe the deviatoric plastic strain had reached a maximum value of $\bar{\gamma}^{\mathrm{p}} = 14.4\%$. Obviously, this strong, local increase in deviatoric plastic deformation was caused by the stress redistribution that balanced the local loss of soil resistance in the suffusion erosion void below the sewer pipe; the specific local shear failure zones in the soil next to the sewer pipe are indicated in Figure. 5.24 by the grey hatched areas.

Figure 5.25 depicts the vertical surface displacement as a function of time for both an elasto-plastic and an elastic sand material. The surface displacement initially grew moderately, but after ~100 days started to increase rapidly due to the formation of the erosion void below the sewer pipe. The surface displacement after 140 days equalled 5.8 and 4.4 mm in the case of the elasto-plastic soil model and

Figure 5.24 Spatial variation of the deviatoric plastic strain invariant $\bar{\gamma}^{\mathrm{p}}$ after the application of the dead weight loading and an erosion process of 140 days under a strong groundwater flow near a defect sewer pipe (second case study). The figure has been reprinted from Scheperboer *et al.* (2022).

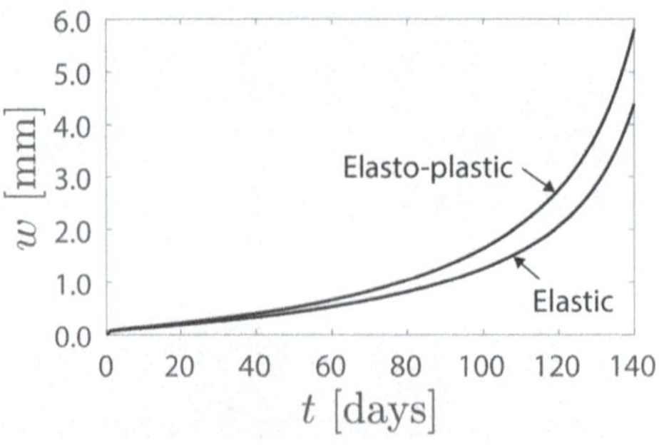

Figure 5.25 Time evolution of the vertical displacement w (in mm) of the ground surface – measured above the centre of the sewer pipe – under a strong groundwater flow near a defect sewer pipe (second case study) for (i) an elasto-plastic sand material and (ii) an elastic sand material. The figure has been reprinted from Scheperboer *et al.* (2022).

the elastic soil model, respectively. The relative difference between these displacements is substantial, that is, 32%, which indicates the importance of accurately modelling the constitutive behaviour of the soil in erosion simulations by means of an elasto-plastic model. Also, it is interesting to notice that the ground surface deflections in Figure 5.25 are much larger than those generated under soil piping erosion, see Figure 5.21.

Additional simulations not presented here have shown that a decrease of the gap width in the sewer pipe by a factor of 40 to a value of 0.5 mm led to a similar surface deflection evolution as illustrated in Figure 5.25 Essentially, the decrease in gap width reduced the flow area at the bottom of the sewer pipe, which was accompanied by an increase in the local flow velocity, such that, as a net effect, the erosion void developed in a more or less comparable fashion. Nevertheless, the value of the surface deflection at 140 days was about 11% smaller, namely 5.2 and 3.9 mm in the case of an elasto-plastic and an elastic sand material, respectively. Note hereby that the assumption of a small gap width of 0.5 mm ignored the fact that the relatively large particles in the sand material cannot flow into the pipe, so that the real surface deflections for this case would be smaller.

Finally, it needs to be emphasized that the remaining stability of the suffosion type of erosion void illustrated in Figure 5.14c is limited, as its bearing capacity was (virtually) generated by the fluidic mixture inside; hence, as soon as the groundwater table lowered and the fluidic mixture left the erosion void, the sewer pipe drops down into the erosion void, whereby the overlying soil structure is likely to collapse into a sinkhole. Accordingly, the abrupt increase in surface settlement registered after 100 days can be considered as a critical warning for catastrophic failure. Since the subsequent surface displacements are in the order of several millimetres, in practice it should be possible to detect these changes with satellite radar interferometry (Chang & Hanssen, 2014; Malinowska *et al.*, 2019), which may help to prevent the eventual collapse into a sinkhole.

The above numerical analyses on soil piping and erosion void formation illustrate that the numerical model developed by Scheperboer *et al.* (2022) realistically predicts the size, location and characteristic time scale of the generated erosion profiles and the deformations of the surrounding soil structure. Hence, the erosion profiles computed by the model may be used as input for a detailed analysis of the local, residual bearing capacity and stress redistribution of buried concrete pipe systems. Additionally, the modelling results may support the early detection of *in situ* subsurface erosion phenomena from ground surface deformations recorded with satellite radar interferometry. The analyses further show that the characteristic time scales of the erosion process and the generated ground surface deflections strongly depend on the geometry and features of the problem, as reflected by the specific hydro-mechanical conditions.

5.7 STRUCTURAL FAILURE OF CONCRETE SEWER PIPES

5.7.1 Introduction

In Dirksen *et al.* (2013a) it is shown that the accuracy and reliability of visual inspection data may be low, with 25% of the damage not being detected. In addition, in Stanic *et al.* (2013) and Stanic *et al.* (2017) it is demonstrated that no evident correlation exists between visual inspection data and the material properties obtained from core samples taken from the sewer pipe. Moreover, an accurate database on the actual condition of sewer systems is lacking (Dirksen *et al.*, 2013a; Wirahadikusumah *et al.*, 2001; van Riel *et al.*, 2016), and the effects of biochemical deterioration and ageing on the structural performance of a sewer pipe are yet not well understood, and therefore difficult to take into account in the assessment procedure (Grengg *et al.*, 2018; Monteny *et al.*, 2000; O'Connell *et al.*, 2010; Stanic *et al.*, 2017).

For the development of a good understanding of the failure behaviour of concrete pipe systems, over the past decades various full-scale field studies (Erdogmus & Tadros, 2009; Hill *et al.*, 1999; McGrath *et al.*, 1999; Nehdi *et al.*, 2016) and laboratory studies (Brachman *et al.*, 2001; MacDougall *et al.*, 2016; Peter *et al.*, 2018; Stanic *et al.*, 2017; Trautmann & O'Rourke, 1985; Younis *et al.*, 2020; Zoladz *et al.*,

1996) have been performed, among which geocentrifuge tests to account for specific interactions with the surrounding soil structure (Rakitin & Xu, 2014; Schofield, 1980; Xu & Shen, 2020). Accordingly, for diverse pipe geometries the bearing strength, deformation behaviour and crack patterns were measured under representative loading and boundary conditions, thereby identifying aspects such as the susceptibility to joint failure and leakage, the most unfavourable loading conditions, and the role of local soil collapse. The experimental failure behaviour of concrete pipes has been complemented and validated by numerical simulations (Kang *et al.*, 2007; Meguid & Dang, 2009; Tan, 2007; Xu *et al.*, 2017), and translated to practical recommendations and design procedures.

In the present section, this line of research is continued by presenting a comprehensive, systematic experimental–numerical study of the structural failure behaviour of unreinforced round and egg-shaped concrete sewer pipes. The main purpose of the study is to obtain thorough insight into the load bearing capacity and failure mechanisms of concrete sewer pipes, and to understand the influence by degradation effects typical of soil erosion, changes in lateral earth pressure, biochemical processes, and ageing. The study considers sewer pipe types most commonly used in the Netherlands, which are (i) two different round pipes with an inner diameter of 400 and 500 mm, respectively and (ii) an egg-shaped pipe with a horizontal inner diameter of 400 mm and a vertical inner diameter of 600 mm. The concrete sewer pipes are subjected to biaxial loading conditions in an experimental set-up customized from the test specifications provided in Stanic *et al.* (2017), whereby the horizontal pressure loading is set equal to 1/3 times the vertical pressure loading. This loading condition is considered to be representative of a concrete sewer pipe embedded in a well-graded sandy gravel and subjected to neutral horizontal earth pressure. The elasto-fracture properties of the concrete are determined from material tests performed on small samples taken from the sewer pipes. The experimental failure responses of the round and egg-shaped sewer pipes are validated by dedicated FEM)analyses. Subsequently, a parameter variation study is carried out using the FEM of the round sewer pipe with an inner diameter of 400 mm. Accordingly, the sensitivity of the overall failure response of the sewer pipe to several parameters is investigated, which are the load contact area (to mimic the effect of soil erosion), the ratio between the applied horizontal and vertical loadings (to analyze the effect by a change in lateral earth pressure), the wall thickness (to simulate the effect of biochemical degradation), and the tensile strength, mode I toughness and Young's modulus of the concrete (to model the effect of biochemical degradation and/or ageing). The outcome of the experimental–numerical study provides a scientific basis and practical guidelines for municipalities in their decision making process regarding the maintenance and replacement of concrete sewer pipe systems. The content of this section is a summary of the work presented in Scheperboer *et al.* (2021).

5.7.2 Experimental programme

The load bearing capacity and failure mechanisms of new concrete sewer pipes are assessed by means of full-scale experiments. The test specimens, experimental set-up, test procedure and measuring devices are specified below.

5.7.2.1 Test specimens

In the experimental programme, three different unreinforced concrete sewer pipes were considered, namely dry-cast round pipes with an inner diameter of 400 and 500 mm, labelled as R400 and R500, respectively, and dry-cast egg-shaped pipes with a horizontal and vertical inner diameter of respectively 400 and 600 mm, labelled as E400/600. The sewer pipes were manufactured by the production companies Kijlstra and De Hamer located in the Netherlands. The strength and stiffness parameters of the specific concrete used were determined by material tests (Scheperboer *et al.*, 2021). The pipes were 1–4 years old and had not been exposed to in-situ conditions. A number of specimens with a manageable length of $\sim$ one metre were sawn from the sewer pipes, whereby the socket and spigot were removed. The specimens were kept moist prior to testing in order to mimic in-situ hydrological conditions. The internal and external surfaces of the pipe specimens were inspected visually to ensure the specimens did not have noticeable defects. For each of the three sewer pipe types the test repeatability was explored by testing six similar pipe specimens. Both the outer and inner specimen height and width and the wall thickness were measured (Scheperboer *et al.*, 2021).

5.7.2.2 Experimental set-up

An overview of the experimental set-up is shown in Figure 5.26. The test frame was composed of steel HEB 300 members, which are sufficiently stiff to marginalize the frame deformations generated during the testing procedure. The horizontal loads were applied by four jacks – two at each side – positioned at the half height of the test specimen, and the vertical load was applied at the top of the test specimen by means of three jacks. The jacks were equally spaced along the specimen length. The ratio between the horizontal and vertical loads was set equal to 1/3. This ratio is representative of a sewer pipe embedded in a well-compacted sandy gravel with an ultimate friction angle f in the range of 40°–45°, for which the coefficient of neutral horizontal earth pressure, in accordance with Jaky's empirical formula, $K_0 = 1 - \sin(f)$, is ~ 1/3 (Verruijt & van Baars, 2007). During the testing the specimen behaviour was recorded by means of two cameras placed at the front side of the specimen. At the top and the two lateral sides of the test specimen, the loading was transferred to the curved outer surface of the specimen by placing a steel member, a steel plate, several plywood boards, an air cushion and a rubber foil between the jacks and the specimen surface. The steel member and plate were used to spread the applied load evenly across the specimen surface, the ply wood boards filled up some space between the steel plate and the test specimen, and the rubber foil served to protect the air cushions from mechanical damage. Once filled with air, the air cushions were observed to accurately follow the curved outer surface of the pipe specimen. For each sewer pipe type, the average contact width of the air cushions was measured from photographs taken prior to failure (Scheperboer *et al.*, 2021).

Throughout the test procedure the relative change in loading contact area turned out to be minor; in specific, the change was less than 5% after the air cushions had been sealed, which allows the loading contact area to be taken as constant. For the round sewer pipes the support at the specimen bottom consisted of an air cushion with a rubber foil, followed by ply wood boards, a steel plate and a steel member. Since the bottom of egg-shaped pipes is flat, for these pipes an adequate contact at the support was simply established with a softboard plate with a thickness of 10 mm, instead of an air cushion.

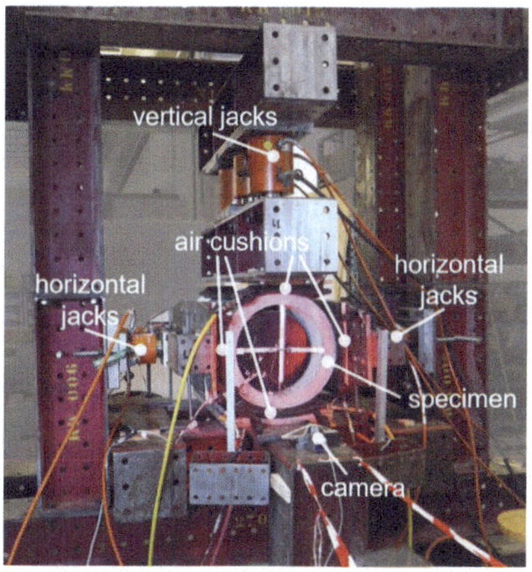

Figure 5.26 Full-scale experimental set-up for a sewer pipe subjected to biaxial loading conditions. The figure has been reprinted from Scheperboer *et al.* (2021).

5.7.2.3 Test procedure

The test protocol adopted for safely and accurately performing the mechanical tests on the sewer pipes is as follows. Firstly, the air cushions at the lateral sides of the pipe were inflated in order to ensure adequate contact between the pipe and the horizontal jacks and to prevent the round pipes from rotating about their length axis. After the cushions had generated a horizontal load of 3 kN, the air cushions at the top and bottom surfaces of the pipe were inflated up to a vertical load of 9 kN, in correspondence with the horizontal to vertical loading ratio of 1/3. Subsequently, all four air cushions were simultaneously inflated further until the horizontal and vertical loads reached the values of 20 and 60 kN, respectively, after which the air cushions were sealed. The inflation of the air cushions was controlled in a manual fashion, whereby the selected horizontal to vertical loading ratio of 1/3 was maintained as close as possible. After the sealing of the air cushions, the jack system was activated and the horizontal and vertical loads were increased up to failure of the pipe specimen. The mechanical testing was performed in a quasi-static, load-controlled fashion, using vertical loading speeds of 6 and 10 kN/min for respectively the round pipe specimens and the egg-shaped pipe specimens. Correspondingly, the test duration was $\sim$ 30–45 min for the round pipes and 45–60 min for the egg-shaped pipes.

5.7.2.4 Measuring devices

The mechanical response was determined by measuring the applied load and the generated deformation at regular time intervals of 1 s. The horizontal and vertical loads were monitored by means of HBM 500 bar pressure sensors that measure the oil pressure inside the tubes activating the jacks. The displacements in horizontal and vertical directions were measured at the front and back sides of the sewer pipe using Solartron SM3 linear variable displacement transducers (LVDTs). The local strains were monitored employing Tokyo Sokki Kenkyojo PL-60-11 strain gauges (SGs). At the top, bottom, left and right sides of the sewer pipe, strain gauges were placed on the inner surface of the pipe, both near the front side and back side. In addition, strain gauges were placed on the front and back surfaces of the sewer pipe, at the top, bottom, left and right sides, at a distance of 10 mm from the outer pipe circumference. The denotations 'left side' and 'right side' hereby relate to a viewpoint taken at the front side of the pipe specimen, so that for the front and back sides the adjectives 'left' and 'right' refer to the same specimen side. Three additional strain gauges were placed at equal mutual distance along the length of the sewer pipe. An HBM X60 two-component adhesive was used to glue the strain gauges to the sewer pipe surfaces. At regular time intervals, photographs were taken of the sewer pipe front surface, which were correlated to the load and deformation recorded at the front side of the specimen.

5.7.3 Numerical model

The experimental results of the concrete sewer pipes are validated by means of 2D plane-strain FEM analyses. The numerical simulations are carried out with the FEM, using the commercial software package ABAQUS Standard. The FEM models are based on 2D plane-strain continuum elements that simulate the elastic (bulk) behaviour of the concrete sewer pipe. The fracture behaviour of concrete is simulated with interface elements endowed with the mixed-mode damage model presented in Cid Alfaro *et al.* (2009), which is implemented in ABAQUS as a user-supplied subroutine (i.e., a UMAT). The essential input parameters of this model are the ultimate fracture strength t^u and the toughness G_c of the concrete material under pure modes I and II loading conditions. The model further contains parameters that (i) allow for including rate-dependent fracture effects and (ii) assist in regularizing fracture responses with complex trajectories towards a physically realistic result. For more details on the formulation and the numerical discretization of this model, the reader is referred to Cid Alfaro *et al.* (2009).

In accordance with the modelling strategy originally proposed in Xu and Needleman (1994), the interface elements are located between all continuum elements, so that the mechanisms of crack bifurcation, crack coalescence and crack branching at arbitrary locations and in arbitrary directions are simulated in an automatic fashion. The accuracy and efficiency of the above approach have been successfully demonstrated in previous fracture studies for a variety of materials. The set of equilibrium equations is solved through an incremental–iterative update procedure, based on an implicit Euler backward scheme combined with a full Newton–Raphson iterative update strategy. The arc-length method is adopted to account for a possible softening and snap-back behaviour in the quasi-static, load-controlled simulations (Riks, 1979).

The geometry, boundary conditions and FEM discretizations of the round and egg-shaped sewer pipes are illustrated in Figure 5.27a and b, respectively. The geometries of the modelled sewer pipes are in close agreement with those used in the experiments. The inner diameters of the modelled geometries are 400 and 500 mm for the round R400 and R500 pipes, and 400 mm/600 mm for the E400/600 egg-shaped sewer pipe, respectively. The wall thickness of the pipes is taken as the average value measured in the experiments, and equals 62 and 72 mm for, respectively, the R400 and R500 pipes, and 70 mm for the upper part of the E400/600 pipe, see Table 5.1. The loads are introduced on the sewer pipe by means of horizontal and vertical surface pressures $\sigma/3$ and σ, respectively, in correspondence with the horizontal to vertical loading ratio of 1/3 used in the experiments. The projected contact width l_c along which the surface pressure is applied equals 250, 280 and 290 mm, for the R400, R500 and E400/600 pipe types, respectively, which is in close correspondence with the widths of the contact zones applied in the testing procedure, see Table 5.1. Accordingly, the total vertical load used in the presentation of the computational results is $F_v = \sigma\, l_c\, \mathrm{L}$, where $\mathrm{L} = 1$ m equals the length of the sewer pipe specimen. The round pipes are restrained against rigid body motions by three roller supports, see Figure 5.27, which are located at the side of the sewer pipe wall that is loaded in compression. The soft board plate used for vertically supporting the egg-shaped pipe has a thickness of 10 mm. As sketched in Figure 5.27b, the soft board is supported in the vertical and horizontal directions by roller supports, and is coherently connected to the pipe specimen. The sewer pipe geometries are discretized by 3-node iso-parametric plane-strain elements equipped with a 1-point Gauss quadrature, whereby the small-strain assumption is adopted. The cracking paths defining the failure response of the sewer pipe are modelled via 4-node interface elements equipped with a 2-point Gauss quadrature. As mentioned above, the interface elements are located between all triangular continuum elements, which allows crack initiation and propagation to be determined by the actual geometry and boundary conditions of the sewer pipe. The detailed cracking paths characterizing the failure response of the sewer pipes are unknown a priori; however, for warranting accurate numerical results, based on the experimental results the areas within which cracks are expected to

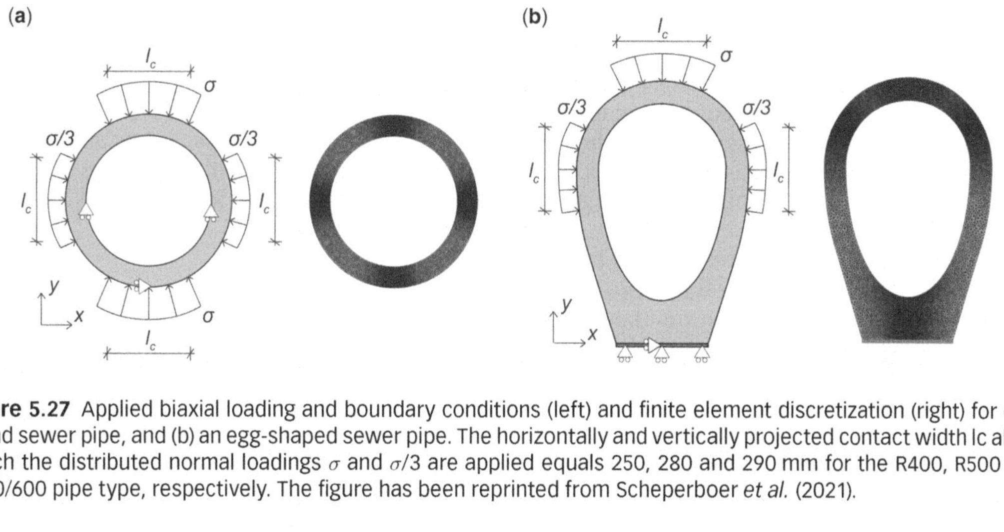

Figure 5.27 Applied biaxial loading and boundary conditions (left) and finite element discretization (right) for (a) a round sewer pipe, and (b) an egg-shaped sewer pipe. The horizontally and vertically projected contact width lc along which the distributed normal loadings σ and $\sigma/3$ are applied equals 250, 280 and 290 mm for the R400, R500 and E400/600 pipe type, respectively. The figure has been reprinted from Scheperboer *et al.* (2021).

emerge are discretized by a relatively fine FEM mesh. The specific number of continuum and interface elements used in the FEM models depends on the actual pipe geometry considered and varies between 34,000–55,000 continuum elements and 50,000–61,000 interface elements, see Figure 5.27 for two examples. Although the spatial discretization of the modelled geometry in principle can have an effect on the location and direction of the cracking path, the FEM meshes used in the present study are sufficiently fine to marginalize this effect. It should be further emphasized that the mesh fineness is bounded by a maximum in order to avoid artificial response contributions related to the use of an elastic interface stiffness K in the traction separation law, see Cid Alfaro *et al.* (2010a) for a detailed discussion of this aspect. The set of equilibrium equations obtained for the FEM models is solved by using an incremental-iterative update approach, which is based on an implicit backward Euler scheme combined with a full Newton–Raphson iterative update procedure. The arc-length method is adopted to robustly simulate a possible snap-back or softening behaviour in the quasi-static, load-controlled simulations (Riks, 1979).

The Young's modulus E_c of the continuum elements describing the elastic behaviour of concrete was determined from uniaxial compression tests, in close conformity with the regulations from Eurocode NEN-EN 12390-13 (European Committee for Standardization, 2013). Accordingly, from each of the three sewer pipe types a total of 12–15 small samples were sawn, whereby the Young's moduli measured for the samples were averaged over the corresponding total number of samples (Scheperboer *et al.*, 2022). The value of the Poisson's ratio ν_c is taken from the Eurocode EN 1992-1-1 (European Committee for Standardization, 2005), in accordance with concrete strength class C50/60. The tensile strength t_1^u , and mode I toughness $G_{I,c}$ used in the interface damage model were determined experimentally from three-point bending tests, with the specimen dimensions and loading procedure based on Hordijk (1991). Accordingly, from each of the three types of sewer pipes 18 concrete notched beam specimens were sawn, whereby the average values of the experimental sample strength and toughness listed were calibrated from dedicated FEM analyses that accurately mimic both the experimental fracture path and the load-displacement response of the beam specimens. The combined experimental–numerical calibration procedure applied here is similar to that recently performed for determining the tensile strength and toughness parameters of oak wood, as described in detail in Luimes *et al.* (2018). For simplicity, the shear strength t_2^u, of concrete is taken equal to the tensile strength t_1^u, and the mode II toughness $G_{II,c}$ is set equal to two times the mode I toughness $G_{I,c}$ (Reinhardt & Xu, 2000). The soft board plate supporting the E400/600 egg-shaped sewer pipe has a Young's modulus $E_{sb} = 3.0$ N/mm^2, which is the average value measured from 8 uniaxial compression tests on soft board samples. Poisson's ratio ν_{sb} of the soft board is set equal to 0.20 (Moarcas & Irle, 1999). In all simulations the elastic stiffness K of the interface damage model was prescribed to have a relatively high value, for which it was confirmed that the overall elastic response of the sewer pipe remained virtually unaffected by this parameter and was determined almost completely by the elastic properties of the surrounding continuum elements. In addition, the material parameters related to the rate-dependency of the kinetic law were chosen such that the fracture response closely approximates the rate-independent limit case. More details on these aspects can be found in previous numerical studies performed with the interface damage model (Cid Alfaro *et al.*, 2009, 2010a, 2010b).

5.7.4 Comparison of experimental and numerical results

5.7.4.1 Load–displacement response of round sewer pipes

Figure 5.28a and b illustrate the simulated load–displacement response (red line) of respectively the round pipes R400 and R500, together with the corresponding experimental responses (black lines). The vertical loading F_v is plotted against the net vertical displacement d_v in Figure 5.28. The comparison related to the horizontal load–displacement response of the pipes is similar to that of the vertical response, and is therefore left out of consideration. It can be seen that the agreement between the numerical and experimental results generally is good. The initial elastic response of the sewer pipes is captured accurately by the FEM models, and the vertical peak load appears to be in good agreement

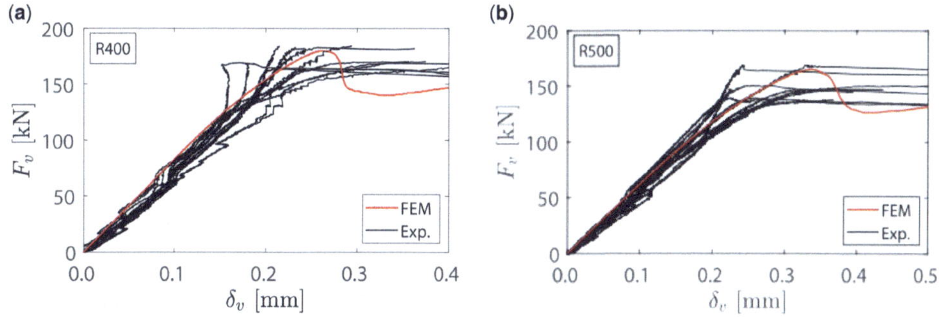

Figure 5.28 Vertical load F_v versus vertical net displacement d_v of (a) the round R400 and (b) the round R500 sewer pipe, as obtained by experiments (black lines) and an FEM simulation (red line). The figure has been reprinted from Scheperboer *et al.* (2021).

with the corresponding experimental peak load. In specific, the numerical vertical peak loads of the R400 and R500 sewer pipes are 180.0 kN and 165.5 kN, which overestimate the average experimental peak loads by 5% and 11%, respectively. After passage of the peak load, the numerical response is characterized by an unstable softening branch, whereby the external load temporarily decreases with increasing displacement, until it reaches a residual value of about 0.8 times the peak load. The experimental response does not show a softening behaviour, due to the fact that the experiments were performed in a load-controlled fashion. Instead, the instability emerging after passing the peak load changed the experimental response from quasi-static to dynamic.

5.7.5 Local strain response of round sewer pipes

Figure 5.29a and b illustrate the experimental (black lines) and numerical (red line) load–strain responses at two specific locations along the inner diameter and near the outer diameter of the R400 pipe, that is, the bottom-inside and the right-outside locations. The correspondence between the experimental and numerical responses generally is good; both the initial linear elastic regime and the response towards the peak load are captured accurately. Since the load–strain responses of the R500 sewer pipe specimens with a larger diameter of 500 mm are comparable to those of the R400 specimens shown in Figure 5.29, with a similar agreement between the experimental and numerical results, the discussion of these curves is omitted here.

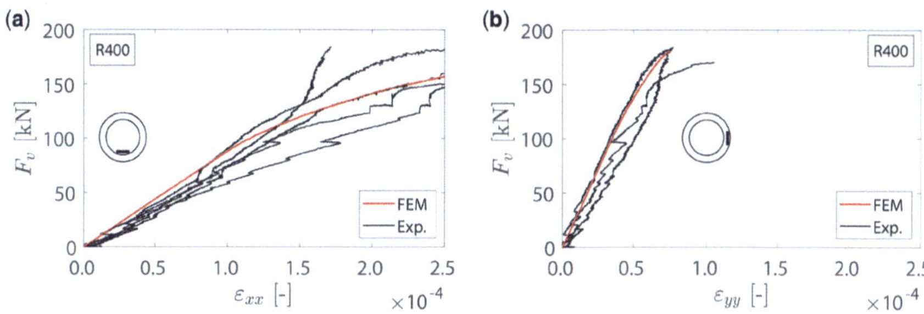

Figure 5.29 Experimental (black lines) and FEM (red line) load–strain response for the round R400 sewer pipe. Vertical load F_v versus the normal strain e_{xx} or e_{yy} as measured at (a) the bottom-inside and (b) the right-outside locations. The corresponding location of the strain gauge is indicated in the inset of the figures. The figure has been reprinted from Scheperboer *et al.* (2021).

5.7.6 Fracture pattern of round sewer pipes

The contour plots in Figure 5.30 show the maximum principal stress in the R400 pipe at four consecutive stages (a)–(d) of the failure process, as indicated in the added load–displacement diagram. Since the macroscopic failure cracks develop in the post-peak regime, the deformation range in the load–displacement diagram needed to be chosen relatively large. For clarification, the deformations of the pipe geometry are magnified in the contour plots by a factor of 50. At stage (a) in the load–displacement diagram the sewer pipe reaches its load bearing capacity, whereby macroscopic failure

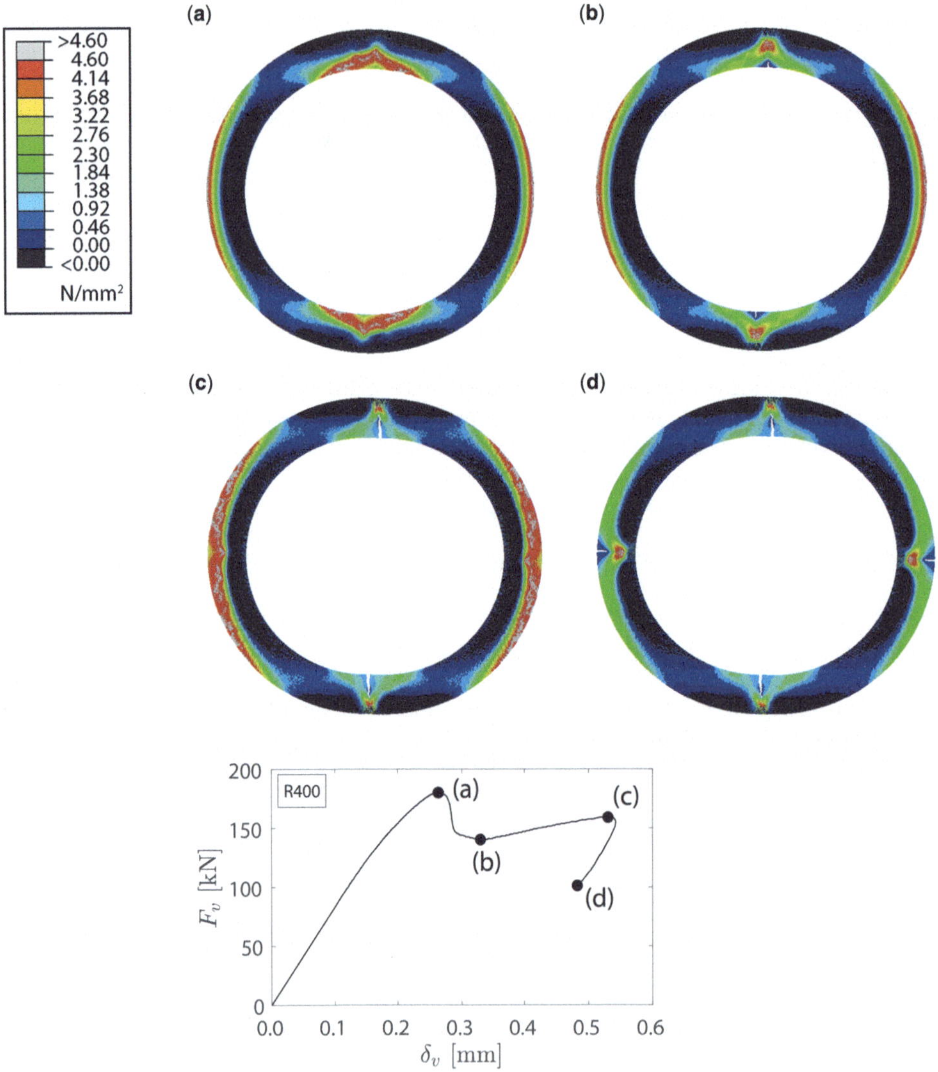

Figure 5.30 Maximum principal stress at crack nucleation (a) and crack propagation (b)–(d) in the round R400 sewer pipe under the applied biaxial loading schematized in Figure 5.27. For clarification, the deformations of the pipe geometry are magnified in the contour plots by a factor of 50. The four failure states (a)-(d) are indicated in the corresponding load–displacement diagram. The figure has been reprinted from Scheperboer *et al.* (2021).

cracks nucleate at the inner top and bottom parts of the sewer pipe. As can be observed from the contour plot in Figure 5.30a, at the locations of these two cracks the maximum principal stress has reached the material tensile strength of 4.6 N/mm^2 (depicted in red). When the two cracks start to propagate, the micro-cracks appearing in the vicinity of these cracks start to close, by which the crack pattern becomes more localized, see Figure 5.30b. Simultaneously, the maximal external load defining the crack nucleation stage drops in magnitude by approximately 20% in accordance with a structural softening behaviour, after which the sewer pipe reaches its residual strength, indicated by stage (b) in the load–displacement diagram. Upon further growth of these cracks, the external load slightly increases with deformation, until reaching stage (c) at which two additional failure cracks nucleate at the outer left and right sides of the sewer pipe, see Figure 5.30c. The development of these failure cracks is illustrated in Figure 5.30d, and causes that the external vertical load substantially decreases under a decrease in the net vertical displacement, in correspondence with reaching stage (d) in the load–displacement diagram. The failure branch from stage (c) to stage (d) commonly is referred to as a 'snap-back behaviour', and preludes catastrophic failure of the sewer pipe. The numerical failure pattern displayed in Figure 5.30d appeared to be in close agreement with the experimental failure pattern (Scheperboer *et al.*, 2021). In addition, the crack sequence presented in Figure 5.30 is in accordance with the experimental and numerical values for the local strains, with the largest tensile strain appearing at the bottom-inside (and top-inside) of the sewer pipe, and the smaller tensile strain emerging at its right-outside (and left-outside) (Scheperboer *et al.*, 2021).

5.7.6.1 Load–displacement response of egg-shaped sewer pipe

In Figure 5.31 the experimental load–displacement response (black lines) of the egg-shaped sewer pipes is compared against the numerical response (red line), by depicting the vertical load F_v versus the vertical net displacement d_v. The FEM response reflects the average of the experimental responses at the front and back sides of the pipe quite well; the vertical ultimate load $F_{v,max}$ following from the FEM simulations equals 363.7 kN, and is 9% larger than the average value of 334.3 kN obtained from the experiments. The local strain response Figure 5.32X.7 shows the applied vertical load F_v as a function of the local normal strains e_{xx} or e_{yy}, for two specific locations along the inner circumference and near the outer circumference of the E400/600 pipe, that is, the bottom-inside and the right-outside locations. The experimental and numerical strain responses are similar, especially along the initial elastic branch of the load–strain diagrams. It can be observed that the response obtained by the plane-strain FEM analysis approximately follows the average of the experimental responses.

The contour plots in Figure 5.33 illustrate the maximum principal stress in the egg-shaped pipe at four successive stages (a)–(d) during the failure process, which are indicated in the added load–displacement diagram. Observe from Figure 5.33a that the first failure crack nucleates at the inner

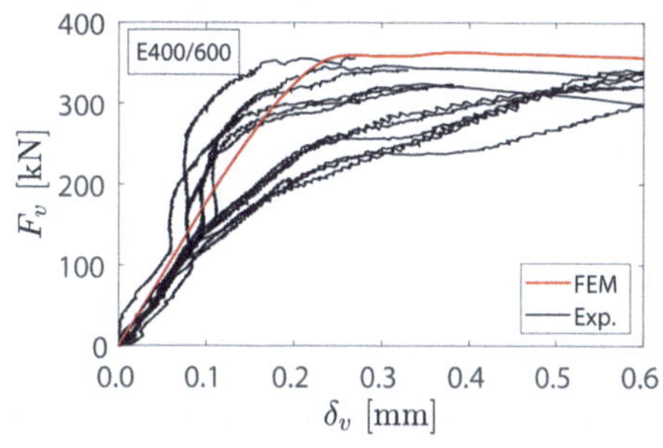

Figure 5.31 Vertical load F_v versus vertical net displacement d_v of the egg-shaped E400/600 sewer pipe, as obtained by experiments (black lines) and an FEM simulation (red line). The figure has been reprinted from Scheperboer *et al.* (2021).

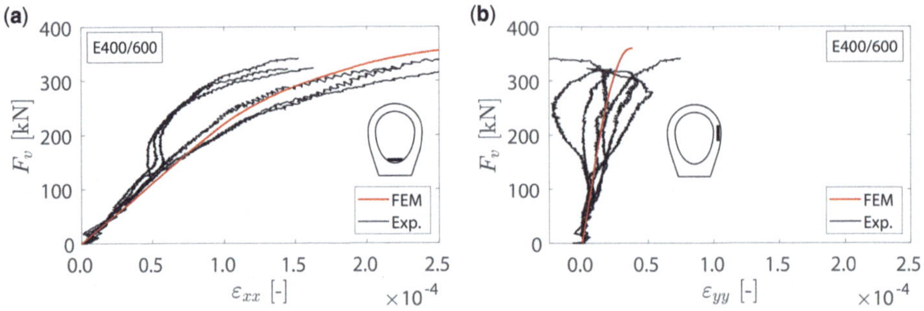

Figure 5.32 Experimental (black lines) and FEM (red line) load–strain response for egg-shaped E400/600 sewer pipe specimens. Vertical load F_v versus the normal strain e_{xx} or e_{yy}, as measured at (a) the bottom-inside and (b) the right-outside locations. The corresponding location of the strain gauge is indicated in the inset of the figures. The figure has been reprinted from Scheperboer *et al.* (2021).

top of the sewer pipe, in correspondence with stage (a) in the load–displacement diagram. When the vertical load is slightly increased towards the ultimate value $F_{v,max}$ at stage (b), this failure crack grows and a second failure crack nucleates at the inner bottom of the sewer pipe, see Figure 5.33b. As indicated in Figure 5.33c, both failure cracks further develop under the appearance of a minor structural softening behaviour from stage (b) to stage (c). Ongoing structural softening towards stage (d) eventually leads to the nucleation and growth of two additional failure cracks, located at the outer top-left and top-right sides of the pipe, see Figure 5.33d, after which the sewer pipe loses its mechanical strength and fails. As for the round pipe, for the egg-shaped pipe the locations of the four numerical failure cracks illustrated in Figure 5.33d are in excellent agreement with those observed experimentally (Scheperboer *et al.*, 2021). In addition, the crack sequence as presented in Figure 5.33 is confirmed by the experimental results for the local strains, with the highest tensile strain emerging at the top-inside of the pipe, followed by the tensile strain at the bottom-inside, and subsequently the tensile strain at the right-outside (and left-outside) of the pipe.

5.7.7 Parameter variation study for the R400 sewer pipe

The good agreement between the experimental and numerical results allows the FEM models to be used for studying the influence of various geometrical and material parameters on the failure response of a sewer pipe. Accordingly, this section treats a parameter variation study for the round R400 sewer pipe, whereby the failure response of the pipe is computed through dedicated FEM simulations that successively consider the influence of (i) the load contact area, (ii) the horizontal to vertical load ratio, (iii) the wall thickness of the sewer pipe, (iv) the tensile strength of concrete, (v) the toughness of concrete and (vi) the Young's modulus of concrete. The load contact area has been included in the parameter variation study since under in-situ conditions the amount of contact of a sewer pipe by the surrounding soil may strongly fluctuate, due to the presence of voids generated by soil erosion (Indiketiya *et al.*, 2019; Meguid & Kamel, 2014; Peter *et al.*, 2018; Scheperboer *et al.*, 2022). Further, the variation of the horizontal to vertical loading ratio provides insight into the sensitivity of the pipes' failure resistance to a change in lateral earth pressure, which in practice may range from active, through neutral, to passive earth pressure. Variations in the lateral earth pressure experienced by a sewer pipe may be caused by near excavations, or by other types of ground work. A reduction in wall thickness can occur under (bio)chemical corrosion, such as biogenic sulphuric acid aggression (Belie *et al.*, 2004; Gu *et al.*, 2018; Vollertsen *et al.*, 2008). This is also the reason that the tensile strength, toughness and/or Young's modulus of the concrete can decrease in time (Bosco *et al.*, 2020; Fan *et al.*, 2010; Ortega *et al.*, 2019; Shaowei *et al.*, 2018). In the parameter variation study, the specific FEM

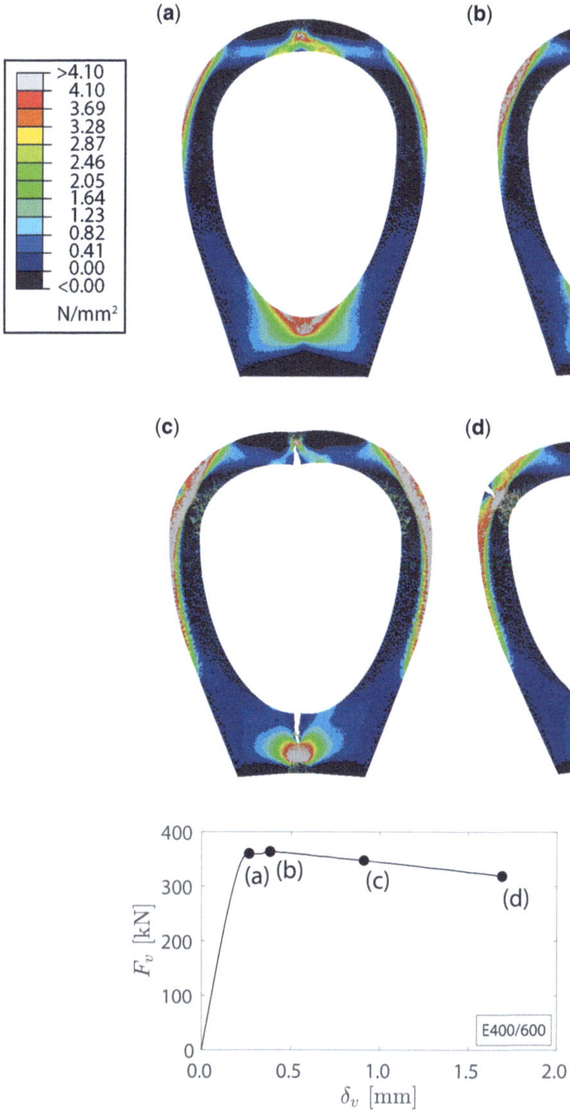

Figure 5.33 Maximum principal stress at crack nucleation (a) and crack propagation (b)–(d) in the egg-shaped E400/600 sewer pipe, as generated under the applied biaxial loading schematized in Figure 5.27. For clarification, the deformations of the pipe geometry are magnified in the contour plots by a factor of 50. The four failure states (a)–(d) are indicated in the corresponding load–displacement diagram. The figure has been reprinted from Scheperboer *et al.* (2021).

model of the R400 pipe used in the experimental validation in Section 5.7.4 will be referred to as the reference case, for which the results will be plotted by using a distinguishable red line or red symbol. The effect of the load contact area on the failure response of the round R400 sewer pipe is studied by using the (dimensionless) relative circumferential contact length $\bar{L}_c$, which represents the proportion of the outer pipe circumference that is in contact with the applied vertical and horizontal loads. The relative circumferential contact length can be expressed in terms of the projected contact width l_c – illustrated in Figure 5.27a – via

$$\bar{L}_c = \frac{4}{\pi}\sin^{-1}\left(\frac{l_c}{D}\right) \quad \text{with} \quad 0 \leq \bar{L}_c \leq 1 \tag{5.10}$$

where D is the outer diameter of the round pipe. For a horizontal to vertical loading ratio of 1/3, at the lower limit $\bar{L}_c = 0$ of zero contact area, the pipe is locally loaded at the top and bottom by a vertical point load F_v, and at the left and right outer sides by a horizontal point load $F_h = F_v/3$. Conversely, at the upper limit $\bar{L}_c = 1$, the complete pipe circumference is in contact with the applied loading, whereby the top and bottom quarter segments of the circular pipe are loaded by a distributed normal load σ, and the left and right quarter segments by a distributed normal load $\sigma/3$. For the reference case with $l_c = 250$ mm and $D = 524$ mm sketched in Figure 5.27a, Eq. (1) results in an intermediate value of $\bar{L}_c = 0.63$. Figure 5.34a depicts the vertical load F_v versus the net vertical displacement d_v of the R400 sewer pipe for a selection of contact lengths in the range $\bar{L}_c \in [0, 1]$, and Figure 5.34b displays the corresponding ultimate failure load $F_{v,max}$ versus the contact length $\bar{L}_c$. It can be observed from Figure 5.34a that under an increasing contact area both the ultimate failure load and the associated vertical displacement increase, and that the elastic resistance against deformation becomes larger. The increase in bearing strength is caused by the fact that under a growing load contact area the pipe becomes predominantly loaded under compression, and therefore becomes less sensitive to fracture. Figure 5.34b shows that the increase in ultimate failure load is about a factor of three in the transition of a sewer pipe with a poor load contact area, $\bar{L}_c \leq 0.2$, to a sewer pipe with a full load contact area, $\bar{L}_c = 1.0$. Since the appearance of erosion voids in the soil structure embedding a sewer pipe typically leads to a more localized load transfer, such voids may substantially reduce the load bearing capacity of the pipe. Hence, to obtain a 'damage tolerant design' for practical sewer systems, the design procedure should account for the presence of voids, in correspondence with a reduced load contact area, $\bar{L}_c < 1$.

The effect of the load contact area $\bar{L}_c$ on the failure response of the three different sewer pipe types is illustrated in Figure 5.35, by plotting the vertical load F_v against the net vertical displacement d_v for (i) the reference case shown in Figure 5.27 and (ii) a sewer pipe with a full load contact area. Note that for the R400 pipe these two load contact scenarios correspond to the curves plotted in Figure 5.34a for $\bar{L}_c = 0.63$ and $\bar{L}_c = 1.0$, respectively. For the egg-shaped pipe, the case of a full load contact area has been realized by applying the vertical pressure σ in the same fashion as for the two round pipes, that is, at the quarter circular segment defining the top part of the egg-shape pipe geometry. In addition, the remaining left and right sides of the egg-shaped pipe were exposed across their full

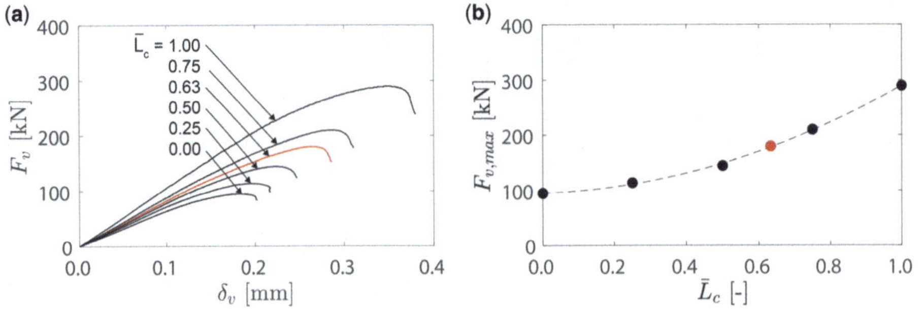

Figure 5.34 Influence of the relative circumferential contact length L_c on the failure response of the round R400 sewer pipe. The results for the reference case shown in Figure 5.27 are depicted in red. The dashed line in the figure for $F_{v,max}$ indicates the best fit to the FEM results. The figure has been reprinted from Scheperboer *et al.* (2021).

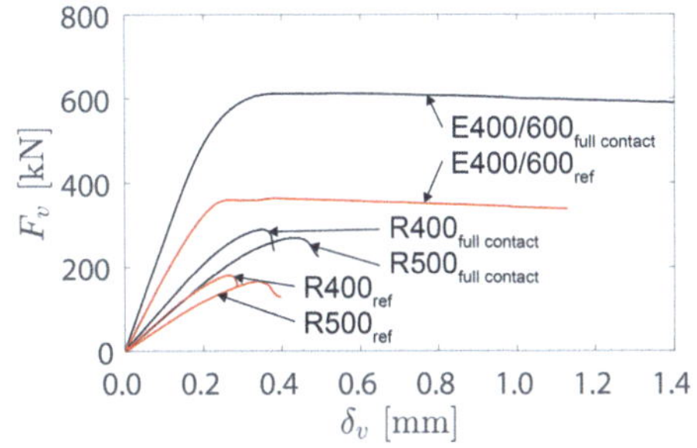

Figure 5.35 Load–displacement response of the R400, R500 and E400/600 sewer pipes with (i) a load contact area corresponding to the reference case shown in Figure 5.27 (red lines) and (ii) a full load contact area (black lines).

length to a horizontal pressure $\sigma/3$, in correspondence with a horizontal to vertical loading ratio of 1/3. It is observed that for all three pipe types the elastic stiffness, as well as the ultimate failure load and the associated vertical displacement, increase with increasing load contact area. The load bearing capacity under full load contact is 289.4, 268.9 and 612.0 kN for the round R400 and R500 pipes and the egg-shaped E400/600 pipe, respectively. In comparison to the respective values of 180.0, 165.5, and 363.7 kN related to the reference case, the relative increase in load bearing capacity is 61%, 62% and 68%, respectively, which thus appears to be similar for the three sewer pipe types.

A mutual comparison of the ultimate failure loads of the three pipe types for the case of a full load contact area shows that the ultimate failure load of the round R500 pipe is a factor of $289.4/268.9 = 1.08$ smaller than that of the round R400 pipe, and a factor of $612.0/268.9 = 2.28$ smaller than that of the egg-shaped E400/600 pipe. Note that these values are comparable to the corresponding values of 1.16 and 2.25 measured in the experiments (whereby, as shown in Figure 5.27, the load contact area is smaller), see Section 3.3.

As explained above, for the construction of Figure 5.34 the load contact area $\bar{L}_c$ has been varied simultaneously at the four specific locations of the external loading applied in the full-scale experiments, that is, at the top, bottom, left and right sides of the pipe specimen. Other studies that consider the influence of soil erosion voids on the mechanical strength of concrete pipes consider alternative profiles of limited load contact area (Meguid & Kamel, 2014; Peter *et al.*, 2018; Scheperboer *et al.*, 2022). Nevertheless, similar to the present study, such studies also demonstrate that a reduction of the load contact area may significantly reduce the load bearing capacity of the pipe.

5.7.7.1 Horizontal to vertical load ratio

The effect of the horizontal to vertical load ratio on the failure response of the R400 sewer pipe is examined by varying the horizontal to vertical load ratio K in the range $K \in [0, 1]$, see Figure 5.36. The value $K = 0$ relates to uniaxial compression. Here, the horizontally distributed load equals zero, and the pipe is effectively subjected to a vertically distributed load, as the net horizontal component of the normal pressure load σ applied at the both the top and bottom of the pipe is also zero, see also Figure 5.27a. In addition, $K = 1$ effectively corresponds to equibiaxial compression, whereby the horizontally and vertically distributed loads are equal to each other. Note that the reference case is in agreement with an intermediate value, $K = 1/3$. It is seen from the load–displacement diagram in Figure 5.36a that under an increasing value of K the ultimate vertical failure load and the vertical displacement at failure first both increase. However, the vertical displacement at failure starts to decrease when K approaches unity, $K = 1$, whereby the post-peak response becomes characterized by a snapback behaviour, instead of a softening behaviour that typifies the post-peak responses for lower values of K. The fracture behaviour at $K = 1$ is also different compared to that for lower values of K; as indicated in the contour plot shown in Figure 5.37, at failure the polar shear stress Σ_{rt} in the thickness direction

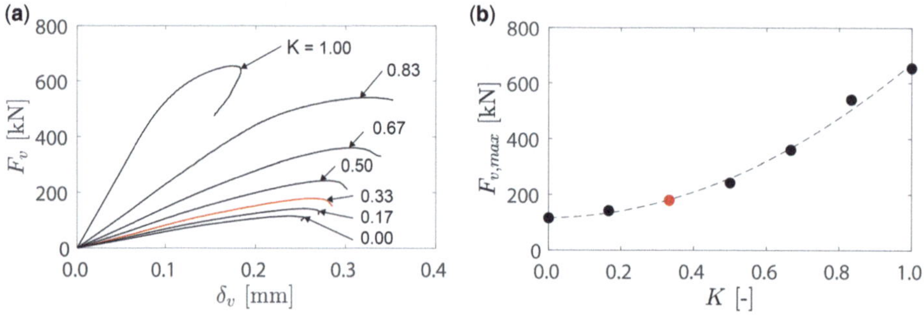

Figure 5.36 Influence of the horizontal to vertical load ratio *K* on the failure response of a round R400 sewer pipe. The results for the reference case shown in Figure 5.27 are depicted in red. The dashed line in the figure for $F_{v,max}$ indicates the best fit to the FEM results. The figure has been reprinted from Scheperboer *et al.* (2021).

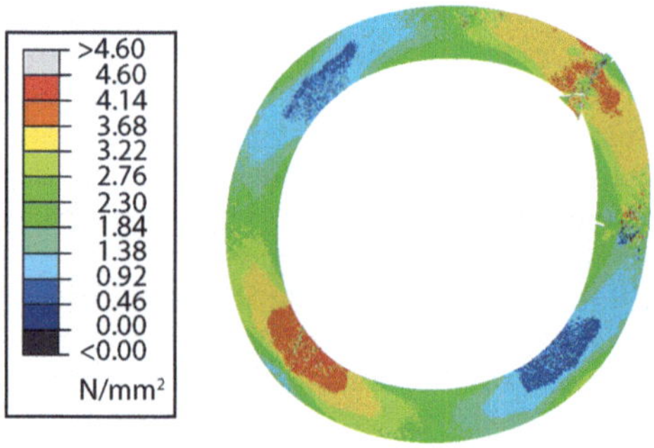

Figure 5.37 Polar shear stress Σ_{rt} (along the pipe thickness direction) at failure under equibiaxial compression, whereby the vertical loading σ sketched in Figure 5.27a equals the horizontal loading, in correspondence with a load ratio $K = 1$. For clarification, the deformations of the pipe are magnified in the contour plot by a factor of 100. The figure has been reprinted from Scheperboer *et al.* (2021).

of the pipe (with the subscripts 'r' and 't' indicating the radial and tangential directions, respectively) has reached the mode II fracture strength $t_u^2 = \pm 4.6$ N/mm² (indicated by the red and blue colours in the contour plot) at locations oriented under an angle of $\pm 45°$ with the horizontal x-axis, that is, at the top-left, top-right, bottom-left and bottom-right sides of the pipe. Accordingly, at the top-right side of the pipe a mode II failure crack develops. The stress redistribution generated under this progressive mode II crack makes that the other three regions with maximum shear stress start to unload, and that a small mode I crack with limited crack length emerges halfway the right specimen side. Although the specimen here effectively fails under mode II cracking, it is emphasized that this fracture behaviour is characterized by the selection of modes I and II fracture strengths of equal magnitude, $t_u^1 = t_u^2 =$ 4.6 N/mm², and thus may change when the ratio between the magnitudes of the mode II and mode I fracture strengths is larger than unity.

Figure 5.37b illustrates the ultimate failure load as a function of the horizontal pressure coefficient K. The reference case with $K = 1/3$ mimics a practical situation of neutral horizontal earth pressure, so that $K < 1/3$ and $K > 1/3$ cover the regimes of active and passive horizontal earth pressure, respectively. The failure resistance of the pipe is the lowest in the regime of active earth pressure, although the ultimate failure load remains fairly constant for $0 \leq K < 1/3$. When entering the regime of passive earth pressure, $K > 1/3$, the failure resistance of the pipe substantially increases under an increasing value of K, eventually reaching an equibiaxial failure load at $K = 1$ that is more than three times higher than the failure load under neutral earth pressure, $K = 1/3$. Hence, for practical pipe design a conservative approach is to determine the pipes' failure resistance in correspondence with critical loading conditions representative of the range of active to neutral earth pressures, $K \leq 1/3$.

5.7.7.2 Wall thickness

The influence of a reduction in wall thickness on the failure response of the R400 sewer pipe is explored by keeping the outer diameter of the pipe fixed and varying the inner diameter. Accordingly, the reduced wall thickness is considered to be the result of a (bio)chemical corrosion process that takes place from the inside of the pipe. The reduced wall thickness is expressed in terms of the relative thickness ratio t/t_{ref}, with t the wall thickness of the corroded pipe, and t_{ref} the wall thickness of the uncorroded pipe corresponding to the reference case studied in Section 5.7.4 The influence of a reduction in wall thickness on the pipes' failure response is illustrated in Figure 5.38. Figure 5.5.38a clearly shows that a decrease in wall thickness leads to a decrease in the ultimate failure load and effective stiffness of the pipe, and further results in an increase of the deformation at which the ultimate failure load is reached. From Figure 5.38b it is observed that a wall thickness decrease by 20% ($t/t_{ref} = 0.8$) already induces a reduction in the ultimate failure load $F_{v,max}$ by almost 40%, while a wall thickness decrease of 50% ($t/t_{ref} = 0.5$) reduces $F_{v,max}$ by somewhat more than 75%. This result emphasizes the importance of accurately monitoring the degradation of the inner surface of sewer pipes during their lifetime for the estimation of the load bearing capacity.

5.7.7.3 Tensile strength, mode I toughness and Young's modulus

The effect on the sewer pipes' failure response by the tensile strength t_1^u, mode I toughness $G_{I,c}$ and Young's modulus E_c of the concrete is shown in Figures 5.39–5.41, respectively.

The material properties have been normalized by a division of the value representative of the reference case discussed in Section 5.7.4. It can be observed that a decrease of each of these material parameters may considerably reduce the load bearing capacity of the pipe. Figures 5.39a and 5.5.40a show that a decrease of the tensile strength t_1^u and mode I toughness $G_{I,c}$ lead to a decrease of the vertical displacement at which the ultimate failure load $F_{v,max}$ is reached. In contrast, Figure 5.41a illustrates that a decrease of the Young's modulus E_c enlarges the displacement associated with the ultimate failure load. In specific, when the relative Young's modulus approaches to zero, $E_c/E_{c,ref} \rightarrow 0$, the displacement associated with the ultimate failure load grows unboundedly towards infinity. Note that the minimal load bearing capacity $F_{v,max} \approx 90$ kN computed near the limit of a zero Young's modulus is comparable to the minimal load bearing capacity calculated for a zero mode I toughness,

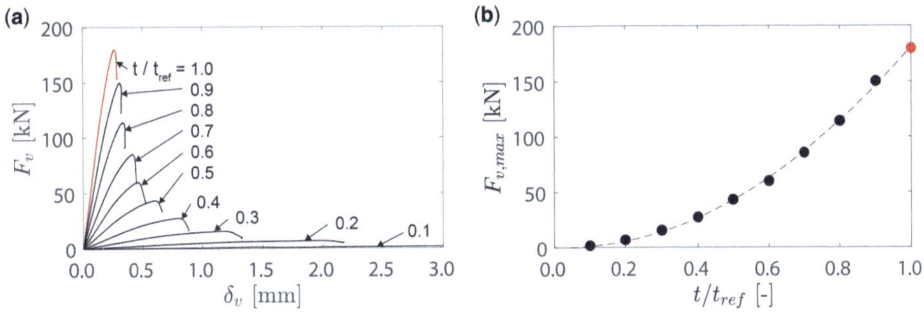

Figure 5.38 Influence of the relative wall thickness t/t_{ref} on the failure response of a round R400 sewer pipe. The results for the reference case shown in Figure 5.27 are depicted in red. The dashed line in the figure for $F_{v,max}$ indicates the best fit to the FEM results. The figure has been reprinted from Scheperboer *et al.* (2021).

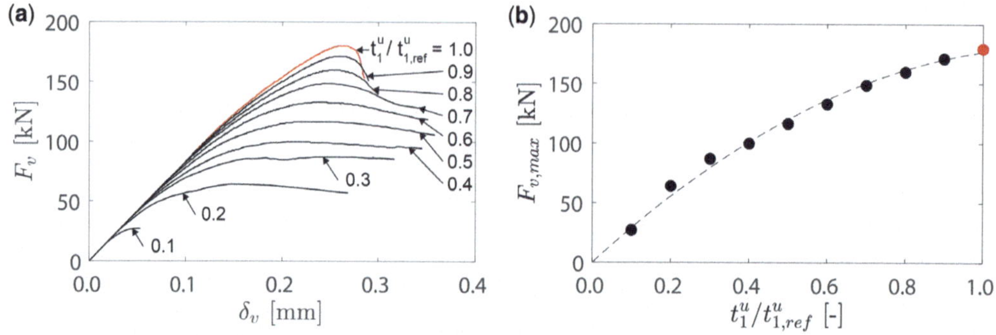

Figure 5.39 Influence of the relative tensile strength $t_1^u/t_{1,\text{ref}}^u$ on the failure response of a round R400 sewer pipe. The results for the reference case shown in Figure 5.27 are depicted in red. The dashed line in the figure for $F_{v,\max}$ indicates the best fit to the FEM results. The figure has been reprinted from Scheperboer *et al.* (2021).

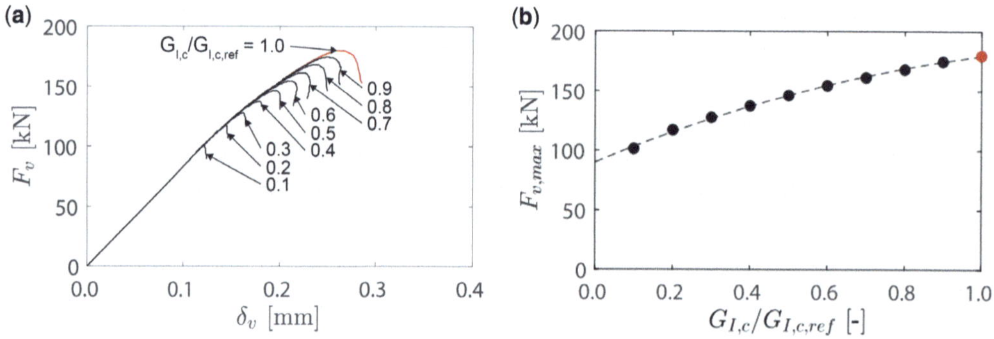

Figure 5.40 Influence of the relative mode I toughness $G_{\text{I,c}}/G_{\text{I,c,ref}}$ the failure response of a round R400 sewer pipe. The results for the reference case shown in Figure 5.27 are depicted in red. The dashed line in the figure for $F_{v,\max}$ indicates the best fit to the FEM results. The figure has been reprinted from Scheperboer *et al.* (2021).

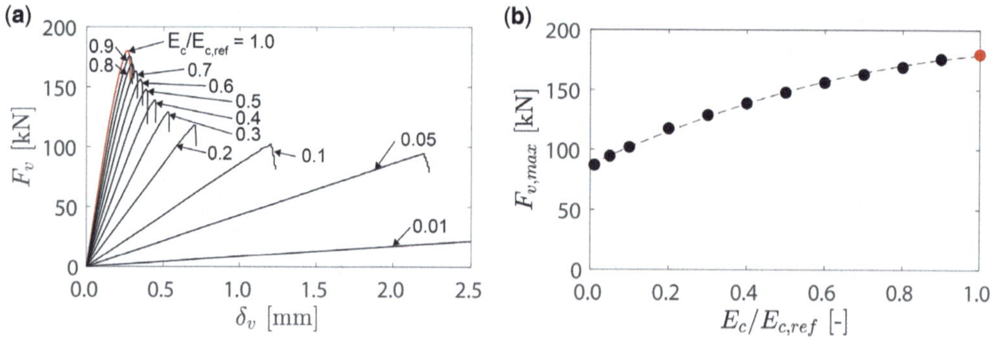

Figure 5.41 Influence of the relative Young's modulus $E_c/E_{c,\text{ref}}$ on the failure response of a round R400 sewer pipe. The results for the reference case shown in Figure 5.27 are depicted in red. The dashed line in the figure for $F_{v,\max}$ indicates the best fit to the FEM results. The figure has been reprinted from Scheperboer *et al.* (2021).

$G_{I,c}/G_{I,c,ref} \rightarrow 0$, see Figures 5.41b and 5.5.40b, respectively. The above results clearly illustrate how the bearing strength of concrete sewer pipes depends on various factors (load contact area, load ratio, wall thickness, tensile strength, toughness, Young's modulus) that can be associated with degradation processes, such as soil erosion, changes in lateral earth pressure, biochemical degradation and ageing. Additional experimental and modelling research is needed to establish accurate relations between these factors and the corresponding degradation process(es), which will allow to directly connect the reduction in bearing strength following from the present computational results to the specific characteristics of a degradation process.

In Luimes *et al.* (2022) additional research results can be found on the structural response and failure mechanisms of 35 *used* concrete sewer pipes, as studied by means of a combined experimental–numerical approach, whereby specific attention focuses on determining the effect by biochemical attack. The degree of degradation of the inner and outer surfaces of the pipes is carefully analyzed by visual inspection, and subsequently categorized into six different surface condition classes. The type and degree of biochemical attack are determined by respectively performing X-ray diffraction analyses and phenolphthalein tests. The sewer pipes, which vary in age, size and geometry, are subjected to biaxial loading conditions in a full-scale test set-up. The experiments show that the material degradation of relatively old sewer pipes can be considerable, and for a large part may be attributed to the process of biogenic sulphide corrosion. This process typically induces a weak, corroded layer at the inside of the pipe, which is characterized by the appearance of exposed granulates and porous mortar between the granulates. The experimental failure responses of the used sewer pipes are compared against numerical results obtained from detailed FEM simulations. It is demonstrated that the failure response of each sewer pipe type is well predicted by the FEM model if the negligible mechanical properties of the biogenic sulphide corrosion layer at the inside of the pipe are accounted for in the simulations. The relatively low concrete tensile strength of old sewer pipes and the reduced effective wall thickness due to biogenic sulphide corrosion cause that the ultimate bearing strength and structural stiffness of these pipes may be more than a factor of two lower than those of completely new sewer pipes. The results of the work in Luimes *et al.* (2022) provide a scientific basis for the decision-making process on sewer rehabilitation and replacement, in particular by revealing and quantifying the influence on the structural bearing strength by biogenic sulphide corrosion.

5.8 LOSS OF HYDRAULIC CAPACITY

5.8.1 Introduction

The hydraulic capacity of (classical) UD systems is determined in the design phase and depends on local climate conditions and is subject to various (local) design rules. In general, the hydraulic capacity is designed to:

- limit the risk and extend of urban flooding to some acceptable limit (basically a political choice), and
- limit the discharge of various degrees of polluted water (either wastewater, stormwater runoff or discharges from combined sewer systems) into either open water courses or groundwater bodies.

The former demand results in selecting dimensions for water conveyance elements in the system (e.g., pipes, weirs, valves and so on) while the latter demand also puts the onus on the available storage capacity in the system. Both the discharge capacities and the storage capacities are subject to processes that decrease them.

5.8.2 Storage capacity

Loss of storage capacity can result from various processes, the most important being subsidence and sedimentation. Subsidence, or more precisely, non-uniform subsidence with subsidence differences over a relatively short distance (i.e., in the order of magnitude of say 100 m) may be the cause of *'lost*

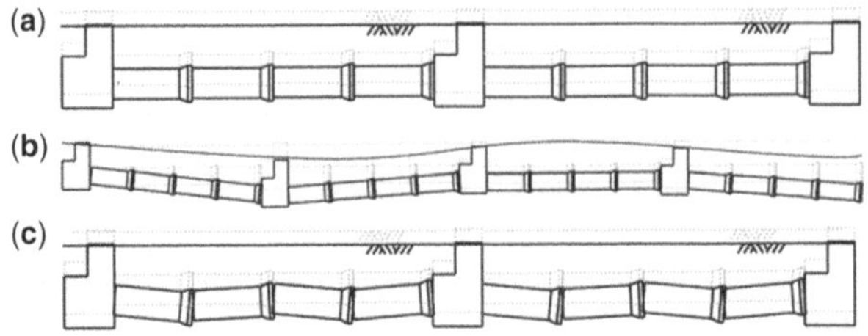

Figure 5.42 Lost storage due to local subsidence Dirksen (2013) (image 3.1 p38)).

storage'. The concept of lost storage is depicted in Figure 5.42. While, as part of the original design and construction it had typically been ensured that no local minima in terms of the invert elevation of the pipes were present, however, due to subsidence of the surrounding soil (e.g., clay or peat) such local minima may have evolved over the course of the lifetime of the system.

These 'sags' form pockets from which water cannot be drained to either an outlet or a pumping station. Especially in sanitary or combined sewer systems the occurrence of such sags is a trigger for a range of processes that, in general, cause deterioration of the functionality of the system. In the first place, not all storage capacity in the sags will be available for storage of water during storm events triggering more frequent occurrence of spills of (diluted) wastewater to receiving water bodies if no corrective action is taken. A second process is the accumulation of sediments in these sags; locally, the transport capacity of solids is substantially reduced, allowing sediments to settle, up to the point at which a new morphological equilibrium is established. During heavy storm events, however, the presence of such accumulated pockets of sediments may result in the 'sudden' release of substantial amounts of pollutants that are flushed out, adding to the environmental impacts of CSO events.

A further contribution to the functional deterioration from subsidence is the fact that in the sags the accumulated sediments, of which a substantial part may consist of biodegradable materials (see e.g., Ashley *et al.*, 2004), act as a reservoir worsening the formation of corrosive substances (e.g., sulphuric attack, see Section 5.5). This process may, in turn, negatively affect the geometry of the pipe (due to concrete corrosion) as well as the hydraulic roughness of the pipe wall. Also, at a catchment scale, subsidence may negatively affect the functioning of the system as whole. Dirksen *et al.* (2012) and Dirksen (2013) reported a clear relation between the subsidence of the sanitary sewer pipes and the occurrence of blockages. The latter may occur when the subsidence reaches a level at which no free water surface is present, see Figure 5.43. In such a case, floating material starts to slowly accumulate at the top of the conduit, until blockage occurs. Even modest settling rates of a few mm/year may trigger substantial problems with sedimentation and/or the accumulation of floating solids (e.g., FOG). Both processes result in a significant reduction of both the storage and the hydraulic conveyance capacity of the system over time. In some extreme cases this may become manifest on time scales of a few months to a few years.

In some cases, siphons are subject to conveyance capacity loss due to the occurrence of three-phase flow phenomena. For example, Figure 5.43 presents several profiles of FOG deposits as measured in siphons in a combined sewer system in the City of Amsterdam, The Netherlands. The brown areas represent the FOG. Although not explicitly observed, a hypothesis of the chain of events responsible for the conditions observed is as follows:

- At the point where the siphon becomes pressurised, floating FOG sticks to the wall of the pipe; these locations vary over time due to the varying dry weather flow discharge.
- In the process, the cross-sectional area decreases resulting in a local increase of the flow velocity.
- This, in turn, increases the amount of air that can be transported into the falling leg of the siphon, allowing for FOG to be deposited deeper into the siphon.

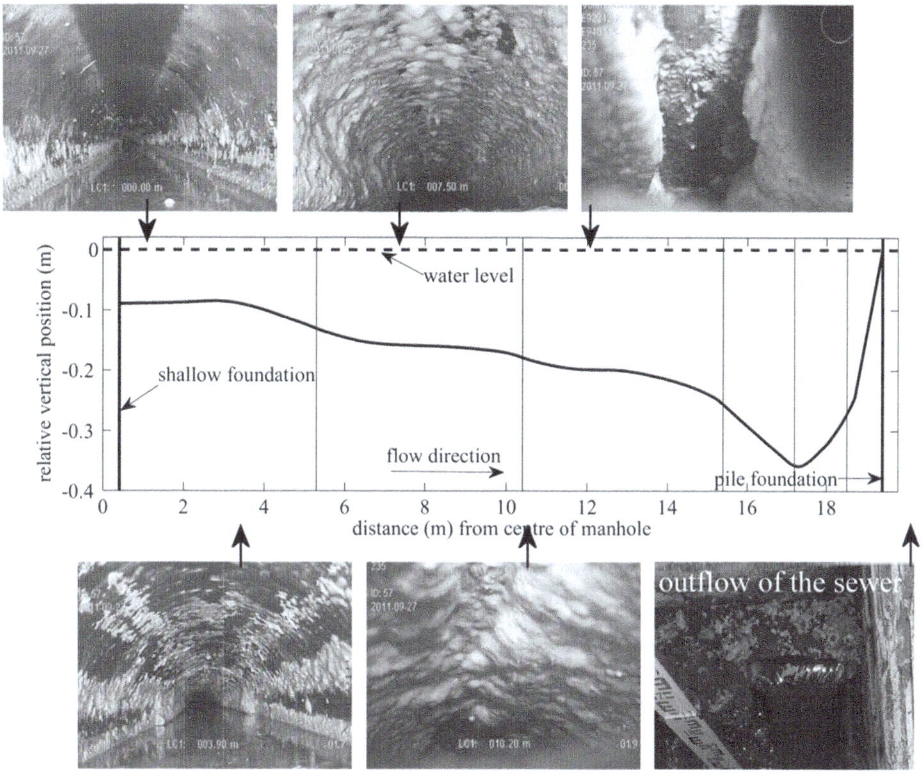

Figure 5.43 Inspection result of a sewer with a filling degree of 100%. The graph represents the vertical cross section of the sewer. (Dirksen *et al.*, 2013b).

- During storm events, air can be transported along the entire length of the siphon causing FOG deposits along its entire length. In one case (I.e., the Entrepotdok siphon in Figure 5.44) this resulted in the creation of an air pocket in the middle of the siphon causing it to move upwards.
- Due to the reduction of the cross-sectional area along a substantial length of the siphons, in addition to the storage capacity, the hydraulic conveyance capacity is significantly reduced as well.

A typical time scale for these processes to develop is about 1 year. Therefore, regular inspection and cleaning of siphons is warranted, especially at locations that are notorious in suffering from FOG (e.g., in the presence of high concentrations of restaurants).

5.8.3 Discharge capacity

5.8.3.1 Introduction

The discharge capacity of a pipe is a function of its length, shape and dimensions of its cross-sectional area as well as the roughness of the pipe wall. Except for the length, all other parameters are subject to change over time. In many design manuals the roughness (or Manning's coefficient) only reflects the type of material of a pipe (e.g., concrete has a rougher texture than for instance PVC or steel) hinting at a difference in hydraulic resistance. Although this may be true for new pipes and/or pipes that are used to transport clean fluids, it is certainly **not** true when a pipe ages and/or when either sediments or gases are present in the transported medium. Given that UD systems typically are expected to accommodate wastewater and/or urban stormwater runoff, it is highly likely that solids and gases **will** be present.

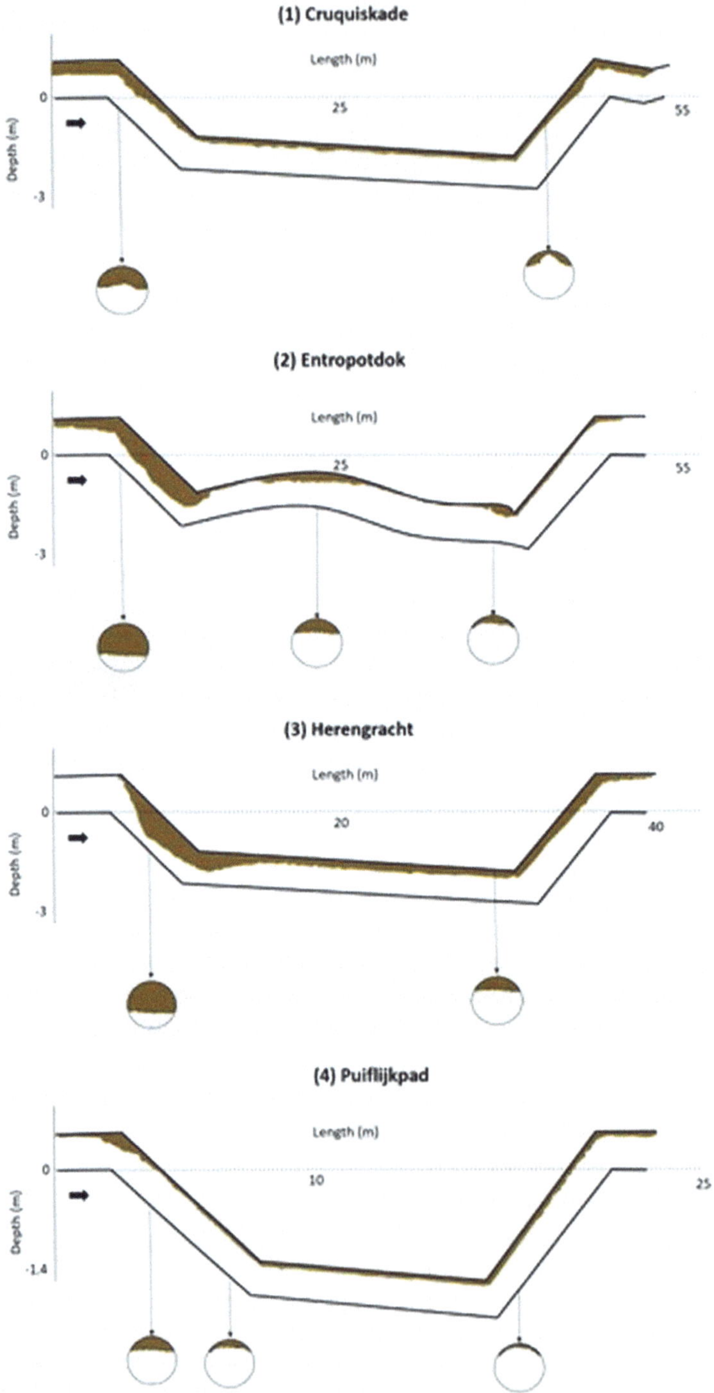

Figure 5.44 FOG deposits in siphons, as measured in Amsterdam, The Netherlands (after de Groot, 2015).

Further, the presence of obstacles (e.g., protruding inlets, root intrusion, debris dumped in the system or remnants from construction activities) may cause significant capacity reductions as does the ingress of soil through cracks or defective joints. Also, a change of the slope, for example, caused by subsidence discussed in the previous section, can results in sedimentation and, in turn, in a reduction of the conveyance capacity.

For pressurized systems, the accumulation of air/gases is a well-known cause of capacity losses that, when the system is pumped, also result in an increased energy consumption. In addition, in such systems the deterioration of the pump is a factor to be reckoned with as well.

5.8.3.2 Obstacles

Van Bijnen (2018) and van Bijnen *et al.* (2016) describe a case study in the Netherlands in which a hydrodynamic model of a small catchment (see Figure 5.45) had been calibrated using several storm events in 2010 and 2015, respectively. For the 2010 scenario, known defects were present in the system (i.e., sediment deposits and root intrusions). These obstacles had been removed in 2012 and the model was calibrated again with additional storm events in 2015. In both cases the system as modelled was supposed to be clean (as is usually done when using a hydrodynamic model). For the perfect case, the residuals (i.e., the differences between measured and modelled water levels) show a Gaussian distribution similar to the uncertainty in the measured values. When the residuals do not present a similar distribution, it indicates the presence of some kind of systematic deviation between the model and the 'real' world, for example, some dimensions are incorrect or wrong invert levels have been used. Such deviations can be eliminated by a careful check of the data describing the structure and geometry of the system in question.

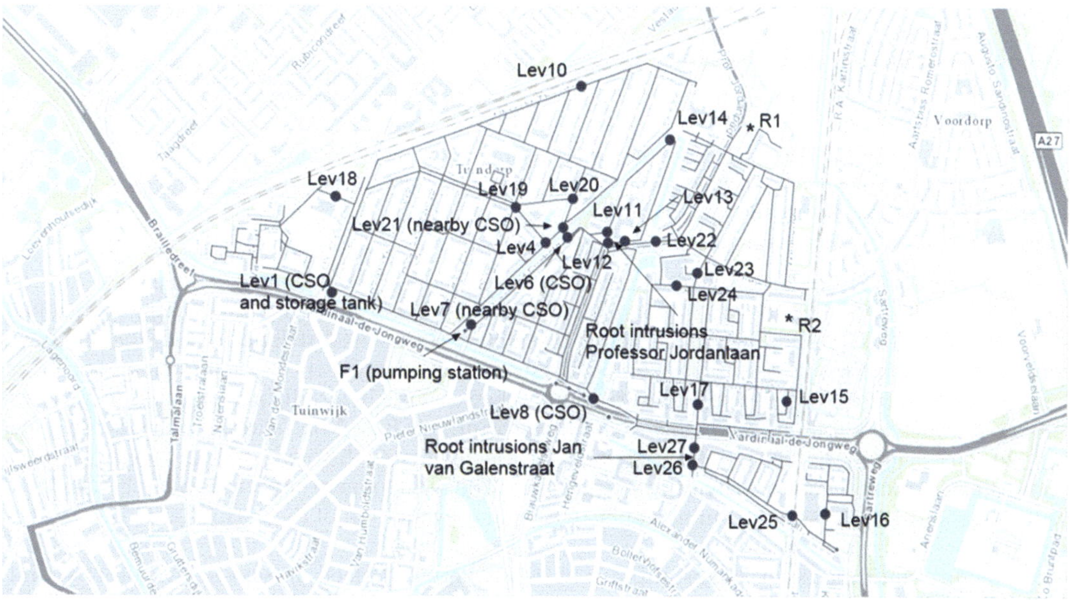

Figure 5.45 Monitoring network 'Tuindorp' catchment. Flows (F1), water levels (Lev1, Lev2, ..., Lev27) and rainfall (R1 and R2) were monitored at several locations in the catchment area. The monitoring network design was based on a combination of hydraulic simulations, reported incidents and observed in-sewer defects. For example, several water level sensors (Lev12 and Lev13 Professor Jordanlaan, Lev23 and Lev24 Troosterlaan, Lev26 and Lev27 Jan van Galenstraat) are installed in the manholes just upstream and downstream of an observed defect. Two tipping-bucket rain gauges (R1 and R2) were used to measure rainfall in 'Tuindorp'. (*Source*: van Bijnen, 2018).

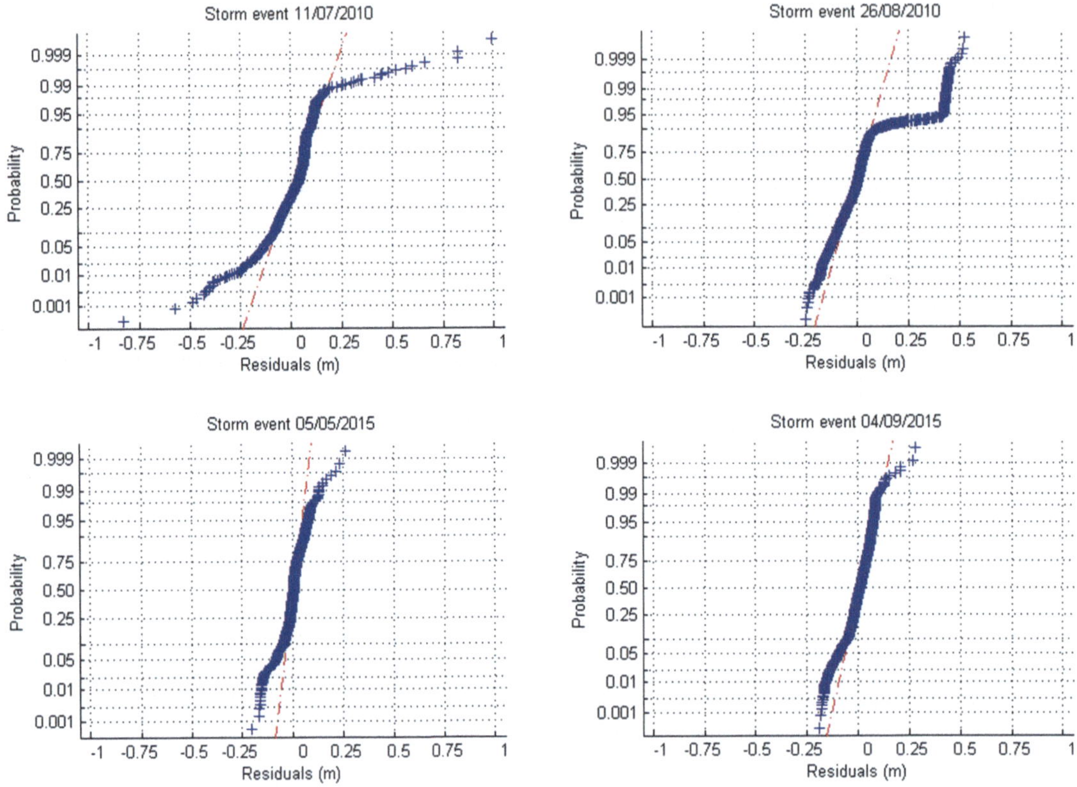

Figure 5.46 Normal probability plots of residuals after calibration of the 'clean' system (storm events 05/05/2015 and 04/09/2015) and the system including defects (storm events 11/07/2010 and 26/08/2010). Residuals of all measuring locations. (*Source*: van Bijnen, 2018).

An analysis of the residuals suggests that for the 2010 scenario the residuals do display a Gaussian distribution with large variations (>0.75 m), see Figure 5.46. Indeed, when analysing the residuals for the (clean) 2015 scenario, they much better resemble a Gaussian distribution with the maximum deviation limited to circa 0.25 m. This example illustrates how the presence of sediment and obstacles in a sewer system can show up when doing a regular check on the hydraulic performance using a well calibrated model as a reference (i.e., hydraulic fingerprinting). Based on further analysis van Bijnen *et al.* (2018) concluded that the risk of citizens being infected due to the increased occurrence of urban flooding (i.e., at more locations, covering a larger area, and with longer time windows) increased with a factor of about 1.5, stressing the importance of a regular evaluation of the performance of a systems as a whole.

Inspections confirmed that a substantial root intrusion was indeed present in the direct vicinity of this monitoring location (Figure 5.47), The geometry of this intrusion, however, was only apparent when the water level exceeded a certain threshold.

5.8.3.3 Air/gas inclusions (pressurized pipes only)

In most designs of hydraulic structures in UD systems as well as during their operation, relatively straight forward one phase flow is implicitly assumed. In practice however, in many cases this is an

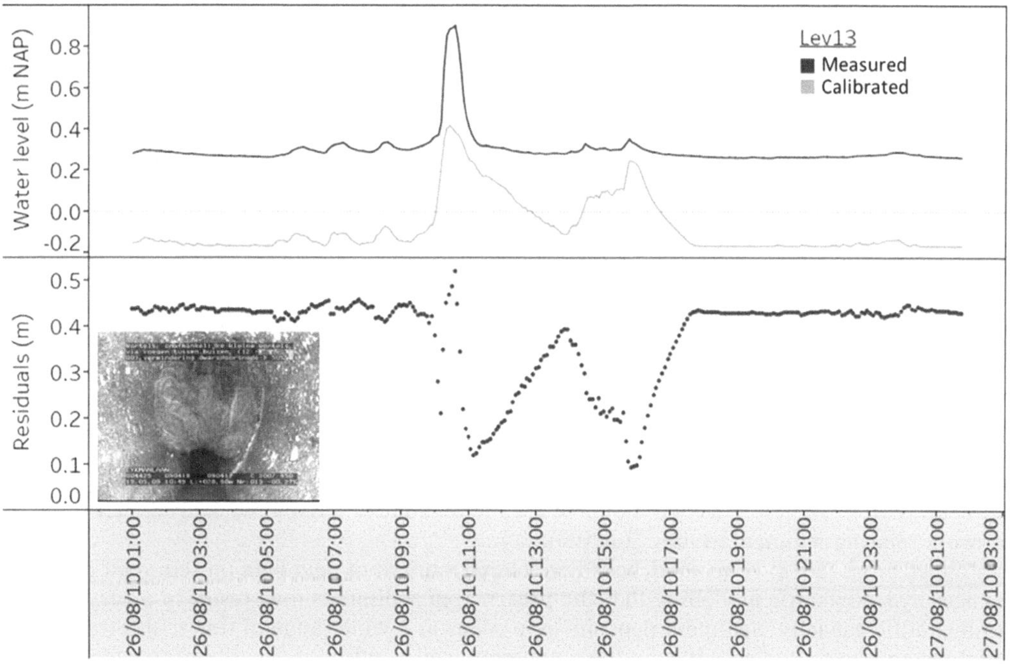

Figure 5.47 Measured and modelled (calibrated) water levels upstream (Lev13) of a root intrusion at the Professor Jordanlaan for the event of 26 August 2010. NAP is reference level. Negative measured and calibrated values represent water levels below this reference. (*Source*: van Bijnen, 2018).

oversimplification, especially when faced with pressurized systems transporting either wastewater or a wastewater/stormwater mixture (as occurs in combined sewer systems) or when siphons are implemented to cross water courses such as canals. The occurrence of gas/air pockets is known to have significant adverse effects on the functionality of such systems (Lubbers, 2007; Pothof, 2011). The formation of air/gas pockets is due to air entrainment in pumping stations, malfunctioning of venting systems or the production of biogas that accumulates at high points in the system (e.g., at the upstream end of siphons). The presence of such gas pockets may give rise to local corrosion in pressure mains, and results in significant loss of hydraulic capacity and energy efficiency. In some cases, even air lock may occur reducing the hydraulic capacity to virtually zero. Kooij *et al.* (2015) present a case study on monitoring the evolution of capacity loss due to air/gas pockets in an extensive system of pumping stations and pressure mains. As discussed in Section 5.7.2 the dynamics of the gas/water/floating material interface may result in substantial FOG deposits in siphons, reducing the hydraulic capacity as well.

The underlying processes and methods to remedy capacity loss due to air/gas pockets are well understood and can be relatively easily quantified using hydrodynamic models. A comprehensive book on the subject is available at IWA Publishing (Tukker *et al.*, 2016), in addition to German guidelines (DWA-A 113; DWA-M 149-9). While the effects of air/gas pockets on the hydraulic capacity as well as the rate of transporting gas/air can well be quantified, the rate at which air/gas accumulates, however is less straight forward to be quantified.

In case of air entrainment in pumping stations, the pump operation and the geometry of the pump sump play a major role. When the formation of biogas is the main contributing factor, the composition of the (waste)water and environmental conditions (i.e., mainly temperature) and how they vary with time play a major role. In practice, the formation of air gas pockets can be avoided by:

- a proper initial design of the system,
- regular checks (e.g., every 6 months) pertaining to the proper functioning of venting valves, and
- monitoring of the discharge, pressure head and energy consumption of the system (Kooij *et al.*, 2015).

The time scales at which changes in hydraulic losses due to the presence of air/gas pockets can be measured range from weeks to years as a function of the contributing processes and the day-to-day operation and management of the system.

5.8.3.4 Sedimentation

Sedimentation is a process that may occur in any UD system and is a key process in the loss of functionality. Accumulation of sediments in pipes reduces the hydraulic capacity (including infiltration for NBS), the volume available for storage and may even lead to complete blockage of elements. Sedimentation has already been discussed in the present chapter in the clogging Section (5.3) and the storage capacity Section (5.8.2). Moreover, a very comprehensive text on sedimentation is provided by Ashley *et al.* (2004). Therefore we will not provide a detailed text here but we will restrict this to some recent work relating sedimentation to functionality.

Van Bijnen *et al.* (2012) concluded, based on a large amount of field data on sediment deposits and subsequent hydrodynamic modelling that the presence of sediments may result in a very substantial loss of hydraulic capacity. Sediment deposits may occur at a wide range of timescales (typical weeks to years) depending on the activities in the catchment, soil conditions, quality of the construction of the system, industry discharges and so on.

5.8.3.5 Ageing of pumps

In many UD systems mechanical sub-systems (e.g., pumping stations, moveable weirs, etc.) are present. These sub-systems have the 'advantage' that malfunctioning becomes more rapidly apparent and pronounced compared to, for instance, a partial loss of the hydraulic capacity of a gravity sewer, or the gradual reduction of the infiltration capacity of NBS. This loss of functionality can be caused by damage of the impeller, which can be the result of wrong pump operations or a poor, initial pump design (e.g., occurrence of cavitation). However, it can also be caused by sustained abrasion due to the presence of for example, sand in wastewater. Such normal ageing becomes manifest in increased energy consumption, vibrations and a decreased discharge. Typically, impellers and bearings are replaced every ten to fifteen years depending on the operational conditions. In extreme cases replacement may be needed every few months; this typically suggests that a pump is being operated beyond its original design intent.

Apart from this 'normal' wear of the impeller which may occur over several years, clogging because of rags or plastic being sucked into the pump may cause sudden failure of the pump (Figure 5.48).

Little literature exists describing deterministic modelling of wastewater pumps (Jensen, 2018) with a few contributions pertaining to statistical models (Korving, 2004; Korving & Ottenhoff, 2008; Korving *et al.*, 2006a). As a result of the statistical analysis of data from pump failures in combined sewer systems, Korving and Ottenhoff (2008) present a positive correlation between pump age and failure rate illustrated by means of a typical 'bathtub' or U-shape. Directly after installation a high failure rate tends to be observed reflecting construction defects and 'start-up' problems. After overcoming this initial phase (which typically lasts about one to two years) a period of about ten years with relatively consistent failure rates can be observed. After this ten-year period the failure rate gradually increases due to material ageing, damage to bearings, impellers and so on. An important factor contributing to the failure rate of pumps, in combined and/or wastewater systems, is the presence of sediments, solids materials and gross solids in the wastewater. Korving and Ottenhoff, (2008) report, based on verbal

Figure 5.48 Clogged pump. (*Source*: Deltares).

information provided by maintenance personnel, that 95% of reported pump failures are (at least partially) related to clogging. In addition to the direct costs of repairs/replacement, malfunctioning of pumps reportedly contributes to substantial (and largely avoidable) spills of sewage via CSOs into receiving waterbodies, with adverse effects on their habitat and human health (Korving *et al.*, 2006b; Giakoumis & Voulvoulis, 2023). A recent article by Hallaij *et al.* (2022) provides a state-of-the-art overview on preventive maintenance of pumps used in civil engineering infrastructure.

5.8.3.6 Scum layers

Formation of layers of floating material in (wastewater) pump sumps may cause pump failure (Duinmeijer, 2020; Korving, 2004). Cleaning out such layers, consisting of FOG (Fat, Oil and Grease) intermixed with floating materials like rags, plastic bottles, and so on., (Figure 5.49) is a costly and unpleasant job

Nieuwenhuis *et al.* (2018) conclude, based on an evaluation of 126 pumping stations in combined sewer systems, that there are three main factors influencing the occurrence of problematic scum layers:

- the average income of the inhabitants within the catchment is negatively correlated to the presence of scum layers,
- the density of restaurants or catering industry within the catchment is positively correlated with the occurrence of scum layers, and
- the intensity of 'mixing', or turbulence in the pump sump is negatively correlated with the occurrence of scum layers.

Apart from being a cause of pump clogging, the presence of scum/FOG layers in pump sumps may block degassing of the wastewater, in the process disabling monitoring systems for the occurrence of potentially hazardous gas mixtures (e.g., H_2S/O_2). When disturbing such a layer unexpected gas may be emitted with potential disastrous health threats for personnel.

Figure 5.49 FOG in pump well. (*Source*: Duinmeijer, 2020).

Only recently, a method was developed to monitor the dynamics of the formation of scum layers (Moreno-Rodenas *et al.*, 2021). So far, the following conclusions have been drawn:

- Choices as to pump management influence the dynamics of the scum layers.
- Changes are seen at small timescales, (i.e., in the order of hours to days) indicating that more regular inspection/cleaning activities are called for than is normally done (one to two times per year).
- There is a clear influence of the occurrence of storm events (in combined sewer systems).

To date, to the authors' knowledge, no attempts to model the dynamics of FOG layers have been reported, likely due to lacking knowledge as to the underlaying processes.

5.9 SEPARATION OF FLOWS

5.9.1 Introduction

Water discharged from urban areas comes in varying qualities, starting with wastewater (discharged from households and industry) that needs to be treated prior to being discharged into the environment. In addition, there is stormwater runoff that may be polluted to varying degrees, depending on the nature of the tributary catchment and the characteristics of the storm events (i.e., reflecting the variability in rain intensity and duration). This stormwater runoff (e.g., when from areas with a heavy traffic load) should get some basic or more sophisticated form of treatment prior to discharge. Further, one should avoid having wastewater leak into groundwater bodies, or vice versa, have groundwater infiltrate either system carrying wastewater or stormwater. The former situation may result in contamination of groundwater, and, by extension, of wells used for producing drinking water. In the latter case, the infiltrating groundwater may consume discharge capacity, dilute the wastewater and may even trigger local scouring of the soil or backfill surrounding the pipes potentially leading to the occurrence of sinkholes and loss of structural stability (see paragraph 5.6). Similarly, one should also avoid directing downspouts from buildings to sanitary sewer systems as this may severely overload the capacity of the system, possibly leading to backwater conditions and flooding of buildings with diluted wastewater.

All combined, keeping flows of different quality separated and avoiding untreated discharge into either surface waters or groundwater bodies necessitates the absence of:

- cross-connections or inadvertent connections from downspouts, and
- the uncontrolled exchange of water between UD systems and surface or groundwater bodies.

5.9.2 Infiltration and exfiltration

Many NBS rely on infiltrating runoff into groundwater bodies. As discussed in Sections 5.2 and 5.3, the clogging process can be characterized by the interception of fine sediments and associated pollutants in the top layers of the infiltration system, which needs to be removed on a regular basis. A further precaution is to ensure no inadvertent cross-connections occur, that is, the discharge of wastewater into NBS should be avoided.

In the case of 'classical' UD systems (i.e., sanitary sewers, storm sewers or combined sewer systems) infiltration and exfiltration can cause a significant loss of functionality.

Causes of infiltration and exfiltration are related to the occurrence of cracks, or faulty joints (e.g., figure 5.50) caused by structural overloading and or subsidence. Generally speaking, infiltration is easier to detect and quantify compared to exfiltration. The reader is referred to Chapter 4 for methods to detect and quantify infiltration and exfiltration.

5.9.3 Leakage (exfiltration) of pressurised systems

5.9.3.1 Introduction

Pressure mains are can be used to transport collected wastewater to Wastewater Treatment Plants. Due to subsidence, traffic loads or corrosion, leakage from these systems may occur. In many cases, the leakage or exfiltration is only manifested when massive soil erosion and damage at the surface become visible (Figure 5.51). Fortunately, the situation illustrated by Figure 5.51 is relatively rare, but it provides a good impression of the risks to the public when sewer pipes are in a poor condition. Often, it has been suspected that such catastrophic event was precipitated by a small leakage due to for example, corrosion. When inspecting pressure mains, the determination of potential 'weak spots' can be facilitated by knowledge of:

- The soil types and indication of uneven subsidence or subsidence rates (e.g., when the soil type changes from peat to sand or clay).
- The location where pressure mains cross main roads (i.e., traffic load) or street car/railway tracks.
- The location where air or gas pockets and associated corrosive conditions (e.g., due to H_2S) may be expected (see Section 5.8.3.3).
- The location where materials were changed used during construction, for example, connections between cast iron and HPDE sections.

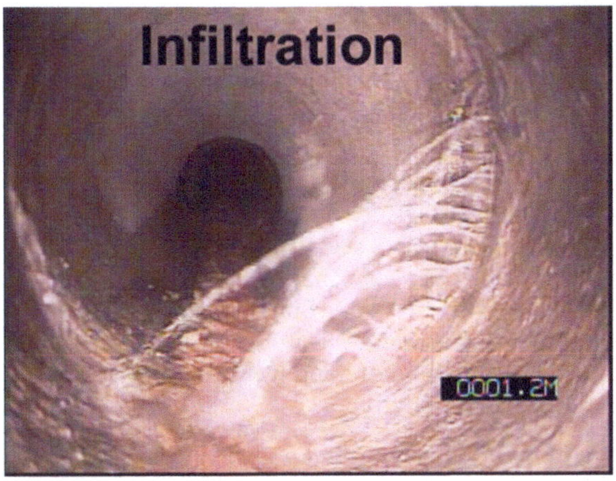

Figure 5.50 Infiltration of groundwater due a defective joint (Deltares).

Figure 5.51 Car is completely 'swallowed' in a sink hole that occurred due to a collapse sewer pressure main. (Photo by C. de Haan).

The time scale of the involved processes varies: subsidence may take years to result in leakage, whereas corrosion can cause serious damage in a matter of months when conditions are unfavourable.

In general leakage can be detected at an early stage using a range of methods:

- Field visits/regular inspection of leak detection devices; also possible by application of drones and so on especially looking for wet areas.
- Low and high pressure testing (either with air or with water) of complete pipeline systems allowing for a rough quantification. This method is used in a lot of countries and described in national guidelines when testing a new pipe and is also very relevant for older pipes-can be applied the same. This method can also be derived in daily operation by use of pressure sensors.
- Focussed electrode leak detection
- Testing of (suspicious) individual joints (both from inside and outside)
- The use of tracer experiments, both gas and fluids. These methods facilitate, to a certain extent, the quantification of the amount of leakage and pinpointing its location, albeit at a relatively low resolution determined by the accessibility of the system.
- External Acoustic methods facilitating the identification of the location; however, very small leaks are still hard to pinpoint. See Almeida *et al.* (2014) and Brennan *et al.* (2017) for leak noise correlations.
- Infrared inspections
- FBG-time domain reflectometry (TDR) (Fatemi Aghda *et al.*, 2018).
- Distributed temperature sensing (DTS), see also Chapter 4.
- Distributed strain sensing (DSS) (see e.g. Zhang *et al.*, 2019).
- Distributed acoustic sensing (DAS) (see e.g. Stajanca *et al.*, 2019).
- Acoustic based methods (e.g., smartball).
- Pigging/other internal inspection in in dry and wet situation; by use of different methods (CCTV, radar, infared, acoustic and conductivity).

It is important to keep in mind is that leakage can also be prevented by searching for local surface or pipe settlement, and deterioration of pipe material and joints and vice versa; leakage is only one aspect and cannot be seen apart from other aspects.

A disadvantage of the TDR, DTS, DAS and DSS methods is that wires need to be installed, preferably during the initial construction. Doing so for existing installations is mostly not an option. Therefore, UD owners are well advised to consider doing so for newly built installations, as the these method allow for a relatively easy and almost continuous monitoring of leakages. Alternatively application of a towed acoustic sensor, like the smartball system, does not required the installation of wires, an additional advantage of such a method is that the presence of air/gas pockets can be detected as well.

In Section 5.9.3.2, a case study is presented that illustrates (combinations of) the processes leading to either leakage, surface damage or collapse. In addition to physical/chemical processes, the following factors contribute as well to incidents concerning pressure mains:

- Incorrect or not up-to date information in the AM systems.
- Damage incurred during construction (see also Section 5.4).
- Damage incurred due to construction activities in the direct vicinity of the pressure main (so-called 'Third party intrusions')

5.9.3.2 A case study on pressurized systems

5.9.3.2.1 General

In the city of Rotterdam, The Netherlands, an extensive system of pressure mains is used to transport wastewater from mainly combined sewer systems to WWPTs. Figure 5.52 shows part of the tracing of one of these pressure mains with some lengths of pipe supported by a pile foundation as a function of the bearing capacity of the soil. The pipeline crosses above the tunnel of the metro and also contains a siphon to pass under a canal. The pressure main contains two high points at which manholes, aerators and vents are present. The system contains aerators and vents at various locations: when the pressure main is not operated, they allow air into the system. In the case of failure, they can cause siphoning of the pipe, through which water may enter from the adjacent river during high river and sea levels. The system was built in 1940 and prior incidents had been reported: a civilian noticed 'water coming out of the street' which was reported to the municipality. It was decided to excavate the pipe, but no problem was identified at the time of the visual inspection. Since the reported water had disappeared, it was decided to rebury the pipe. A few weeks later a similar call came in; this time it was noticed that the street had subsided by a few centimetres, sand has flushed out of the construction.

5.9.3.2.2 Environmental aspects

The pipeline runs through an urban area with a lot of infrastructure and a variety of street level changes: an intersection with a waterway where the pipeline is near a bridgehead and a passage of three dykes, one of which is a primary flood defence. In addition, the pipeline crosses a metro tunnel and many cables and other underground infrastructure (e.g., water supply and gas supply). Taken all these conditions and circumstances into consideration, the system is to be regarded as a hotspot in terms of consequences of failure and should therefore be scrutinized.

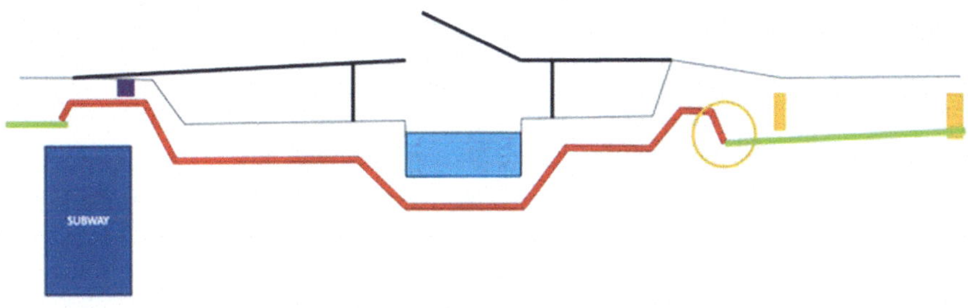

Figure 5.52 Schematic impression of the vertical tracing of a pressure main, red = pile foundation, green = no pile foundation and dark blue = subway, orange = manhole, the circle indicates the location of the incident.

5.9.3.2.3 Emergency plan

The municipality has a process plan in which roles and responsibilities are laid out; however, it is recognized that improvement may be needed, see the discussion of a recent event below under 'Failure history'. Important points for attention are the moment of signalling the field operations staff and the availability of a file with information about the environment (i.e., the pipes and foundation method).

5.9.3.2.4 Failure history

Due to uneven settlement, the joint at the transition from the not supported to the supported pressure main had failed (see Figures 5.52–5.54). This is an example (occurred in 2013) of failure occurring at the transition from cast iron elements. Similar incidents with this type of pipes had occurred in the past. In almost all cases, (uneven) settlement was suspected to be the primary cause. In a few cases, it decreased the quality of the joint; the rope used in the seal between pipe sections had dried out. The cast iron pipe itself, however, had not failed (i.e., no cracks or holes due to corrosion were observed).

Over the course of time uneven subsidence of the soil in the direct vicinity of the pressure main had occurred (as illustrated in Figure 5.55).

This case illustrates the added value of monitoring the evolution of known, or suspected, failure mechanisms for critical infrastructure. The city of Rotterdam monitors the performance of its pressure mains in considerable detail as it pertains to hydraulic performance (see e.g., Korving, 2004 or Duinmeijer, 2020). Monitoring of the (rate of) subsidence may add to the early detection of potential critical situations and allow for preventive action. A further important lesson is that citizen calls that may somehow be related to critical infrastructure ask for more than normal attention, as the initial call related to 'water on the street' was, in retrospect, an early warning of a joint giving way, which had not been noticed initially. Monitoring such events can provide valuable information for future cases and may help shape protocols for tracking down locations where deterioration processes are occurring.

Figure 5.53 Aerial photo of the situation, the red and green line indicates the alignment of the pressure main, the blue line indicates the alignment of the underground metro, the circle indicates the location of the incident. (Courtesy by C. de Haan).

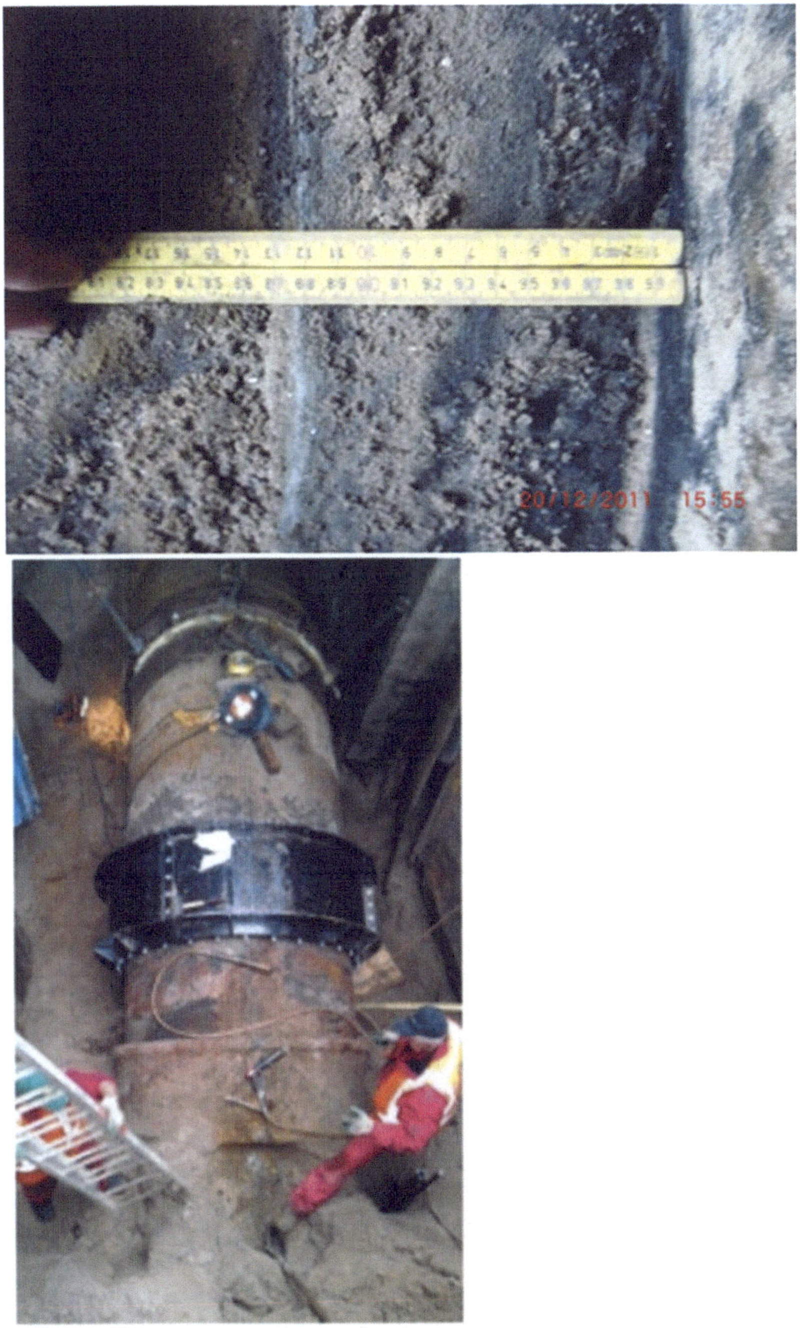

Figure 5.54 Example of a joint that has given way. The pipes were pushed about 13 cm apart leading to a leaking joint. (Photos by C. de Haan).

Figure 5.55 Observed subsidence of the soil, the blue line indicates the location of the pressure main, the red area indicates the location of the incident, the coloured dots indicate the observed subsidence rate of the soil (red = between −46 and −10 mm/year, the green dots between −3 and + 4 mm/year the blue dots between + 11 and + 20 mm/year) As can be seen the differences in subsidence rate in the indicated area are quite high: up to 50 mm/year).

5.9.4 Wrong connections

Misconnections, or wrong connections, cross-connections, or illicit connections are terms indicating the occurrence of premises or locations where wastewater is (unintentionally) discharged wastewater into the public stormwater system, or vice versa, where stormwater runoff is discharged into the sanitary sewer system. In both cases the functionality of the systems is corrupted.

In many cases stormwater runoff is either directly, or after a simple or more sophisticated purification step, discharged into open water courses, is infiltrated to replenish aquifers, or is used as (usually non-potable) water source. As discussed in Section 6.4 many stormwater systems display a significant wastewater signature. Needless to say, the discharge of wastewater or polluted runoff into such systems directly affects its water quality and may give rise to odour complaints from residents as stormwater systems are normally not equipped with odour control measures. In addition, it may negatively impact public health where this stormwater is used for other purposes, see Section 6.2. The case of discharging stormwater runoff into a sanitary sewer system has negative repercussions as well, albeit primarily affecting flood risks and increasing the operational costs of transport and treatment. To a certain extent, allowing some stormwater runoff into a sanitary system at its most upstream terminals to ensure frequent flushing velocities may be good practice; however, when the contributing area increases above a certain threshold, regular flooding of the sanitary system may occur. This is obviously not in compliance with one of the main premises of sanitary system practice, that is, preventing people from getting in direct contact with wastewater. The occurrence of wrong connections is in most cases due to human errors or negligence. During construction of new developments or reconstruction activities mistakes are made in connecting downspouts and storm service connections to sanitary pipes and vice versa. Despite efforts to track down wrong connections after construction, estimates of percentages of wrong connections vary enormously between case studies with values from 1% up to 25% reported (Ellis & Butler, 2015). Without regular preventative actions this percentage is likely to increase over time due to reconstruction activities within homes and apartment buildings. These reconstruction activities are in many cases DIY projects; hence, the application/enforcement of professional standards and quality checks may be largely lacking. To the authors' knowledge no data have being published on the rate at which wrong connections increase, but it is likely to be an unpredictable process as reconstruction activities and people's mobility (linked to reconstruction) depend on a wide range of (economical) factors. A plethora of methods has been developed and is being applied to detect and localise misconnections:

- DTS (Hoes *et al.*, 2009) for wastewater discharged into storm water sewers.
- Smoke methods (for stormwater into wastewater systems, Gokhale & Graham, 2004).
- Acoustic methods (for stormwater into wastewater systems, SAHARA system, Rizzo, 2010).
- Conductivity (Deffontis *et al.*, 2013).
- Chemical analysis (Panasiuk *et al.*, 2015).
- Tracers (Lepot *et al.*, 2016).
- Infrared camera system (Lepot *et al.*, 2017).

More recently, the application of a network of simple low-cost sensors was reported as a means for detecting illicit connections using temperature and conductivity sensors (Shi *et al.*, 2022).

5.10 STATISTICAL MODELLING OF AGEING PROCESSES

5.10.1 Introduction

Since the 1980s many organizations managing UD systems have created large scale inventories of the technical status of their assets using visual inspection (mainly through CCTV techniques). As discussed in Chapter 4, guidelines and norms exist to apply standardized methods to classify observed defects using video footage. Over the last few decades some organizations have managed to build a comprehensive database of those elements in their UD systems that have been inspected multiple times over the course of time, potentially allowing them to identify the speed of the deterioration processes.

It is implicitly assumed that the conditions between inspections were identical and that all processes happened at a constant process speed over time. In addition, note that defect classification, as for example, defined in the EN 13508-2 (2011) has no quantifiable relation to loss of functionality.

A rather abundant literature has been dedicated to sewer deterioration modelling since the end of the 1990s. Sound reviews of these research works may be found in Ana and Bauwens (2010) and El-Housni *et al.* (2018). Two broad categories of modelling approaches can be distinguished: physical and probabilistic-statistical models.

5.10.2 Physical deterioration models

Physical deterioration models mainly focus, at the pipe element scale, on the physico-chemical processes of deterioration, involving varying combinations of load stresses and pipe material corrosion. Physical deterioration models can be either fully deterministic or include a probabilistic aspect as in the limit state concept. The real-world application of physical models is however limited by the impossibility to get precise measurements of the physico-chemical conditions within the entire sewer network and, consequently, by our ignorance of the deterioration factor values needed for the calculations. Research activities in the field of physical deterioration models are nevertheless far from being useless since they can provide guidance as to the choice and design of explanatory factors in the creation of the probabilistic-statistical models. Previous studies emphasized cross-relationships effects involving the pipeline geometry (e.g., the length, height and width of elements) and environmental factors such as the location (e.g., under roadway versus under a sidewalk or green space), the pipe depth and the traffic intensity.

5.10.3 Probabilistic-statistical deterioration models

Probabilistic-statistical deterioration models seek to explain, and then to predict, the probability that a given sewer element with given characteristics (i.e., their age, material, size, depth, slope, effluent type and so on) belongs to a condition class. These models formally depend on parameters that should be, for operational applications, statistically estimated from sewerage utility data.

Probabilistic-statistical deterioration models can, in turn, be divided into two main types, reflecting the particular role assigned to time, and more precisely the age of the sewer pipes, allowing one to distinguish:

- classification models, for which the sewer age is a deterioration factor like any other, and
- stochastic processes that structurally depend on the sewer age by considering the deterioration speed.

These two approaches are illustrated in the remainder of this section with the help of typical examples composed of multinomial logistic regression (MLR) models and Markov chains (MC), respectively.

5.10.4 Multinomial logistic regression models

Multinomial logistic regression (MLR), as presented in the reference book 'Generalized Linear Models' (McCullagh & Nelder, 1989), provides the canonical framework for studying phenomena formalized by a random variable (RV), the distribution of which is multinomial. The condition class assigned to a given sewer pipe following its inspection is considered as a random variable denoted Y, which takes its possible values as per a discrete set of integers $J = \{1,2,\ldots,C\}$, which represent the various condition classes, ordered from best (e.g., 1 = 'as good as new' to worst (e.g., C = 'rehabilitation urgently needed'). The condition class probabilities $p_j(\boldsymbol{Z}) = \Pr[Y = j \mid \boldsymbol{Z}]$, $j \in J$ are assumed to depend on pipe characteristics (or functions of them), gathered within the 'covariate' vector $\boldsymbol{Z}$, which are then considered as 'deterioration factors'.

Since the condition classes are ordered it is preferable, according to McCullagh and Nelder (1989), to consider the cumulative probabilities $\pi_j(\boldsymbol{Z}) = \Pr[Y \le j \mid \boldsymbol{Z}]$, $j \in J$, *with* $\pi_C(\boldsymbol{Z}) = 1$. The condition class probabilities are then calculated by their difference: $p_j(\boldsymbol{Z}) = \pi_j(\boldsymbol{Z}) - \pi_{j-1}(\boldsymbol{Z})$, which ensures $\sum_{j\in J} p_j(\boldsymbol{Z}) = 1$.

The key feature of MLR consists of mapping $\pi_j(\boldsymbol{Z}) \in]0,1[$ onto $g(\pi_j(\boldsymbol{Z})) \in]-\infty, +\infty[$, by using the so-called 'logit' transform $g(\pi) = \ln(\pi/(1-\pi))$, which allows one to adopt the linearity hypothesis: $g(\pi_j(\boldsymbol{Z})) = \alpha_j + \boldsymbol{Z}^T\beta$, where $j \in \{1,2,\ldots,C-1\}$. The vectors $\boldsymbol{\alpha} = (\alpha_j)_{j \in J}$ and β are regression parameters to be estimated from observed data; the α_j's estimates must satisfy the inequality constraints $\alpha_1 < \alpha_2 < \ldots < \alpha_{C-1}$ to ensure $p_j(\boldsymbol{Z}) \in]0,1[$, by preventing the graphs of the $\pi_j(\boldsymbol{Z})$'s from intersecting. This theoretical construct is typical of generalized linear models in that:

- unlike simple linear models, it is not the response RV value itself that is modelled but the probability of this value; this has the crucial consequence that parameter estimation cannot proceed by minimizing a sum of squared deviations between observed and predicted response values; and
- the linear effect of covariates on the response RV is ensured via an adequate transformation.

Additionally, the analytical form $g(\pi_j(\boldsymbol{Z})) = \alpha_j + \boldsymbol{Z}^T\beta$ makes the covariate effects independent of the condition class while the regression intercepts are assumed to solely depend on it (i.e., 'parallel regression' hypothesis).

As discussed earlier, estimating the model parameters $\theta = (\alpha, \beta)$ cannot be conducted using a least squares minimization method. A more general estimation method is needed, based on the 'maximum likelihood' concept. This involves defining a likelihood function $L(\theta|E)$ that quantifies how likely a parameter value θ is, given a set of n observations (i.e., n sewer segment inspections) $E = \{(y_i, \boldsymbol{Z}_{i}): i = 1,\ldots,n\}$, where y_i stands for the Y value at observation i. 'Maximum Likelihood Estimation' consists of numerically determining the 'optimal' value θ_{opt} that maximizes $L(\theta\,|E)$. The adequate MLR likelihood function is simply defined as the product of probabilities $L(\theta|E) = \prod_{i=1,\ldots,n} \Pr[Y = y_i|\, \boldsymbol{Z}_i]$. In practice, rather the natural logarithm $\ln L(\theta|E)$ is considered since it involves a summation which is numerically more tractable than a product. According to the authors' experience, the so-called 'Nelder-Mead simplex' method, proposed by Nelder and Mead (1965) can be used to maximize the MLR log-likelihood, although a wide spectrum of optimisation algorithms is presently available, see for example, Bazaraa *et al.* (2013).

The definition of $L(\theta|E)$ above assumes, however, that the n observations are mutually independent of each other. In the case where the data set E concerns pipes inspected repeatedly, this independence assumption is no longer valid, and the model can be refined by adding a pipe-specific random effect to the linear combination $\boldsymbol{Z}^{\mathrm{T}}\beta$; the MLR model is then turned into a 'mixed' MLR (i.e., considering a mixture of fixed and random effects), which requires an adapted likelihood maximization method such as the so-called 'Expectation-Maximization' method, proposed by Dempster *et al.* (1977).

Classification tools such as the MLR presented above are generally considered efficient for predicting which sewer pipelines are likely to be the most deteriorated within the short term (i.e., within the next year or next two years) that can then be used in scheduling inspection activities. These models are however not designed to explicitly account for the increase in deterioration as a function of the ageing of the assets. They can therefore not be used in performing mid- to long-range simulations of the evolution of the condition of a set of sewer segments that compose a sewerage network, given that they are subject to various imposed constraints such as a pipe renewal budget and the interaction with the rehabilitation strategy of other, third-party infrastructure (particularly roadways). Markov chains were developed to fill this gap; they are discussed in the next section.

5.10.5 Markov chains

Following previous research works dedicated to deterioration modelling of engineering structures such as bridges and roadways, the mathematical theory of probabilistic Markov chains was recognized from the beginning as a relevant approach (Madanat *et al.*, 1995).

In the context of sewer deterioration, the theoretical development of Markov chains can be seen as an adaptation of the MLR model with the aim of endowing it with appropriate properties expected of ageing models.

A Markov chain is a stochastic process formalized by a random variable (RV) $Y(t)$ that structurally depends on time t, namely the age of the sewer main. It should however be emphasized that, for a fixed age, the variable has the same multinomial distribution as that presented in the case of the classification models: the RV $Y(t)$ takes its values on a discrete state space, the elements of which are usually formalized by a set of integers, such as $J = \{1,2,\ldots,C\}$ (C is in usual applications a small finite integer, such as 4 or 5, but it could also be theoretically considered as infinite). Markov chains have the characteristic property of being memoryless, which means that the state reached at the next time step depends only on the current state (i.e., having a 'Markovian' property).

The time dimension can either be discrete, that is, $t \in \mathbb{N}$, in which case the MC is said to be a 'Discrete Time Markov Chain' (DTMC), or continuous, that is, $t \in \mathbb{R}_+$, and the MC is then called a 'Continuous Time Markov Chain' (CTMC). These two classes of Markov chains will be presented successively, with for each an example of construction usable in practice. At the practical level, for both MC classes, the parameter calibration is confronted with the scarcity of sewer inspection data, leading to the impossibility of precisely dating the transitions between condition states. It is then pertinent to design a calibration procedure rather based on the state probabilities and build a likelihood function such as that used in MLR.

5.10.6 DTMC

The state of a DTMC is formalized at any age t by the state probability vector: $\boldsymbol{p}(t) = (\Pr[Y(t) = j])_{j \in J}$, and its evolution over time, that is, when the system ages, it is governed in the most general case by a time-dependent square $C \times C$ matrix of transition probabilities $\boldsymbol{Q}(t) = (q_{ij}(t))$, with $q_{ij}(t) = \Pr[Y(t) = j \,|Y(t-1) = i]$ which only depends on the current state $Y(t)$, and not on the MC process trajectory $Y(s)$, $s \in \{0,1,\ldots, t-1\}$. If the transition probabilities do not depend on time, the MC is said to be 'homogeneous' (HDTMC), and 'non-homogeneous' (NHDTMC) otherwise.

In the rather general case of a NHDTMC, the development of condition state probabilities is governed by the relationship: $\boldsymbol{p}^{\mathrm{T}}(t+s) = \boldsymbol{p}^{\mathrm{T}}(t)\prod_{k=1,\ldots,s}\boldsymbol{Q}(t+k), \forall\, t \in \mathbb{N}, \forall\, s \in \mathbb{N}^*$. In the HDTMC case, this relationship simplifies to $\boldsymbol{p}^{\mathrm{T}}(t+s) = \boldsymbol{p}^{\mathrm{T}}(t)\boldsymbol{Q}^{s}$.

A practical DTMC construct can empirically start from the observation that the graph of the cumulative probability $\pi_j(t; \boldsymbol{Z}) = \Pr[Y(t) \leq j \mid \boldsymbol{Z}]$ *vs* age t has most often a decreasing sigmoid shape. A simple transition matrix $\boldsymbol{Q}(t)$ can be designed by assuming that, in a unit time step $\Delta t = 1$, a sewer pipe can only either stay in the same condition state j, with probability $q_j(t+1)$, or jump into the adjacent more deteriorated state $j+1$, with complementary probability $1\text{-}q_j(t+1)$, which is calculated as $[\pi_j(t)\text{-}\pi_j(t+1)]/p_j(t)$. This simple DTMC is illustrated by Figure 5.40. Usual sigmoid functions lead to an age-dependent transition matrix (NHDTMC). An example of such a construct is the GompitZ model presented by Le Gat (2008); the analytical form is $\pi_j(t; \boldsymbol{Z},u) = \exp[-\exp(\alpha_j + t\exp(\boldsymbol{Z}^{\mathrm{T}}\beta + u))]$, designed to prevent the cumulative probability curves from intersecting, and the condition state from improving with age. It also accounts for the pipe-specific frailty via the random effect coefficient u; this model can also be viewed as a generalized linear mixed model.

5.10.7 CTMC

In the case of a CTMC the transition probabilities are replaced by transition rates, which are either time-dependent $\tau_{ij}(t) = \mathrm{E}[\mathrm{d}Y(t) = j - i \,|Y(t) = i]$, in the case of a 'non-homogeneous' CTMC (NHCTMC, see Figure 5.56), or independent of time τ_{ij}, in the case of a 'homogeneous' CTMC (HCTMC).

A construction like that of the GompitZ model is possible in continuous time, and is the subject of ongoing research work, which is briefly discussed here. It is more practical in continuous time to consider transition rates independent of pipe age, and then to build a HCTMC, as illustrated by Figure 5.57. In this kind of construct, the deterioration rates can be designed in a 'proportional hazard' manner, as $\eta_j = \exp(\alpha_j + \boldsymbol{Z}^{\mathrm{T}}\beta)$. It is easy to account for pipe decommissioning by adding an absorbing 'decommissioned' state, fed by decommission rates ρ_j, which are the product of inspection rates and conditional probabilities of decommissioning given the pipe was observed in this or that condition state. Considering such an absorbing 'decommissioned' state allows for accounting in a 'natural' manner for the right censoring of the deterioration process due to pipe decommissioning, which rarefies the observation of the most deteriorated state. As the decommissioned state is unobserved at the pipe inspection, the condition probabilities to be considered when building the likelihood function

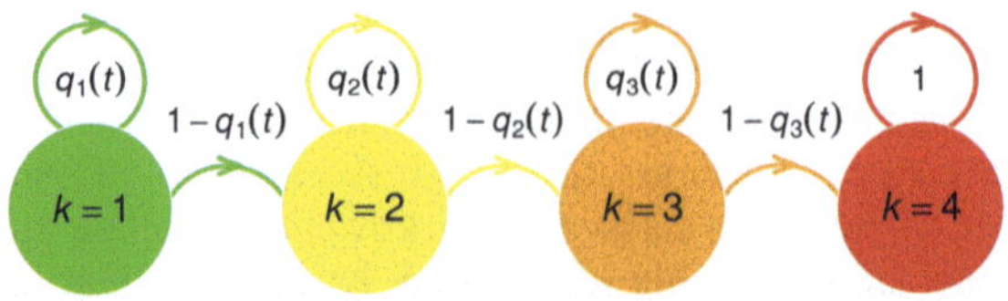

Figure 5.56 Illustration of a simple NHDTMC with $C = 4$ possible condition states, the 4th being absorbing.

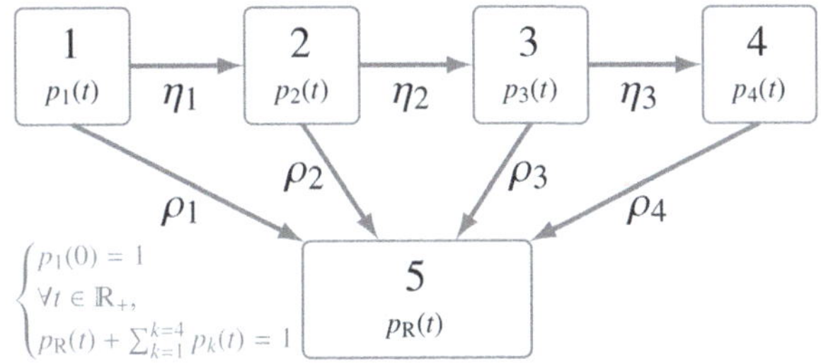

Figure 5.57 Illustration of a simple HCTMC with $C = 4$ possible condition states, and a 5th state that stands for decommissioned pipes and is absorbing.

are $p_j^*(t;\mathbf{Z}) = p_j(t;\mathbf{Z})/\sum_{k=1,C} p_k(t;\mathbf{Z})$, which allows one to account for the selective survival phenomenon. These questions of informative censoring of pipe inspections, and pipe cohort truncation due to the selective survival phenomenon, are discussed in more detail in Chapter 8 of this book.

5.11 APPLICATION OF MACHINE LEARNING BASED MODEL

Artificial intelligence-based models, such as machine learning models, for sewer deterioration modelling differ from statistical models as they do not require any assumptions on the model structure (Tscheikner-Gratl *et al.*, 2019). These models are known to be data-driven, as the mathematical relationship between the factors affecting the deterioration and condition class of the sewer pipe are constructed by 'learning' the deterioration process from data obtained during sewer inspection (Hawari *et al.*, 2020). The strength of machine learning models over statistical models is their capability to handle complex relationships between variables that are difficult to describe using statistical models (Ana & Bauwens, 2010). On the other hand, machine learning-based models have the disadvantage of being 'black boxes', meaning that the internal processes are somewhat unknown (Tu, 1996). Additionally, machine learning-based models requires a substantial amount of computational power and have a high demand of data to be trained on. The latter has an impact on utilities who lack condition data of their sewer systems, typically smaller utilities (e.g. Chen *et al.*, 2022; Jenkins *et al.*, 2015; Kabir *et al.*, 2020).

As one of the goals of a sewer deterioration models is to predict the condition of sewer pipes that have not yet been inspected by using inspection data from other pipes (Harvey and McBean, 2014b), the machine learning algorithms need to be able to learn from a set of input with a given output. Within the machine learning field of supervised learning, this is known as classification (Géron, 2019). In sewer deterioration modelling, the input will be the desired physical, environmental or operational factors that could describe the deterioration process (Hawari *et al.*, 2017), and the output would be the corresponding condition class, based on the inspected data used in the model.

Several algorithms for doing classification problems exist the most common are described in some detail by Géron (2019). Regarding sewer deterioration modelling, many of these algorithms have been applied in various studies, from the simple structures as decision trees (e.g. Harvey and McBean, 2014a) and random forest (e.g. Rokstad and Ugarelli, 2015), to the more complex structures of artificial neural networks (Atambo *et al.*, 2022). Before looking deeper into the use of the different algorithms for sewer deterioration modelling, the most commonly utilized algorithms will be briefly introduced by their strengths and weaknesses.

Géron (2019) gives a detailed description of decision trees, and their main strength is that they are powerful and capable of fitting complex datasets, and they are able to perform both classification and regression tasks. The most common way of implementing decision trees is by the algorithm CART, which is an abbreviation for alassification and regression tree and is developed by Breiman *et al.* (1984). Additionally, to being powerful, decision trees are also easy to understand and to use. However, the main issues when using decision trees is that they are very sensitive to small changes in the dataset used for training (Pedregosa *et al.*, 2011), and they are known to be easily overfitted to the data (Kotsiantis, 2013).

The random forest algorithm is developed by Breiman (2001), and is based on grouping a number of decision trees in an ensemble which reduces the chance of overfitting due to the law of large numbers. The algorithm grows multiple decision trees, known as a forest, where the class that gets the most votes at the end is the resulting prediction. By using these votes, it is possible to calculate the probability of belonging to a given class (Rokstad and Ugarelli, 2015). The random feature of the algorithm is implemented by using so-called bagging or bootstrap aggregating (Breiman, 1996, 2001). This is done by training each decision tree of the forest on a randomly selected subset of the training data, and a random selection of input variables are used at each node of the tree.

Another commonly used algorithm is support vector machine (SVM), based on the work of Cortes and Vapnik (1995). The concept of the algorithm is to map the input variables onto a high dimensional feature space, where a linear decision surface is constructed. To do this mapping, so-called kernel functions are used (Shawe-Taylor & Cristianini, 2004). Typical available kernel functions are linear, polynomial and radial basis function. The goal of SVM is to find the optimal hyperplane that maximizes the separation between the classes, but also generalize well, that is reduce overfitting (Cortes & Vapnik, 1995). SVM are very well suited for classification on complex, but small- or medium-sized datasets, but they are very sensitive to adjustments of the hyperparameters (Géron, 2019).

Artificial Neural Network (ANN) is a type of algorithm that tries to mimic the performance of the human nervous system (Al-Barqawi & Zayed, 2008). The composition of ANNs is consisting of artificial neurons that are connected to each other in different layers, with the aim of mimicking the ability of the human brain to recognize patters and using these to predict outcomes based on its observations. The neurons are linked together with connections having a certain weight (Fausett, 1994). For a given input, the associated weights are summed up and the neurons send a signal to identify the activation function, when the sum of weights reaches a certain threshold value. The most common neural network used for deterioration modelling is the back-propagation neural network (BPNN) (Hawari *et al.*, 2020; Tran *et al.*, 2007; Zeng *et al.*, 2023). Figure 5.58 shows the layout of a BPNN used for sewer deterioration modelling, where each circle represents a neuron, the lines indicate where the connection weights are located. The goal of the BPNN is to tune the connection weights so that the error between predicted condition class and actual condition class is minimized.

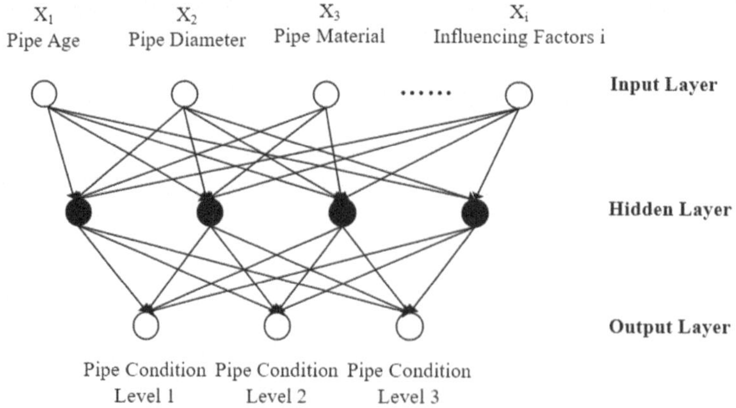

Figure 5.58 Schematic presentation of a BPNN for sewer deterioration modelling. (After Zeng *et al.* (2023).

The strength of ANNs is that they are very powerful and scalable, making them suitable for large and complex machine learning tasks (Géron, 2019). Still, they have their limitation as the underlying processes are hidden, making it difficult to really understand how the model works (Tran *et al.*, 2007).

To address the predictive power of a machine learning classifier, several metrics exist (Géron, 2019). A visual representation is often provided by presenting the predicted condition versus the actual condition in a confusion matrix. This gives an overview of the amounts of correct and incorrect predictions. The goal of a classifier is to predict good pipes as good and bad pipes as bad, meaning that the number of incorrect predictions should be as close to zero as possible (Harvey & McBean, 2014b). Accuracy is commonly used as a performance metric, which is defined as the number of correct predictions divided by the total number of predictions. Still, using accuracy solely to measure the performance is unsuitable for imbalanced datasets, where a high accuracy does not necessarily mean that the model is good at predicting (Géron, 2019; Harvey & McBean, 2014b). As this often is the case for sewers, where the number of pipes in good condition is higher than the number of pipes in bad condition, other measures should be used. A commonly used method is to plot the True Positive Rate (TPR) as a function of the False Positive Rate (FPR) by using different probability cut-offs (Géron, 2019). TPR is calculated as the number of correctly predicted good pipes over the total number of good pipes, while FPR is the number of bad pipes predicted as good over the total number of bad pipes. The resulting plot is known as the receiver operating characteristic (ROC) curve and is shown in Figure 5.59. A measure that can be extracted from the ROC curve is the area under the curve, known as AUC, where a score equal to 1 is a perfect model and a score equal to 0.5 is a random model. An AUC score above 0.7 is by Hosmer and Lemeshow (2000) considered a good model.

All the mentioned machine learning models have been applied for sewer deterioration modelling in several studies over the years. Harvey and McBean (2014a) compared two deterioration models in their study, one Decision Tree model and one SVM model, using data from the municipality of Guelph in Ontario, Canada. The decision tree model had an accuracy of 76%, and an AUC score of 0.78, which was a significantly higher accuracy than the one of the SVM model (58%), while the AUC score was more similar between the two models. Syachrani *et al.* (2013) also used a Decision Tree model, which they compared with a regression model and an artificial neural network on data from Johnson County Wastewater in Kansas, USA. The findings of the study were that the decision tree model consistently outperformed the other models.

As earlier mentioned, decision trees are models who are easily overfitted, and the use of random forest has been more usual. Harvey and McBean (2014b) used a random forest model with different

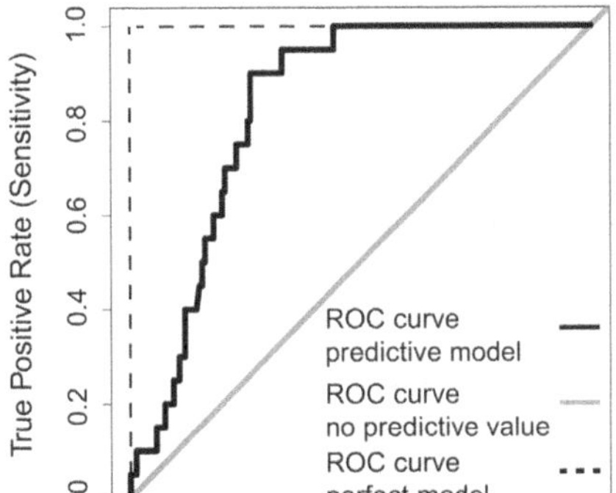

Figure 5.59 Example of ROC curve. (After Harvey & McBean, 2014b).

probability cut-offs (i.e., threshold for when a pipe is classified as bad or good), on the dataset from Guelph, achieving accuracies of 72%, 74% and 89%. All models gave the same AUC score, but the latter model favourited classifying pipes as good, failing to find the bad pipes. By intuition, having a deterioration model that can find the bad pipes is more useful than one who fails to find them. This states the fact that using a single performance measure when evaluating a model is unfavourable. Rokstad and Ugarelli (2015) compared a random forest model with the statistical GompitZ model (Le Gat, 2008) on sewer data from Oslo, Norway, where the random forest model gave the best predictive accuracy on the network level. Caradot *et al.* (2018) also compared the same two models on sewer data from Berlin, Germany. Their study showed that random forest performed better in terms of measures such as False Negative Rate, FPR, and their own defined measures K_{Network} and K_{Pipe}. Laakso *et al.* (2018) used a logistic regression model and a random forest model on data from the southern Finland, with the former model giving an accuracy of 56% and the latter an accuracy of 62%. Nguyen *et al.* (2022) compared 17 different algorithms on sewer data from the city of Ålesund in Norway, where the random forest model was ranked the best based on the TOPSIS ranking method. The random forest also yielded the highest accuracy of around 78%, and the highest AUC score (0.776).

Mashford *et al.* (2011) used SVM models with varying input variables on sewer data from Adelaide, South Australia. In general, the models gave high accuracy, up to 91% at most. Sousa *et al.* (2014) used SVM in comparison with a logistic regression model and an ANN on sewer data from Lisbon, Portugal. The ANN model gave the best accuracy of 78.5%, with the SVM model having an accuracy of 71.1%. Both models still performed better than the logistic regression model, which had an accuracy of 65.9%. The study of Hernández *et al.* (2021) looked into optimizing SVM models in the cities of Bogota and Medellin in Colombia. Models for both cities yielded fairly accurate results, giving the best results when optimized for the K_{pipe} measure described in Caradot *et al.* (2018).

The use of ANNs has also been tested in various other studies. In the early twenty-first century, Najafi and Kulandaivel (2005) created an ANN-based sewer deterioration model on data from the city of Atlanta, United States. The model gave a high accuracy score, but it was trained on a relatively small dataset. More recently, Atambo *et al.* (2022) compared a multinomial logistic regression model against an ANN model on data from the city of Dallas, United States. The ANN model outcompeted the regression model in terms of accuracy, with a score of 85% over 75%. Nguyen *et al.* (2022) tested three different ANN-based models of the 17 in total they used on the sewer data from Ålesund, Norway. The three ANN models had one, two and three hidden layers respectively, and used different activation functions. Still, they did not rank high on the TOPSIS ranking, with accuracy ranging between approximately 67% and 72%, and AUC scores between ~ 0.65–0.69.

Even though several algorithms have been tested in various studies, sig datasets of different size and geographical location, no algorithm came out as better than the other in general, analogously described by the 'No free lunch' theorem (Wolpert & Macready, 1997). The ranges of accuracy, AUC score or other performance measures varies both between the algorithms, and also within each algorithm.

5.12 CONCLUSIONS AND OUTLOOK

As illustrated in the previous sections, a huge amount of physical and (bio)chemical processes lay at the heart of the loss of functionality of UD systems. In most cases, the loss of functionality occurs gradually and slowly develops over time; however, sudden and catastrophic collapse may occur occasionally. In most cases this is the result of a combination of processes undermining the stability of a core element: an unexpected (traffic) mechanical load may trigger a sudden collapse. A similar thing may be said with respect to the occurrence of sinkholes (which sometimes can be linked to the collapse of pipes or severe leaking of underground pipes). Although the processes tend to be well understood and described at some length and detail in the (scientific) literature, this does not imply that deterioration can be predicted or modelled to the extent of producing reliable forecasts on for example, the timing of a collapse or the rate at which the hydraulic capacity of a given conduit will decay. The main reasons for this are:

- Many (deterministic) models need a lot of process parameters that should be determined separately for each case, which in most cases is a non-viable option, also given the substantial heterogeneousness of for example, material and soil parameters.
- The initial conditions are often not known to the detail needed for reliable modelling.
- The boundary conditions and their development over time are largely unknown and sometimes inherently unknowable, unless monitored over the life span of an element (e.g., stormwater runoff loadings, the composition of wastewater, traffic loads, changes in the daily management of the system and so on).

For long-term assessments on relatively large cohorts of elements (i.e., at the scale of an entire catchment or urban area) statistical/data driven models based on condition class data seem to present a useable tool, at least at a large spatial scale and mid- to long-term planning (i.e., >five years). These models, however, lose their predictive power when applied at the scale of a single element or for a small number of elements and short-term 'prediction' horizons. This implies in practical situations, when a decision must be made to replace or rehabilitate a given element, no tools are available to reliably estimate a 'deadline' for action. In such a case, a sound understanding of the processes and their characteristic timescales responsible for the given situation comes at hand. The latter highlights the real value of detailed models of the processes responsible for deterioration: appreciating the concept 'model' as a means to understand the processes and as tool to get a 'feeling' for the process characteristics instead of expecting exact 'predictions'. These models can be of value in decision making, while the understanding of the processes involved is crucial in deciding what and how to monitor to keep track of a developing situation and/or formulate actions to remove or mitigate the cause of the process or slow the deterioration process down. For example, forced degassing combined with forced ventilation and a proper material selection are measures taken during design and operation to reduce/avoid bio sulphuric corrosion.

Regular monitoring of the hydraulic performance of UD systems using high-quality hydrodynamic models as a reference has shown to be able to identify capacity loss due to, for example, sedimentation or the occurrence of obstacles. Further, process knowledge and estimates of the temporal and spatial scales at which these occur can be instrumental to identify locations in systems that are sensitive to certain deterioration processes. This, in turn, can be used as a guide for monitoring. Generally speaking, organizations responsible for UDAM are encouraged to incorporate monitoring options when building new systems, like, for example, wiring pressure mains to allow for regular leakage testing using distributed sensing technologies.

REFERENCES

Alabtah F. G., Mahdi E. and Eliyan F. F. (2021). The use of fiber reinforced polymeric composites in pipelines: A review. *Composite Structures*, **276**, 114595, https://doi.org/10.1016/j.compstruct.2021.114595

Alakayleh Z., Fang X. and Clement T. P. (2019). A comprehensive performance assessment of the modified Philip–Dunne infiltrometer. *Water*, **11**, 1881, https://doi.org/10.3390/w11091881

Al-Barqawi H. and Zayed T. (2008). Infrastructure management: integrated AHP/ANN model to evaluate municipal water mains' performance. *Journal of Infrastructure Systems*, **14**(4), 305–318. Scopus, https://doi.org/10.1061/(ASCE)1076-0342(2008)14:4(305)

Allahverdi A. and Skvara F. (2000). Acidic corrosion of hydrated cement based materials. *Ceramics-Silik' aty*, **44**(2000), 114–120.

Allaire S. E., Roulier S. and Cessna A. J. (2009). Quantifying preferential flow in soils: a review of different techniques. *Journal of Hydrology*, **378**, 179–204, https://doi.org/10.1016/j.jhydrol.2009.08.013

Allroggen N. and Tronicke J. (2015). Attribute-based analysis of time-lapse ground-penetrating radar data. *Geophysics*, **81**, H1–H8, https://doi.org/10.1190/geo2015-0171.1

Almeida F., Brennan M., Joseph P., Whitfield S., Dray S. and Paschoalini A. (2014). On the acoustic filtering of the pipe and sensor in a buried plastic water pipe and its effects on leak detection: An experimental investigation. *Sensors*, 2014, **14**(3), 5595–5610, https://doi.org/10.339/s10305595

Al-Rubaei A. M., Viklander M. and Blecken G. T. (2015). Long-term hydraulic performance of stormwater infiltration systems. *Urban Water Journal*, **12**(8), 660–671, https://doi.org/10.1080/1573062X.2014.949796

Ana E. V. and Bauwens W. (2010). Modelling the structural deterioration of urban drainage pipes: the state-of-the-art in statistical methods. *Urban Water Journal*, **7**(1), 47–59, https://doi.org/10.1080/15730620903447597

Anderson A. E., Weiler M., Alila Y. and Hudson R. O. (2009a). Dye staining and excavation of a lateral preferential flow network. *Hydrology and Earth System Sciences*, **13**, 935–944, https://doi.org/10.5194/hess-13-935-2009

Anderson A. E., Weiler M., Alila Y. and Hudson R. O. (2009b). Subsurface flow velocities in a hillslope with lateral preferential flow. *Water Resources Research*, **45**, https://doi.org/10.1029/2008WR007121

Angulo-Jaramillo R., Bagarello V., Iovino M. and Lassabatere L. (2016). Infiltration Measurements for Soil Hydraulic Characterization. Springer International Publishing, Switzerland (eBook), https://doi.org/10.1007/978-3-319-31788-5

Angulo-Jaramillo R., Bagarello V., Di Prima S., Gosset A., Iovino M. and Lassabatere L. (2019). Beerkan estimation of soil transfer parameters (BEST) across soils and scales. *Journal of Hydrology*, **576**, 239–261, https://doi.org/10.1016/j.jhydrol.2019.06.007

Ankeny M., Kaspar T. and Horton R. (1988). Design for an automated tension infiltrometer. *Soil Science Society of America Journal*, **52**(3), 893–895. http://dx.doi.org/10.2136/sssaj1988.03615995005200030054x

Anton-Prinet C., Mur G., Gay M., Audouin L. and Verdu J. (1999). Change of mechanical properties of rigid poly (vinylchloride) during photochemical ageing. *Journal of Materials Science*, **34**, 379–384, https://doi.org/10.1023/A:1004482328203

Archer N. A. L., Quinton J. N. and Hess T. M. (2002). Below-ground relationships of soil texture, roots and hydraulic conductivity in two-phase mosaic vegetation in South-east Spain. *Journal of Arid Environments*, **52**(4), 535–553, https://doi.org/10.1006/jare.2002.1011

Arnold J. C. (2003). 6.06 Environmental effects on crack growth in polymers. *Comprehensive Structural Integrity*, **6**, 281–319, https://doi.org/10.1016/B0-08-043749-4/06123-1

Ashely R. M., Bertrand-Krajewski J.-L., Hvitved-Jacobsen T. and Verbanck, M. (eds) (2004). Solids in Sewers. Characteristics, Effects and Control of Sewer Solids and Associated Pollutants. Scientific and Technical report No. 14. IWA Publishing, London, ISBN 1 900222914.

Atambo D. O., Najafi M. and Kaushal V. (2022). Development and comparison of prediction models for sanitary sewer pipes condition assessment using multinomial logistic regression and artificial neural network. *Sustainability (Switzerland)*, **14**(9), 5549. Scopus, https://doi.org/10.3390/su14095549

Attiogbe E. K. and Rizkalla S. H. (1988). Response of concrete to sulfuric acid attack. *ACI Materials Journal*, **85**, 481–488.

Azzout Y., Barraud S., Cres F.-N. and Alfakih E. (1994). Techniques alternatives en assainissement pluvial: choix, conception, réalisation et entretien, Technique & Documentation, Lavoisier.

Balades J.-D., Legret M. and Madiec H. (1995). Permeable pavements: pollution management tools. *Water Science & Technology*, **32**(1), 49–56, https://doi.org/10.2166/wst.1995.0012

Barraud S., Gonzalez-Merchan C., Nascimento N., Moura P. and Silva A. (2014). A method for evaluating the evolution of clogging: application to the Pampulha Campus infiltration system (Brazil). *Water Science & Technology*, **69**(6), 1241–1248, https://doi.org/10.2166/wst.2013.819

Bazaraa M. S., Sherali H. D. and Shetty C. M. (2013) Nonlinear Programming: Theory and Algorithms. John Wiley & Sons, Inc., New Jersey, USA, ISBN-13: 978-0-471-48600-8

Bedell J.-P., Saulais M. and Delolme C. (2013). Rôle de la végétation sur l'évolution des charactéristiques physico-chimiques des sédiments déposés dans un bassin d'infiltration des eaux pluviales. *Etude Gest. Sols*, **20**, 27–38.

Bedell J.-P., Hechelski M., Saulais M. and Lassabatere L. (2021). Are acts of selective planting and maintenance drivers for vegetation change in stormwater systems? A case study of two infiltration basins. *Ecol. Eng.*, **172**, 106400, https://doi.org/10.1016/j.ecoleng.2021.106400

Belie N. D., Monteny J., Beeldens A., Vincke E., van Gemert D. and Verstraete W. (2004). Experimental research and prediction of the effect of chemical and biogenic sulfuric acid on different types of commercially produced concrete sewer pipes. *Cement and Concrete Research*, **34**, 2223–2236, https://doi.org/10.1016/j.cemconres.2004.02.015

Bell J. P. (1976). Neutron Probe Practice. Institute of Hydrology, Wallingford, UK, Rep., **19**, 1–63.

Benjamin P. (1980). The influence of the extrusion process on the quality of unplasticized polyvinyl chloride (UPVC) pressure pipe. *Journal of Vinyl and Additive Technology*, **2**, 254–258, https://doi.org/10.1002/vnl.730020412

Bergman M., Hedegaard M. R., Petersen M. F., Binning P., Mark O. and Mikkelsen P. S. (2011). Evaluation of two stormwater infiltration trenches in central Copenhagen after 15 years of operation. *Water Science &Technology*, **63**(10), 2279–2286, https://doi.org/10.2166/wst.2011.158

Bernatek A. (2015). The influence of piping on mid-mountain relief: a case study from the polish Bieszczady Mts. (Eastern Carpathians). *Carpathian J Earth Environ Sci*, **10**, 107–120.

Bernatek-Jakiel A. and Poesen J. (2018). Subsurface erosion by soil piping: significance and research needs. *Earth-Science Reviews*, **185**, 1107–1128, https://doi.org/10.1016/j.earscirev.2018.08.006

Bernatek-Jakiel A., Vannoppen W. and Poesen J. (2017). Assessment of grass root effects on soil piping in sandy soils using pinhole test. *Geomorphology*, **295**, 563–571, https://doi.org/10.1016/j.geomorph.2017.08.027

Beryani A., Goldstein A., Al-Rubaei A. M., Viklander M., Hunt W. and Blecken G.-T. (2021). Survey of the operational status of twenty-six urban stormwater biofilter facilities in Sweden. *Journal of Environmental Management*, **297**(2021), 113375, https://doi.org/10.1016/j.jenvman.2021.113375

Beven K. and Germann P. (1982). Macropores and water flow in soils. *Water Resources Research*, **18**, 1311–1325, https://doi.org/10.1029/WR018i005p01311

Beven K. and Germann P. (2013). Macropores and water flow in soils revisited. *Water Resources Research*, **49**, 3071–3092, https://doi.org/10.1002/wrcr.20156

Bielefeldt A., Gutierrez-Padilla M. G. D., Ovtchinnikov S., Silverstein J. and Hernandez M. (2010). Bacterial kinetics of sulfur oxidizing bacteria and their biodeterioration rates of concrete sewer pipe samples. *Journal of Environmental Engineering*, **136**, 731–738, https://doi.org/10.1061/(ASCE)EE.1943-7870.0000215

Bishop S., Isaac D. H., Hinksman P. and Morrissey P. (2000). Environmental stress cracking of poly (vinyl chloride) in alkaline solutions. *Polymer Degradation and Stability*, **70**, 477–484, https://doi.org/10.1016/S0141-3910(00)00144-0

Blake G. R. and Hartge K. H. (1986). Particle density. In: Methods of Soil Analysis, John Wiley & Sons, Ltd, New Jersey, USA, pp. 377–382, https://doi.org/10.2136/sssabookser5.1.2ed.c14

Blecken G.-T., Hunt W. F., Al-Rubaei A. M., Viklander M. and Lord W. G. (2017). Stormwater control measure (SCM) maintenance considerations to ensure designed functionality. *Urban Water Journal*, **14**(3), 278–290, https://doi.org/10.1080/1573062X.2015.1111913

Blecken G., Beryani A., Al-Rubaei A., Viklander M., Goldstein A. and Hunt W. (2021). (Review evaluation of the functionality of 26 stormwater biofilters (*org. title in Swedish: Oversiktlig utvardering av funktionaliteten av 26 dagvattenbiofilter.*) https://www.diva-portal.org/smash/get/diva2:1841857/FULLTEXT01.pdf

Bolderman R., van der Meulen M. and Op de Beek R. M. J. (2003). Riolen van beton, State of the art 2003. Vereniging van Producenten van Betonleidingsystemen VPB, Woerden, the Netherlands.

Bonelli S., Jommi C. and Sterpi D. (eds) (2018). Internal erosion in Earthdams, Dikes and Levees. Proceedings of EWG-IE 26th annual meeting 2018, Milan, Italy.

Bonneau J., Fletcher T., Costelloe J., Poelsma P., James R. and Burns M. (2018). Where does infiltrated stormwater go? Interactions with vegetation and subsurface anthropogenic features. *Journal of Hydrology*, **567**, 121–132, https://doi.org/10.1016/j.jhydrol.2018.10.006

Bosch D. D. and West L. T. (1998). Hydraulic conductivity variability for Two sandy soils. *Soil Science Society of America Journal*, **62**, 90–98, https://doi.org/10.2136/sssaj1998.03615995006200010012x

Bosco E., Claessens R. J. M. A. and Suiker A. S. J. (2020). Multi-scale prediction of chemo-mechanical properties of concrete materials through asymptotic homogenization. *Cement and Concrete Research*, **128**, 105929, https://doi.org/10.1016/j.cemconres.2019.105929

Bouarafa S., Lassabatere L., Lipeme-Kouyi G. and Angulo-Jaramillo R. (2019). Hydrodynamic characterization of sustainable urban drainage systems (SuDS) by using beerkan infiltration experiments. *Water*, **11**, 660, https://doi.org/10.3390/w11040660

Bouwer H. (2002). Artificial recharge of groundwater: hydrogeology and engineering. *Hydrogeology Journal*, **10**, 121–142, https://doi.org/10.1007/s10040-001-0182-4

Brachman R. W. I., Moore I. D. and Rowe R. K. (2001). The performance of a laboratory facility for evaluating the structural response of small-diameter buried pipes. *Canadian Geotechnical Journal*, **38**, 260–275, https://doi.org/10.1139/t00-102

Breen J. (1993). Environmental stress cracking of PVC and PVC-CPE. *Journal of Materials Science*, **28**, 3769–3776, https://doi.org/10.1007/BF00353177

Breen J. (1994). Environmental stress cracking of PVC and PVC-CPE. *Journal of Materials Science*, **29**, 39–46, https://doi.org/10.1007/BF00356570

Breen J. (1995). Environmental stress cracking of PVC and PVC-CPE. *Journal of Materials Science*, **30**, 5833–5840, https://doi.org/10.1007/BF00356729

Breen J. (2006). Expected Lifetime of Existing PVC Water Systems: Summary. TNO, Eindhoven.

Breiman L. (1996). Bagging predictors. *Machine Learning*, **24**(2), 123–140. Scopus, https://doi.org/10.1023/A:1018054314350

Breiman L. (2001). Random forests. *Machine Learning*, **45**(1), 5–32, https://doi.org/10.1023/A:1010933404324

Breiman L., Friedman J., Olshen R. and Stone C. J. (1984). *Classification and Regression Trees* (First), https://doi.org/10.1201/9781315139470

Brennan M., Karimi M., Almeida F., Kroll de Lima F., Ayala P., Obata D., Paschoalini A. and Kessissoglou N. (2017). On the role of vibro-acoustics in leak detection for plastic water distribution pipes. *Procedia Engineering*, **199**(2017), 1350–1355, https://doi.org/10.1016/j.proeng.2017.09.350

Brooks R. H. and Corey, A. T. (1964). Hydraulic properties of porous media. Hydrological papers Colorado state University. Fort Collins, Colorado (USA).

Bruand A. and Coquet Y. (2005). Les sols et le cycle de l'eau 345.

Buckingham E. (1907). United States. Bureau of Soils. Studies on the movement of soil moisture. Washington, Govt. Print. Off.

Burdine N. T. (1953). Relative Permeability Calculations from Pore Size Distribution Data. *Transactions of the American Institute of Mining & Metallurgical Engineers*, **5**, 71–78, https://doi.org/10.2118/225-g

Butters G. (ed.) (1982). Particulate Nature of PVC: Formation, Structure and Processing. Applied Science Publishers, London.

Caradot N., Riechel M., Fesneau M., Hernandez N., Torres A., Sonnenberg H., Eckert E., Lengemann N., Waschnewski J. and Rouault P. (2018). Practical benchmarking of statistical and machine learning models for predicting the condition of sewer pipes in Berlin, Germany. *Journal of Hydroinformatics*, ***20***(5), 1131–1147. Scopus, https://doi.org/10.2166/hydro.2018.217

Caradot N., Sampaio P. R., Guilbert A. S., Sonnenberg H., Parez V. and Dimova V. (2020). Using deterioration modelling to simulate sewer rehabilitation strategy with low data availability. *Water Science & Technology*, **83**(3), 631–640, https://doi.org/10.2166/wst.2020.604

Cardarelli F. (2008). Polymers and elastomers. In: Materials Handbook: A Concise Desktop Reference, 2nd edn, F. Cardarelli (ed.), Springer, London, pp. 691–750, https://doi.org/10.1007/978-1-84628-669-8_11

Cassel D. K. and Nielsen D. R. (1986). Field capacity and available water capacity. In: Methods of Soil Analysis, John Wiley & Sons, Ltd, New Jersey, USA, pp. 901–926, https://doi.org/10.2136/sssabookser5.1.2ed.c36

Castiglione P., Shouse P. J., Mohanty B. P. and van Genuchten M. T. (2005). Analysis of temperature effects on tension infiltrometry of low permeability materials. *Vadose Zone Journal*, **4**, 481–487, https://doi.org/10.2136/vzj2004.0134

Cerdà A. (1996). Seasonal variability of infiltration rates under contrasting slope conditions in southeast Spain. *Geoderma*, **69**, 217–232, https://doi.org/10.1016/0016-7061(95)00062-3

Cerdà A. (1997). Seasonal changes of the infiltration rates in a Mediterranean scrubland on limestone. *Journal of Hydrology*, **198**, 209–225, https://doi.org/10.1016/S0022-1694(96)03295-7

Chang L. and Hanssen R. F. (2014). Detection of cavity migration and sinkhole risk using radar interferometric time series. *Remote Sensing of Environment*, **147**, 56–64, https://doi.org/10.1016/j.rse.2014.03.002

Chaoui K., Chudnovsky A. and Moet A. (1987). Effect of residual stress on crack propagation in MDPE pipes. *Journal of Materials Science*, **22**, 3873–3879, https://doi.org/10.1007/BF01133334

Chen T. Y.-J., Vladeanu G., Yazdekhasti S. and Daly C. M. (2022). Performance evaluation of pipe break machine learning models using datasets from multiple utilities. *Journal of Infrastructure Systems*, ***28***(2). Scopus, https://doi.org/10.1061/(ASCE)IS.1943-555X.0000683

Cheng Q., Tang C.-S., Xu D., Zeng H. and Shi B. (2021). Water infiltration in a cracked soil considering effect of drying-wetting cycles. *Journal of Hydrology*, **593**, 125640, https://doi.org/10.1016/j.jhydrol.2020.125640

Cherqui F., James R., Poelsma P., Burns M. J., Szota C., Fletcher T. and Bertrand-Krajewski J.-L. (2020). A platform and protocol to standardise the test and selection low-cost sensors for water level monitoring. *H2 Open Journal*, **3**(1), 437–456, https://doi.org/10.2166/h2oj.2020.050

Childs J. L., Wallender W. W. and Hopmans J. W. (1993). Spatial and seasonal variation of furrow infiltration. *Journal of Irrigation and Drainage Engineering*, **119**, 74–90, https://doi.org/10.1061/(ASCE)0733-9437(1993)119:1(74)

Cho K. S. and Mori T. (1995). A newly isolated fungus participates in the corrosion of concrete sewer pipes. *Water Sci. Technol.*, **31**, 263–271.

Choi B. H., Zhou Z., Chudnovsky A., Stivala S. S., Sehanobish K. and Bosnyak C. P. (2005). Fracture initiation associated with chemical degradation: observation and modeling. *International Journal of Solids and Structures*, **42**, 681–695, https://doi.org/10.1016/j.ijsolstr.2004.06.028

Chopra M., Kakuturu S., Ballock C., Spence J. and Wanielista M. (2010). Effect of rejuvenation methods on the infiltration rates of pervious concrete pavements. *Journal of Hydrologic Engineering*, **15**(6), 426–433, https://doi.org/10.1061/(ASCE)HE.1943-5584.0000117

Chossat J.-C. (2005). La mesure de la conductivité hydraulique dans les sols. Choix des méthodes, TEC&DOC. ed. Lavoisier.

Cid Alfaro M. V., Suiker A. S. J., de Borst R. and Remmers J. J. C. (2009). Analysis of fracture and delamination in laminates using 3D numerical modelling. *Engineering Fracture Mechanics*, **76**, 761–780, https://doi.org/10.1016/j.engfracmech.2008.09.002

Cid Alfaro M. V., Suiker A. S. J. and de Borst R. (2010a). Transverse failure behavior of fiber-epoxy systems. *Journal of Composite Materials*, **44**, 1493–1516, https://doi.org/10.1177/0021998309360941

Cid Alfaro M. V., Suiker A. S. J., Verhoosel C. V. and de Borst R. (2010b). Numerical homogenization of cracking processes in thin fibre-epoxy layers. *European Journal of Mechanics – A/Solids*, **29**, 119–131, https://doi.org/10.1016/j.euromechsol.2009.09.006

CIRIA (2007). The SuDS manual CIRIA C697 London. UK.

Concialdi P., Di Prima S., Bhanderi H. M., Stewart R. D., Abou Najm M. R., Lal Gaur M., Angulo-Jaramillo R. and Lassabatere L. (2020). An open-source instrumentation package for intensive soil hydraulic characterization. *Journal of Hydrology*, **582**, 124492, https://doi.org/10.1016/j.jhydrol.2019.124492

Cortes C. and Vapnik V. (1995). Support-Vector networks. *Machine Learning*, ***20***(3), 273–297. Scopus, https://doi.org/10.1023/A:1022627411411

Cossais N., Thomas A., Cherqui F., Morison P., Bos D., Martouzet D., Sibeud E., Honegger A., Lavau S. and Fletcher T. (2017). Understanding the challenges of managing SUDS to maintain or improve their perform.

County of San Diego (2014). Low Impact Development Handbook-Stormwater Management Strategies. https://www.sandiegocounty.gov/content/dam/sdc/pds/docs/LID_Handbook_2014.pdf (accessed 07/04/2023)

Dagenais D., Brisson J. and Fletcher T. (2018). The role of plants in bioretention systems: does the science underpin current guidance? *Ecological Engineering*, **120**, 532–545, https://doi.org/10.1016/j.ecoleng.2018.07.007

Daily W., Ramirez A., LaBrecque D. and Nitao J. (1992). Electrical resistivity tomography of vadose water movement. *Water Resources Research*, **28**, 1429–1442, https://doi.org/10.1029/91WR03087

Daily W., Ramirez A., Binley A. and LaBrecque D. (2005). Electrical Resistance Tomography? Theory and Practice, in: Near-Surface Geophysics, Investigations in Geophysics. Society of Exploration. Geophysicists, pp. 525–550, https://doi.org/10.1190/1.9781560801719.ch17

Danielson R. E. and Sutherland P. L. (1986). Porosity. In: Methods of Soil Analysis, John Wiley & Sons, Ltd, New Jersey, USA, pp. 443–461, https://doi.org/10.2136/sssabookser5.1.2ed.c18

Davies J. P., Clarke B. A., Whiter J. T. and Cunningham R. J. (2001). Factors influencing the structural deterioration and collapse of rigid sewer pipes. *Urban Water*, **3**, 73–89, https://doi.org/10.1016/S1462-0758(01)00017-6

Davis J. L., Nica D., Shields K. and Roberts D. J. (1998). Analysis of concrete from corroded sewer pipe. *International Biodeterioration & Biodegradation*, **42**, 75–84, https://doi.org/10.1016/S0964-8305(98)00049-3

Davis A., Traver R., Hunt W. and Lee R. (2012). Hydrologic performance of bioretention storm-water control measures. *Journal of Hydrologic Engineering*, **5**, 17.

Deb S. K. and Shukla M. K. (2012). Variability of hydraulic conductivity due to multiple factors. *American Journal of Environmental Sciences*, **8**, 489–502, https://doi.org/10.3844/ajessp.2012.489.502

Dechesne M., Barraud S. and Bardin J.-P. (2002). Performance of stormwater infiltration basins on the long term. Proc of the 9th ICUD. https://doi.org/10.1061/40644(2002)35

Dechesne M., Barraud S. and Bardin J.-P. (2005). Experimental assessment of stormwater infiltration basin evolution. *Journal of Environmental Engineering*, **131**(7), 1090–1098, https://doi.org/10.1061/(ASCE)0733-9372(2005)131:7(1090)

Deffontis S., Breton A., Vialle C., Montrejaud-Vignoles M., Vignoles C. and Sablayrolles C. (2013). Impact of dry weather discharges on annual pollution from a separate storm sewer in Toulouse, France. *Science of the Total Environment*, **452–453**, 394–403, https://doi.org/10.1016/j.scitotenv.2013.03.014

de Groot A. C. (2015). A study on the fouling process in inverted siphons over time. Respository TU dElft: https://repository.tudelft.nl/islandora/object/uuid%3A6262a3e6-f408-4cd5-84e5-db388e127cd6 (visited 25 August 2023).

Dempster A. P., Laird N. M. and Rubin D. B. (1977). Maximum likelihood from incomplete data via the EM algorithm. *Journal of the Royal Statistical Society. Series B*, **39**(1), 1–38.

Di Prima S., Lassabatere L., Bagarello V., Iovino M. and Angulo-Jaramillo R. (2015). Testing a new automated single ring infiltrometer for Beerkan infiltration experiments. *Geoderma*, **262**, 20–34, https://doi.org/10.1016/j.geoderma.2015.08.006

Di Prima S., Winiarski T., Angulo-Jaramillo R., Stewart R. D., Castellini M., Abou Najm M. R., Ventrella D., Pirastru M., Giadrossich F., Capello G., Biddoccu M. and Lassabatere L. (2020). Detecting infiltrated water and preferential flow pathways through time-lapse ground-penetrating radar surveys. *Science of The Total Environment*, **726**, 138511, https://doi.org/10.1016/j.scitotenv.2020.138511

Di Prima S., Fernandes G., Marras E., Giadrossich F., Stewart R. D., Abou Najm M. R., Winiarski T., Mourier B., Angulo-Jaramillo R., Comegna A., del Campo A. and Lassabatere L. (2023). Evaluating subsurface flow connectivity in a pine-covered hillslope with stemflow infiltration and ground-penetrating radar surveys. *Journal of Hydrology*, **620**, 129527, https://doi.org/10.1016/j.jhydrol.2023.129527

Dirksen J., Baars E., Langeveld G. and Clemens F. (2012). Settlement as a driver for sewer rehabilitation. *Water Science & Technology*, **66**(7), 1534–1539, https://doi.org/10.2166/wst.2012.347

Dirksen J. (2013). Monitoring ground settlement to guide sewer asset management. PhD thesis, Delft University of Technology. ISBN 978-94-6108-474 3. https://repository.tudelft.nl/islandora/search/author%3A%22Dirksen%2C%20J.%22?f%5B0%5D=mods_genre_s%3A%22doctoral%5C%20thesis%22

Dirksen J., Clemens F., Korving H., Cherqui F., Le Gauffre P., Ertl T., Plihal H., Mueller K. and Snaterse C. (2013a). The consistency of visual sewer inspection data. *Structure and Infrastructure Engineering*, **9**(3), 214–228, https://doi.org/10.1080/15732479.2010.541265

Dirksen J., Baars E., Langeveld J. G. and Clemens F. H. L. R. (2013b). The impact of fat, oil and grease on the performance of sanitary systems. Proc. Of the 7th int. Conf. On Sewer Processes and Networks (SPN7), Sheffield, UK.

Doolittle J. A. and Brevik E. C. (2014). The use of electromagnetic induction techniques in soils studies. *Geoderma*, **223–225**, 33–45, https://doi.org/10.1016/j.geoderma.2014.01.027

Drake J. and Bradford A. (2013). Assessing the potential for restoration of surface permeability for permeable pavements through maintenance. *Water Science & Technology*, **68**(9), 1950–1958, https://doi.org/10.2166/wst.2013.450

Duinmeijer S. P. A. (2020). On the free vortex driven motion of buoyant particles. PhD thesis, TU Delft. ISBN 978-946366-271-0, https://doi.org/10.4233/uuid:a33fa2a9-f347-40a3-96be-51e880018974

Duinmeijer S. P. A. and Clemens F. H. L. R. (2021). An experimental study on the motion of buoyant particles in the free-surface vortex flow. *Journal of Hydraulic Research*, **59**, 947–962, https://doi.org/10.1080/00221686.2020.1845827

DWA-A 113. Hydraulic dimensioning and testing of sewage ressure mains (org. Title in German: Hydraulische Dimensionierung und Leistungsnachweis von Abwasserdrucksystemen).

DWA-M 149-9. Assessement of the condition of drainage systems outside buildings, part 9 : inspection and manitenance of sewage pressure mains.(*org title in German : Zustandserfassung und -beurteilung von Entwässerungssystemen außerhalb von Gebäuden – Teil 9: Inspektion und Wartung von Abwasserdruckleitungen*).

El-Housni H., Ouellet M. and Duchesne S. (2018). Identification of most significant factors for modelling deterioration of sewer pipes. *Canadian Journal of Civil Engineering*, **45**(3), 215–226, https://doi.org/10.1139/cjce-2015-0293

Ellis B. and Butler D. (2015). Surface water sewer misconnections in England and Wales: pollution sources and impacts. *Science of the Total Environment*, **526**, 98–109, https://doi.org/10.1016/j.scitotenv.2015.04.042

Emerson C., Wadzuk B. and Traver R. (2010). Hydraulic evolution and total suspended solids capture of an infiltration trench. In *Hydrological processes*, **24**(8), 1008–1014, https://doi.org/10.1002/hyp.7539

EN 13508-2 (2011). Investigation and assessment of drain and sewer systems outside buildings – Part 2: Visual inspection coding system (European Standard). European Committee for Standardization, Brussels, Belgium.

Erdogmus E. and Tadros M. K. (2009). Behavior and Design of Buried Concrete Pipe-Phase 2. Nebraska Department of Roads, Lincoln, NE, p. 124.

Eriksen K. (2003). Thaumasite attack on concrete at marbjerg waterworks. *Cement and Concrete Composites*, **25**, 1147–1150, https://doi.org/10.1016/S0958-9465(03)00151-3

Erickson A., Weiss P. and Gulliver J. (2013). Optimizing stormwater treatment practices. A handbook of assessment and maintenance. Springer New York, NY, ISBN 978-1-899-9401-1/978-1-4614-623-1. https://doi.org/10.1007/978-1-4614-4624-8

European Committee for Standardization (2005). NEN-EN 1992-1-1 – Eurocode 2: Design of concrete structures – Part 1-1: General rules and rules for buildings.

European Committee for Standardization (2013). NEN-EN 12390-13 – Testing hardened concrete – Part 13: Determination of secant modulus of elasticity in compression.

Fan Y. F., Hu Z. Q., Zhang Y. Z. and Liu J. L. (2010). Deterioration of compressive property of concrete under simulated acid rain environment. *Construction and Building Materials*, **24**, 1975–1983, https://doi.org/10.1016/j.conbuildmat.2010.04.002

Fan B., Liu X., Zhu Q., Qin G., Li J., Lin H. and Guo L. (2020). Exploring the interplay between infiltration dynamics and critical zone structures with multiscale geophysical imaging: a review. *Geoderma*, **374**, 114431, https://doi.org/10.1016/j.geoderma.2020.114431

Fannin R. J. and Slangen P. (2014). On the distinct phenomena of suffusion and suffosion. *Geotechnique Letters*, **4**, 289–294, https://doi.org/10.1680/geolett.14.00051

Farny J. A. and Kerkhoff B. (1977). Diagnosis and Control of Alkali-aggregate Reactions in Concrete, Portland Cement Association Skokie, IL.

Fassman E. A., Simcock R. and Wang S. (2013). *Media Specification for Stormwater Bioretention Devices*, Auckland UniServices for Auckland Council. technical report. TR2013/011.

Fatemi Aghda S., Ganjalipour K. and Nabiollahi K. (2018). Comparison of performance inclinometer casing and TDR technique. *Journal of Applied Geophysics*, **150**, 182–194, https://doi.org/10.1016/j.appgeo.2018.01.022

Faure B. and Lemaire F. (1990). Nettoyage et décolmatage des enrobés drainants en milieu urbain. *Revue générale des routes et aérodromes*, N°677.

Fausett L. V. (1994). Fundamentals of Neural Networks: Architectures, Algorithms, and Applications. Prentice-Hall, New Jersey, USA.

Fernandes G., Roques O., Lassabatere L., Sarles L., Venisseau A., Marchand P. and Bedell J.-P. (2022). Influence of edaphic conditions and persistent organic pollutants on earthworms in an infiltration basin. , **304**, 119192, https://doi.org/10.1016/j.envpol.2022.119192

Fillot L. A., Hajji P., Gauthier C. and Masenelli-Varlot K. (2006). U-PVC gelation level assessment, part 1: comparison of different techniques. *Journal of Vinyl and Additive Technology*, **12**, 98–107, https://doi.org/10.1002/vnl.20077

Fletcher T., Shuster W., Hunt W. F., Butler D., Scott A., Trowsdale S., Barraud S., Semandeni-Davies A., Bertrand-Krajewski J.-L., Mikkelsen P. S., Rivard G., Uhl M., Dagenais D. and Viklander M. (2015). SUDS, LID, BMPs, WSUD and more – The evolution and application of terminology surrounding urban drainage. *Urban Water Journal*, **12**(7), 525–542, https://doi.org/10.1080/1573062X.2014.916314

Foster M., Fell R. and Spannagle M. (2000). The statistics of embankment dam failures and accidents. *Canadian Geotechnical Journal*, **37**, 1000–1024, https://doi.org/10.1139/t00-030

Fowdar H., Payne E., Deletic A., Zhang K. and McCarthy D. (2022). Advancing the sponge city agenda: evaluation of 22 plant species across a broad range of life forms for stormwater management. *Ecological Engineering*, **175**, 106501, https://doi.org/10.1016/j.ecoleng.2021.106501

Fragoudakis R. (2017). Failure concepts in fiber reinforced plastics. In: Failure Analysis and Prevention, IntechOpen, https://doi.org/10.5772/intechopen.71822

Fujiyama M. and Kondou M. (2004). Effect of degree of polymerization on gelation and flow processability of poly (vinyl chloride). *Journal of Applied Polymer Science*, **92**, 1915–1938, https://doi.org/10.1002/app.20181

Gardner W. H. (1986). Water Content, in: Methods of Soil Analysis. John Wiley & Sons, Ltd, New Jersey, USA, pp. 493–544, https://doi.org/10.2136/sssabookser5.1.2ed.c21

Gautier A. (1998). Contribution à la connaissance du fonctionnement d'ouvrages d'infiltration d'eau de ruissellement pluvial urbain. PhD INSA de Lyon, Villeurbanne, France.

Gerke H. H., Germann P. and Nieber J. (2010). Preferential and unstable flow: from the pore to the catchment scale. *Vadose Zone Journal*, **9**, 207–212, https://doi.org/10.2136/vzj2010.0059

Géron A. (2019). Hands-on Machine Learning with Scikit-Learn, Keras, and TensorFlow, 2nd edition, O'Reilly Media, California, USA.

Giakoumis T. and Voulvoulis N. (2023). Combined sewer overflows: relating event dureation monitoring to wastewater systems' capacity in England. *Environmental Science Water Research & Technology*, **9**, 707–722, https://doi.org/10.1039/d2ew00637e

Gladchenko A. N., Shevelya I. V., Kiyanitsa E. V. and Derkach V. V. (1997). Wear of working members of extruders for processing polymers. *Chemical and Petroleum Engineering*, **33**, 367–370, https://doi.org/10.1007/BF02416718

Glaser B., Jackisch C., Hopp L. and Klaus J. (2019). How meaningful are plot-scale observations and simulations of preferential flow for catchment models? *Vadose Zone Journal*, **18**, 1–18, https://doi.org/10.2136/vzj2018.08.0146

Goh H. W., Zakaria N. A., Lau T. L., Foo K. Y., Chang C. K. and Leow C. S. (2017). Mesocosm study of enhanced bioretention media in treating nutrient rich stormwater for mixed development area. *Urban Water Journal*, **14**(2), 134–142, https://doi.org/10.1080/1573062X.2015.1076861

Gokhale S. and Graham J. A. (2004). A new development in locating leaks in sanitary sewers. *Tunnelling and Underground Space Technology*, **19**(1), 85–96, https://doi.org/10.1016/j.tust.2003.08.003

Gonzalez-Merchan C., Barraud S., Le Coustumer S. and Fletcher T. (2012). Monitoring of clogging evolution in the stormwater infiltration system and determinant factors. *European Journal of Environmental and Civil Engineering*, **16**(1), 34–67, https://doi.org/10.1080/19648189.2012.682457

Gonzalez-Merchan C., Barraud S. and Bedell J. P. (2014). Influence of spontaneous vegetation in stormwater infiltration system clogging. *Environmental Science and Pollution Research*, **21**(8), 5419–5426, https://doi.org/10.1007/s11356-013-2398-y

Goutaland D., Winiarski T., Angulo-Jaramillo R., Lassabatere L., Bièvre G., Buoncristiani J.-F., Dubé J.-S., Mesbah A. and Cazalets H. (2007). Hydrogeophysical study of the heterogeneous unsaturated zone of a stormwater infiltration basin, pp. 173–192.

Green W. H. and Ampt G. A. (1911). Studies on soil physics. *The Journal of Agricultural Science*, **4**, 1–24, https://doi.org/10.1017/S0021859600001441

Green T. R., Ahuja L. R. and Benjamin J. G. (2003). Advances and challenges in predicting agricultural management effects on soil hydraulic properties. *Geoderma, Quantifying Agricultural Management Effects on Soil Properties and Processes*, **116**, 3–27.

Gregory J., Dukes M., Miller G. and Jones P. (2005). Analysis of double-ring infiltration techniques and development of a simple automatic water delivery system. *Applied Turfgrass Science*, **2**, 1–7, https://doi.org/10.1094/ATS-2005-0531-01-MG

Grengg C., Mittermayr F., Baldermann A., Böttcher M. E., Leis A., Koraimann G., Grunert P. and Dietzel M. (2015). Microbiologically induced concrete corrosion: a case study from a combined sewer network. *Cement and Concrete Research*, **77**, 16–25, https://doi.org/10.1016/j.cemconres.2015.06.011

Grengg C., Mittermayr F., Ukrainczyk N., Koraimann G., Kienesberger S. and Dietzel, M. (2018). Advances in concrete materials for sewer systems affected by microbial induced concrete corrosion: A review. *Water Research*, **134**, 341–352, https://doi.org/10.1016/j.watres.2018.01.043

Gu L., Visintin P. and Bennett T. (2018). Evaluation of accelerated degradation test methods for cementitious composites subject to sulphuric acid attack: application to conventional and alkali activated concretes. *Cement and Concrete Composites*, (87), 187–204, https://doi.org/10.1016/j.cemconcomp.2017.12.015

Guo L., Chen J. and Lin H. (2014). Subsurface lateral preferential flow network revealed by time-lapse ground-penetrating radar in a hillslope. *Water Resources Research*, **50**, 9127–9147, https://doi.org/10.1002/2013WR014603

Haile T. M., Hobiger G., Kammerer G., Allabashi R., Schaerfinger B. and Fuerhacker M. (2016). Hydraulic performance and pollutant concentration profile in a stormwater runoff filtration systems. *Water Air Soil Pollution*, **227**(1), 34, https://doi.org/10.1007/s11270-015-2736-4

Hallaji S. M., Fang Y. and Winfrey B. K. (2022). Predictive maintenance of pumps in civil infrastructure: state-of-the-art, challenges and future directions. *Automation in Constriction*, **134**, 104049, https://doi.org/10.1016/j.autcon.2021.104049

Hallett P. D. (2008). A brief overview of the causes, impacts and amelioration of soil water repellency – a review. *Soil and Water Research*, **3**, S21–S29, https://doi.org/10.17221/1198-swr

Hallett P. D., Nunan N., Douglas J. T. and Young I. M. (2004). Millimeter-scale spatial variability in soil water sorptivity. *Soil Science Society of America Journal*, **68**, 352–358, https://doi.org/10.2136/sssaj2004.3520

Hamel P. and Fletcher T. D. (2014). The impact of stormwater source-control strategies on the (low) flow regime of urban catchments. *Water Science & Technology*, **69**(4), 739–745, https://doi.org/10.2166/wst.2013.772

Hart T. D. (2017). Root-enhanced Infiltration in Stormwater Bioretention Facilities in Portland, Oregon. Dissertations and Theses, Portland state University. Paper 3468, https://doi.org/10.15760/etd.5352

Harvey R. R. and McBean E. A. (2014a). Comparing the utility of decision trees and support vector machines when planning inspections of linear sewer infrastructure. *Journal of Hydroinformatics*, ***16***(6), 1265–1279. Scopus, https://doi.org/10.2166/hydro.2014.007

Harvey R. R. and McBean E. A. (2014b). Predicting the structural condition of individual sanitary sewer pipes with random forests. *Canadian Journal of Civil Engineering*, **41**(4), 294–303. Scopus, https://doi.org/10.1139/cjce-2013-0431

Hatt B. E., Lewis J. F., Fletcher T. D. and Deletic A. (2007). Insights from the design, construction and operation of an experimental biofiltration system. 13th International Rainwater Catchment Systems Conference and 5th International Water Sensitive Urban Design.

Hatt B. E., Fletcher T. D. and Deletic A. (2009). Hydrologic and pollutant removal performance of stormwater biofiltration systems at the field scale. *Journal of Hydrology*, **365**(3–4), 310–321, https://doi.org/10.1016/j.jhydrol.2008.12.001

Hawari A., Alkadour F., Elmasry M. and Zayed T. (2017). Simulation-based condition assessment model for sewer pipelines. *Journal of Performance of Constructed Facilities*, **31**(1). Scopus, https://doi.org/10.1061/(ASCE)CF.1943-5509.0000914

Hawari A., Alkadour F., Elmasry M. and Zayed T. (2020). A state of the art review on condition assessment models developed for sewer pipelines. *Engineering Applications of Artificial Intelligence*, **93**, 103721, https://doi.org/10.1016/j.engappai.2020.103721

Haws N. W., Das B. S. and Rao P. S. C. (2004). Dual-domain solute transfer and transport processes: evaluation in batch and transport experiments. *Journal of Contaminant Hydrology*, **75**, 257–280, https://doi.org/10.1016/j.jconhyd.2004.07.001

Henderson V. and Tighe S. (2011). Evaluation of pervious concrete pavement permeability renewal maintenance methods at field sites in Canada. *Canadian Journal of Civil Engineering*, **38**(12), 1404–1413.

Hernández N., Caradot N., Sonnenberg H., Rouault P. and Torres A. (2021). Optimizing SVM models as predicting tools for sewer pipes conditions in the two main cities in Colombia for different sewer asset management purposes. *Structure and Infrastructure Engineering*, **17**(2), 156–169. Scopus, https://doi.org/10.1080/15732479.2020.1733029

Hill J. J., Kurdziel J. M., Nelson C. R., Nystrom J. A. and Sondag M. S. (1999). Minnesota department of transportation overload field tests of standard and standard installation direct design reinforced concrete pipe installations. *Transportation Research Record: Journal of the Transportation Research Board*, **1656**, 64–72, https://doi.org/10.3141/1656-09

Hillel D. (1998). Environmental Soil Physics: Fundamentals, Applications, and Environmental Considerations. Elsevier, Amsterdam, The Netherlands.

Hoes O. A. C., Schilperoort R. P. S., Luxemburg W. M. J., Clemens F. H. L. R. and van de Giessen, N. C. (2009). Locating illicit connections in storm water sewers using fiber-optic distributed temperature sensing. *Water Research*, **43**(20), 5187–5197, https://doi.org/10.1016/j.watres.2009.08.020

Hordijk D. A. (1991). Local approach to fatigue of concrete. Ph.D. thesis, Delft University of Technology, Delft, The Netherlands.

Horton R. E. (1941). An approach toward a physical interpretation of infiltration-Capacity1. *Soil Science Society of America Journal*, **5**, 399–417, https://doi.org/10.2136/sssaj1941.036159950005000C0075x

Horton R. E. (1945). Erosional development of streams and their drainage basin; hydrophysical approach to quantitative morphology. *Geological Society of America Bulletin*, **56**, 275–370, https://doi.org/10.1130/0016-7606(1945)56[275:EDOSAT]2.0.CO;2

Hosmer D. W. and Lemeshow S. (2000). Applied Logistic Regression. Wiley Publishing, New Jersey, USA.

Hunt W. F., Greenway M., Moore T. C., Brown R. A., Kennedy S. G., Line D. E. and Lord W. G. (2011). Constructed stormwater wetland installation and maintenance: are we getting it right? *Journal of Irrigation and Drainage Engineering*, **137**(8), 469–474, https://doi.org/10.1061/(ASCE)IR.1943-4774.0000326

Hunter R. P. and Bowman E. T. (2018). Visualisation of seepage-induced suffusion and suffosion within internally erodible granular media. *Geotechnique*, **68**, 918–930, https://doi.org/10.1680/jgeot.17.P.161

Hussain I., Hamid S. H. and Khan J. H. (1995). Polyvinyl chloride pipe degradation studies in natural environments. *Journal of Vinyl and Additive Technology*, **1**, 137–141, https://doi.org/10.1002/vnl.730010305

Hutař P., Ševčík M., Nahlik L., Pinter G., Frank A. and Mitev I. (2011). A numerical methodology for lifetime estimation of HDPE pressure pipes. *Engineering Fracture Mechanics*, **78**, 3049–3058, https://doi.org/10.1016/j.engfracmech.2011.09.001

Hutař P., Ševčík M., Frank A., Nahlik L., Kučera J. and Pinter G. (2013). The effect of residual stress on polymer pipe lifetime. *Engineering Fracture Mechanics*, **108**, 98–108, https://doi.org/10.1016/j.engfracmech.2013.04.014

Hutchinson J. M. (1995). Physical aging of polymers. *Progress in Polymer Science*, **20**, 703–760, https://doi.org/10.1016/0079-6700(94)00001-I

Hvitved-Jacobsen T., Vollertsen J. and Nielsen A. H. (2013). Sewer processes. Microbial and chemical process engineering of sewer networks, second edition. eBook, ISBN 9780429110986, https://doi.org/10.1201/b14666

Indiketiya S., Jegatheesan P., Rajeev P. and Kuwano R. (2019). The influence of pipe embedment material on sinkhole formation due to erosion around defective sewers. *Transportation Geotechnics*, **19**, 110–125, https://doi.org/10.1016/j.trgeo.2019.03.001

Ismail N., Nonaka T., Noda S. and Mori T. (1993). Effect of carbonation on microbial corrosion of concretes. *Doboku Gakkai Ronbunshu*, **20**(474), 133–138, https://doi.org/10.2208/jscej.1993.474_133

Jačka L., Pavlásek J., Kuráž V. and Pech P. (2014). A comparison of three measuring methods for estimating the saturated hydraulic conductivity in the shallow subsurface layer of mountain podzols. *Geoderma*, **219–220**, 82–88, https://doi.org/10.1016/j.geoderma.2013.12.027

Janson L. E. (2003). Plastics Pipes for Water Supply and Sewage Disposal, 4th ed., Borealis, Stenungsund.

Jenkins L., Gokhale S. and McDonald M. (2015). Comparison of pipeline failure prediction models for water distribution networks with uncertain and limited data. *Journal of Pipeline Systems Engineering and Practice*, **6**(2). Scopus, https://doi.org/10.1061/(ASCE)PS.1949-1204.0000181

Jensen A. L. (2018). Characterisation of rag properties and their transport in wastewater pumps. PhD Thesis, Aalborg University. ISBN (online): 978-87-7210-211-5, https://doi.org/10.5278/vbn.phd.eng.00049

Jeong H. Y., Jun S.-C., Cheon J.-Y. and Park M. (2018). A review on clogging mechanisms and managements in aquifer storage and recovery (ASR) applications. *Geosciences Journal*, **22**, 667–679, https://doi.org/10.1007/s12303-017-0073-x

Jiang Y, Zevenbergen C. and Ma Y. (2018). Urban pluvial flooding and stormwater management: A contemporary review of Chia's challenges and "sponge" cities strategy. *Environmental Science and Policy*, **80**(2018), 132–143, https://doi.org/10.1016/j.envsci.2017.11.016

Johansson L. and Tornell B. (1987). Initiation of fractures in rigid PVC pipes by soft particles. *Journal of Vinyl and Additive Technology*, **9**, 103–107, https://doi.org/10.1002/vnl.730090304

Jones J. A. A. (2010). Soil piping and catchment response. *Hydrological Processes*, **24**, 1548–1566, https://doi.org/10.1002/hyp.7634

Kabir G., Tesfamariam S., Hemsing J. and Sadiq R. (2020). Handling incomplete and missing data in water network database using imputation methods. *Sustainable and Resilient Infrastructure*, **5**(6), 365–377, https://doi.org/10.1080/23789689.2019.1600960

Kang J., Parker F. and Yoo C. H. (2007). Soil-structure interaction and imperfect trench installations for deeply buried concrete pipes. *Journal of Geotechnical and Geoenvironmental Engineering*, **133**, 277–285, https://doi.org/10.1061/(ASCE)1090-0241(2007)133:3(277)

Kauffman G. J., Belden A. C., Vonck K. J. and Homsey A. R. (2009). Link between impervious cover and base flow in the white clay creek wild and scenic watershed in Delaware. *Journal of Hydrologic Engineering*, **14**(4), 324–334, https://doi.org/10.1061/(ASCE)1084-0699(2009)14:4(324)

Kluge B., Markert A., Facklam M., Sommer H., Kaiser M., Pallasch M. and Wessolek G. (2018). Metal accumulation and hydraulic performance of bioretention systems after long-term operation. *Journal of Soils & Sediments*, **18**(2), 431–441, https://doi.org/10.1007/s11368-016-1533-z

Knapen A., Poesen J., Govers G., Gyssels G. and Nachtergaele J. (2007). Resistance of soils to concentrated flow erosion: a review. *Earth-Science Reviews*, **80**, 75–109, https://doi.org/10.1016/j.earscirev.2006.08.001

Kodešová R., Šimůnek J., Nikodem A. and Jirků V. (2010). Estimation of the dual-permeability model parameters using tension disk infiltrometer and Guelph permeameter. *Vadose Zone Journal*, **9**, 213–225, https://doi.org/10.2136/vzj2009.0069

Kooij C., Muhle S., Clemens F., Pothof I. and Blokzijl F. (2015). Performance Indicators for Complex Wastewater Pumping Stations and Pressure Mains, https://doi.org/10.4108/icst.iniscom.2015.258385

Koppe E., Schneider F. J. A., Londero A. L., de Queiroz R., Buligon L. and Minella J. P. G. (2022). Soil water infiltration evaluation from punctual to hillslope scales. *Environmental Monitoring and Assessment*, **194**, 300, https://doi.org/10.1007/s10661-022-09893-x

Korving J. L. (2004). Probabilistic assessment of the performance of combined sewer systems. PhD Thesis, Delft University of Technology. ISBN: 90-9018056-7.

Korving J. L. and Ottenhoff E. C. (2008). Analysis of the causes of pump failure and differences of failure characteristics. *Water Science & Technology*, **57**(8), 1271–1276, https://doi.org/10.2166/wst.2008.304

Korving J. L., Clemens F. and van Noortwijk J. (2006a). Statistical modelling of the serviceability of sewage pumps. *Journal of Hydraulic Engineering*, **132**(10), 1076–1085, https://doi.org/10.1061/(ASCE)0733-9429(2006)132:10(1076)

Korving J. L., Geise M. and Clemens F. (2006b). Failure of sewage pumps: statistical modelling and impact assessment. *Water Science & Technology*, **54**(6–7), 119–126, https://doi.org/10.2166/wst.2006.577

Kotsiantis S. B. (2013). Decision trees: A recent overview. *Artificial Intelligence Review*, **39**(4), 261–283, https://doi.org/10.1007/s10462-011-9272-4

Kuriyama T., Narisawa I., Shina R. and Kotaki M. (1998). Effects of morphology on the fracture toughness of PVC-U pipe. *Journal of Vinyl and Additive Technology*, **4**, 164–168, https://doi.org/10.1002/vnl.10035

Kutílek M. and Nielsen D. R. (2015). Soil. Springer Netherlands, Dordrecht, https://doi.org/10.1007/978-94-017-9789-4

Laakso T., Kokkonen T., Mellin I. and Vahala R. (2018). Sewer condition prediction and analysis of explanatory factors. *Water (Switzerland)*, **10**(9), 1239. Scopus, https://doi.org/10.3390/w10091239

Lassabatere L., Di Prima S., Bouarafa S., Iovino M., Bagarello V. and Angulo-Jaramillo R. (2019). BEST-2K method for characterizing dual-permeability unsaturated soils with ponded and tension infiltrometers. *Vadose Zone Journal*, **18**, 1–20, https://doi.org/10.2136/vzj2018.06.0124

Le Coustumer S. and Barraud S. (2007). Long-term hydraulic and pollution retention performance of infiltration systems. *Water Science & Technology*, **55**(4), 235–243, https://doi.org/10.2166/wst.2007.114

Le Coustumer S., Fletcher T. D., Deletic A., Barraud S. and Lewis J. F. (2009). Hydraulic performance of biofilter systems for stormwater management: influences of design and operation. *Journal of Hydrology*, **376**(1–2), 16–23, https://doi.org/10.1016/j.jhydrol.2009.07.012

Le Coustumer S., Fletcher T. D., Deletic A., Barraud S. and Poelsma P. (2012). The influence of design parameters on clogging of stormwater biofilters: a large-scale column study. *Water Research*, **46**(20), 6743–6752, https://doi.org/10.1016/j.watres.2012.01.026

Le Gat Y. (2008). Modelling the deterioration process of drainage pipelines. *Urban Water Journal*, **5**(2), 97–106, https://doi.org/10.1080/15730620801939398

Lepot M., Hernandez B., Cedillo S., Marzloff I., Goes B., Schellart A., Verlinden J. and Clemens F. (2016). Use of numbered and coloured 3D printed pills a a new, low-cost robust and easy method to inspect mis-,cross-, and illicit connections in separate sewers. Proc. of NOVATECH 2016, Lyon, France.

Lepot M., Makris K. and Clemens F. (2017). Detection and quantification of lateral, illicit connections and infiltration in sewers with infra-red camera: conclusions after a wide experimental plan. *Water Research*, **122**, 679–691, https://doi.org/10.1016/j.watres.2017.06.030

Lindsey G., Roberts L. and Page W. (1992). Inspection and maintenance of infiltration facilities. *J. Soil and Water Conservation*, **47**(6), 481–486.

Lubbers C. L. (2007). On gas pockets in wastewater pressure mains and their effect on hydraulic performance. Delft University of Press, IOS Press, Amsterdam. ISBN: 978-1-58603-789-5.

Luimes R. A., Suiker A. S. J., Verhoosel C. V., Jorissen A. J. M. and Schellen H. L. (2018). Fracture behaviour of historic and new oak wood. *Wood Science and Technology*, **52**, 1243–1269, https://doi.org/10.1007/s00226-018-1038-6

Luimes R. A., Scheperboer I. C., Suiker A. S. J., Bosco E. and Clemens F. H. L. R. (2022). Effect of biochemical attack on the mechanical performance of used concrete sewer pipes. *Construction and Building Materials*, **346**, 128390, https://doi.org/10.1016/j.conbuildmat.2022.128390

Luimes R. A., Scheperboer I. C., Suiker A. S. J., Bosco E. and Clemens F. H. L. R. (2023). Assessment of chemo-mechanical degradation of concrete sewer pipes through an integrated experimental approach. *Engineering Structures*, **293**, 116590, https://doi.org/10.1016/j.engstruct.2023.116590

MacDougall K., Hoult N. A. and Moore I. D. (2016). Measured load capacity of buried reinforced concrete pipes. *ACI Structural Journal*, **113**, 63–74, https://doi.org/10.14359/51688059

Madanat S., Mishalani R. and Ibrahim W. H. W. (1995). Estimation of infrastructure transition probabilities from condition rating data. *Journal of Infrastructure Systems*, **1**(2), 120–125, https://doi.org/10.1061/(ASCE)1076-0342(1995)1:2(120)

Makris K. F., Langeveld J. G. and Clemens F. H. L. R. (2021). Extensive testing on PVC sewer pipes identifying the factors that affect their operational lifetime. *Journal of Structure and Infrastructure Engineering*, **18**, 1–13, https://doi.org/10.1080/15732479.2021.1907601

Malinowska A. A., Witkowski W. T., Hejmanowsky R., Chang L., van Leijen F. J. and Hanssen R. F. (2019). Sinkhole occurrence monitoring over shallow abandoned coal mines with satellite-based persistent scatterer interferometry. *Engineering Geology*, **262**, 105336, https://doi.org/10.1016/j.enggeo.2019.105336

Marquez-Penaranda J. F., Sanchez-Silva M., Husserl J. and Bastidas-Arteaga E. (2015). Effects of biodeterioration on the mechanical properties of concrete. *Materials and Structures*, **49**(2015), 4085–4099.

Marsalek J., Watt W. and Anderson B. (2006). Trace metal levels in sediments deposited in urban stormwater management facilities. *Water Science & Technology*, **53**(2), 175–183, https://doi.org/10.2166/wst.2006.051

Mashford J., Marlow D., Tran D. and May R. (2011). Prediction of sewer condition grade using support vector machines. *Journal of Computing in Civil Engineering*, **25**(4), 283–290. Scopus, https://doi.org/10.1061/(ASCE)CP.1943-5487.0000089

McCullagh P. and Nelder J. A. (1989). Generalized Linear Models, 2nd edn. Chapman & Hall, London, UK, https://doi.org/10.1201/9780203753736

McGrath T. J., Selig E. T., Webb M. C. and Zoladz G. V. (1999). Pipe Interaction with the Backfill Envelope. US Department of Transportation, McLean, VA, p. 269.

Medjnoun A. and Bahar R. (2016). Shrinking–swelling of clay under the effect of hydric cycles. *Innovative Infrastructure Solutions*, **1**, 46, https://doi.org/10.1007/s41062-016-0043-6

Meguid M. A. and Dang H. K. (2009). The effect of erosion voids on existing tunnel linings. *Tunnelling and Underground Space Technology*, **24**, 278–286, https://doi.org/10.1016/j.tust.2008.09.002

Meguid M. A. and Kamel S. (2014). A three-dimensional analysis of the effects of erosion voids on rigid pipes. *Tunnelling and Underground Space Technology*, **43**, 276–289, https://doi.org/10.1016/j.tust.2014.05.019

Melbourne Water (2023). guidelines. https://www.melbournewater.com.au/building-and-works/developer-guides-and-resources/guidelines-drawings-and-checklists/guidelines (07/04/2023).

Mittermayr F., Rezvani M., Baldermann A., Hainer S., Breitenbücher P., Juhart J., Graubner C.-A. and Proske T. (2015). Sulfate resistance of cement-reduced eco-friendly concretes. *Cement and Concrete Composites*, **55**, 364–373, https://doi.org/10.1016/j.cemconcomp.2014.09.020

Moarcas O. and Irle M. (1999). Determination of Poisson's ratio for particleboard in pure bending. *Wood Science & Technology*, (33), 439–44, https://doi.org/10.1007/s002260050128

Moffat R., Fannin R. J. and Garner S. J. (2011). Spatial and temporal progression of internal erosion in cohesionless soil. *Canadian Geotechnical Journal*, **48**, 399–412, https://doi.org/10.1139/T10-071

Moghri M., Garmabi H. and Akbarian M. (2003). Effect of processing parameters on fusion and mechanical properties of a twin-screw extruded rigid PVC pipe. *Journal of Vinyl and Additive Technology*, **9**, 81–89, https://doi.org/10.1002/vnl.10067

Monteny J., Vincke E., Beeldens A., De Belie N., Taerwe L., Van Gemert D. and Verstraete W. (2000). Chemical, microbiological, and in situ test methods for biogenic sulfuric acid corrosion of concrete. *Cement and Concrete Research*, **30**, 623–634, https://doi.org/10.1016/S0008-8846(00)00219-2

Moreno-Rodenas A. M., Duinmeijer S. P. A. and Clemens-Meyer F. (2021). Deep learning based monitoring of FOG layer dynamics in wastewater pumping stations. *Water Research*, **202**, 117482, https://doi.org/10.1016/j.watres.2021.117482

Morris P. H., Graham J. and Williams D. J. (1992). Cracking in drying soils. *Canadian Geotechnical Journal*, **29**, 263–277, https://doi.org/10.1139/t92-030

Mubarak I., Mailhol J. C., Angulo-Jaramillo R., Ruelle P., Boivin P. and Khaledian M. (2009). Temporal variability in soil hydraulic properties under drip irrigation. *Geoderma*, **150**, 158–165, https://doi.org/10.1016/j.geoderma.2009.01.022

Najafi M. and Kulandaivel G. (2005). *Pipeline condition prediction using neural network models*. 767–781. Scopus, https://doi.org/10.1061/40800(180)61

Nehdi M. L., Mohamed, N. and Soliman A. (2016). Investigation of buried full-scale SFRC pipes under live loads. *Construction and Building materials*, (102), 733–742, https://doi.org/10.1016/j.conbuildmat.2015.10.203

Nelder J. A. and Mead R. (1965). A simplex method for function minimization. *The Computer Journal*, **7**, 308–313, https://doi.org/10.1093/comjnl/7.4.308

Neville A. (2004). The confused world of sulfate attack on concrete. *Cement and Concrete Research*, **34**(2004), 1275–1296, https://doi.org/10.1016/j.cemconres.2004.04.004

Nguyen L. V., Bui D. T. and Seidu R. (2022). Comparison of machine learning techniques for condition assessment of sewer network. *IEEE Access*, **10**, 124238–124258, https://doi.org/10.1109/ACCESS.2022.3222823

Nielsen D., Biggar J. and Erh K. (1973). Spatial variability of fieldmeasured soil-water properties. *Hilgardia*, **42**, 215–259, https://doi.org/10.3733/hilg.v42n07p215

Nieuwenhuis E., Post J., Duinmeijer A., Langeveld J. and Clemens F. (2018). Statistical modelling of Fat, Oil and grease (FOG) depoits in wastewater pump sumps. *Water Research*, **135**, 155–167, https://doi.org/10.1016/j.watres.2018.02.026

Nimmo J. R., Schmidt K. M., Perkins K. S. and Stock J. D. (2009). Rapid measurement of field-saturated hydraulic conductivity for areal characterization. *Vadose Zone Journal*, **8**, 142–149, https://doi.org/10.2136/vzj2007.0159

North Carolina Department of Environmental Quality (NCDEQ) (2018). Stormwater Design Manual: https://www.deq.nc.gov/about/divisions/energy-mineral-and-land-resources/stormwater/stormwater-program/stormwater-design-manual (07/04/2023).

O'Connell M., McNally C. and Richardson M. G. (2010). Biochemical attack on concrete in wastewater applications: a state of the art review. *Cement and Concrete Composites*, **32**, 479–485, https://doi.org/10.1016/j.cemconcomp.2010.05.001

Olsson L., Barbosa H., Bhadwal S., Cowie A., Delusca K., Flores-Renteria D., Hermans K., Jobbagy E., Kurz W., Li D., Sonwa D. J. and Stringer L. (2019). Land degradation. In: Climate Change and Land: an IPCC Special Report on Climate Change, Desertification, Land Degradation, Sustainable Land Management, Food Security, and Greenhouse gas Fluxes in Terrestrial Ecosystems, P. R. Shukla, J. Skea, E. Calvo Buendia, V. Masson-Delmotte, H.-O. Pörtner, D. C. Roberts, P. Zhai, R. Slade, S. Connors, R. van Diemen, M. Ferrat, E. Haughey, S. Luz, S. Neogi, M. Pathak, J. Petzold, J. Portugal Pereira, P. Vyas, E. Huntley, K. Kissick, M. Belkacemi and J. Malley (eds), Cambridge University Press, Cambridge, UK, pp. 345–436, https://doi.org/10.1017/9781009157988.006

Oral H. V., Carvalho P., Gajewska M., Ursino N., Masi F., Hullebusch E. D., van Kazak J. K., Exposito A., Cipolletta G., Andersen T. R., Finger D. C., Simperler L., Regelsberger M., Rous V., Radinja M., Buttiglieri G., Krzeminski P., Rizzo A., Dehghanian K., Nikolova M. and Zimmermann M. (2020). A review of nature-based solutions for urban water management in European circular cities: a critical assessment based on case studies and literature. *Blue-Green Systems*, **2**(1), 112–136, https://doi.org/10.2166/bgs.2020.932

Ortega J., García-Vera V., Solak A. and Tenza-Abril A. (2019). Pore structure degradation of different cement mortars exposed to sulphuric acid. *Appl. Sci.*, **9**, 1–13.

Panasiuk O., Hedström A., Marsalek J., Ashley R. M. and Viklander M. (2015). Contamination of stormwater by wastewater: a review of detection method. *Journal of Environment Management*, **152**, 241–250, https://doi.org/10.1016/j.jenvman.2015.01.050

Parker C. D. (1945). The corrosion of concrete. *Australian Journal of Experimental Biology and Medical Science*, **23**, 81–90, https://doi.org/10.1038/icb.1945.13

Paus K. H., Morgan J., Gulliver J. S., Leiknes T. and Hozalski R. M. (2014). Assessment of the hydraulic and toxic metal removal capacities of bioretention cells after 2 to 8 years of service. *Water, Air, and Soil Pollution*, **225**(1), 1–12.

Pedregosa F., Varoquaux G., Gramfort A., Michel V., Thirion B., Grisel O., Blondel M., Prettenhofer P., Weiss R., Dubourg V., Vanderplas J., Passos A., Cournapeau D., Brucher M., Perrot M. and Duchesnay É (2011). Scikit-learn: machine learning in python. *Journal of Machine Learning Research*, **12**, 2825–2830. Scopus.

Peter J. M., Chapman D., Moore I. D. and Hoult N. (2018). Impact of soil erosion voids on reinforced concrete pipe responses to surface loads. *Tunnelling and Underground Space Technology*, **82**, 111–124, https://doi.org/10.1016/j.tust.2018.08.003

Petrucci G. (2012). La diffusion du contrôle à la source des eaux pluviales urbaines : confrontation des pratiques à la rationalité hydrologique. PhD, Ecole des Ponts et Chaussées- Université Paris Est, Paris France.

Philip J. R. (1957). The theory of infiltration: 4 sorptivity and algebraic infiltration equations. *Soil Science*, **84**, 257–264, https://doi.org/10.1097/00010694-195709000-00010

Poduška J., Hutař P., Kučera J., Frank A., Sadilek J., Pinter G. and Nahlık L. (2016). Residual stress in polyethylene pipes. *Polymer Testing*, **54**, 288–295, https://doi.org/10.1016/j.polymertesting.2016.07.017

Poesen J. (2018). Soil erosion in the anthropocene: research needs. *Earth Surface Processes and Landforms*, **43**, 64–84, https://doi.org/10.1002/esp.4250

Pothof I. W. M. (2011). Co-current air-water flow in downward sloping pipes. PhD thesis, Delft University of Technology. ISBN: 978-90-8957-018-5. https://www.bing.com/search?q=Pothof%2C+I.W.M.+(2011)+co-current+air-water+flow+in+downward+sloping+pipes&cvid=1b3f4467f6b640889296b2a58b129d8f&aqs=edge..69i57.5508j0j4&FORM=ANAB01&PC=U531

Prima S. D., Stewart R. D., Najm M. R. A., Roder L. R., Giadrossich F., Campus S., Angulo-Jaramillo R., Yilmaz D., Roggero P. P., Pirastru M. and Lassabatere L. (2021,). BEST-WR: an adapted algorithm for the hydraulic characterization of hydrophilic and water-repellent soils. *Journal of Hydrology*, **603**, 126936, https://doi.org/10.1016/j.jhydrol.2021.126936

Proton A. (2008). Etude hydraulique des tranchées de rétention / infiltration. PhD, INSA Lyon, Villeurbanne, France, 299 p.

Rabinovitch E. B. and Summers J. W. (1992). The effect of physical ageing on properties of rigid polyvinyl chloride. *Journal of Vinyl Technology*, **14**, 126–130, https://doi.org/10.1002/vnl.730140303

Rajabipour F., Giannini E., Dunant C., Ideker J. H. and Thomas M. D. A. (2015). Alkali–silica reaction: current understanding of the reaction mechanisms and the knowledge gaps. *Cement and Concrete Research*, **76**, 130–146, https://doi.org/10.1016/j.cemconres.2015.05.024

Rakitin B. and Xu M. (2014). Centrifuge modeling of large-diameter underground pipes subjected to heavy traffic loads. *Canadian Geotechnical Journal*, **51**, 353–368, https://doi.org/10.1139/cgj-2013-0253

Reinhardt H. W. and Xu S. (2000). A practical testing approach to determine mode II fracture energy for concrete. *Int. J. Fract.*, **105**, 107–125, https://doi.org/10.1023/A:1007649004465

Reynolds W. D. and Lewis J. K. (2012). A drive point application of the Guelph permeameter method for coarse-textured soils. *Geoderma*, **187–188**, 59–66, https://doi.org/10.1016/j.geoderma.2012.04.004

Riks E. (1979). An incremental approach to the solution of snapping and buckling problems. *Int. Journal of Solids and Structures*, **15**(7), 529–551, https://doi.org/10.1016/0020-7683(79)90081-7

Rizzo P. (2010). Water and Wastewater Pipe Non-destructive Evaluation and Health Monitoring: A Review. *Advances in Civil Engineering*, **2010**, 818597, https://doi.org/10.1155/2010/818597

Robeson L. M. (2013). Environmental stress cracking: a review. *Polymer Engineering & Science*, **53**, 453–467, https://doi.org/10.1002/pen.23284

Robinson M. and Dean T. J. (1993). Measurement of near surface soil water content using a capacitance probe. *Hydrological Processes*, **7**, 77–86, https://doi.org/10.1002/hyp.3360070108

Rodrigues-Iturbe I., Febres De Power B., Sharif M. B. and Georgakakos K. P. (1989). Chaos in rainfall. *Water Resources Research*, **25**(7), 1667–1675, https://doi.org/10.1029/WR025i007p01667

Rokstad M. M. and Ugarelli R. M. (2015). Evaluating the role of deterioration models for condition assessment of sewers. *Journal of Hydroinformatics*, **17**(5), 789–804. Scopus, https://doi.org/10.2166/hydro.2015.122

Rooyackers F. A. M., Bosco E., Suiker A. S. J. and Clemens F. H. L. R. (2022). A chemo-mechanical model for biogenic sulphide corrosion of concrete. *Cement and Concrete Research*, **160**, 106809, https://doi.org/10.1016/j.cemconres.2022.106809

Saulais M. (2011). Colonisation végétale des bassins d'infiltration et de rétention : caractérisation de la flore et évolution des caractéristiques physico-chimiques de l'horizon de surface végétalisé. PhD, INSA, Lyon, France. https://theses.hal.science/tel-00715802.

Saulais M., Bedell J. P. and Delolme C. (2011). Cd, Cu and Zn mobility in contaminated sediments from an infiltration basin colonized by wild plants: The case of Phalaris arundinacea and Typha.

Savija B. and Lukovic M. (2016). Carbonation of cement paste: understanding, challenges, and opportunities. *Construction and Building Materials*, **117**, 285–301, https://doi.org/10.1016/j.conbuildmat.2016.04.138

Scheperboer I. C., Luimes R. A., Suiker A. S. J., Bosco E. and Clemens F. H. L. R. (2021). Experimental-numerical study on the structural failure of concrete sewer pipes. *Tunnelling and Underground Space Technology*, **116**, 104075, https://doi.org/10.1016/j.tust.2021.104075

Scheperboer I. C., Suiker A. S. J., Bosco E. and Clemens F. H. L. R. (2022). A coupled hydro-mechanical model for subsurface erosion with analysis of soil piping and void formation. *Acta Geotechnica*, **17**, 4769–4798, https://doi.org/10.1007/s11440-022-01479-8

Schofield A. N. (1980). Cambridge geotechnical centrifuge operations. *Geotechnique*, **30**, 227–268, https://doi.org/10.1680/geot.1980.30.3.227

Scholten F. L., van der Stok E., Gerets B., Wenzel M. and Boege M. (2016). Residual quality of excavated UPVC gas and water distribution pipelines. Proceedings of the 18th International Conference on Plastics Pipes, Berlin, Germany.

Shaowei H., Qiyong Q. and Nina G. (2018). Effect of acid corrosion on crack propagation of concrete beams. *S˜adhaña*, **43**, 1–23.

Shawe-Taylor J. and Cristianini N. (2004). Kernel Methods for Pattern Analysis. Cambridge University Press, Cambridge, UK, https://doi.org/10.1017/CBO9780511809682

Shi B., Catsamas S., Delectic B., Wang M., Bach P. M., Lintern A., Delectic A. and McCarthy D. T. (2022). Illicit discharge detection in stormwater drains using an arduino-based low-cost sensor network. *Water Science and Technology*, **85**(5), 1372–1383, https://doi.org/10.2166/wst.2022.034

Siegmann A., Buchman A. and Kenig S. (1981). Residual stresses in polymers. II. Their effect on mechanical behaviour. *Polymer Engineering and Science*, **21**, 997–1002, https://doi.org/10.1002/pen.760211503

Siegmann A., Buchman A. and Kenig S. (1982). Residual stresses in polymers I: The effect of thermal history. *Polymer Engineering and Science*, **22**, 40–47, https://doi.org/10.1002/pen.760220107

Silva A., Nascimento N., Seidl M. and Vieira L. (2010). SWITCH in Belo Horizonte, Brazil: infiltration and detention systems for more sustainable stormwater control in Belo Horizonte. *Reviews of Environmental Science and Biotechnology*, **9**, 7–13, https://doi.org/10.1007/s11157-010-9196-5

Šimůnek J., Jarvis N. J., van Genuchten M. T. and Gärdenäs A. (2003). Review and comparison of models for describing non-equilibrium and preferential flow and transport in the vadose zone. *Journal of Hydrology*, **272**, 14–35, https://doi.org/10.1016/S0022-1694(02)00252-4

Šimůnek J., Šejna M., Saito H. and Van Genuchten M. T. (2008). The HYDRUS-1D software package for simulating the movement of water, heat, and multiple solutes in variably saturated media. *HYDRUS Softw. Ser.*, **3**.

Šimůnek J., van Genuchten M. Th. and Šejna M. (2016). Recent developments and applications of the HYDRUS computer software packages. *Vadose Zone Journal*, **15**, 1–25, https://doi.org/10.2136/vzj2016.04.0033

Siriwardene N., Deletic A. and Fletcher T. D. (2007). Clogging of stormwater gravel infiltration systems and filters: insights from a laboratory study. *Water Research*, **41**(7), 1433–1440, https://doi.org/10.1016/j.watres.2006.12.040

Sivakumar B., Berndtsson J., Olsson J. and Jinno K. (2009). Evidence of chaos in the rainfall-runoff process. *Hydrological Sciences Journal*, **46**(1), 131–145, https://doi.org/m10.1080/02626660109492805

Skorobogatov A. (2014). Hydrological Functionality of Plants and Its Application to Stormwater Management. Master's thesis, University of Calgary, Calgary, Canada. Retrieved from https://prism.ucalgary.ca, https://doi.org/10.11575/PRISM/26403

Smakhtin V. U. (2001). Low flow hydrology: a review. *Journal of Hydrology*, **240**, 147–186, https://doi.org/10.1016/S0022-1694(00)00340-1

Sousa V., Matos J. P. and Matias N. (2014). Evaluation of artificial intelligence tool performance and§§ uncertainty for predicting sewer structural condition. *Automation in Construction*, **44**, 84–91. Scopus, https://doi.org/10.1016/j.autcon.2014.04.004

Stajanca P., Chruscicki S., Homann T., Seifert S., Sschmidt D. and Habib A. (2019). Detection of leak-induced pipeline vibrations using fiber optic distributed acoustic sensing. *Sensors*, **18**, 2841, https://doi.org/10.3390/s18092842

Stanic N., de Haan C., Tirion M., Langeveld J. G. and Clemens F. H. L. R. (2013). Comparison of core sampling and visual inspection for assessment of concrete sewer pipe condition. *Water Science and Technology*, **67**, 2458–2466, https://doi.org/10.2166/wst.2013.138

Stanic N., Langeveld J., Salet T. and Clemens F. (2017). Relating the structural strength of concrete sewer pipes and material properties retrieved from core samples. *Structure and Infrastructure Engineering*, **13**, 637–651, https://doi.org/10.1080/15732479.2016.1187631

Struik L. C. E. (1977). Physical aging in amorphous polymers and other materials. Doctoral dissertation, Delft University of Technology, The Netherlands. Retrieved from http://resolver.tudelft.nl/uuid:941d2af6-903a-4260-9953-2efb4cb38d2e

Syachrani S., Jeong H. S. ". and Chung C. S. (2013). Decision tree–based deterioration model for buried wastewater pipelines. *Journal of Performance of Constructed Facilities*, **27**(5), 633–645, https://doi.org/10.1061/(ASCE)CF.1943-5509.0000349

Talesnick M. and Baker R. (1999). Investigation of the failure of a concrete-lined steel pipe. *Geotechnical and Geological Engineering*, **17**, 99–121, https://doi.org/10.1023/A:1008913408452

Tan Z. (2007). Nonlinear finite element study of deteriorated sewers including the influence of erosion voids. Master's thesis, Queen's University, Kingston, Ontario, Canada.

Toronto and Region Conservation Authority (TRCA) (2016). Low Impact Development Stormwater Management Practice Inspection and Maintenance Guide. https://sustainabletechnologies.ca/app/uploads/2016/08/LID-IM-Guide-2016-1.pdf (07/04/2023).

Tran H., Perera B. and Ng A. (2007). Neural Network Based Prediction Models For Structural Deterioration of Urban Drainage Pipes. MODSIM07 – Land, Water and Environmental Management: Integrated Systems for Sustainability, Proceedings.

Trautmann C. H. and O'Rourke T. D. (1985). Lateral force-displacement response of buried pipe. *Journal of Geotechnical Engineering*, **111**, 1077–1092, https://doi.org/10.1061/(ASCE)0733-9410(1985)111:9(1077)

Tscheikner-Gratl F., Caradot N., Cherqui F., Leitão J. P., Ahmadi M., Langeveld J. G., Le Gat Y., Scholten L., Roghani B., Rodríguez J. P., Lepot M., Stegeman B., Heinrichsen A., Kropp I., Kerres K., Almeida M. D. C., Bach P. M., Moy de Vitry M., Sá Marques A., … Clemens F. (2019). Sewer asset management–state of the art and research needs. *Urban Water Journal*, **16**(9), 662–675. Scopus, https://doi.org/10.1080/1573062X.2020.1713382

Tu J. V. (1996). Advantages and disadvantages of using artificial neural networks versus logistic regression for predicting medical outcomes. *Journal of Clinical Epidemiology*, **49**(11), 1225–1231. Scopus, https://doi.org/10.1016/S0895-4356(96)00002-9

Tukker M., Kooij C. and Pothof I. (2016). Hydraulic Design and Management of Wastewater Transport Systems. IWA Publishing, London, UK, https://doi.org/10.2166/9781780407814

Utsi E. C. (2017). Ground Penetrating Radar: Theory and Practice. Butterworth-Heinemann.

Valdelfener M., Barraud S., Sibeud E., Bacot L., Perrin Y., Jourdain F. and Marmonier P. (2018). Do sustainable drainage systems favour mosquito proliferation in cities compared to stormwater networks? *Urban Water Journal*, **16**(6), 436–443, https://doi.org/10.1080/1573062X.2018.1523442

Van Bijnen M. (2018). The impact of sewer condition on the performance of sewer systems. PhD thesis, Delft University of Technology. ISBN 978-94-6233-987-3.

Van Bijnen M., Korving H. and Clemens F. (2012). Impact of sewer condition on urban flooding: an uncertainty analysis based on field observations and Monte Carlo simulations on full hydrodynamic models. *Water Science and Technology*, **65**, 2219–2227, https://doi.org/10.2166/wst.2012.134

Van Bijnen M., Korving H., Langeveld J. and Clemens F. (2016). Calibration of hydrodynamic model driven sewer maintenance. *Structure and Infrastructure Engineering*, **13**, 1167–1185, https://doi.org/10.1080/15732479.2016.1247287

Van Bijnen M., Korving H., Langeveld J. and Clemens F. (2018). Quantitative impact assessment of sewer condoition on health risk. *Water*, **10**(3), 245, https://doi.org/10.3390/w10030245

Van Duin B. (2023). Private communication on clogging of geotextiles in practical UDAM applications.

Van Duin B., Brown C., Chu A., Marsalek J. and Valeo C. (2008). Characterization of long-term solids removal and clogging processes in two types of permeable pavement under cold climate conditions. 11th International Conference on Urban Drainage, Edinburgh, pp. 1–10.

Van Genuchten M. T. (1980). A closed-form equation for predicting the hydraulic conductivity of unsaturated soils. *Soil Science Society of America Journal*, **44**, 892–898, https://doi.org/10.2136/sssaj1980.03615995004400050002x

van Riel W. A. P. (2016). On decision-making for sewer replacement. PhD thesis Delft University of Technology. ISBN:978-94-6233-327-7, https://doi.org/10.4233/uuid:92b10448-795d-43ac-8071-d779af9d374d

van Riel W., Langeveld J., Herder P. and Clemens F. (2016). The influence of information quality on decision making for networked infrastructure management. *Structure and Infrastructure Engineering*, **13**(6), 1–13, https://doi.org/10.1080/15732479.2016.1187633

van Schaik N. L. M. B., Hendriks R. F. A. and van Dam J. C. (2010). Parameterization of macropore flow using dye-tracer infiltration patterns in the SWAP model. *Vadose Zone Journal*, **9**, 95, https://doi.org/10.2136/vzj2009.0031

Verachtert E., Maetens W., Van Den Eeckhaut M., Poesen J. and Deckers J. (2011). Soil loss rates due to piping erosion. *Earth Surface Processes and Landforms*, **36**, 1715–1725, https://doi.org/10.1002/esp.2186

Verruijt A. and van Baars S. (2007). Soil Mechanics. VSSD, Delft, the Netherlands.

Virahsawmy H., Stewardson M., Vietz G. and Fletcher T. (2014). Factors that affect the hydraulic performance of raingardens: implications for design and maintenance. *Water Science & Technology*, **69**(5), 982–988, https://doi.org/10.2166/wst.2013.809

Visser H. A. (2009). Residual lifetime assessment of uPVC gas pipes, PhD Thesis University of Twente, https://doi.org/10.3990/1.9789036529587.

Visser H. A., Bor T. C., Wolters M., Warnet L. L. and Govaert L. E. (2011). Influence of physical aging on impact embrittlement of uPVC pipes. *Plastics, Rubber and Composites*, **40**, 201–212, https://doi.org/10.1179/1743289810Y.0000000021

Vollertsen J., Nielsen A. H., Jensen H. S., Wium-Andersen T. and Hvitved-Jacobsen T. (2008). Corrosion of concrete sewers – the kinetics of hydrogen sulfide oxidation. *Science of The Total Environment*, **394**, 162–170, https://doi.org/10.1016/j.scitotenv.2008.01.028

Walthan T., Bell F. G. and Culshaw M. G. (2005). Sinkholes and Subsidence – Karst and Cavernous Rocks in Engineering and Construction. Springer-Verlag, Berlin-Heidelberg-New York.

Wang Z., Wu Q. J., Wu L., Ritsema C. J., Dekker L. W. and Feyen J. (2000). Effects of soil water repellency on infiltration rate and flow instability. *Journal of Hydrology*, **231–232**, 265–276, https://doi.org/10.1016/S0022-1694(00)00200-6

Wang J., Huang H., Xie X. and Bobet A. (2014). Void-induced liner deformation and stress redistribution. *Tunnelling and Underground Space Technology*, **40**, 263–276, https://doi.org/10.1016/j.tust.2013.10.008

Watlet A., Kaufmann O., Triantafyllou A., Poulain A., Chambers J. E., Meldrum P. I., Wilkinson P. B., Hallet V., Quinif Y., Van Ruymbeke M. and Van Camp M. (2018). Imaging groundwater infiltration dynamics in the karst vadose zone with long-term ERT monitoring. *Hydrology and Earth System Sciences*, **22**, 1563–1592, https://doi.org/10.5194/hess-22-1563-2018

Werey C., Cherqui F., Le Nouveau N., Rodriguez F., Sibeud E., Joannis C. and Barraud S. (2016). Asset management of Best Management Practices: a new story for urban stormwater management (Gestion patrimoniale des techniques alternatives : une nouvelle histoire à écrire pour la gestion des eaux pluviales en ville). 9th International Conference Novatech, Lyon, France.

Whalley W. R. (1993). Considerations on the use of time-domain reflectometry (TDR) for measuring soil water content. *Journal of Soil Science*, **44**, 1–9, https://doi.org/10.1111/j.1365-2389.1993.tb00429.x

Willard L. L., Wynn-Thompson T., Krometis L. H., Neher T. P. and Badgley B. D. (2017). Does it pay to be mature? Evaluation of bioretention cell performance seven years postconstruction. *J. Environ. Eng.*, **143**(9), 04017041, https://doi.org/10.1061/(ASCE)EE.1943-7870.0001232

Winston R., Al-Rubaei A., Blecken G.-T., Viklander M. and Hunt W. (2016). Maintenance measures for preservation and recovery of permeable pavement surface infiltration rate – the effects of street sweeping, vacuum cleaning, high pressure washing, and milling. *Journal of Environmental Management*, **169**, 132–144, https://doi.org/10.1016/j.jenvman.2015.12.026

Wirahadikusumah R., Abraham D. and Iseley T. (2001). Challenging issues in modeling deterioration of combined sewers. *Journal of Infrastructure Systems*, **7**, 77–84, https://doi.org/10.1061/(ASCE)1076-0342(2001)7:2(77)

Wolpert D. H. and Macready W. G. (1997). No free lunch theorems for optimization. *IEEE Transactions on Evolutionary Computation*, **1**(1), 67–82, https://doi.org/10.1109/4235.585893

Xu X. P. and Needleman A. (1994). Numerical simulations of fast crack growth in brittle solids. *Journal of the Mechanics and Physics of Solids*, **42**, 1397–1434, https://doi.org/10.1016/0022-5096(94)90003-5

Xu M. and Shen D. (2020). The influence of erosion voids on the longitudinal behaviour of a jointed large-diameter reinforced concrete pipeline. *Tunnelling and Underground Space Technology*, **103**, 103494, https://doi.org/10.1016/j.tust.2020.103494

Xu M., Shen D. and Rakitin B. (2017). The longitudinal response of buried large-diameter reinforced concrete pipeline with gasketed bell-and-spigot joints subjected to traffic loading. *Tunnelling and Underground Space Technology*, **64**, 117–132, https://doi.org/10.1016/j.tust.2016.12.020

Yang J., Yin Z. Y., Laouafa F. and Hicher P. Y. (2019). Internal erosion in dike-on-foundation modeled by a coupled hydromechanical approach. *International Journal for Numerical and Analytical Methods in Geomechanics*, **43**, 663–683, https://doi.org/10.1002/nag.2877

Yang F., Fu D., Zevenbergen C. and Rene E. A. (2022). Comprehensive review on the long-term performance of stormwater biofiltration systems (SBS): operational challenges and future directions. *Journal of Environmental Management*, **302**, 113956, https://doi.org/10.1016/j.jenvman.2021.113956

Yasuda N., Tsukada K. and Asakura T. (2017). Elastic solutions for circular tunnel with void behind lining. *Tunnelling and Underground Space Technology*, **70**, 274–285, https://doi.org/10.1016/j.tust.2017.08.032

Yilmaz D., Lassabatere L., Deneele D., Angulo-Jaramillo R. and Legret M. (2013). Influence of carbonation on the microstructure and hydraulic properties of a basic oxygen furnace slag. *Vadose Zone Journal*, **12**, 1–15, https://doi.org/10.2136/vzj2012.0121

Yilmaz D., Di Prima S., Stewart R. D., Abou Najm M. R., Fernandez-Moret D., Latorre B. and Lassabatere L. (2022). Three-term formulation to describe infiltration in water-repellent soils. *Geoderma*, **427**, 116127, https://doi.org/10.1016/j.geoderma.2022.116127

Younis A., Ramadan A., Wong L. and Nehdi M. (2020). New rational test for reinforced-concrete pipe eliminating subjective crack-width criteria. *Structures*, (28), 2507–2522, https://doi.org/10.1016/j.istruc.2020.10.076

Zaqout T., Ólöf Andradóttir H. and Arnalds Ó (2022). Infiltration capacity in urban areas undergoing frequent snow and freeze-thaw cycles: implications on sustainable urban drainage systems. *Journal of Hydrology*, **607**, 127495, https://doi.org/10.1016/j.jhydrol.2022.127495

Zeng X., Wang Z., Wang H., Zhu S. and Chen S. (2023). Progress in drainage pipeline condition assessment and deterioration prediction models. *Sustainability*, **15**(4), Article 4, https://doi.org/10.3390/su15043849

Zhang S., Liu B. and He J. (2019). Pipeline deformation monitoring using distributed fiber optical sensir. *Measurement*, **133**, 208–213, https://doi.org/10.1016/j.measurement.2018.10.021

Zhao K., Xu Q., Liu F., Xiu D. and Ren X. (2020a). Field monitoring of preferential infiltration in loess using time-lapse electrical resistivity tomography. *Journal of Hydrology*, **591**, 125278, https://doi.org/10.1016/j.jhydrol.2020.125278

Zhao S., Jia Y., Gong J., Niu C., Su H., Gan Y. and Liu H. (2020b). Spatial variability of preferential flow and infiltration redistribution along a rocky-mountain hillslope, northern China. *Water,* **12**, 1102, https://doi.org/10.3390/w12041102

Zoladz G. V., McGrath T. J. and Selig E. T. (1996). Laboratory tests of buried pipe installation procedures. *Transportation Research Record: Journal of the Transportation Research Board,* **1541**, 86–96, https://doi.org/10.1177/0361198196154100111

doi: 10.2166/9781789063059_0231

Chapter 6

From condition-based to service-based strategies

Steve Auger[1], *Liam Carson*[2], *Frédéric Cherqui*[3], *Shamsuddin Daulat*[4], *Bert van Duin*[5,6], *Norman F. Neumann*[2], *Jeroen Langeveld*[7], *David Lembcke*[1], *Bardia Roghani*[4] *and Franz Tscheikner-Gratl*[4]

[1]*Integrated Watershed Management, Lake Simcoe Region Conservation Authority, 120 Bayview Parkway, Newmarket, Ontario, Canada L3Y 3W3*
[2]*University of Alberta, Edmonton, Alberta, Canada*
[3]*University of Lyon, INSA-LYON, Université Claude Bernard Lyon 1, DEEP, F-69621, F-69622, Villeurbanne, France*
[4]*NTNU – Norwegian University of Science and Technology, Trondheim, Norway*
[5]*City of Calgary, Canada*
[6]*University of Calgary, Canada and University of Alberta, Edmonton, Canada*
[7]*Faculty of Civil Engineering and Geoscience, Department of Water Management, Delft University of Technology, Delft, The Netherlands*

ABSTRACT

Condition assessment often serves as the primary, and at times, the exclusive factor driving the prioritization of rehabilitation requirements. While this methodology signifies a substantial departure from a 'run-to-failure' strategy, it does possess inherent limitations. Notably, it has been demonstrated to exhibit constrained efficiency, potentially resulting in the refurbishment of assets with minimal risks or inconsequential impacts. An evolved and more sophisticated perspective on asset management involves a comprehensive evaluation of the functions delivered by infrastructure elements, whether they pertain to drainage pipes or other stormwater control measures. By harmonizing the state of assets with the contextual stakes and vulnerabilities within the specific territory or region under the purview of the utility manager responsible for the upkeep of the drainage network, a more precise targeting of rehabilitation necessities can be achieved. This precision, in turn, culminates in a notable enhancement of the system's overall performance. This holistic approach, commonly referred to as a risk-based strategy, furnishes an inclusive framework for optimizing location strategies. This optimization hinges on the prioritization of rehabilitation requisites through a meticulous multi-criteria analysis. This chapter delves into the foundational functionalities inherent in urban drainage systems, coupled with their associated services. Subsequently, the succeeding section elucidates the shift from a condition-centred methodology to a performance-centric approach. A series of illustrative case studies follow, providing real-world context to the concepts discussed.

Keywords: decision aid, impact, multi-criteria, performance, prioritization, service, stake, risk, vulnerability

6.1 INTRODUCTION AND LAYOUT OF THE CHAPTER

Chapter 4 of this book is dedicated to investigating the condition and performance of the drainage assets. Results of these investigations are generally expressed in different categories (i.e., quantitative and qualitative) of defects, failures or condition, usually reflecting regional practices, guidelines and standards. They provide critical information, often including already some kind of asset ranking as to the decision on when and where to rehabilitate and to which extent, thereby defining future rehabilitation needs. In theory, any asset in poor condition should be rehabilitated. In practice, however, due to budget constraints (or economic efficiencies), only a percentage of the assets in poor condition will be rehabilitated. In this regard, the question arises: how to decide on whether to rehabilitate certain assets while not others? In practice, many protocols, guidelines and standards implicitly assume the application of a condition-based strategy for urban drainage systems (see Chapter 3). Such an approach has the advantage that guidance is available to describe the condition of an element within a system. However, (i) it largely ignores the impact of a (group of) element(s) with respect to the performance level of the system as whole, (ii) does not consider cascading effects of failure and (iii) last, but not least, assumes that the currently applied methods of condition classes relate to some physical meaning. Condition class-based strategies have found wide application in current asset management (AM) practice given their well-structured and easy to apply methodology. That said, in view of their shortcomings, they are increasingly challenged by service-based (and, by extension risk-based) approaches. This chapter provides an in-depth presentation and discussion of condition-based versus service-based strategies covering the entire life cycle of our urban drainage (UD) infrastructure.

First, it is important to understand that there is a difference between service life and technical life expectancy. In literature, various terms can be found relevant to the life of assets such as service life, technical service life, economic service life (Vanier, 2001), design life, remaining and end-of-asset life (Marlow *et al.*, 2009), remaining life expectancy (Ana & Bauwens, 2010), and remaining service life (Snider & McBean, 2021). Regardless of the individual terms and their definitions, their central tenet reflects economic or service-delivery considerations. That is, it seeks to answer the question: is-it economically rational to retain or replace the asset? Other terms that are disconnected from the costs reflect the 'pure' asset life such as the end of physical asset life or obsolescence. Marlow *et al.* (2009) define the end of physical asset life as *'when the asset is physically derelict and non-functioning'*, and obsolescence as *'when the asset is obsolete because of changes in technology, regulatory requirements, or performance criteria'*. For decision-making in infrastructure AM, three additional terms are of interest: (i) design life, (ii) end-of-asset life and (iii) remaining life expectancy. Design life can be seen as 'the period over which the asset is designed to be available for use and able to provide the required level of service at an acceptable cost and level of risk'; end-of-asset life as 'the time at which a significant (capital rather than operational) investment is (or should be) made' (Marlow *et al.*, 2009). After the investment has been made and intervention is complete, the asset starts a new design life. Marlow *et al.* (2009) define the remaining life as 'the time left before a significant capital intervention is required'. So, in general, economic life refers to accounting services and is based on the depreciation period. Service life is the period in which an asset fulfils its requirements without unexpected costs due to service interruptions or repair (Vanier & Rahman, 2004). The technical life expectancy, on the other hand, is the period during which the asset maintains its physical integrity (which could be different from the design life). Assets reaching their technical life expectancy may not need to be immediately rehabilitated as their structural condition may have no major impact on the service delivered by the larger infrastructure system that they are part of. For example, a stormwater pipe or a nature-based solution may be heavily damaged or collapsed but its rehabilitation may be postponed because it is either located in a zone where flooding does not have any major consequence, or its malfunctioning is 'buffered out' by the performance of the network it is part of. Conversely, an asset could be rehabilitated before the end of its technical lifetime because its present condition negatively impacts the performance of the larger infrastructure system. A key approach when dealing with a limited budget is thus to focus on the performance or service delivered as well instead of solely considering asset condition.

Therefore, how long an in-service asset can be properly used is one of the core discussions in infrastructure AM. Assessing or modelling (as discussed in Chapter 5) the remaining life of a pipeline can be a complex issue given that these assets have different characteristics such as material, diameter, length, thickness and so on., that affect their life. Adding to the complexity is the fact that various operational and environmental conditions affect the longevity of assets. Since the end of the physical life of an asset is coupled with system or subsystem failure which can range from minor consequences to major catastrophic events, the end-of-asset life can be decided well before it reaches its end-of-physical life. The issue of 'repairability' versus 'non-repairability' of assets adds even more complexity to the remaining asset life (Marlow *et al.*, 2009), as some assets can be repaired, and hence have their life prolonged by maintenance activities. As a result, the end-of-asset life is therefore usually a decision made by the asset owners, typically reflecting economic evaluations. Sustainability evaluation is not yet fully embedded into the asset life definition of UD systems as it adds two more dimensions into the end-of-asset life decision-making: (i) environmental and (ii) societal assessments. If assets are prematurely replaced based solely on economic evaluations, this can pose an extra burden on the environment. However, this burden can be mitigated if the assets are maintained and used for a longer period, thereby generating less waste. The societal evaluation is the impact of the end-of-asset life on human health and safety. For example, breakage of a sewer pipe (i.e., end-of-physical asset life) can result in sewage spilling into the open urban environment posing a serious risk to human health. Accordingly, a balance between the three dimensions of sustainability is vital. Although numerous studies (e.g., Akhtar *et al.*, 2015; Fathollahi and Coupe, 2021; Filion *et al.*, 2004; Vahidi *et al.*, 2016) have evaluated the sustainability aspects of UD assets in terms of life cycle approaches, asset life definitions in literature still largely reflect economic considerations.

All this added complexity requires, when taken seriously, a risk-based approach that includes several layers of risks by coupling the probabilities of failure (i.e., a drop in the condition or performance to an unacceptable level) with several dimensions of consequences (i.e., environmental, economic and human health related). In this chapter, we will first consider the performance of the system, relating the objectives of UD with their functional requirements and the impacts of dysfunctions. Pipes will be differentiated from nature-based solutions (NBS). Prioritizing rehabilitation needs based on performance requires the consideration of risk-based and impact-based approaches, coupled with multi-criteria analysis. Strategies and concepts will be explained and then illustrated. Case-studies are introduced in the last section to discover full-scale applications.

6.2 FUNCTIONALITY OF THE SYSTEM AND ITS COMPONENTS IN RELATION TO THE SERVICEABILITY OF THE SYSTEM

As discussed in Chapter 1, our UD systems are not one uniform entity in terms of functionality but represent rather a hybrid system, consisting of parts with very different characteristics reflecting local topography, design philosophies, management approaches and preferences. These parts can be divided into groups reflecting (i) the piped network (and to a degree overland conveyance as well) or (ii) stormwater control measures (SCM).

6.2.1 The case of piped networks

The functionality of a piped network, consisting not only of the pipes themselves but rather the connected structures, manholes and machinery necessary to keep the system operating, is often described as 'collect and convey'. This view, however, is too narrow and not in line with the objectives given by, for instance, the EN 752 (2008), which defines four objectives for sewers and drains:

(1) Public health and safety.
(2) Occupational health and safety.
(3) Environmental protection.
(4) Sustainable development.

With respect to the above four objectives functional requirements as shown in Table 6.1 can be defined. These range from the more familiar functions, such as flood protection and flow conveyance to functions that historically have not received a lot of attention, such as aesthetics, odours or safety.

To quantify the contribution of the condition of an asset to an objective (as listed above), an important (but not exclusive) component is the condition assessment. No definitive list of condition indicators can be presented as this will depend on the objective chosen. However, two different approaches to aid in choosing indicators are proposed:

- The EN 752 (2008) recommends four specific conditions for consideration:
  - The hydraulic condition allows the assessment of whether a pipe continues to be able to convey the design flow, which could have been reduced because of infiltration, illicit connections, or surface deterioration.
  - The environmental condition allows the assessment of whether the pipe condition could lead to water escaping and polluting the adjacent soil or downstream water bodies, mostly because of exfiltration.
  - The structural condition considers all defects that could affect the structural integrity of the pipe ultimately leading to collapse.
  - The operational condition allows the assessment of how the condition of a pipe may lead to an increased number or intensity of operational activities because of, among others, blockages, pumping station failures, and gravel, sand, or sediment deposits.

Table 6.1 Relationship between objectives and functional requirements.

Clause No	Public Health and Safety	Occupational Health and Safety	Environmental Protection	Sustainable Development
Protection from flooding	XXX	XX	XXX	–
Maintainability	XX	XXX	XX	XX
Protection of receiving surface waters	XXX	X	XXX	XX
Protection of groundwater	XXX	–	XXX	XXX
Prevention of odours and toxic, explosive and corrosive gases	XXX	XXX	XXX	XXX
Prevention of noise and vibration	XX	XXX	X	X
Sustainable use of products and materials	–	–	XX	XXX
Sustainable use of energy	–	–	XX	XXX
Structural integrity and design life	XXX	XXX	XXX	XXX
Maintaining the flow	XXX	–	XXX	X
Watertightness	XXX	X	XXX	XX
Not endangering adjacent structures and utility services	XXX	XXX	X	XX
Inputs quality*	XX	XXX	XXX	XX

Source: Adapted from (EN 752, 2008). XXX is high, XX is medium, X is low and – is unrelated.

*The drain and sewer system can be designed to receive both domestic and non-domestic wastewater inputs. According to EN 752 (2008), inputs quality corresponds to the 'quality of the non-domestic inputs', and such quality 'shall be controlled so that they do not compromise the integrity of the fabric of the system or its function or constitute a danger for the environment'.

- A more exhaustive list of condition indicators (called dysfunctions) has been proposed by Le Gauffre *et al.* (2007a). These indicators assess if the pipe condition could lead to:
    - ABR – ongoing degradation from abrasion,
    - BLO – blockage,
    - COL – risk of collapse,
    - COR – ongoing corrosion,
    - CSO – excessive spillage,
    - EXF – exfiltration (seepage loss),
    - INF – infiltration or exfiltration,
    - FLO – flooding,
    - HYD – decrease in hydraulic capacity (flow reduction),
    - ROO – ongoing degradation from root intrusion,
    - SAN – sand silting, and
    - SPD – Destabilization of ground-pipe system.

Depending on the situation, the list proposed by Le Gauffre *et al.* (2007a) may be too detailed, or the list proposed by EN 752 (2008) too limited to connect the individual contribution to a specific function (or dysfunction).

Criteria that can be used to rank assets for rehabilitation activities can be of many different natures, ranging from technical to social to environmental aspects (e.g., Tahmasebi Birgani *et al.*, 2013; Tarigan *et al.*, 2018). For instance, a spill in highly urbanized areas may seriously threaten public health (Carriço *et al.*, 2012). Le Gauffre *et al.* (2007a) proposed an exhaustive set of rehabilitation criteria for pipes based on the following impacts:

- DAB – damage to (important) buildings (including seepage of water into basements),
- NUH – nuisance of a hydraulic nature (e.g., service interruptions and flooding),
- OCP – excessive treatment plant operating costs,
- OCS – excessive network operation costs (including the cost of shortened lifetime of equipment),
- POG – pollution of ground and groundwater,
- POL – pollution of surface water resources, due to overflows, spills or disturbance to treatment system processes,
- SLC – shortened lifetime costs, along with the surplus costs associated with remedial actions,
- TRA – traffic disruption (e.g., because of the needs of sewer operations such as pipe cleaning).

A comparison of the two above-mentioned approaches (EN 752, 2008; Le Gauffre *et al.*, 2007a) is presented in Table 6.2.

6.2.2 The case of stormwater control measures

SCMs are fundamentally different from drains and sewer pipe networks. They are increasingly called NBS but have inherited various names depending on the country of origin, the water-related focus and the underlying functions or concepts (Fletcher *et al.*, 2015): alternative techniques, best management practices, green infrastructure, to name only a few. Accordingly, over the last six decades, drainage systems have evolved and become sophisticated beyond the traditional characteristics of piped networks by considering and addressing a growing number of objectives, as shown in Figure 6.1.

This growing number of objectives stems from the need for urban water management to adapt to our evolving urban developments. The driving forces of these changes are numerous and diverse: social (e.g., increasing expectations regarding levels of service), societal (e.g., increasing complexity of regulations and institutions), environmental (e.g., climate change and its consequences), technological (e.g., including new monitoring capabilities), economic (e.g., global markets, cost sharing among stakeholders), and most importantly, the urbanization process itself which has had a significant impact on the urban water cycle (Chocat *et al.*, 2007). The pressures exerted on our urban areas,

Table 6.2 Requirements of EN 752 (2008) related to the French RERAU approach (Le Gauffre *et al.*, 2007a) dysfunctions and impacts (Ahmadi *et al.*, 2014a). RERAU ('REhabilitation des Reseaux d'Assainissement urbain' or urban sewer networks rehabilitation) is a French sewer AM approach (tools and methods) dedicated to pipe condition assessment, inspection prioritization and rehabilitation programme elaboration.

Functional Requirement of EN752 (2008)	Related RERAU Dysfunction	Related RERAU Impacts
Protection from flooding	HYD, BLO, FLO	NUH
Maintainability	–	–
Protection of receiving surface waters	INF, HYD, BLO, CSO	POL, POG
Protection of groundwater	EXF	POL, POG
Prevention of odours and toxic, explosive and corrosive gases	–	NUH
Prevention of noise and vibration	–	–
Sustainable use of products and materials	–	–
Structural integrity and design life	SPD, ROO, COR, ABR	SLC
Maintaining the flow	BLO	POL, POG, NUH, OCS, OCP
Water tightness	INF, EXF	POL, POG
Not endangering adjacent structures and utility services	COL	DAB, TRA
Inputs quality*	–	–

*The drain and sewer system can be designed to receive both domestic and non-domestic wastewater inputs. According to EN 752 (2008), inputs quality corresponds to the 'quality of the non-domestic inputs', and such quality 'shall be controlled so that they do not compromise the integrity of the fabric of the system or its function or constitute a danger for the environment'.

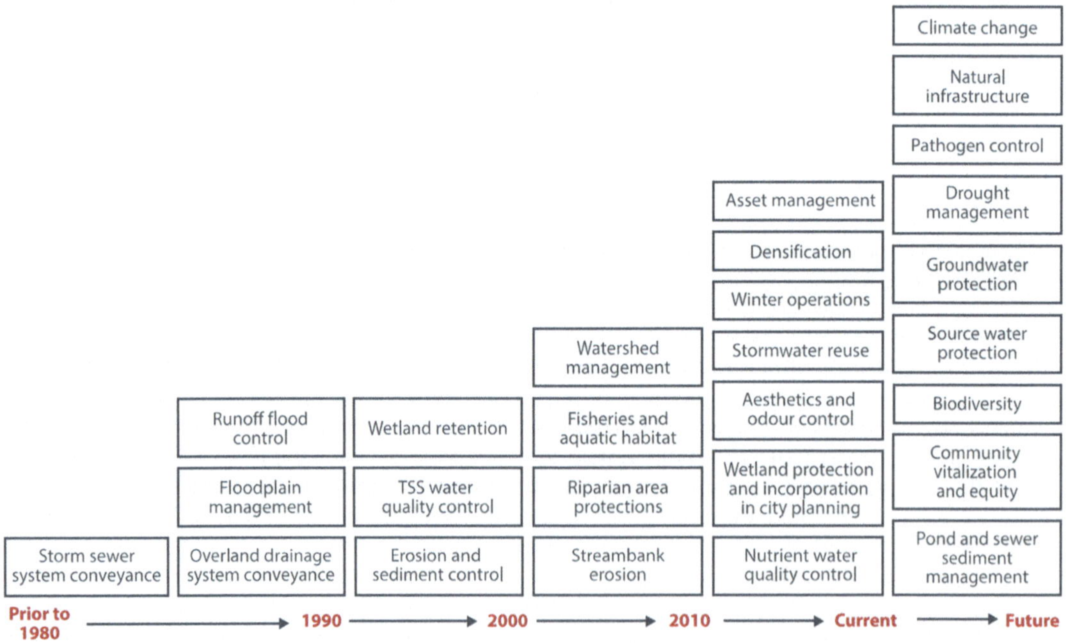

Figure 6.1 Increasing integration and sophistication of stormwater management considerations for the city of Calgary (2023).

Figure 6.2 Generic representation of services delivered by the sustainable urban water management system (Belmeziti *et al.* 2015).

on the associated urban water systems, and on the natural environment have necessitated the consideration of different ways of managing the urban water cycle. Belmeziti *et al.* (2015) proposed an exhaustive representation of the services delivered by sustainable urban water management systems, see Figure 6.2.

The representation of Figure 6.2 is suitable for application to SCMs and can be used to describe their functionality. In this respect, Cherqui *et al.* (2013) described the performance indicators that should be assessed when considering SCMs, as illustrated in Figure 6.3. Figures 6.2 and 6.3 are complementary in that the former presents a global overview of SCMs as a system that is integrated within sustainable urban water management, while the latter misses the integration within the urban landscape but focuses more on a specific system: it is thus better suited for the evaluation of individual assets.

6.3 FROM CONDITION TO PERFORMANCE ASSESSMENT

Figure 6.4 proposes a representation of AM approaches dedicated to prioritizing the rehabilitation of sewer assets (adapted from Ahmadi *et al.*, 2014b). Figure 6.4 can also be adapted to SCMs. The

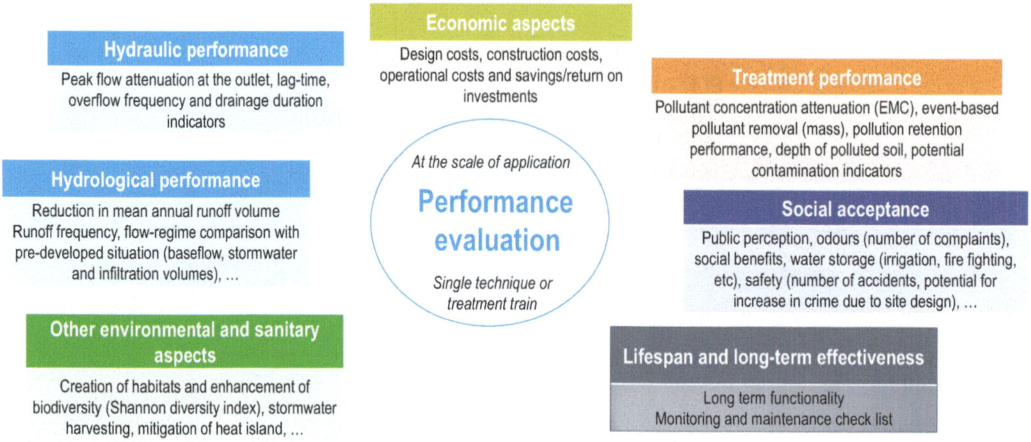

Figure 6.3 Performance evaluation of stormwater control measures (Cherqui *et al.*, 2013).

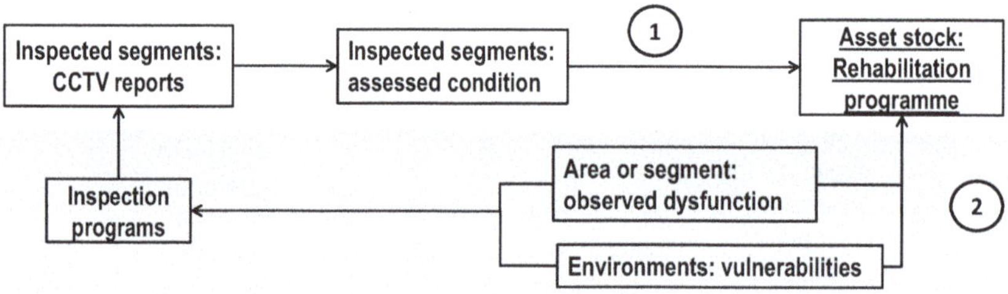

Figure 6.4 Two approaches for prioritizing assets to be rehabilitated (adapted from Ahmadi *et al.*, 2014b).

output of this approach mostly depends on the level of knowledge of the utility manager and the quality of the asset database. The simplest practice is to inspect a portion of the asset stock, then determine the condition grade of the inspected pipe based on the corresponding CCTV report, and finally, conduct a rehabilitation programme for those sewers having the worst condition rating (1). As discussed previously, this approach is too limited because the poor condition itself may not be sufficient motivation. In an evolution of this approach (2), observed dysfunctions and the network's environment variables are combined with assessed condition grades (i.e., a risk or impact-based approach): only the assets responsible for considerable negative impacts or risks are at the top of the list for rehabilitation.

Chapter 4 exhaustively presents the methods dedicated to assessing the condition of a pipe or a SCM. To move on from condition assessments, it is needed to combine the condition of the pipe with an impact-based or risk-based approach. In this section, the key concepts to enable such a risk-based approach are presented. Another way of illustrating such an approach is shown in Figure 6.5, inspired by ISO 31000 (2018) on risk management.

6.3.1 Impact-based or risk-based approaches: definitions

For budgetary reasons, it is impossible to rehabilitate all assets that are not in perfect condition (and it would not make sense to do so). Assets that need to be rehabilitated should be in a 'poor condition' AND this 'poor condition' should contribute to an '(existing) serious impact' or lead to a '(future) serious risk'.

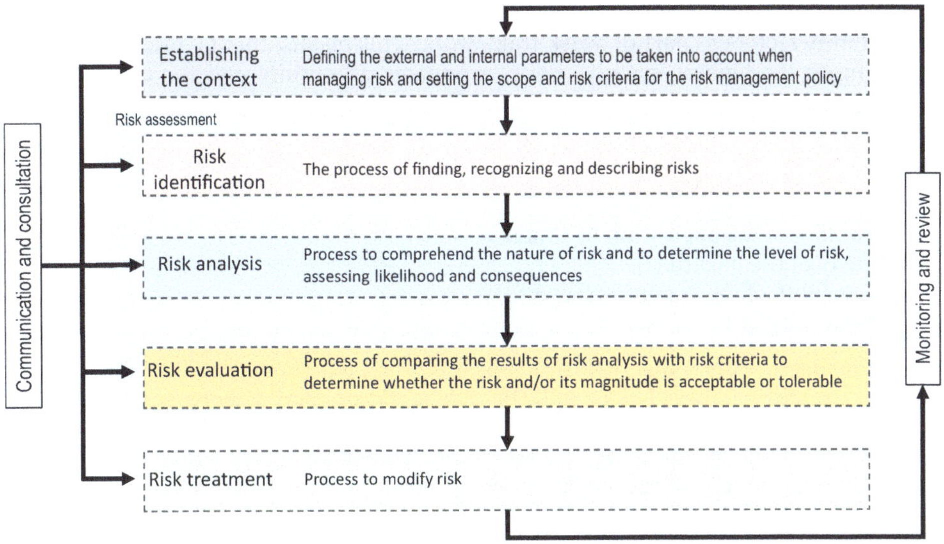

Figure 6.5 An overview of a risk management approach (inspired from ISO 31000, 2018).

There is a subtle nuance between impact-based approaches and risk-based approaches. Impact-based approaches are designed to target the assets responsible for an existing impact (Figure 6.6).

In risk-based approaches, the impact has not happened (yet), and one targets those assets whose poor condition will most probably lead to unwanted consequences: for example, as shown in Figure 6.6, cracks in a sewer pipe could lead to combined sewer overflows (CSOs), because of groundwater infiltration. Such CSOs, when they happen, will most probably pollute the downstream aquatic environment, thus leading to increased fish mortality. As part of impact-based approaches, one observes an existing impact, and the objective is to find the pipe(s) responsible for this impact: for example, as shown in Figure 6.6, when one observes fish mortality due to CSOs one may arrange a CCTV campaign within the upstream pipes to detect the responsible pipe(s) (i.e., the ones with the highest infiltration). Both approaches (i.e., impact-based and risk-based) are required to prioritize

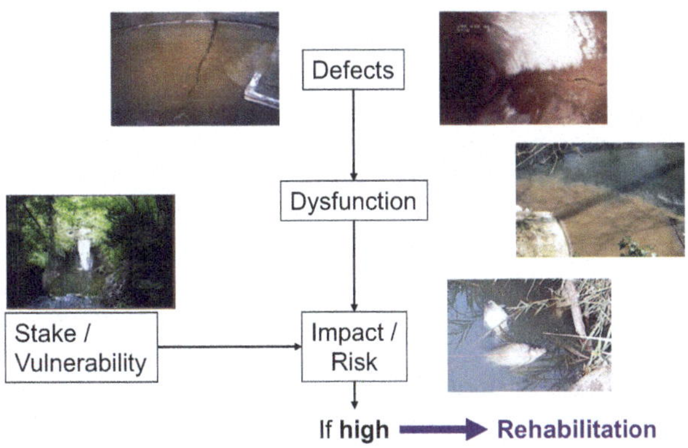

Figure 6.6 Risk or impact-based approach: for example, sewer pipes are targeted for rehabilitation because their condition leads to a risk of pollution (or existing pollution) of a water body.

rehabilitation needs. For both approaches, it is highly recommended to combine condition grades with stakes/vulnerability factors as part of a risk framework. This means, that by prioritizing the renewal of assets leading to the highest negative risks or impacts, the utility and the community it serves will avoid high economic, social and environmental costs. Accordingly, risks can be defined as the combination of the consequences of failure and the likelihood or probability that such a failure will occur (ISO 24765, 2017; ISO 55000, 2014). As stated by Tira (1997), risk can be described as a function of four components (R = f(P,I,V,E)) including:

- The probability of the apparition of a hazard (P).
- The intensity of the hazard (I).
- The vulnerability of the exposed element (V).
- The element value (E).

If we consider for instance the risk of an overflow or spill event, the hazard depends on the probability that the pipe will overflow or spill (P), the overflow or spill volume and the type of effluent (I). The risk depends on the hazard and the element at risk: does it overflow or spill directly into a basement or outside of the dwelling (V) and does that element constitute a high value or a low value (E)? Most of the literature considers only three components (see e.g. Chen *et al.*, 2020) by merging V (vulnerability) and E (stake).

Generally, consequences comprise social, economic and environmental impacts that may occur because of a failure in the system. For determining a pipe's failure probability, several variables should be used, including the pipe's age, material, length, depth, diameter, slope, number of failures and soil type. A comprehensive list of risk factors associated with drainage systems can be found in literature (e.g., Ba *et al.*, 2021; Davies *et al.*, 2001; Laakso *et al.*, 2018). It is noteworthy that those assets with the potential to significantly impact the delivery of network objectives are called 'critical assets' (ISO 55000, 2014) which should preferably never fail. Nevertheless, there will always be some likelihood of failure. This implies the necessity to accept some level of risk as its elimination is neither feasible nor affordable. However, the degree of risk should be kept as low as reasonably practicable (Frangopol, 2011).

6.3.2 Existing literature on the risk approach

As AM is moving away from reactive strategies, risk-based approaches are increasingly getting traction over purely impact-driven ones. A vast amount of research has been conducted in the field of risk-based AM, resulting in the development of a multitude of tools and methods. For example, Table 6.3 presents references that combine the assessment (or prediction) of a condition with other indicators for sewer pipes, under a risk-based framework. In a more recent attempt, Ba *et al.* (2021) proposed a comprehensive risk assessment framework for drainage pipes, taking human, management, pipeline attributes and environmental factors into account while calculating the probability of accidents (see Figure 6.7). However, to estimate the consequences, only a limited number of them were considered and assessed in detail. Publications to date have mainly focused on piped networks, while the very few studies available for SCMs have a distinct focus on flood risk reduction (e.g., Sörensen & Emilsson, 2019). The last decade has also seen the emergence of software solutions (e.g., Debères *et al.*, 2012; Halfawy *et al.*, 2008) dedicated to risk-based approaches, combining pipe condition or CCTV reports (when the tool has a pipe assessment protocol) with a broad range of indicators related to the surroundings of the pipe (which have often been extracted from a municipality's geospatial information system). It should be emphasized that there are some complexities involved in determining the direct costs of repair or replacement activities, due to the method selection process. Even more uncertainty, difficulty, and subjectivity are encountered during the estimation of indirect costs of pipeline failure in monetary terms as they are intangible in nature (Salman & Salem, 2012). Accordingly, one major impediment to the adoption of such methods and tools is the lack of data, although approaches with valuations from survey data have been tested (Rozan *et al.*, 2017). In an urban environment, it is also

Table 6.3 Selected risk-based approaches regarding sewer asset management.

Reference	Impacts/Risk Considered	Decision Support Method	Case Study
Anbari *et al.* (2017)	Wastewater inflow to river, lake or water distribution network, overflow and service disruption	Weighted average and fuzzy inference	Tehran, Iran
Baah *et al.* (2015)	Roadway type, intersecting a railway, proximity to hospital or school, proximity to river, proximity to park or recreational areas	Weighted sum using a geographical information system (ArcGIS)	'Mid-sized community', Ontario, Canada
Debères *et al.* (2012)	Groundwater and soil quality, deterioration, operating costs, flooding, traffic and urban disturbance	Outranking method (ELECTRE TRI) and a simple procedure called 'thresholds method'	Caen-la-Mer, France
Egger and Maurer (2015)	Rainfall variability, sewer deterioration, preference uncertainty, socio-economic and land use development, groundwater contamination, sewer failure and collapse, flooding weighted by location and cost increase	Multi-criteria decision analysis (MCDA)	Two small towns, Kanton of Zurich, Switzerland
Elsawah *et al.* (2016)	Type of soil, number of road lanes, land use and function of the pipe	Risk matrices	Montreal, Canada
Ennaouri and Musandji (2013)	Type of soil, traffic, groundwater, exfiltration, infiltration and hydraulic capacity	Analytic hierarchy process (AHP)	Saint-Hyacinthe, Canada
Hahn *et al.* (2002)	Human health, environmental, commerce and traffic impacts	Bayesian belief network	Input from a US national group of experts from both the public and private sectors
Halfawy *et al.* (2008)	Sewer type, sewer function, land use and road classification	Simple weighing	City of Regina, Canada
Korving *et al.* (2009)	Cost of environmental damage due to CSOs and investment costs	Optimization	De Hoven catchment, The Netherlands
Kuliczkowska (2016)	Water table level, intensity of road traffic, backflow or overflow of wastewater	Simple weighing	Poland
Lee *et al.* (2021)	12 factors to consider: physical factors (e.g., diameter, length and depth), social factors (e.g., commercial or residential land uses) and environmental factors (e.g., proximity to roads, river and forest)	Risk matrices	Seoul, South Korea
Rozan *et al.* (2017)	Intangible goods and inhabitants' preferences	Assessment of inhabitants' willingness to pay	Strasbourg, France
Salman and Salem (2012)	Proximity to water sources, pipe function, landslide potential, roadway type and building type	Weighted scoring and fuzzy inference	Cincinnati, USA
Tscheikner-Gratl *et al.* (2014)	Flood risk and opportunities connected to urban development	Weighted scoring system	'Medium-sized Alpine city', Austria
Vladeanu and Matthews (2019)	14 factors, including pipe characteristics and demographics, are integrated into a hierarchical model to calculate one consequence-of-failure score.	AHP	Northwestern Louisiana, USA

Source: adapted from Tscheikner-Gratl *et al.* (2019).

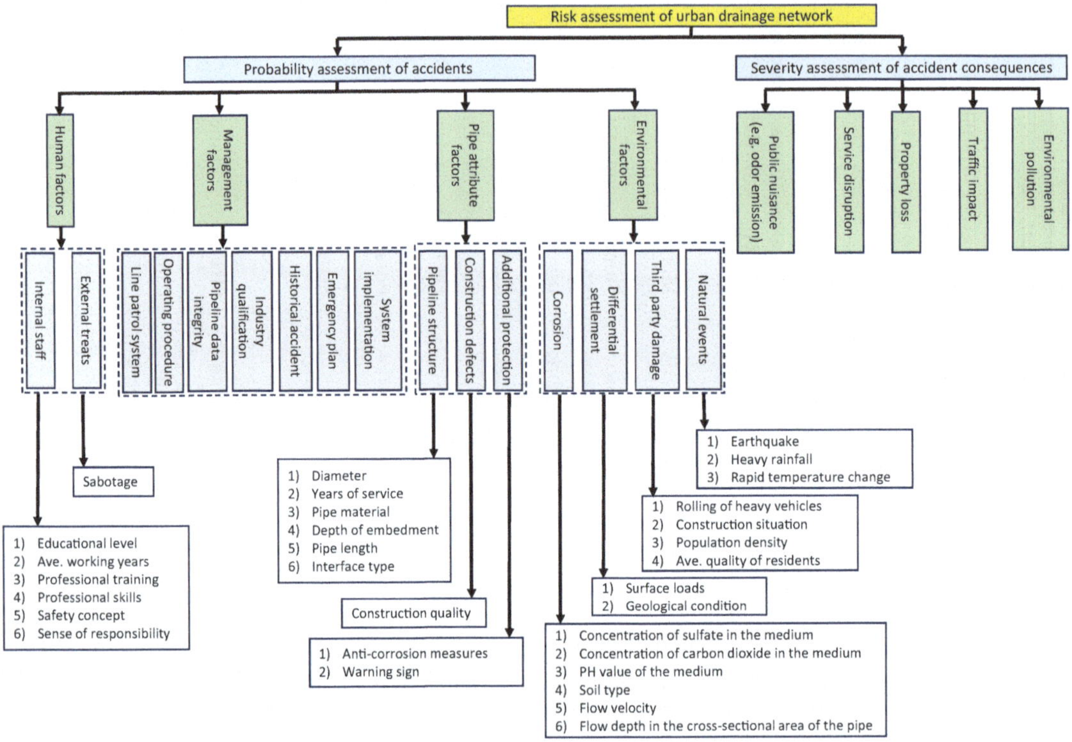

Figure 6.7 Hierarchy of risk factors of drainage pipelines (Ba *et al.*, 2021).

important to note that risk-based approaches that focus solely on one type of infrastructure may miss parts of the overall picture (e.g., the influence on the risks of other adjacent types of infrastructure and vice versa). Examples of approaches trying to incorporate those risks into an integrated approach and the limitations and challenges of such attempts are described by Daulat *et al.* (2022).

6.3.3 Risk assessment

Another important aspect of a risk framework is its evaluation (see Figure 6.5). The aim of any risk evaluation tool is to ensure that the decision process is transparent, well documented, based on the best available knowledge and reflects a common understanding among stakeholders. In this regard, risk evaluation methods are typically based on a combination of indicators by means of a limited number of 'if … then …' rules. One of the most famous and simplest risk evaluation structures is the risk matrix. Basically, it is a simple two-dimensional map of probability and consequences of failure ratings corresponding to each risk (see Figure 6.8).

For both probability and impact ratings, a risk matrix is generally developed using a Likert scale and divided into various discrete risk exposure zones, such as high, medium, and low (Aven, 2017). For instance, if an asset failure triggers high consequences, but has a low probability of failure, this can result in a low risk of failure (see Figure 6.8). Categories rather than numerical values are preferred by both risk specialists and laymen for understandability as well as visualization of the uncertainties in risk statements (Duijm, 2015). In his paper, Duijm (2015) proposed a probability consequence diagram with continuous scales for likelihood and consequence as an alternative to the risk matrix with discrete categories. Such representation can address some disadvantages of discrete

Consequences

Likelihood of failure	Insignificant	Minor	Moderate	Major	Catastrophic
Very Likely	Low	Moderate	High	Very High	Very High
Likely	Low	Moderate	Moderate	High	Very High
Possible	Low	Low	Moderate	High	High
Unlikely	Very Low	Low	Low	Moderate	High
Rare	Very Low	Very Low	Low	Moderate	High

Figure 6.8 Example of a risk matrix.

risk matrices (such as risk ties), but it should be noted that the risk appetite of a decision-maker or an organization is reflected in the partitioning scheme such that different stakeholders (organizations) perceive risk criticality differently (Ruan *et al.*, 2015). The French RERAU method (Le Gauffre *et al.*, 2007a) uses 4 × 4 tables defining how to combine couples of indicators, each one being evaluated on a four-grade ordinal scale.

A review of earlier discussions on risk matrices highlights the following aspects to take into consideration:

- The consistency between risk matrices and quantitative measures, and, consequently, the appropriateness of decisions based on them (Cox, 2008; Levine, 2012).
- The use of a subjective classification of consequence and probability (Cox, 2008; IEC 31010, 2019; Smith *et al.*, 2009).
- The definition of risk scores and the relationship between them and the scaling (linear or logarithmic) of the categories (Franks & Maddison, 2006; Levine, 2012; Ni *et al.*, 2010; Ruge, 2004).
- The limited resolution of risk matrices – leading to 'risk ties' (Cox, 2008; Ni *et al.*, 2010).
- The aggregation of scenarios, consequences, and risks (Franks & Maddison, 2006; IEC 31010, 2019).
- The issues associated with using universal or corporate-wide risk matrices (IEC 31010, 2019).
- The uncertainties related to indicators (components of risk) assessment and the need to consider them (Le Gauffre & Cherqui, 2009).

Rehabilitation based on risk criteria allows for a degree of compensating for uncertainties associated with the condition grading methods by combining them with other types of information (e.g., observed overflows or infiltration); however, rehabilitation criteria will inevitably contain uncertainties linked to data and to knowledge or decision rules. Le Gauffre and Cherqui (2009) proposed to convert crisp aggregation tables into fuzzy aggregation rules associated with the use of fuzzy indicators. Fuzzy indicators allow dealing with the uncertainty linked to raw data (e.g., a score calculated after visual inspection, based on reported defects), or linked to calibration processes (e.g., the uncertainty linked to the hypotheses used for defining a cut-off value). Moreover, fuzzy rules (i.e., rules with several conclusions) allow one to address imprecise human reasoning during the combining phase of complementary indicators into a more synthetic indicator. Examples provided tend to demonstrate that using a fuzzy framework, instead of crisp indicators and crisp rules, improves the risk assessment process by providing finer and more robust conclusions. To see more practical examples in the field of UDAM, see Salman and Salem (2012) and Anbari *et al.* (2017). On the other hand, a major drawback of using fuzzy indicators is the inherent complexity of such an approach and thus the difficult appropriation by the utility manager: tools must be proposed to facilitate the use of this framework.

6.3.4 Multi-criteria decision analysis

AM often combines multiple risk assessments when considering the different impacts and the associated hazards. For example, EN 752 (2008) lists a series of requirements for drainage and sewer systems that directly correspond to impacts (see Table 6.1). Ahmadi *et al.* (2014a) also proposed a list of risks or impacts for drainage and sewer systems. To calculate the relative importance of the assessed risks, multi-criteria decision analysis (MCDA) has been widely applied to structures to support such types of decisions in infrastructure AM. The philosophy behind this analysis is to rank (or classify) solutions based on a set of different (and sometimes difficult and almost incomparable) criteria. As such, MCDA is, by definition, a structured approach to measuring the performance of multiple alternatives against a set of (often conflicting) criteria as part of the decision-making process (Achillas *et al.*, 2013; Hajkowicz & Collins, 2007). For example, when considering a list of SCMs with varying conditions and different risk dimensions (e.g., pollution of groundwater or flooding a basement), one should have a method to rank rehabilitation priorities and to justify this ranking. A common limitation of multi-criteria decision analysis applications in UDAM is that usually only aggregation and evaluation models are used. As with multi-criteria analysis and indicator assessment, the need for consideration of the decision support process and elicitation and inclusion of decision-maker preferences is often overlooked (Tscheikner-Gratl *et al.*, 2019).

Still, methods in this category can assist in the decision-analysis process when more than one criterion, or attribute, is being evaluated at the same time (Lai *et al.*, 2008). Here, each criterion can be given a relative weight, using decision analysis tools. Applying the assigned weights and combining the weighted performance ratings, the alternatives are ranked considering the selected criteria. By utilizing these methods, complex decision-making processes can be made more transparent, auditable and analytically rigorous (Roy, 2016). A variety of MCDA approaches exist with different levels of complexity and based on different assumptions. The available methods can be categorized into three schools of thought (Løken, 2007):

- Value measurement models: each alternative is assigned a numerical score. Moreover, each criterion is assigned a weight 'w', which indicates its importance (e.g., weighted sum model, analytic hierarchy process).
- Goal, aspiration, and reference level models: these models measure how well alternatives achieve predetermined goals (e.g., VIKOR (Mardani *et al.*, 2016), TOPSIS (Birgani & Yazdandoost, 2018)).
- Outranking models: these methods compare alternatives pairwise, determining the strength of preference for one over another (e.g., ELECTRE (Le Gauffre *et al.*, 2007b), PROMETHEE (Behzadian *et al.*, 2010)).

Despite the development of diverse methods, there is still a knowledge gap about how the selection of a decision-support approach itself affects the outcome. Tscheikner-Gratl *et al.* (2017) provided an overview of the various methods that have been developed for application in an integrated rehabilitation management scheme and analyzed their suitability for urban water systems. The study investigated whether different methods lead to different priorities when rehabilitating street sections with their underlying networks. Their findings revealed that due to different score scales, weights and distributions of scores within the criteria, not all methods provide the same ranking. However, the decision support method has less influence on the outcomes than the definition of the criteria and the related score scales. MCDA methods are only part of the decision (i.e., they are decision-aid tools): the final decision relies on the humans themselves (and thus on the method settings), as discussed in Chapter 7. Independently of the method used, the user must be familiar with its machinations and strengths but also its weaknesses. A well-applied and fully understood simpler method will provide a better defendable and explainable result than a poorly applied, more sophisticated method with similar data availability.

MCDA methods have been widely used for various purposes in the field of planning and management of UD systems. For example, Saadat Foomani and Malekmohammadi (2020) presented a method for locating SCM systems at optimal locations. Three systems were selected for this study: infiltration trenches, vegetated swales and bioretention areas. The first step was to identify criteria that were effective in finding locations for these systems. Multiple layers were combined using fuzzy logic and the analytical hierarchical process in GIS to select optimal locations for these three systems. A final location was determined by considering the specific land uses in the region. A study by Loc *et al.* (2017) assessed the applicability of SCMs as flood control measures, in Ho Chi Minh City, Vietnam, using a multi-criteria decision analysis. They applied four different procedures: Borda count, pair-wise voting, range of value, and analytic hierarchy process. Despite some differences, the integrated results generally agreed (not surprisingly) that urban green space is the most beneficial flood control measure in that area. Furthermore, Bakhshipour *et al.* (2021) developed a generic solution within an MCDA platform to (1) facilitate the optimization of hybrid (de)centralized UD infrastructure involving multiple decisions with conflicting objectives (i.e., reliability, resilience, sustainability, and construction costs); (2) examine the relationship between performance indicators and system configuration; and (3) reduce conflicts between optimization analysts and decision makers through different stages of planning. First, all those hybrid UD systems were designed using multi-objective optimization. Finally, the solutions were comprehensively analyzed and ranked. As demonstrated by the results, the proposed framework can handle multiple decisions, objectives, and indicators for a complex optimization problem in a reasonable amount of time while providing realistic solutions. This and very similar problems have been addressed in recent literature (e.g., Lin *et al.*, 2021; Wang, 2022).

6.4 CASE STUDIES

6.4.1 Deploying the INDIGAU decision support system in Caen-la-Mer urban community, France

This case study is based on an article by Debères *et al.* (2012). It presents an implementation of the INDIGAU AM toolbox based on the French national RERAU methodology (Le Gauffre *et al.*, 2007a). This toolbox allows a utility manager or consultant to define a rehabilitation programme using CCTV reports and other indicators such as economic, social or environmental indicators. The INDIGAU toolbox was applied to a length of about 99 km of sewers (covering about 10% of the network) within the Caen-la-Mer urban community in France.

As part of the first step, raw CCTV data were assessed (with inspection reports coded using the EN 13508-2 (2011) coding system) and entered into the GIS. The assessment was used to generate sewer dysfunction indicators scores, using interpretation models, on a four-grade scale (grade 1 corresponding to the lowest risk). The calibration of the interpretation model against an expert-opinion database removed the risk of single-human bias (Cherqui *et al.*, 2008).

A set of rehabilitation criteria was chosen in discussion with the utility manager (i.e., the City of Caen-la-Mer). These criteria are presented in Table 6.4

Each criterion combined (at least) a dysfunction and an impact, with their definitions adapted to the context of the case study. Their importance (corresponding to the weights) was defined by stakeholders through multi-criteria analysis. Here, we present three examples: one environmental criterion (R/PON1/EXF), one economic criterion (R/CXR3/INF) and one social criterion (R/TRA1/EFF).

For the environmental criterion R/PON1/EXF, three basic indicators were employed as shown in Figure 6.9. First, a basic indicator representing the water tightness deficiency for exfiltration estimated by visual inspection (EXF4) and the risk factors for exfiltration (EXF2) were combined (Figure 6.10) resulting in the risk of exfiltration (EXF6). Then this composite indicator was combined with the vulnerability of the soil and groundwater to pollution (V-SN-PON).

Table 6.4 Rehabilitation criteria descriptions used to prioritize sections to be rehabilitated and their importance weights.

Abbreviation	Criterion Description, Considering the Consequences of:	Weight
R/PON1/EXF	Exfiltration on groundwater and soil quality deterioration	0.17
R/PON2/HYD	Reduction in section hydraulic capacity on groundwater quality deterioration	0.09
R/CXS/INF	Infiltration on operating costs of the WWTP	0.05
R/CXR1/ENS	Silting on cleansing cost	0.05
R/CXR2/BOU	Blockage contributing to an extra cost of network operation	0.13
R/CXR3/INF	Infiltration on network operation surplus cost	0.14
R/DEB/ENS	Siltation leading to flooding risks	0.05
R/DEB/BOU	Blockage leading to flooding risks	0.05
R/DEB/HYD	Reduction in section hydraulic capacity on flooding risks	0.05
R/TRA1/EFF	Sewer possible collapse on traffic and urban disturbance	0.22

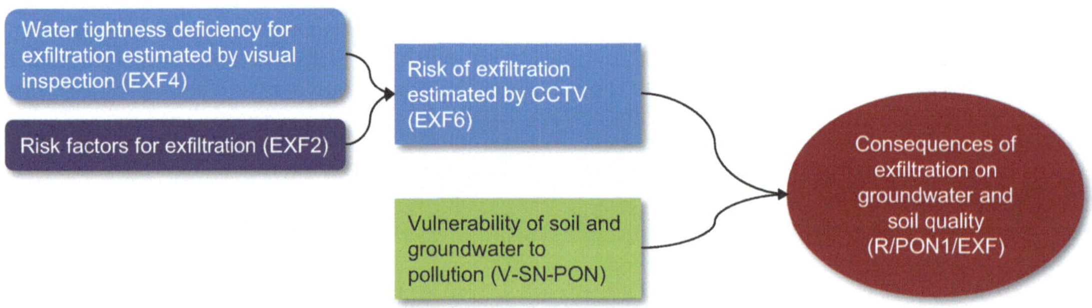

Figure 6.9 Consequences of exfiltration on groundwater and soil quality criterion construction procedure.

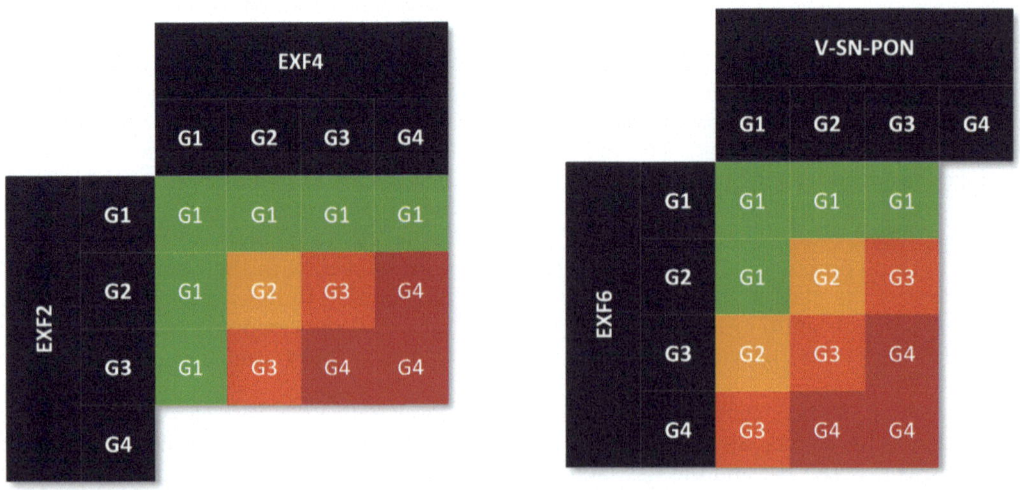

Figure 6.10 EXF6 and R/PON1/EXF evaluation matrix.

For instance, when assigning Grade G1 to EXF6 (i.e., one of the sublevel indicators from EXF2 or EXF4 is equal to G1), means that there is no pollution risk due to exfiltration because the segment is in good condition, or the segment environment does not allow any exfiltration to occur. As mentioned above, most of the work is concerned translating the existing data into a useful format using the INDIGAU toolbox. The translation procedure for assessing the EXF2 criterion into ordinal levels is illustrated hereafter. All pipe segments according to their position with respect to the groundwater level were regrouped. They were classified into Group 1 when the groundwater usually reaches the segment, Group 2 when it occasionally reaches the segment, and finally Group 3 when groundwater never reaches the segment (Figure 6.10). As the depth of all segments was not available for the entire sewer system that had been inspected, a systematic rule was applied to decide in which group each segment would be classified based on the groundwater level (highest level observed): G1 corresponds to groundwater levels between 0 and 1 m below the ground surface, G2 for a level between 1 and 5 m below the ground surface and G3 for a level more than 5 m below the ground surface (Figure 6.11).

Concerning V-SN-PON, the area was divided into three ordinal grades: G3 was applied when the segment was in a protected area; G2 was applied when the presence of groundwater can lead to a bigger pollution area and finally, G1 was applied when groundwater does not exist, or it is deep enough to be ignored.

Figure 6.12 shows an example of an economic criterion which considered the consequences of infiltration on the network operation surplus cost R/CXR3/INF. The pumping cost was calculated by using the monitoring within the network for gauging the infiltration water volume. Subsequently, this cost vulnerability indicator was combined with an indicator representing the infiltration volume in the catchment. This latter was also combined with the risk of infiltration estimated by the CCTV indicator.

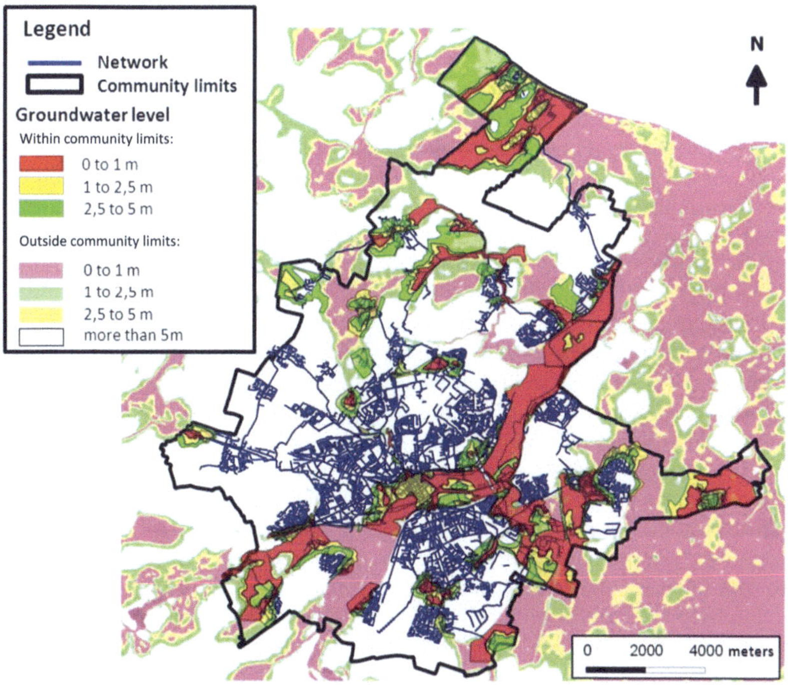

Figure 6.11 Groundwater level map for the studied area.

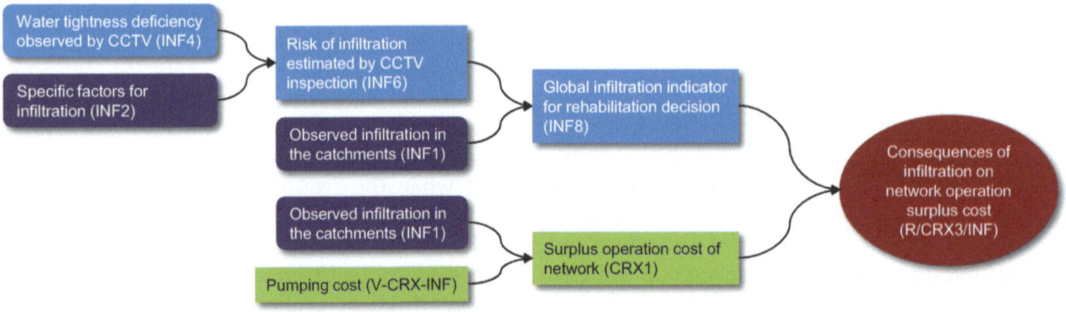

Figure 6.12 Consequences of infiltration on network operation surplus cost criterion construction procedure.

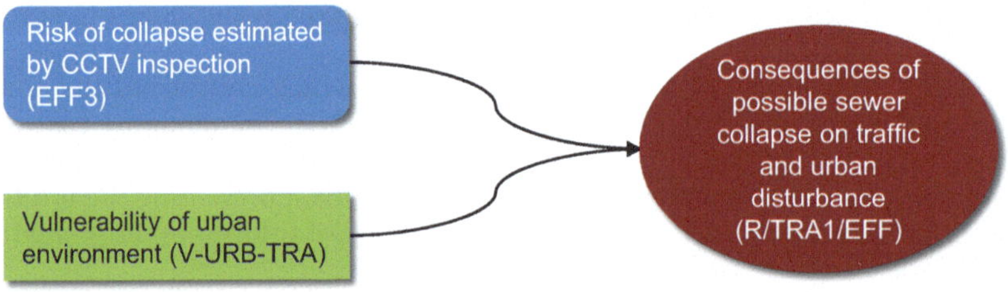

Figure 6.13 Consequences of possible sewer collapse on traffic and urban disturbance criterion construction procedure.

The social R/TRA/EFF criterion considered the consequences of a possible pipe collapse with respect to traffic and urban disturbance (Figure 6.13). It was defined as a combination of six indicators as per the RERAU methodology. However, a simplified version had to be implemented in this case study. The simplification was necessitated due to the lack of data on the interaction between the pipes and the ground (i.e., the ground-pipe complex), chemical attacks or abrasion. The risk factor for sewer collapse was also neglected for the same reason. This simplification was agreed upon with the stakeholders.

To prioritize the segments to be rehabilitated, two multi-criteria analysis modules have been integrated into the INDIGAU toolbox: ELECTRE TRI (Rogers *et al.*, 2000; Yu, 1992) and a simple procedure called the 'thresholds method' (Cherqui *et al.*, 2010). This method uses two threshold values (α, β) to define three priority levels: Green, Orange and Red (successively from low to high priority). Table 6.5 shows an example of an assessment with the threshold method.

For instance, with $\alpha = 15\%$ and $\beta = 30\%$, the sewer segment assessed in Table 6.5 was assigned a 'Red' priority level (i.e., since $W3 + W4 = 0.43 > 0.30$). α and β can control the number of segments in each priority level: α controlled the number of 'green' segments and β controlled the proportion of 'red' and 'orange' segments. The simplicity of this approach made it understandable and helped the stakeholders to limit the number of segments to be prioritized.

A multi-criteria analysis helped to prioritize the sections in terms of rehabilitation needs from 'most urgent' to 'not-needed' ones. No constraint was given by the utility in relation to the various unknowns that can help reformulating a rehabilitation programme like:

- Duration of works.
- Annual rehabilitation budget.
- A specific length of network to be rehabilitated.

Table 6.5 Assessment of a Segment with the 'Threshold Method'.

Criterion	Grade	Weight	Synthesis	Rules to Define Priority Level
R/CXR3/INF	G4	0.14	W4 = 0.14	[(W4 > α) or
R/CXS/INF	G3	0.05		(W3 + W4 > β)] => Red
R/DEB/HYD	G3	0.05	W3 = 0.32	
R/TRA1/EFF	G3	0.22		(W3 + W4 > α) => Orange
R/PON2/HYD	G2	0.09	W2 = 0.09	
R/PON1/EXF	G1	0.17		
R/CXR1/ENS	G1	0.05		
R/CXR2/BOU	G1	0.13	W1 = 0.45	
R/DEB/ENS	G1	0.05		
R/DEB/BOU	G1	0.05		

Using the above approach, 5.5 and 7.7 km of the inspected network were found to correspond to the red and orange categories, which translated into 6% and 8% of the inspected network, respectively (Figure 6.14). Table 6.6 illustrates the results of the multi-criteria analysis. As mentioned above, over 10% of the assets were inspected with 13% to be rehabilitated in short and medium order.

6.4.2 Public health approaches to managing microbial quality of urban stormwater: Alberta, Canada

Although stormwater management is critical for flood mitigation, erosion control and pollution management, the water 'produced' as runoff has traditionally been viewed as a nuisance – a water source generally unfit for societal use. However, for jurisdictions facing water scarcity, climate change (e.g., drought), and rapid population growth, stormwater is increasingly being seen as an important and essential resource for promoting urban resiliency (see Figure 6.1). This shift in thinking, from 'nuisance-to-resource', has profound implications in terms of how we view stormwater management and assessing the condition of these assets – that is, there is an inherent and fundamental need to protect a resource, but by contrast, there is no need to protect a nuisance. Consequently, source protection is recognized as an emerging and important element in the condition-based management of stormwater infrastructure, particularly for jurisdictions using stormwater to supplement potable water demands. Reciprocally, the requirement to protect a resource obligates that stormwater and its assets be managed around a set of water quality criteria congruent to societal risk. In such cases, public health standards emerge as paramount importance in managing the quality of stormwater, assessing infrastructure conditions and strategically planning environmental services around these assets.

Importantly, storm events and human disease are inextricably linked, and weather is a well-known determinant affecting the prevalence of waterborne disease outbreaks (Guzman-Herrador *et al.*, 2015). A robust association exists between extreme rainfall events (e.g., defined as the 90th percentile of average precipitation), and outbreaks of waterborne gastrointestinal disease (Cann *et al.*, 2013; Curriero *et al.*, 2001; Guzman-Herrador *et al.*, 2015). In a landmark study by Curriero *et al.* (2001), 68% of gastrointestinal waterborne disease outbreaks between 1948 and 1994 in the United States were preceded by precipitation events above the 80th percentile and 51% of outbreaks were preceded by the 90th percentile. Recent studies have also found a correlation between precipitation and the incidence of respiratory, skin, and vector-borne diseases (Gleason *et al.*, 2016; Kambali *et al.*, 2021; Liang & Messenger, 2018). For example, the prevalence of Legionnaire's Disease, a respiratory disease caused by *Legionella pneumophila*, has also been associated with increased precipitation (De Giglio *et al.*, 2019; Gleason *et al.*, 2016; Mitsui *et al.*, 2021). Likewise, *Pseudomonas aeruginosa*, an

Figure 6.14 Segments and their need to rehabilitate according to their risk level.

Table 6.6 Multi-criteria analysis results.

Total Length (m)	Inspected Length (m)	Green (m)	Orange (m)	Red (m)	Inspection Ratio (%)	Total Rehabilitation Length Ratio (%)	Inspected Rehabilitation Length Ratio (%)
931,633	99,467	86,310	7741	5416	10.68	0.58	5.45

opportunistic pathogen causing skin, eye, ear, wound and soft tissue infections, as well as pneumonia, is a common and abundant microbial constituent of stormwater (Lee *et al.*, 2020; Olivieri *et al.*, 1977) and the prevalence of disease caused by this microbe has also been linked to precipitation (Ramos *et al.*, 2013; Villedieu *et al.*, 2018). Importantly, a broad spectrum of human diseases appears to become more prevalent during, and after, storm events.

This, by no means, implicates urban stormwater as a direct cause of this increased risk of disease in humans. In fact, direct human exposure to stormwater is relatively limited within the context of the urban environment (i.e., occupational exposures, urban flooding scenarios, etc.). Rather, it is the mobilization, re-distribution and concentration of pathogens in storm runoff, and the accumulation of that runoff in a receptor environment (i.e., a pond, river, lake, etc.), that is the fundamental harbinger of increased risk of disease. Urban stormwater acts as a conveyance medium for transporting pathogens to aquatic receptor bodies where human exposures subsequently occur, and consequently, microbial risks related to stormwater are largely contingent upon indirect exposures and transmission pathways – that is, the pathways by which pathogens (or their products [e.g., toxins]) traffic into the human population, the most notable being stormwater-impacted drinking water, recreational water, or irrigation water sources. Additionally, nutrients from stormwater runoff can also accentuate the growth of endogenous opportunistic saprozoic pathogens in stormwater (i.e., *Pseudomonas* spp.) (McLellan *et al.*, 2015), potentially explaining why a correlation exists between precipitation and human diseases other than gastroenteritis. Nutrient-loading from stormwater can also contribute to the growth of cyanobacteria, leading to harmful algal blooms, and which can subsequently affect drinking water, recreational water and natural water sources (e.g., lakes, rivers and even urban storm ponds themselves) (Brewton *et al.*, 2022; Lusk & Toor, 2016).

As such, stormwater is recognized as one of the most dominant sources of urban pollution affecting water quality. Urban-generated storm runoff is considered the primary source of impairment of coastal waters in the United States (Arnone & Perdek Walling, 2007) and contamination with microbes is considered the primary cause of this impaired water quality. Inland waters are also heavily contaminated by urban stormwater runoff, and microbes are a major concern (Arnone & Perdek Walling, 2007). Consequently, as a basic public health premise, stormwater should be considered as 'risky' in terms of human exposure, regardless of drainage design (i.e., CSO versus non-CSO), and due to the fact that stormwater often fails to meet public health-related water quality guidelines/standards (Ahmed *et al.*, 2019; Converse *et al.*, 2011; Hart *et al.*, 2020; Lee *et al.*, 2020; Parker *et al.*, 2010; Sauer *et al.*, 2011; Schreiber *et al.*, 2019; Sidhu *et al.*, 2012; Steele *et al.*, 2018).

In terms of traditional microbial water quality indicators, faecal indicator bacteria (FIB), such as *Enterococcus* spp. and *Escherichia coli (E. coli)*, have been found in high concentrations in both CSO (Hachad *et al.*, 2022; McGinnis *et al.*, 2018; Olds *et al.*, 2018) and non-CSO urban stormwater systems (Ahmed *et al.*, 2019; Converse *et al.*, 2011; Hart *et al.*, 2020; Lee *et al.*, 2020; Parker *et al.*, 2010; Sauer *et al.*, 2011; Schreiber *et al.*, 2019; Sidhu *et al.*, 2012; Steele *et al.*, 2018). As a result, stormwater does not always meet water quality standards or guidelines for safe water (e.g., for recreation, irrigation) (Ahmed *et al.*, 2019; Converse *et al.*, 2011; Hart *et al.*, 2020; Lee *et al.*, 2020; Parker *et al.*, 2010; Sauer *et al.*, 2011; Schreiber *et al.*, 2019; Sidhu *et al.*, 2012; Steele *et al.*, 2018). For example, in segregated stormwater systems, Converse *et al.* (2011) observed that *E. coli* concentrations exceeded single sample limits of 400 MPN/100 mL in 4 of 5 studied sites, while greater than 85% of samples exceeded single sample limits for *Enterococcus*, with these observations consistent with studies done elsewhere (Parker *et al.*, 2010; Sauer *et al.*, 2011; Schreiber *et al.*, 2019). Indeed, faecal indicator bacteria concentrations (such as for *E. coli*) can vary as low as 1 $\log_{10}$ MPN/100 mL (Parker *et al.*, 2010) to as high as 6.5 $\log_{10}$ MPN/100 mL (Sauer *et al.*, 2011) in segregated storm sewer systems. Similarly, estimates of *E. coli* can range from as low as 1.4 $\log_{10}$ MPN/100 mL (Hachad *et al.*, 2022) to as high as 6 $\log_{10}$ MPN/100 mL (Olds *et al.*, 2018) in CSO systems, though mean concentrations generally hover between 3 and 5 $\log_{10}$ MPN/100 mL, and typically violate ambient water quality criteria in the vast majority of samples (Hachad *et al.*, 2022; McGinnis *et al.*, 2018; Olds *et al.*, 2018).

The public health risks associated with poor quality urban stormwater also depend on the sources of faecal pollution impacting drainage systems. Microbial pathogens (i.e., viruses, bacteria, protozoa, and parasitic worms) transmitted via faecal–oral routes in humans are typically well-adapted to infect and colonize other humans, and consequently, exposure to stormwater (or stormwater receptor bodies) polluted with human faeces represents a high-risk exposure scenario for public health. Many studies on stormwater, particularly CSOs, have demonstrated significant health risks associated with exposure to stormwater-contaminated recreational waters (Sales-Ortells & Medema, 2015; Wade *et al.*, 2003, 2006, 2010) irrigation water (Lim *et al.*, 2015) and even drinking water (Jalliffier-Verne *et al.*, 2015). Many studies have also demonstrated that human faeces are a dominant source of microbial pollution in urban stormwater systems, even those systems separate from sanitary sewerage (Beaudry, 2019; Hart *et al.*, 2020; Sauer *et al.*, 2011; Sidhu *et al.*, 2012; Steele *et al.*, 2018). Sources of human faecal pollution in non-CSO stormwater systems can originate from residential/commercial cross connections, sewerage infrastructure failures (e.g., leaky sewer pipes located in common utility trenches with storm drains), illegal discharges (e.g., septic dumping from recreational vehicles), and/or vulnerable or homeless persons/encampments lacking the provision of waste management. In one of the most comprehensive assessments regarding sources of pollution found in stormwater systems separate from sanitary, Sauer *et al.* (2011), observed that among 45 stormwater outfalls (consisting of 845 samples) monitored over 4 years in Milwaukee, Wisconsin, that human faecal pollution could be detected in all outfalls, suggesting that human faecal contamination may be a ubiquitous and chronic problem in urban stormwater in general. The prevalence of human faecal pollution across all 45 sites ranged from 11% to 100% of samples. In another study by Hart *et al.* (2020) that assessed water quality in separated stormwater sewers of Beaufort, North Carolina, 71% (65 of 92) samples taken from two outfalls were positive for human sewage. Indeed, research carried out in Alberta, Canada, has found similar results (Carson *et al.*, 2023; Shahirnia *et al.*, 2023). Estimating the actual amount of human sewage contaminating stormwater can be done by comparing levels of microbial source tracking markers (e.g., HF183) present in municipal sewage (i.e., 7–8 $\log_{10}$ copies/100 mL – see Mayer *et al.*, 2018) to those observed in stormwater. Sauer *et al.*, (2011) found the concentration of these markers as high as 4–6 $\log_{10}$ copies/100 mL, suggesting that even in non-CSO drainage networks raw human sewage can comprise as much as 10% of stormwater flows. By contrast, a study by Hart *et al.* (2020), found the average concentration of HF183 in stormwater outfall was between 1 and 2 $\log_{10}$ copies/100 mL, suggesting 0.01% of stormwater flows were made up of human sewage. Similarly, high prevalence (73%–98%) but low average concentrations (2–3 $\log_{10}$ copies/100 mL) of human sewage were found by Lee *et al.* (2020) and Steele *et al.* (2018) in non-CSO systems. Once again, the Alberta research in this area supports these findings, noting the wide variation in the concentration and occurrence of human faecal pollution in stormwater systems in Alberta, Canada (Carson *et al.*, 2023; Shahirnia *et al.*, 2023), with variation attributable to parameters such as drainage network characteristics (e.g., residential vs. commercial), age and integrity of infrastructure, and meteorological conditions (Shahirnia *et al.*, 2023). From an AM perspective, Carson *et al.* (2023), as well as others (Ahmed *et al.*, 2020; Hachad *et al.*, 2022), have used these tools to track human faecal pollution signatures within storm drain networks to pinpoint where (and why) human faecal pollution may be intruding into storm drains (e.g., identifying cross-connections). Once identified these sites can subsequently be targeted for remediation. *This approach provides an example of strategic AM centered around protecting stormwater as a resource.*

Humans are not the only contributor to microbial pollutants in urban stormwater. Regardless of their design, all urban stormwater systems are subject to contamination from animal faeces, and several studies have demonstrated that urban stormwater often contains faecal contamination from domestic pets (Monteiro *et al.*, 2021; Olson *et al.*, 2021), as well as urban wildlife (Olson *et al.*, 2021; Steele *et al.*, 2018). As an example, dog faeces is a common contaminant of urban stormwater (Sales-Ortells & Medema, 2015; Steele *et al.*, 2018), an observation that is not surprising given that the American Veterinary Medical Association (AVMA) and the Canadian Animal Health Institute (CAHI) estimate that more than 76 million and 8.1 million dogs inhabit the U.S. and Canada (AVMA, 2018; CAHI, 2021), respectively, with the vast majority residing in urban centers. Dogs are reservoirs for a range of zoonotic pathogens including *Giardia* (Bouzid *et al.*, 2015), *Campylobacter* (Fox *et al.*, 1988; Montgomery *et al.*, 2018) and *Salmonella* (Reimschuessel *et al.*, 2017). In terms of wildlife, aquatic birds, such as geese and gulls have also been shown to be major contributors of faecal pollution to urban stormwater drainage systems (Staley *et al.*, 2018; Steele *et al.*, 2018; Williams *et al.*, 2022), and can often be infected with relatively high levels of *Campylobacter* and *Salmonella* (Antilles *et al.*, 2021; Lu *et al.*, 2011; Moré *et al.*, 2017; Moriarty *et al.*, 2011; Russo *et al.*, 2021). For example, up to 59%

of gulls can carry *Campylobacter* (Moriarty *et al.*, 2011), and can shed up to 7.1 $\log_{10}$ *Campylobacter* per gram of faeces (Lévesque 2000). Consequently, although human sources of pollution are often considered to be riskier than animal sources, the risks from animal faeces in stormwater are, by no means, negligible (Brown *et al.*, 2017; Schoen & Ashbolt, 2010; Soller *et al.*, 2010).

In an eloquent risk assessment study by Sales-Ortells and Medema (2014) examining human health risks associated with urban flooding, risks associated with gastrointestinal illness were greatest for individuals playing in pluvial flood waters from CSOs, as well as swimming and rowing in a river impacted by effluents. However, these researchers also found that individuals playing in pluvial flood water from stormwater drains that were separate from sanitary sewers were also at higher risk for illness due to water quality that often violated the European Directorate's acceptable criteria for excellent water quality for recreation and bathing. In a study by Lee *et al.* (2020) examining stormwater systems separate from sanitary, diverse sources of faecal pollution were observed in urban stormwater systems in Ohio, including faecal contamination from ruminants, humans, dogs, and birds (the most dominant being ruminants), and microbiome analysis revealed the potential presence of pathogenic *Salmonella* strains in stormwater samples. Similarly, Beaudry (2019) observed that in stormwater systems separate from sanitary, the most dominant faecal source of contamination came from humans (i.e., 114 of 533 samples [27%] taken from 13 sites across three different storm ponds), followed by gulls (11%), geese (2%), dogs (2%), ruminants (2%) and muskrats (1%). At one particular site, 93% of stormwater samples were shown to be contaminated with human wastes, results that were similar to those of Sauer *et al.* (2011) who found that at some stormwater outfalls, 100% of samples contained evidence of human faeces. Beaudry (2019) also demonstrated that bacterial pathogens could be routinely detected in these stormwater samples using molecular methods (i.e., DNA amplification using quantitative polymerase chain reaction [qPCR]), with 25% of all stormwater samples containing *Arcobacter butzleri*, 8% contaminated with Shiga toxin-producing *E. coli*, 4% with *Campylobacter* and 1% containing *Salmonella*. Spatiotemporal variation in pathogen occurrence was also observed, and at some sites almost half of all stormwater samples surveyed were positive for *A. butzleri* by qPCR. Ahmed *et al.* (2019), provide an excellent comprehensive review of studies on stormwater in terms of sources of microbial pollution impacting urban stormwater systems and pathogen occurrence in these systems.

The emerging focus of using stormwater as a resource to promote urban resiliency has resulted in several studies focused on risk assessment. Various QMRA studies have been carried out on non-potable uses of stormwater, with several suggesting considerable risks of illness from microbial pathogens from stormwater used for irrigation, recreation, toilet flushing, and clothes washing, and when water treatment is not carefully considered (Murphy *et al.*, 2017; Petterson *et al.*, 2016; Sales-Ortells & Medema, 2015; Schoen *et al.*, 2017). Recreation may be of particular concern, when considering the strong epidemiological evidence for stormwater associated illness in users (such as swimmers and surfers) of recreational beaches impacted by stormwater runoff (Colford *et al.*, 2007, 2012; Haile *et al.*, 1999; Soller *et al.*, 2017).

Stormwater use guidelines exist in Canada, Australia and in several U.S. states, though they also vary widely in their individual scope, criteria and guiding principles (AHS, 2021; Los Angeles County Department of Public Health [LACDPH], 2016; National Academies of Science, Engineering, & Medicine (NASEM), 2016; Natural Resource Management Ministerial Council [NRMMC] *et al.*, 2009; Sharvelle *et al.*, 2017; Texas Water Development Board [TWDB], 2006). While most of these guidelines use some measure of FIB enumeration, they again vary on both the concentration set as criteria, as well as what FIB are used (LACDPH, 2016; NASEM, 2016). For example, varying guidelines in California suggest using total coliforms, faecal coliforms and/or *E. coli* (LACDPH, 2016; NASEM, 2016). Recommended FIB concentrations are also dependent on the jurisdiction, as well as competing guidelines within jurisdictions such as those set out by Title 22 legislation and/or modified NSF350 requirements in California (LACDPH, 2016; NASEM, 2016). For some jurisdictions stormwater use guidelines do not include many allowable end-uses, while in other jurisdictions the criteria are extrapolated to be used across multiple end-uses which all may have potentially different risks (LACDPH, 2016; NASEM, 2016; Schoen *et al.*, 2017; TWDB, 2006).

In Canada, the Province of Alberta has adopted a risk-based, fit-for-purpose approach for the use of alternative water sources such as stormwater. In this context stormwater protection is paramount to resource management, and subsequently linked to treatment requirements to ensure that water quality is acceptable for the intended end-use, and to the user, the public, and the environment. The approach follows that proposed by Sharvelle *et al.* (2017). This fit-for-purpose approach minimizes the cost of (unnecessary) water treatment, the optimization of energy usage, and potentially minimizes the creation of other wastes (e.g., by-products of treatment) (AHS, 2021). Alberta's Public Health Guidelines for Water Reuse and Stormwater Use are based on the principles of a water safety plan (WSP) aimed at comprehensively assessing and managing risks from source to end-use (WHO, 2005). The risk assessment process begins with established health-based targets, which set an annual tolerable infection risk from a set of reference pathogens. The Province of Alberta has adopted two health-based targets from the United States Environmental Protection Agency (US EPA): the one in 100 voluntary infection risk for recreational water and the one in 10 000 involuntary infection risk for potable water (Schoen *et al.*, 2017; US EPA 2003, 2012). The health-based targets have been used to set system performance targets, referred to as $\log_{10}$ reduction targets (LRT), derived using reverse QMRA and calculated for each end-use of stormwater (WHO, 2011). The system performance targets may be met through treatment and/or exposure controls which have been assigned $\log_{10}$ reduction credits (LRC) (based on peer-reviewed literature), and which ultimately require the LRC to meet or exceed the LRT requirements for the water to be used for an intended purpose.

The LRTs required for stormwater use are also linked to the source of contamination impacting stormwater. As outlined previously, studies have demonstrated that upwards of 10% of stormwater flows in non-CSO systems can be made up of raw municipal sewage (Sauer *et al.*, 2011), originating from broken sewer lines, illegal discharges, cross-connections, and vulnerable populations. Since many of these sources of human faecal contamination will not be detected through a visual sanitary survey, the potential risks should be mitigated through treatment or other management practices. In Alberta's risk-based guidelines, a default assumption is that urban stormwater contains 10% (or a 10^{-1} dilution) of unintentional raw human sewage, and consequently, the LRT values are based on this worst-case-scenario – that is, it is important that they adhere to the precautionary principle in standard risk assessment practice. However, although human sewage may be ubiquitous in urban stormwater, the levels of human sewage contamination tend to be far lower (e.g., 0.1% or lower [$<10^{-3}$ dilution of human sewage]). In cases where it can be demonstrated that urban stormwater contains less than 0.1% of raw human sewage (i.e., based on the microbial source tracking methods described above), the LRT values within the guidelines shift by at least two orders of magnitude. This ultimately translates into less treatment requirements and long-term operational maintenance of these systems. Consequently, a utility manager is incentivized to assess the condition of the stormwater asset and protect stormwater from intrusion of human sewage contamination, and by so doing, is confident that the lower LRT performance targets meet public health guidelines and expectations.

The practical implementation of these recently derived Public Health Guidelines is currently being debated in 'Communities-of-Practice' involving industry, municipalities, regulators and academic professionals. Of specific interest is the interpretation of these guidelines in the case of greenfield development where no upfront monitoring data is available. Given the advances in modern construction techniques and materials, it is questioned whether adoption of a 10^{-1} dilution ratio and associated more stringent treatment requirements is excessive. However, adequate inspection will be vital since only a few inadvertent cross-connections may raise the contamination level from the target 10^{-3} to an actual 10^{-1} dilution ratio across time and space. In addition, ongoing monitoring of the performance of the system will be paramount to ensure the desired dilution ratio continues to be met as the stormwater system ages.

6.4.3 Prioritizing the sewer pipe asset for maintenance activities based on a fuzzy multidimensional risk assessment method: case study of Tehran, Iran

This case study is part of a PhD thesis entitled 'Developing a risk-based multi-criteria decision-making model to prioritize sewer pipes for rehabilitation activities' (Roghani, 2020). Here, the effectiveness of classical risk analysis was enhanced by incorporating fuzzy logic and MCDM. To be more comprehensive, the risk of collapse of the pipes was viewed as a multi-dimensional parameter related to impacts on the environment, traffic disturbance, and loss of quality of life. The proposed method was applied to central parts of the sewer network of the City of Tehran, the capital of Iran (Figure 6.15). At the time of the study, there were 45 km of sewer pipes in the district of interest, most of which had been built within the last 15 years. There was little information about the condition of the pipes (e.g., there was no visual inspection data available) and the utility faced challenges in determining which parts should be inspected/rehabilitated and in what order. Since decision-support tools were not available at the time of this research, sections tended to be chosen at random, resulting in unnecessary maintenance and increased costs. While this study focused on pipes with a diameter greater than 400 mm, a portion of the network with smaller diameters was also examined (named NW in Figure 6.15) to demonstrate that the proposed methodology can work for pipes of all sizes.

Four phases were outlined: (1) data collection, (2) identification of factors contributing to risk components, (3) calculation of risk at pipe level based on pre-defined fuzzy membership functions (FMFs) and (4) clustering sewer pipes according to their defuzzified values and creating risk maps.

Criteria that were used to estimate the probability of structural failure and different categories of failure consequences are presented in Table 6.7. A triangular fuzzy number (TFN) was used to describe the effects of the value of each factor on the risk of sewer pipe failure. For instance, a visual representation of suggested TFNs for 'Pipe depth' is presented in Figure 6.16. The classification of parameter values was based on suggestions in prior studies. For example, according to Sousa *et al.* (2007), a pipe depth between 2 and 5 m has the least impact on the structural failure of sewer pipes when compared to deeper pipes.

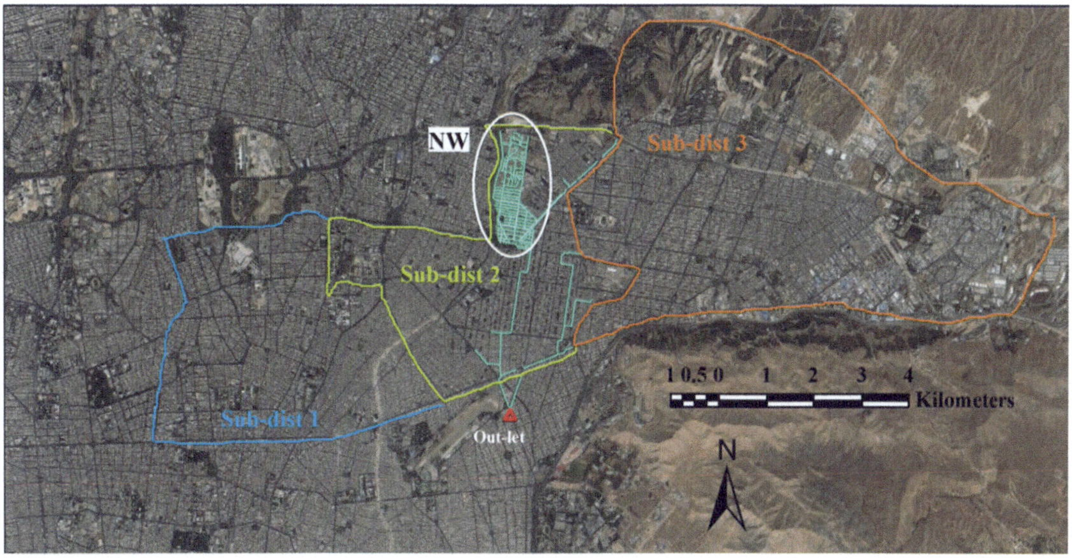

Figure 6.15 Studied area (blue, green and orange lines denote sub-districts number 1, 2 and 3, respectively; NW: region with smaller diameters than 400 mm) (Tehran wastewater company, 2019).

Table 6.7 Considered factors for the risk assessment of sewer pipes' collapse.

Likelihood of Overall Structural Failure (LoF)			**Consequences of Failure (CoF)**		
Construction Factor	**Local External Factor**	**Other Factors**	**Natural Environment**	**Traffic and Road**	**Life Quality and Public Expenditure**
Pipe depth Installation condition Pipe material Pipe length to diameter ratio Operating conditions	Road type Groundwater level Soil type Root penetration Produced amount of H2S	Pipe age Wastewater type	Soil contamination Groundwater contamination	Parking loss Decrease the life span of road asphalt Traffic density Service interruption of critical buildings	Drop in people's work efficiency and comfort due to odour, noise, and so on. Damage to nearby buildings

Source: adapted from Roghani *et al.* (2023).

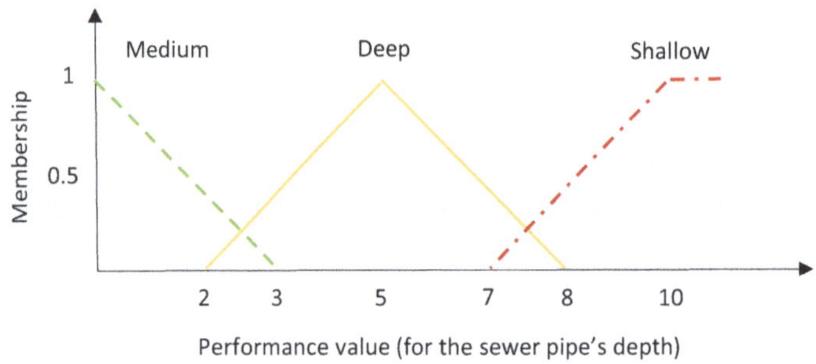

Figure 6.16 Suggested fuzzy membership function for sewer pipe depth effects on the risk of its structural failure (adapted from Roghani *et al.*, 2023).

Upon developing a questionnaire, experts from different backgrounds and disciplines were consulted to assess the relative importance of different risk factors on the sewer pipe failure (Table 6.7), using paired comparisons. According to Fuzzy AHP results, the most critical factors on the likelihood of failure (LoF) were wastewater type, pipe depth and road type. In terms of the consequence of failure (CoF), soil contamination, service interruptions of critical buildings and a decrease in workers' efficiency and comfort were evaluated as the three main consequences. Following that, individual risk values (RVs) were calculated by multiplying LoF with CoF.

The risk of environmental contamination due to pipe collapse was the most serious, with nearly half the total length of the pipes clustered in risk group number 3, extremely critical (Figure 6.17). A major reason could be the coarse-grained soil type in this region; therefore, when a pipe collapses, the resulting pollution plume can easily spread. From Figure 6.17, nearly 60% of the pipes are in cluster number 1 with regards to the risk of annoyance of life quality and increased public expenditure. Generally, the RV of dead-end and main collective branches is higher since the collapse of those branches will cause wastewater drainage problems for nearly all customers. Based on traffic disturbance and road damage risk assessments, less than 30% of the pipes were considered extremely critical (Figure 6.17) as the local roads normally have little traffic flow and the wider roads have enough space for passing traffic if a pipeline collapse occurs.

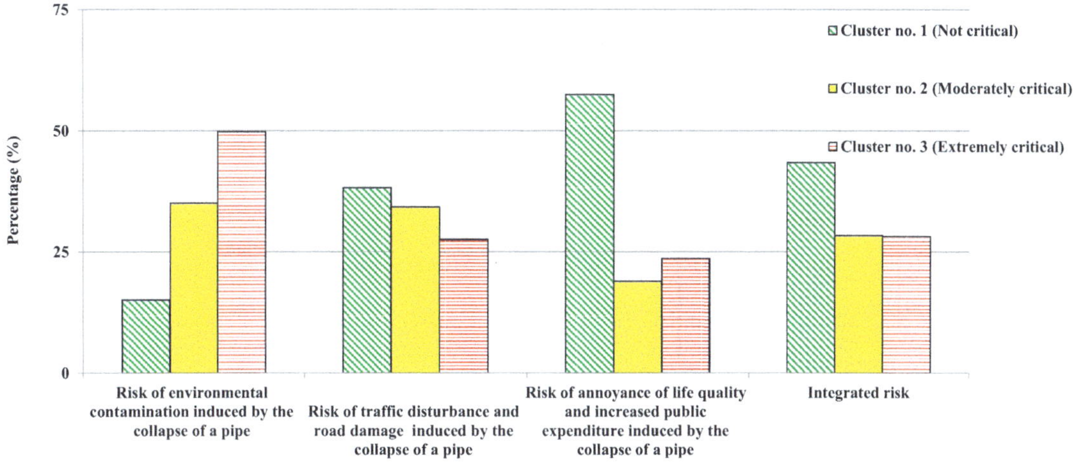

Figure 6.17 Distribution of pipes' length in each cluster (%) for individual and integrated risk of sewer pipe collapse (adapted Roghani *et al.*, 2023).

To calculate the integrated risk at an individual segment level, each individual risk was assigned a fuzzy weight with a value between (1, 1, 3) and (7, 9, 11), which varied to reflect RV impact. To illustrate, a maximum weight of (7, 9, 11) was assigned to very high RVs (i.e., extremely critical) and a minimum weight of (1, 1, 3) was assigned to very low RVs (i.e., not critical). According to the results, integrating different risk indexes changed the criticality of a pipeline for rehabilitation activities: approximately 30% of the length of the pipes were placed in the third cluster, extremely critical.

6.4.4 Precondition for asset management, experience from the Netherlands

The concept of AM (as described in Chapter 2) is based on the potential to successfully apply the well-known Deming cycle (Plan-Do-Check-Act – Deming, 1982), which is also called Plan-Do-Study-Act (Moen, 2009). Considering service-based AM, this learning cycle can only successfully be applied if five general preconditions for effective control are met, see Figure 6.18 (De Leeuw, 1974; van Riel *et al.*, 2014). The control paradigm of De Leeuw (1974) is a conceptual and general framework for describing optimal control of systems, which has been applied previously to sewer AM by van Riel *et al.* (2014). The system (i.e., a pipe, sewer network, stormwater control measure or urban area) is 'controlled' by the controller (e.g., an object manager, network operator, urban planner) using the information on the condition and performance of the controlled system. The controller also uses contextual information related to the surroundings of the UD system (which could be another layer in Figure 6.18) when defining the control actions. Finally, this environment affects the performance of the controlled system (e.g., pipe conditions could be affected by heavy traffic), and the performance of the controlled system affects the receiving waterbodies (e.g., CSO spills can hamper the use of urban canals for recreational purposes). The following sections discuss whether the state of the art is sufficient to apply AM to infrastructure networks and to network components.

6.4.4.1 Condition 1: asset manager has an objective and evaluation mechanism to check whether the goals are met

The ultimate goals of the sewer infrastructure are to protect public health and safety and to contribute to urban water management that sustains urban life at the required level. At this high conceptual level, sewer asset managers tend to neither have the criteria nor the data to evaluate their contribution to these ultimate goals. As such, one could argue that Condition 1 cannot be met.

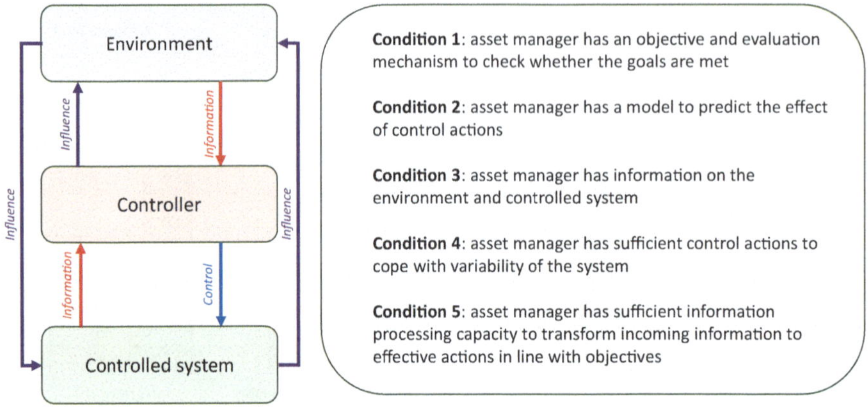

Figure 6.18 Control paradigm of De Leeuw (1974) and the five preconditions for effective control (adapted from van Riel *et al.*, 2014).

However, at a lower conceptual level, key performance indicators (KPIs) have often been developed that enable sewer asset managers to assess the hydraulic performance of sewers as infrastructure networks and to assess the structural condition of individual pipes. Despite shortcomings in both hydraulic monitoring networks and inspection techniques, it can be argued that Condition 1 can be met in practice for traditional sewer systems. For the overland drainage component of the dual drainage system as well as SCMs; however, there is still a lack of available techniques and data to assess their performance as well as to determine their condition. Moreover, SCMs typically deliver a wide range of services, such as contributing to the livability of an area, which is very hard to quantify.

6.4.4.2 Condition 2: asset manager has a model to predict the effect of control actions

State-of-the-art hydrodynamic models predict the effect of measures within infrastructure networks on the hydraulic performance of the networks. In contrast, sewer water quality models are significantly less developed and may not be adequate to model the effect of measures on water quality of, for example, CSO or SSO spills due to a lack of knowledge pertaining to the transport processes of sewer solids. Water quality models describing the physical processes taking place in sewers, such as the WATS model (wastewater aerobic/anaerobic transformations in Sewers (Hvitved-Jacobsen, 2002), are capable of adequately simulating the effect of measures to control the formation of H_2S in sewers. Hardly any models exist to simulate the effect of measures to improve the structural integrity of individual components, such as pipes and manholes. Available statistical deterioration models have been mostly developed to model the deterioration of sewer cohorts and not of individual assets. Consequently, the second precondition can only be partially met.

6.4.4.3 Condition 3: asset manager has information on the environment and controlled system

Information with respect to the environment, which is the urban fabric for sewer networks and the infrastructure network for sewer objects, is typically widely available when it comes to static information, such as the layout of the road network or the urban canals or the structure of the sewer network, although the incorporation of revisions in asset databases is notoriously slow. However, when the information becomes more dynamic, it becomes harder for the sewer asset manager to keep all information from all sources up-to-date, despite the current potential of open data and data sharing.

6.4.4.4 Condition 4: asset manager has sufficient control actions to cope with variability of the system

Sewer asset managers typically have a long list of potential measures at their disposal: measures can be applied at the level of the infrastructure network as a whole or can be related to rehabilitation of individual components/objects (for the latter, please refer to Chapter 9). Both sets of measures are adequate to address the traditional issues related to the hydraulic performance of the system and to maintaining the structural integrity of the assets. The existing measures have been proven to be inadequate to address new challenges, such as the need for climate change adaptation, and a whole range of SCM have been deemed necessary in both public space and on private property. For these newer measures, which are often positioned outside of the traditional sewer network, sewer asset managers often lack the ability, skills, and responsibility to take on or enforce them. In other words: adequate control actions are available for the more traditional rehabilitation needs and hydraulic network performance, but the measures are still inadequate to deal with issues at the level of the urban fabric.

6.4.4.5 Condition 5: asset manager has sufficient information processing capacity to transform incoming information into effective actions in line with objectives

In line with the statements with respect to Condition 4, it can be argued that regarding the hydraulic performance of networks and rehabilitation needs of objects, sewer asset managers have sufficient information processing capacity to convert the incoming information into effective actions. For more sophisticated or more detailed problems, the education level of and resources available to many sewer asset managers tend to be inadequate, and they must rely on consultants to provide them with sufficient information processing capacity. For very advanced issues, research institutes or universities may be requested to provide this expertise and capacity. The discussion of the extent to which the state of the art is sufficient to professionally apply sewer AM shows several shortcomings and/or challenges. However, it has become clear that an asset manager relies on a long list of different types of information.

6.4.5 Lake Simcoe watershed's maintenance prioritization methodology developed for conventional stormwater management features

The Lake Simcoe Region Conservation Authority (LSRCA) in partnership with the Ontario Ministry of the Environment, Conservation, and Parks (MECP), has developed a maintenance prioritization methodology for conventional stormwater management (SWM) features. This prioritization methodology was informed by initial stakeholder interviews and peer reviewed at several intervals in 2018–2019. Cross-jurisdictional uptake efforts are ongoing for this methodology through the sustainable technologies evaluation programme – water inspection and maintenance training in Ontario. This methodology ranks maintenance and repair needs to support decisions made for municipal resource allocation to address issues. It also provides the municipal SWM inspection and maintenance programme teams the opportunity to re-evaluate these rankings at a set frequency, respecting the dynamic nature of annual budget revisions and addressing continued maintenance issues.

This methodology allows for adaptation to suit the needs of each municipality within the recommended evaluation criteria (i.e., (i) sediment accumulation; (ii) pond conditions; (iii) past clean-out, outlet assessment; (iv) upstream and downstream areas, public) (Figure 6.19). Additional categories may also be added, as needed. The maximum total score that can be given is 158. For dry SWM ponds, the scoring criteria categories have been simplified for categories (i) sediment accumulation and (iii) past clean-out, outlet assessment, resulting in a maximum total score of 108.

The scoring and associated weights for each category including sub-criteria reflect the significance a certain item may have on the performance of a SWM pond, any risk to the public, and downstream receiving water. These refined prioritization methods have been tested by LSRCA staff against

	Evaluation Score (Not Weighted)			Evaluation Score (Weighted)			
Category	Magna Centre - Newmarket Pond 96	John Clarkson Park - Caledon Pond E2		Magna Centre - Newmarket Pond 96		John Clarkson Park - Caledon Pond E2	
				Weighted Score	Percent of Total Potential Score	Weighted Score	Percent of Total Potential Score
1. Sediment Accumulation	4	3		60	38%	30	19%
2. Pond Conditions	6	6		12	8%	40	25%
3. Past Clean-out, Outlet Assessment	8	6		8	5%	6	4%
4. Upstream & Downstream Areas, Public	3	9		5	3%	15	9%
5. Additional	0	0		0	0%	0	0%
Not Weighted	Magna Centre - Newmarket Pond 96	John Clarkson Park - Caledon Pond E2	Weighted	Magna Centre - Newmarket Pond 96		John Clarkson Park - Caledon Pond E2	
Total Scores	21	24	Total Scores	85	54%	91	58%

Figure 6.19 Prioritization score examples for conventional SWM ponds. Notes: (1) additional categories may be added to suit specific municipal needs (e.g., Health and safety, financial, permitting requirements). (2) Highest potential score: 158 (wet feature) or 108 (dry feature) considering all relevant evaluation categories prescribed in the default spreadsheet.

previously ranked maintenance, repair and/or clean-out recommendations for SWM ponds in several municipalities in the watershed. The higher the maintenance prioritization score for each category along with the overall total, the greater the maintenance and/or repair need(s) identified. The scoring summary outlining results from all four categories supports the overall prioritization rankings for short term decisions within the municipal budget cycle, as well as longer term planning. Municipalities have the option to add categories (e.g., health and safety, financial) with associated criteria or modify existing categories within the maintenance prioritization scoring spreadsheet. The weighting for each criterion may be modified to suit the municipal team's preference as well.

The integration of this prioritization methodology as part of the overall objective to support municipalities with improving their SWM inspection, maintenance and monitoring programmes is critical for success. If managed and acted on effectively, the maintenance prioritization methodology will ensure SWM features are functioning effectively, aligned with new provincial linear infrastructure legislation, protect valued ecosystem components, and ensure public safety remains a top priority.

6.5 CONCLUSION

Condition assessment is often the primary, or sometimes the sole, driver for prioritizing rehabilitation needs. While this approach is a significant improvement over a 'run-to-failure' strategy, it has its limitations: it has been proven to provide limited efficiency as it may lead to the rehabilitation of assets with minimal risks or insignificant impacts. A more advanced approach to AM therefore involves identifying the services provided by infrastructure, whether drainage networks or SCM. By cross-referencing the asset condition with the stakes and vulnerabilities in the territory or region where the utility manager is tasked with maintaining and managing the drainage network, he can better target rehabilitation needs, leading to a significant improvement in the system's performance. This approach, commonly known as a risk-based approach, offers a comprehensive way to optimize asset location strategies by prioritizing rehabilitation needs through a multi-criteria analysis.

As discussed in this chapter, various multi-criteria decision aid methods demonstrate remarkable diversity, catering to the specific needs and contexts of different decision-making scenarios. These methods offer solutions that range from simple and intuitive techniques to more complex methodologies capable of handling intricate and uncertain situations, allowing for a higher level of sophistication in decision-making processes. This adaptability ensures that decision-makers can effectively utilize the most appropriate method, aligning with the complexity and nuances of their unique AM or problem-solving challenges. However, selecting the right multi-criteria decision aid method presents several challenges. On one hand, opting for a simple method may provide ease of understanding, but it might overlook essential aspects and complexities of the problems at hand, leading to suboptimal decisions. On the other hand, selecting a complex method may better capture the intricacies of the problem, but its difficulty to comprehend and apply could hinder effective decision-making. Striking the right balance between simplicity and comprehensiveness is crucial, as it ensures decision-makers can make informed and optimal choices, tailored to the specific needs and nuances of the decision-making scenario. Additionally, the diversity of available methods adds to the complexity of the choice, as decision-makers must carefully assess which method aligns best with their unique problem-solving challenges. Despite their diversity, the input parameters of these methods often exert a greater influence on the outcome than the choice of the method itself and users must pay careful attention to their setting.

REFERENCES

Achillas C., Moussiopoulos N., Karagiannidis A., Banias G. and Perkoulidis G. (2013). The use of multi-criteria decision analysis to tackle waste management problems: a literature review. *Waste Management & Research: The Journal for a Sustainable Circular Economy*, **31**, 115–129, https://doi.org/10.1177/0734242X12470203

Ahmadi M., Cherqui F., Massiac J.-C. de, Werey C., Lagoutte S. and Gauffre P. L. (2014a). Condition grading for dysfunction indicators in sewer asset management. *Structure and Infrastructure Engineering*, **10**, 346–358, https://doi.org/10.1080/15732479.2012.756916

Ahmadi M., Cherqui F., Massiac J.-C. D. and Gauffre P. L. (2014b). From sewer inspection programmes to rehabilitation needs: research and results related to data quality and availability with the support of numerical experiment. *European Journal of Environmental and Civil Engineering*, **18**, 1145, https://www.tandfonline.com/doi/abs/10.1080/19648189.2014.893212

Ahmed W., Hamilton K., Toze S., Cook S. and Page D. (2019). A review on microbial contaminants in stormwater runoff and outfalls: potential health risks and mitigation strategies. *Science of The Total Environment*, **692**, 1304–1321, https://doi.org/10.1016/j.scitotenv.2019.07.055

Ahmed W., Payyappat S., Cassidy M., Harrison N. and Besley C. (2020). Sewage-associated marker genes illustrate the impact of wet weather overflows and dry weather leakage in urban estuarine waters of Sydney, Australia. *Science of The Total Environment*, **705**, 135390, https://doi.org/10.1016/j.scitotenv.2019.135390

Akhtar S., Reza B., Hewage K., Shahriar A., Zargar A. and Sadiq R. (2015). Life cycle sustainability assessment (LCSA) for selection of sewer pipe materials. *Clean Technologies and Environmental Policy*, **17**, 973–992, https://doi.org/10.1007/s10098-014-0849-x

Alberta Health Services (AHS) (2021). Public health guidelines for water reuse and stormwater use. https://open.alberta.ca/publications/public-health-guidelines-water-reuse-stormwater-use

American Veterinary Medical Association (AVMA) (2018). U.S. pet ownership statistics. Retrieved from https://www.avma.org/resources-tools/reports-statistics/us-pet-ownership-statistics

Ana E. V. and Bauwens W. (2010). Modeling the structural deterioration of urban drainage pipes: the state-of-the-art in statistical methods. *Urban Water Journal*, **7**, 47–59, https://doi.org/10.1080/15730620903447597

Anbari M. J., Tabesh M. and Roozbahani A. (2017). Risk assessment model to prioritize sewer pipes inspection in wastewater collection networks. *Journal of Environmental Management*, **190**, 91–101, https://doi.org/10.1016/j.jenvman.2016.12.052

Antilles N., García-Bocanegra I., Alba-Casals A., López-Soria S., Pérez-Méndez N., Saco M., González-Solís J. and Cerdà-Cuéllar M. (2021). Occurrence and antimicrobial resistance of zoonotic enteropathogens in gulls from Southern Europe. *Science of The Total Environment*, **763**, 143018, https://doi.org/10.1016/j.scitotenv.2020.143018

Arnone R. D. and Perdek Walling J. (2007). Waterborne pathogens in urban watersheds. *Journal of Water and Health*, **5**, 149–162, https://doi.org/10.2166/wh.2006.001

Aven T. (2017). Improving risk characterisations in practical situations by highlighting knowledge aspects, with applications to risk matrices. Reliab. Eng. Syst. Saf., special section: applications of probabilistic graphical models in dependability. *Reliability Engineering & System Safety*, **167**, 42–48, https://doi.org/10.1016/j.ress.2017.05.006

Ba Z., Fu J., Liang J., Liang K. and Wang M. (2021). Risk assessment method of drainage network operation based on fuzzy comprehensive evaluation combined with analytic network process. *Journal of Pipeline Systems Engineering and Practice*, **12**, 04021009, https://doi.org/10.1061/(ASCE)PS.1949-1204.0000543

Baah K., Dubey B., Harvey R. and McBean E. (2015). A risk-based approach to sanitary sewer pipe asset management. *Science of The Total Environment*, **505**, 1011–1017, https://doi.org/10.1016/j.scitotenv.2014.10.040

Bakhshipour A. E., Dittmer U., Haghighi A. and Nowak W. (2021). Toward sustainable urban drainage infrastructure planning: a combined multiobjective optimization and multi-criteria decision-making platform. *Journal of Water Resources Planning and Management*, **147**, 04021049, https://doi.org/10.1061/(ASCE)WR.1943-5452.0001389

Beaudry M. (2019). From nuisance to resource: understanding microbial sources of contamination in urban stormwater-impacted bodies of water intended for water reuse activities. MSc thesis, University of Alberta, Edmonton. https://era.library.ualberta.ca/items/6f834aa3-56d3-4829-8c9e-5b294d95b11c

Behzadian M., Kazemzadeh R. B., Albadvi A. and Aghdasi M. (2010). PROMETHEE: a comprehensive literature review on methodologies and applications. *European Journal of Operational Research*, **200**(1), 198–215, https://doi.org/10.1016/j.ejor.2009.01.021

Belmeziti A., Cherqui F., Tourne A., Granger D., Werey C., Gauffre P. L. and Chocat B. (2015). Transitioning to sustainable urban water management systems: how to define expected service functions? *Civil Engineering and Environmental Systems*, **32**, 316–334, https://doi.org/10.1080/10286608.2015.1047355

Birgani Y. T. and Yazdandoost F. (2018). An integrated framework to evaluate resilient-sustainable urban drainage management plans using a combined-adaptive MCDM technique. *Water Resources Management*, **32**(8), 2817–2835, https://doi.org/10.1007/s11269-018-1960-2

Bouzid M., Halai K., Jeffreys D. and Hunter P. R. (2015). The prevalence of *Giardia* infection in dogs and cats, a systematic review and meta-analysis of prevalence studies from stool samples. *Veterinary Parasitology*, **207**, 181–202, https://doi.org/10.1016/j.vetpar.2014.12.011

Brewton R. A., Kreiger L. B., Tyre K. N., Baladi D., Wilking L. E., Herren L. W. and Lapointe B. E. (2022). Septic system–groundwater–surface water couplings in waterfront communities contribute to harmful algal blooms in southwest Florida. *Science of The Total Environment*, **837**, 155319, https://doi.org/10.1016/j.scitotenv.2022.155319

Brown K. I., Graham K. E., Soller J. A. and Boehm A. B. (2017). Estimating the probability of illness due to swimming in recreational water with a mixture of human- and gull-associated microbial source tracking markers. *Environmental Science: Processes & Impacts*, **19**, 1528–1541, https://doi.org/10.1039/C7EM00316A

Calgary, City of (2023). Stormwater Management Strategy, report from the City of Calgary, 33p., online: https://www.calgary.ca/water/stormwater.html (last accessed 31.07.23)

Canadian Animal Health Institute (CAHI) (2021). 2020 pet population figures. Retrieved from https://cahi-icsa.ca/news/2020-canadian-pet-population-figures-released

Cann K. F., Thomas D. R., Salmon R. L., Wyn-Jones A. P. and Kay D. (2013). Extreme water-related weather events and waterborne disease. *Epidemiology and Infection*, **141**, 671–686, https://doi.org/10.1017/S0950268812001653

Carriço N., Covas D. I. C., Almeida M. C., Leitão J. P. and Alegre H. (2012). Prioritization of rehabilitation interventions for urban water assets using multiple criteria decision-aid methods. *Water Science and Technology*, **66**, 1007–1014, https://doi.org/10.2166/wst.2012.274

Carson L. R., Goodman C., van Duin B. and Neumann N. F. (2023). Application of a microbial and pathogen source tracking 'toolbox' to identify infrastructure problems in stormwater drainage networks: a case study. Manuscript submitted for publication.

Chen T. Y.-J., Washington V. N., Aven T. and Guikema S. D. (2020). Review and evaluation of the J100-10 risk and resilience management standard for water and wastewater systems. *Risk Analysis*, **40**, 608–623, https://doi.org/10.1111/risa.13421

Cherqui F., Werey C., Ibrahim M. and Gauffre P. L. (2008). CCTV inspection of sewer segments: calibration of performance indicators based on experts' opinions. 11th International Conference on Urban Drainage, Edinburgh, Scotland, UK, pp. 11.

Cherqui F., Le Gauffre P., Werey C., Ibrahim M. and Christol J. (2010). L4 v1 – Formulation des critères de réhabilitation et modèle(s) de synthèse multicritère. (In French), available on: https://gestionpatrimoniale.files.wordpress.com/2017/05/indigau_l4_formulation-criteres_synthese-multicritere.pdf

Cherqui F., Granger D., Métadier M., Fletcher T. D., Barraud S., Lalanne P. and Litrico X. (2013). Indicators Related to BMP Performances: Operational Monitoring Propositions. Presented at the Novatech, Lyon, France.

Chocat B., Ashley R., Marsalek J., Matos M. R., Rauch W., Schilling W. and Urbonas B. (2007). Toward the sustainable management of urban storm-water. *Indoor and Built Environment*, **16**, 273–285, https://doi.org/10.1177/1420326X07078854

Colford J. M., Wade T. J., Schiff K. C., Wright C. C., Griffith J. F., Sandhu S. K., Burns S., Sobsey M., Lovelace G. and Weisberg S. B. (2007). Water quality indicators and the risk of illness at beaches with nonpoint sources of fecal contamination. *Epidemiology*, **18**, 27–35, https://doi.org/10.1097/01.ede.0000249425.32990.b9

Colford J. M., Schiff K. C., Griffith J. F., Yau V., Arnold B. F., Wright C. C., Gruber J. S., Wade T. J., Burns S., Hayes J., McGee C., Gold M., Cao Y., Noble R. T., Haugland R. and Weisberg S. B. (2012). Using rapid indicators for *Enterococcus* to assess the risk of illness after exposure to urban runoff contaminated marine water. *Water Research*, **46**, 2176–2186, https://doi.org/10.1016/j.watres.2012.01.033

Converse R. R., Piehler M. F. and Noble R. T. (2011). Contrasts in concentrations and loads of conventional and alternative indicators of fecal contamination in coastal stormwater. *Water Research*, **45**, 5229–5240, https://doi.org/10.1016/j.watres.2011.07.029

Cox L. A. (2008). What's wrong with risk matrices? *Risk Analysis*, **28**, 497–512, https://doi.org/10.1111/j.1539-6924.2008.01030.x

Curriero F. C., Patz J. A., Rose J. B. and Lele S. (2001). The association between extreme precipitation and waterborne disease outbreaks in the United States, 1948–1994. *American Journal of Public Health*, **91**, 1194–1199, https://doi.org/10.2105/AJPH.91.8.1194

Daulat S., Rokstad M. M., Klein-Paste A., Langeveld J. and Tscheikner-Gratl F. (2022). Challenges of integrated multi-infrastructure asset management: a review of pavement, sewer, and water distribution networks. *Structure and Infrastructure Engineering*, **20**(4), 1–20, https://doi.org/10.1080/15732479.2022.2119480

Davies J. P., Clarke B. A., Whiter J. T. and Cunningham R. J. (2001). Factors influencing the structural deterioration and collapse of rigid sewer pipes. *Urban Water*, **3**, 73–89, https://doi.org/10.1016/S1462-0758(01)00017-6

Debères P., Ahmadi M., De Massiac J.-C., Le Gauffre P., Cherqui F., Werey C. and Nirsimloo K. (2012). Deploying a sewer asset management strategy using the indigau decision support system. Proceedings of 4th Leading Edge Conference on Strategic Asset Management (LESAM), Mühlheim, Germany, pp. 11.

De Giglio O., Fasano F., Diella G., Lopuzzo M., Napoli C., Apollonio F., Brigida S., Calia C., Campanale C., Marzella A., Pousis C., Rutigliano S., Triggiano F., Caggiano G. and Montagna M. T. (2019). *Legionella* and legionellosis in touristic-recreational facilities: influence of climate factors and geostatistical analysis in Southern Italy (2001–2017). *Environmental Research*, **178**, 108721, https://doi.org/10.1016/j.envres.2019.108721

De Leeuw A. C. J. (1974). Systeemleer en organisatiekunde. Een onderzoek naar mogelijke bijdragen van de systeemleer tot een integrale organisatiekunde. PhD thesis, Eindhoven University of Technology.

Deming W. E. (1982). Quality, Productivity and Competitive Position. MIT Press, Cambridge.

Duijm N. J. (2015). Recommendations on the use and design of risk matrices. *Safety Science*, **76**, 21–31, https://doi.org/10.1016/j.ssci.2015.02.014

Egger C. and Maurer M. (2015). Importance of anthropogenic climate impact, sampling error and urban development in sewer system design. *Water Research*, **73**, 78–97, https://doi.org/10.1016/j.watres.2014.12.050

Elsawah H., Bakry I. and Osama M. (2016). Decision support model for integrated risk assessment and prioritization of intervention plans of municipal infrastructure. *Journal of Pipeline Systems Engineering and Practice*, **7**, 04016010, https://doi.org/10.1061/(ASCE)PS.1949-1204.0000245

EN 752 (2008). Drain and Sewer Systems Outside Buildings (European Standard). European Committee for Standardization, Brussels, Belgium.

EN 13508-2 (2011). Investigation and Assessment of Drain and Sewer Systems Outside Buildings – Part 2: Visual Inspection Coding System (European Standard). European Comittee for Standardization, Brussels, Belgium.

Ennaouri I. and Musandji F. (2013). New integrated condition-assessment model for combined storm-sewer systems. *Journal of Water Resources Planning and Management*, **139**, 53–64, https://doi.org/10.1061/(ASCE)WR.1943-5452.0000217

Fathollahi A. and Coupe S. J. (2021). Life cycle assessment (LCA) and life cycle costing (LCC) of road drainage systems for sustainability evaluation: quantifying the contribution of different life cycle phases. *Science of The Total Environment*, **776**, 145937, https://doi.org/10.1016/j.scitotenv.2021.145937

Filion Y. R., MacLean H. L. and Karney B. W. (2004). Life-cycle energy analysis of a water distribution system. *Journal of Infrastructure Systems*, **10**, 120–130, https://doi.org/10.1061/(ASCE)1076-0342(2004)10:3(119)

Fletcher T. D., Shuster W., Hunt W. F., Ashley R., Butler D., Arthur S., Trowsdale S., Barraud S., Semadeni-Davies A., Bertrand-Krajewski J.-L., Mikkelsen P. S., Rivard G., Uhl M., Dagenais D. and Viklander M. (2015). SUDS, LID, BMPs, WSUD and more – The evolution and application of terminology surrounding urban drainage. *Urban Water Journal*, **12**(7), 525–542, https://doi.org/10.1080/1573062X.2014.916314

Fox J. G., Claps M. C., Taylor N. S., Maxwell K. O., Ackerman J. I. and Hoffman S. B. (1988). *Campylobacter jejuni/coli* in commercially reared beagles: prevalence and serotypes. *Lab Animal Science*, **38**(3), 262–265.

Frangopol D. M. (2011). Life-cycle performance, management, and optimisation of structural systems under uncertainty: accomplishments and challenges. *Structure and Infrastructure Engineering*, **7**, 389–413, https://doi.org/10.1080/15732471003594427

Franks A. P. and Maddison T. (2006). A simplified method for the estimation of individual risk. *Process Safety and Environmental Protection*, **84**, 101–108, https://doi.org/10.1205/psep.04287

Gleason J. A., Kratz N. R., Greeley R. D. and Fagliano J. A. (2016). Under the weather: legionellosis and meteorological factors. *EcoHealth*, **13**, 293–302, https://doi.org/10.1007/s10393-016-1115-y

Guzman-Herrador B. R., De Blasio B. F., MacDonald E., Nichols G., Sudre B., Vold L., Semenza J. C. and Nygård K. (2015). Analytical studies assessing the association between extreme precipitation or temperature and drinking water-related waterborne infections: a review. *Environmental Health*, **14**, 29, https://doi.org/10.1186/s12940-015-0014-y

Hachad M., Lanoue M., Vo Duy S., Villlemur R., Sauvé S., Prévost M. and Dorner S. (2022). Locating illicit discharges in storm sewers in urban areas using multi-parameter source tracking: field validation of a toolbox composite index to prioritize high risk areas. *Science of The Total Environment*, **811**, 152060, https://doi.org/10.1016/j.scitotenv.2021.152060

Hahn M. A., Palmer R. N., Merrill M. S. and Lukas A. B. (2002). Expert system for prioritizing the inspection of sewers: knowledge base formulation and evaluation. *Journal of Water Resources Planning and Management*, **128**, 121–129, https://doi.org/10.1061/(ASCE)0733-9496(2002)128:2(121)

Haile R. W., Witte J. S., Gold M., Cresssey R., McGee C., Millikan R. C., Glasser A., Harawa N., Ervin C., Harmon P., Harper J., Dermand J., Alamillo J., Barrett K., Nides M. and Wang G.-Y. (1999). The health effects of swimming in ocean water contaminated by storm drain runoff. *Epidemiology*, **10**, 355–363, https://doi.org/10.1097/00001648-199907000-00004

Hajkowicz S. and Collins K. (2007). A review of multiple criteria analysis for water resource planning and management. *Water Resources Management*, **21**, 1553–1566, https://doi.org/10.1007/s11269-006-9112-5

Halfawy M. R., Dridi L. and Baker S. (2008). Integrated decision support system for optimal renewal planning of sewer networks. *Journal of Computing in Civil Engineering*, **22**, 360–372, https://doi.org/10.1061/(ASCE)0887-3801(2008)22:6(360)

Hart J. D., Blackwood A. D. and Noble R. T. (2020). Examining coastal dynamics and recreational water quality by quantifying multiple sewage specific markers in a North Carolina estuary. *Science of The Total Environment*, **747**, 141124, https://doi.org/10.1016/j.scitotenv.2020.141124

Hvitved-Jacobsen T. (2002). Sewer Processes: Microbial and Chemical Process Engineering of Sewer Networks. CRC PRESS, Washington, DC.

IEC 31010 (2019). Risk Management – Risk Assessment Techniques (International Standard). International Electrotechnical Commission, Geneva, Switzerland.

ISO 55000 (2014). Asset Management – Overview, Principles and Terminology (International Standard). International Organization for Standardization, Geneva, Switzerland.

ISO 24765 (2017). Systems and Software Engineering – Vocabulary (International Standard). International Organization for Standardization, Geneva, Switzerland.

ISO 31000 (2018). Risk Management – Guidelines. International Organization for Standardization, Geneva, Switzerland.

Jalliffier-Verne I., Leconte R., Huaringa-Alvarez U., Madoux-Humery A.-S., Galarneau M., Servais P., Prévost M. and Dorner S. (2015). Impacts of global change on the concentrations and dilution of combined sewer overflows in a drinking water source. *Science of The Total Environment*, **508**, 462–476, https://doi.org/10.1016/j.scitotenv.2014.11.059

Kambali S., Quinonez E., Sharifi A., Shahraki A. H., Kumar N., Dushyantha J. and Mirsaeidi M. (2021). Pulmonary nontuberculous mycobacterial disease in Florida and association with large-scale natural disasters. *BMC Public Health*, **21**, 2058, https://doi.org/10.1186/s12889-021-12115-7

Korving H., Noortwijk J. M. V., Gelder P. H. A. J. M. V. and Clemens F. H. L. R. (2009). Risk-based design of sewer system rehabilitation. *Structure and Infrastructure Engineering*, **5**, 215–227, https://doi.org/10.1080/15732470601114299

Kuliczkowska E. (2016). Risk of structural failure in concrete sewers due to internal corrosion. *Engineering Failure Analysis*, **66**, 110–119, https://doi.org/10.1016/j.engfailanal.2016.04.026

Laakso T., Kokkonen T., Mellin I. and Vahala R. (2018). Sewer condition prediction and analysis of explanatory factors. *Water*, **10**, 1239, https://doi.org/10.3390/w10091239

Lai E., Lundie S. and Ashbolt N. J. (2008). Review of multi-criteria decision aid for integrated sustainability assessment of urban water systems. *Urban Water Journal*, **5**, 315–327, https://doi.org/10.1080/15730620802041038

Lee S., Suits M., Wituszynski D., Winston R., Martin J. and Lee J. (2020). Residential urban stormwater runoff: a comprehensive profile of microbiome and antibiotic resistance. *Science of The Total Environment*, **723**, 138033, https://doi.org/10.1016/j.scitotenv.2020.138033

Lee J., Park C. Y., Baek S., Han S. H. and Yun S. (2021). Risk-Based prioritization of sewer pipe inspection from infrastructure asset management perspective. *Sustainability*, **13**, 7213, https://doi.org/10.3390/su13137213

Le Gauffre P. and Cherqui F. (2009). Sewer Rehabilitation Criteria Evaluated by Fusion of Fuzzy Indicators. Presented at the LESAM 2009, Miami, USA, p. 16.

Le Gauffre P., Joannis C., Vasconcelos E., Breysse D., Gibello C. and Desmulliez J.-J. (2007a). Performance indicators and multi-criteria decision support for sewer asset management. *Journal of Infrastructure Systems*, **13**, 105–114, https://doi.org/10.1061/(ASCE)1076-0342(2007)13:2(105)

Le Gauffre P., Haidar H., Poinard D., Laffréchine K., Baur R. and Schiatti M. (2007b). A multi-criteria decision support methodology for annual rehabilitation programs of water networks. *Computer-Aided Civil and Infrastructure Engineering*, **22**(7), 478–488, https://doi.org/10.1111/j.1467-8667.2007.00504.x

Lévesque B. (2000). Study of the bacterial content of ring-billed gull droppings in relation to recreational water quality. *Water Research*, **34**, 1089–1096, https://doi.org/10.1016/S0043-1354(99)00266-3

Levine E. S. (2012). Improving risk matrices: the advantages of logarithmically scaled axes. *Journal of Risk Research*, **15**, 209–222, https://doi.org/10.1080/13669877.2011.634514

Liang S. Y. and Messenger N. (2018). Infectious diseases after hydrologic disasters. *Emergency Medicine Clinics of North America*, **36**, 835–851, https://doi.org/10.1016/j.emc.2018.07.002

Lim K.-Y., Hamilton A. J. and Jiang S. C. (2015). Assessment of public health risk associated with viral contamination in harvested urban stormwater for domestic applications. *Science of The Total Environment*, **523**, 95–108, https://doi.org/10.1016/j.scitotenv.2015.03.077

Lin S.-H., Zhao X., Wu J., Liang F., Li J.-H., Lai R.-J., Hsieh J.-C. and Tzeng G.-H. (2021). An evaluation framework for developing green infrastructure by using a new hybrid multiple attribute decision-making model for promoting environmental sustainability. *Socio-Economic Planning Sciences*, **75**, 100909, https://doi.org/10.1016/j.seps.2020.100909

Loc H. H., Duyen P. M., Ballatore T. J., Lan N. H. M. and Das Gupta A. (2017). Applicability of sustainable urban drainage systems: an evaluation by multi-criteria analysis. *Environment Systems and Decisions*, **37**, 332–343, https://doi.org/10.1007/s10669-017-9639-4

Løken E. (2007). Use of multi-criteria decision analysis methods for energy planning problems. *Renewable and Sustainable Energy Reviews*, **11**, 1584–1595, https://doi.org/10.1016/j.rser.2005.11.005

Los Angeles County Department of Public Health (2016). Guidelines for alternate water sources: Indoor and outdoor non-potable uses. http://www.publichealth.lacounty.gov/EH/docs/permit/guidelines-alternate-water-sources.pdf

Lu J., Ryu H., Santo Domingo J. W., Griffith J. F. and Ashbolt N. (2011). Molecular detection of *Campylobacter* spp. in California Gull (*Larus californicus*) excreta. *Applied and Environmental Microbiology*, **77**, 5034–5039, https://doi.org/10.1128/AEM.00018-11

Lusk M. G. and Toor G. S. (2016). Biodegradability and molecular composition of dissolved organic nitrogen in urban stormwater runoff and outflow water from a stormwater retention pond. *Environmental Science & Technology*, **50**, 3391–3398, https://doi.org/10.1021/acs.est.5b05714

Mardani A., Zavadskas E. K., Govindan K., Amat Senin A. and Jusoh A. (2016). VIKOR Technique: a systematic review of the state of the art literature on methodologies and applications. *Sustainability*, **8**(37), 37, https://doi.org/10.3390/su8010037

Marlow D., Davis P., Trans D., Beale D. and Burn S. (2009). Remaining Asset Life: A State of the Art Review. WERF, Alexandria, VA, USA.

Mayer R. E., Reischer G. H., Ixenmaier S. K., Derx J., Blaschke A. P., Ebdon J. E., Linke R., Egle L., Ahmed W., Blanch A. R., Byamukama D., Savill M., Mushi D., Cristóbal H. A., Edge T. A., Schade M. A., Aslan A., Brooks Y. M., Sommer R., Masago Y., Sato M. I., Taylor H. D., Rose J. B., Wuertz S., Shanks O. C., Piringer H., Mach R. L., Savio D., Zessner M. and Farnleitner A. H. (2018). Global distribution of human-associated fecal genetic markers in reference samples from six continents. *Environmental Science & Technology*, **52**, 5076–5084, https://doi.org/10.1021/acs.est.7b04438

McGinnis S., Spencer S., Firnstahl A., Stokdyk J., Borchardt M., McCarthy D. T. and Murphy H. M. (2018). Human Bacteroides and total coliforms as indicators of recent combined sewer overflows and rain events in urban creeks. *Science of The Total Environment*, **630**, 967–976, https://doi.org/10.1016/j.scitotenv.2018.02.108

McLellan S. L., Fisher C., J J. and Newton R. (2015). The microbiome of urban waters. International Microbiology, **18**, 141–149, https://doi.org/10.2436/20.1501.01.244

Mitsui M., Ito A., Ishida T., Tachibana H., Nakanishi Y., Yamazaki A. and Washio Y. (2021). Increased risk of *Legionella* pneumonia as community-acquired pneumonia after heavy rainfall in 2018 in west Japan. *Journal of Infection and Chemotherapy*, **27**, 1429–1435, https://doi.org/10.1016/j.jiac.2021.05.018

Moen R. (2009). Foundation and History of the PDSA Cycle. Available from: https://www.praxisframework.org/files/pdsa-history-ron-moen.pdf (Last access 31.07.23)

Monteiro S., Queiroz G., Ferreira F. and Santos R. (2021). Characterization of stormwater runoff based on microbial source tracking methods. *Frontiers in Microbiology*, **12**, 674047, https://doi.org/10.3389/fmicb.2021.674047

Montgomery M. P., Robertson S., Koski L., Salehi E., Stevenson L. M., Silver R., Sundararaman P., Singh A., Joseph L. A., Weisner M. B., Brandt E., Prarat M., Bokanyi R., Chen J. C., Folster J. P., Bennett C. T., Francois Watkins L. K., Aubert R. D., Chu A., Jackson J., Blanton J., Ginn A., Ramadugu K., Stanek D., DeMent J., Cui J., Zhang Y., Basler C., Friedman C. R., Geissler A. L., Crowe S. J., Dowell N., Dixon S., Whitlock L., Williams I., Jhung M. A., Nichols M. C., De Fijter S. and Laughlin M. E. (2018). Multidrug-resistant *Campylobacter jejuni* outbreak linked to puppy exposure – United States, 2016–2018. *MMWR. Morbidity and Mortality Weekly Report*, **67**, 1032–1035, https://doi.org/10.15585/mmwr.mm6737a3

Moré E., Ayats T., Ryan P. G., Naicker P. R., Keddy K. H., Gaglio D., Witteveen M. and Cerdà-Cuéllar M. (2017). Seabirds (laridae) as a source of *Campylobacter* spp., *Salmonella* spp. and antimicrobial resistance in South Africa. *Environmental Microbiology*, **19**, 4164–4176, https://doi.org/10.1111/1462-2920.13874

Moriarty E., Karki N., Mackenzie M., Sinton L., Wood D. and Gilpin B. (2011). Faecal indicators and pathogens in selected New Zealand waterfowl. *New Zealand Journal of Marine and Freshwater Research*, **45**, 679–688, https://doi.org/10.1080/00288330.2011.578653

Murphy H. M., Meng Z., Henry R., Deletic A. and McCarthy D. T. (2017). Current stormwater harvesting guidelines are inadequate for mitigating risk from *Campylobacter* during nonpotable reuse activities. *Environmental Science & Technology*, **51**, 12498–12507, https://doi.org/10.1021/acs.est.7b03089

National Academies of Science, Engineering, and Medicine (2016). Using graywater and stormwater to enhance local water supplies: an assessment of risks, costs, and benefits, https://doi.org/10.17226/21866

Natural Resource Management Ministerial Council, Environment Protection and Heritage Council, & National Health and Medical Research Council (2009). Australian Guidelines for water recycling: Managing health and environmental risks (phase 2) stormwater harvesting and reuse. https://www.waterquality.gov.au/sites/default/files/documents/water-recycling-guidelines-stormwater-23.pdf

Ni H., Chen A. and Chen N. (2010). Some extensions on risk matrix approach. *Safety Science*, **48**, 1269–1278, https://doi.org/10.1016/j.ssci.2010.04.005

Olds H. T., Corsi S. R., Dila D. K., Halmo K. M., Bootsma M. J. and McLellan S. L. (2018). High levels of sewage contamination released from urban areas after storm events: a quantitative survey with sewage specific bacterial indicators. *PLOS Medicine*, **15**, e1002614, https://doi.org/10.1371/journal.pmed.1002614

Olivieri V. P., Kruse C. W., Kawata K. and Smith J. E. (1977). Microorganisms in Urban Stormwater. Municipal Environmental Research Laboratory, U.S. Environmental Protection Agency, Cincinnati, OH.

Olson E., Hargiss C. L. M. and Norland J. (2021). *Escherichia coli* levels and microbial source tracking in stormwater retention ponds and detention basins. *Water Environment Research*, **94**(1), 1–16, https://doi.org/10.1002/wer.1675

Parker J. K., McIntyre D. and Noble R. T. (2010). Characterizing fecal contamination in stormwater runoff in coastal North Carolina, USA. *Water Research*, **44**, 4186–4194, https://doi.org/10.1016/j.watres.2010.05.018

Petterson S. R., Mitchell V. G., Davies C. M., O'Connor J., Kaucner C., Roser D. and Ashbolt N. (2016). Evaluation of three full-scale stormwater treatment systems with respect to water yield, pathogen removal efficacy and human health risk from faecal pathogens. *Science of The Total Environment*, **543**, 691–702, https://doi.org/10.1016/j.scitotenv.2015.11.056

Ramos G. P., Rocha J. L. and Tuon F. F. (2013). Seasonal humidity may influence *Pseudomonas aeruginosa* hospital-acquired infection rates. *International Journal of Infectious Diseases*, **17**, e757–e761, https://doi.org/10.1016/j.ijid.2013.03.002

Reimschuessel R., Grabenstein M., Guag J., Nemser S. M., Song K., Qiu J., Clothier K. A., Byrne B. A., Marks S. L., Cadmus K., Pabilonia K., Sanchez S., Rajeev S., Ensley S., Frana T. S., Jergens A. E., Chappell K. H., Thakur S., Byrum B., Cui J., Zhang Y., Erdman M. M., Rankin S. C., Daly R., Das S., Ruesch L., Lawhon S. D., Zhang S., Baszler T., Diaz-Campos D., Hartmann F. and Okwumabua O. (2017). Multilaboratory survey to evaluate *Salmonella* prevalence in diarrheic and nondiarrheic dogs and cats in the United States between 2012 and 2014. *Journal of Clinical Microbiology*, **55**, 1350–1368, https://doi.org/10.1128/JCM.02137-16

Rogers M., Bruen M. and Maystre L. Y. (2000). ELECTRE And Decision Support. Methods and Applications in Engineering and Infrastructure Investment. Kluwer Academic, Dordrechet, The Netherlands.

Roghani B. (2020). Developing a risk-based multi-criteria decision making model to prioritize sewer pipes for rehabilitation activities. PhD thesis, University of Tehran, College of Engineering, Faculty of Civil Engineering, Tehran, Iran.

Roghani B., Tabesh M. and Cherqui F. (2023). A fuzzy multidimensional risk assessment method for sewer asset management. *International Journal of Civil Engineering*, **22**, 1–17, https://doi.org/10.1007/s40999-023-00888-4

Roy B. (2016). Paradigms and challenges. In: Multiple Criteria Decision Analysis: State of the Art Surveys, International Series in Operations Research & Management Science, S. Greco, M. Ehrgott and J. R. Figueira (eds.), Springer, New York, NY, Vol. **233**, pp. 19–39, https://doi.org/10.1007/978-1-4939-3094-4_2

Rozan A., Rulleau B. and Werey C. (2017). Assessing preferences for sewer network asset management in France. *International Journal of Environmental Technology and Management*, **20**, 163–182, https://doi.org/10.1504/IJETM.2017.10010689

Ruan X., Yin Z. and Frangopol D. M. (2015). Risk matrix integrating risk attitudes based on utility theory. *Risk Analysis*, **35**, 1437–1447, https://doi.org/10.1111/risa.12400

Ruge B. (2004). Risk matrix as tool for risk assessment in the chemical process industries. In: Probabilistic Safety Assessment and Management, C. Spitzer, U. Schmocker and V. N. Dang (eds.), Springer, London, pp. 2693–2698, https://doi.org/10.1007/978-0-85729-410-4_431

Russo T. P., Pace A., Varriale L., Borrelli L., Gargiulo A., Pompameo M., Fioretti A. and Dipineto L. (2021). Prevalence and antimicrobial resistance of enteropathogenic bacteria in yellow-legged gulls (Larus michahellis) in southern Italy. *Animals*, **11**, 275, https://doi.org/10.3390/ani11020275

Saadat Foomani M. and Malekmohammadi B. (2020). Site selection of sustainable urban drainage systems using fuzzy logic and multi-criteria decision-making. *Water and Environment Journal*, **34**, 584–599, https://doi.org/10.1111/wej.12487

Sales-Ortells H. and Medema G. (2014). Screening-level microbial risk assessment of urban water locations: a tool for prioritization. *Environmental Science & Technology*, **48**, 9780–9789, https://doi.org/10.1021/es5020407

Sales-Ortells H. and Medema G. (2015). Microbial health risks associated with exposure to stormwater in a water plaza. *Water Research*, **74**, 34–46, https://doi.org/10.1016/j.watres.2015.01.044

Salman B. and Salem O. (2012). Risk assessment of wastewater collection lines using failure models and criticality ratings. *Journal of Pipeline Systems Engineering and Practice*, **3**, 68–76, https://doi.org/10.1061/(ASCE)PS.1949-1204.0000100

Sauer E. P., VandeWalle J. L., Bootsma M. J. and McLellan S. L. (2011). Detection of the human specific *Bacteroides* genetic marker provides evidence of widespread sewage contamination of stormwater in the urban environment. *Water Research*, **45**, 4081–4091, https://doi.org/10.1016/j.watres.2011.04.049

Schoen M. E. and Ashbolt N. J. (2010). Assessing pathogen risk to swimmers at non-sewage impacted recreational beaches. *Environmental Science and Technology*, **44**(7), 2286–2291, https://doi.org/10.1021/es903523q

Schoen M. E., Ashbolt N. J., Jahne M. A. and Garland J. (2017). Risk-based enteric pathogen reduction targets for non-potable and direct potable use of roof runoff, stormwater, and greywater. *Microbial Risk Analysis*, **5**, 32–43, https://doi.org/10.1016/j.mran.2017.01.002

Schreiber C., Heinkel S.-B., Zacharias N., Mertens F.-M., Christoffels E., Gayer U., Koch C. and Kistemann T. (2019). Infectious rain? Evaluation of human pathogen concentrations in stormwater in separate sewer systems. *Water Science and Technology*, **80**, 1022–1030, https://doi.org/10.2166/wst.2019.340

Shahirnia M., Vidrio-Sahagún C. T., He J., Valeo C., van Duin B., Beaudry M. and Neumann N. F. (2023). Land use and rainfall influences on bacteria levels and sources in stormwater ponds. Manuscript submitted for publication.

Sharvelle S., Ashbolt N., Clerico E., Holquist R., Levernz H. and Olivieri A. (2017). Risk-based Framework for the Development of Public Health Guidance for Decentralized non-Potable Water Systems. Prepared by the National Water Research Institute for the Water Environment & Reuse Foundation. Alexandria, VA, Vol. **2017**, pp. 3799–3809. WE&RF Project No. SIWM10C15, https://www.nwri-usa.org/_files/ugd/632dc3_8831385f1c2f4bb1b2976b06719832ae.pdf?index=true

Sidhu J. P. S., Hodgers L., Ahmed W., Chong M. N. and Toze S. (2012). Prevalence of human pathogens and indicators in stormwater runoff in Brisbane, Australia. *Water Research*, **46**, 6652–6660, https://doi.org/10.1016/j.watres.2012.03.012

Smith E. D., Siefert W. T. and Drain D. (2009). Risk matrix input data biases. *Systems Engineering*, **12**, 344–360, https://doi.org/10.1002/sys.20126

Snider B. and McBean E. A. (2021). Combining machine learning and survival statistics to predict remaining service life of watermains. *Journal of Infrastructure Systems*, **27**, 04021019, https://doi.org/10.1061/(ASCE)IS.1943-555X.0000629

Soller J. A., Schoen M. E., Bartrand T., Ravenscroft J. E. and Ashbolt N. J. (2010). Estimated human health risks from exposure to recreational waters impacted by human and non-human sources of faecal contamination. *Water Research*, **44**, 4674–4691, https://doi.org/10.1016/j.watres.2010.06.049

Soller J. A., Schoen M., Steele J. A., Griffith J. F. and Schiff K. C. (2017). Incidence of gastrointestinal illness following wet weather recreational exposures: harmonization of quantitative microbial risk assessment with an epidemiologic investigation of surfers. *Water Research*, **121**, 280–289, https://doi.org/10.1016/j.watres.2017.05.017

Sörensen J. and Emilsson T. (2019). Evaluating flood risk reduction by urban blue-green infrastructure using insurance data. *Journal of Water Resources Planning and Management*, **145**, 04018099, https://doi.org/10.1061/(ASCE)WR.1943-5452.0001037

Sousa V., Silva M., Veigas T., Matos J., Martins J. and Teixeira A. (2007). Technical management of sewer networks: a simplified decision tool. 2nd Leading Edge Conference on Strategic Asset Management (LESAM), Lisbon, Portugal.

Staley Z. R., Boyd R. J., Shum P. and Edge T. A. (2018). Microbial source tracking using quantitative and digital pcr to identify sources of fecal contamination in stormwater, river water, and beach water in a great lakes area of concern. *Applied and Environmental Microbiology*, **84**, e01634, https://doi.org/10.1128/AEM.01634-18

Steele J. A., Blackwood A. D., Griffith J. F., Noble R. T. and Schiff K. C. (2018). Quantification of pathogens and markers of fecal contamination during storm events along popular surfing beaches in San Diego, California. *Water Research*, **136**, 137–149, https://doi.org/10.1016/j.watres.2018.01.056

Tahmasebi Birgani Y., Yazdandoost F. and Moghadam M. (2013). Role of resilience in sustainable urban storm water management. *Journal of Hydraulic Structures*, **1**, 44, https://jhs.scu.ac.ir/article_10074.html

Tarigan A. P. M., Rahmad D., Sembiring R. A. and Iskandar R. (2018). An application of the AHP in water resources management: a case study on urban drainage rehabilitation in Medan City. *IOP Conference Series: Materials Science and Engineering*, **309**, 012096, https://doi.org/10.1088/1757-899X/309/1/012096

Texas Water Development Board (TWDB) (2006). Rainwater Harvesting Potential and Guidelines for Texas. https://www.twdb.texas.gov/innovativewater/rainwater/doc/RainwaterCommitteeFinalReport.pdf

Tira M. (1997). Pianificare la città sicura, Edizioni Librerie Dedalo, Roma, Italy.

Tscheikner-Gratl F., Mikovits C., Rauch W. and Kleidorfer M. (2014). Adaptation of sewer networks using integrated rehabilitation management. *Water Science and Technology*, **70**, 1847–1856, https://doi.org/10.2166/wst.2014.353

Tscheikner-Gratl F., Egger P., Rauch W. and Kleidorfer M. (2017). Comparison of multi-criteria decision support methods for integrated rehabilitation prioritization. *Water*, **9**, 68, https://doi.org/10.3390/w9020068

Tscheikner-Gratl F., Caradot N., Cherqui F., Leitão J. P., Ahmadi M., Langeveld J. G., Le Gat Y., Scholten L., Roghani B., Rodríguez J. P., Lepot M., Stegeman B., Heinrichsen A., Kropp I., Kerres K., Almeida M. do C., Bach P. M., Vitry M. M. de, Marques A. S., Simões N. E., Rouault P., Hernandez N., Torres A., Werey C., Rulleau B. and Clemens F. (2019). Sewer asset management – state of the art and research needs. *Urban Water Journal*, **16**, 662–675, https://doi.org/10.1080/1573062X.2020.1713382

United States Environmental Protection Agency (2003). Long Term 2 Enhanced Surface Water Treatment Rule (EPA 815-D-03-009). Office of Water, United States Environmental Protection Agency, Washington, DC, USA.

United States Environmental Protection Agency (2012). Recreational water quality criteria (820-F-12-058). https://www.epa.gov/sites/default/files/2015-10/documents/rwqc2012.pdf

Vahidi E., Jin E., Das M., Singh M. and Zhao F. (2016). Environmental life cycle analysis of pipe materials for sewer systems. *Sustainable Cities and Society*, **27**, 167–174, https://doi.org/10.1016/j.scs.2016.06.028

Vanier D. J. (2001). Why industry needs asset management tools. *Journal of Computing in Civil Engineering*, **15**, 35–43, https://doi.org/10.1061/(ASCE)0887-3801(2001)15:1(35)

Vanier D. J. and Rahman S. (2004). MIIP report: a primer on municipal infrastructure asset management. National Research Council of Canada, Ottawa, Ontario, Canada, https://doi.org/10.4224/20377283

van Riel W., Langeveld J. G., Herder P. M. and Clemens F. H. L. R. (2014). Intuition and information in decision-making for sewer asset management. *Urban Water Journal*, **11**, 506–518, https://doi.org/10.1080/1573062X.2014.904903

Villedieu A., Papesh E., Weinberg S. E., Teare L., Radhakrishnan J. and Elamin W. F. (2018). Seasonal variation of *Pseudomonas aeruginosa* in culture positive otitis externa in South East England. *Epidemiology and Infection*, **146**, 1811–1812, https://doi.org/10.1017/S0950268818001899

Vladeanu G. J. and Matthews J. C. (2019). Consequence-of-failure model for risk-based asset management of wastewater pipes using AHP. *Journal of Pipeline Systems Engineering and Practice*, **10**, 04019005, https://doi.org/10.1061/(ASCE)PS.1949-1204.0000370

Wade T. J., Pai N., Eisenberg J. N. S. and Colford J. M. (2003). Do U.S. environmental protection agency water quality guidelines for recreational waters prevent gastrointestinal illness? A systematic review and meta-analysis. *Environmental Health Perspectives*, **111**, 1102–1109, https://doi.org/10.1289/ehp.6241

Wade T. J., Calderon R. L., Sams E., Beach M., Brenner K. P., Williams A. H. and Dufour A. P. (2006). Rapidly measured indicators of recreational water quality are predictive of swimming-associated gastrointestinal illness. *Environmental Health Perspectives*, **114**, 24–28, https://doi.org/10.1289/ehp.8273

Wade T. J., Sams E., Brenner K. P., Haugland R., Chern E., Beach M., Wymer L., Rankin C. C., Love D., Li Q., Noble R. and Dufour A. P. (2010). Rapidly measured indicators of recreational water quality and swimming-associated illness at marine beaches: a prospective cohort study. *Environmental Health*, **9**, 66, https://doi.org/10.1186/1476-069X-9-66

Wang R. (2022). Fuzzy-based multi-criteria analysis of the driving factors and solution strategies for green infrastructure development in China. *Sustainable Cities and Society*, **82**, 103898, https://doi.org/10.1016/j.scs.2022.103898

Williams N. L. R., Siboni N., Potts J., Campey M., Johnson C., Rao S., Bramucci A., Scanes P. and Seymour J. R. (2022). Molecular microbiological approaches reduce ambiguity about the sources of faecal pollution and identify microbial hazards within an urbanised coastal environment. *Water Research*, **218**, 118534, https://doi.org/10.1016/j.watres.2022.118534

World Health Organization (2005). Water safety plans: managing drinking-water quality from catchment to consumer (WHO/SDE/WSH/05.06). https://apps.who.int/iris/bitstream/handle/10665/42890/WHO_SDE_WSH_05.06_eng.pdf?sequence=1&isAllowed=y

World Health Organization (2011). Chapter 3: health-based targets. In: Guidelines for Drinking-Water Quality, Geneve, Switzerland, 4th edn, pp. 35–44. https://cdn.who.int/media/docs/default-source/wash-documents/water-safety-and-quality/dwq-guidelines-4/gdwq4-with-add1-chap3.pdf?sfvrsn=85204bc2_3

Yu W. (1992). ELECTRE Tri: Methodological Aspects and User Manual. Paris Dauphine University, Paris, 80p (in French).

doi: 10.2166/9781789063059_0271

Chapter 7

Decision-making in urban drainage asset management

Lisa Scholten[1] and Aad Oomens[2]

[1]Faculty of Technology, Policy and Management, Delft University of Technology, Jaffalaan 5, 2628 BX Delft, The Netherlands
[2]Waterschap De Dommel, Bosscheweg 56, 5283 WB Boxtel, The Netherlands

ABSTRACT

Decision-making is at the core of urban drainage asset management (UDAM), but its importance is often underestimated, leading to a lack of improvement of decision quality in practice. Therefore, our objective is to present fundamental concepts and theories of decision-making from literature and compare them with real-world experiences of observing, supporting, and participating in UDAM decisions in the Netherlands. The observations are contrasted against selected observations from other nations to illustrate the potential impact of key factors on decision-making processes and outcomes. From this, we observe that despite the available UDAM literature and experiences suggesting otherwise, decision-making in UDAM practice tends to focus on information acquisition, cognitive processing, and judgmental processes. This can lead to known decision biases such as protection of mindset and following fragmented, path-dependent processes influenced by formal and informal structures or institutions. To improve decision-making in UDAM, it is necessary to look beyond optimization of existing assets within the pre-existing technical paradigm and instead work toward aligning it with governing structures and processes for effective decision-making at a system level. While the existing evidence – although limited and mostly anecdotal – is compelling, it does not allow for generalization or validation of theoretical propositions against practical findings and vice versa. We therefore see a need for strengthened efforts into a more systematic study of current UDAM practices that incorporates existing theories and empirical insights on decision-making from several disciplines. This will foster accumulation of knowledge and mutual learning to enhance the research and practice of UDAM decision-making.

Keywords: urban drainage asset management, decision-making, cognitive heuristics and biases, multi-actor problem-solving, decision quality, governance gaps, process blunders.

7.1 WHY A CLOSER LOOK AT DECISION-MAKING IN URBAN DRAINAGE ASSET MANAGEMENT?

In organizational and public life, the governance, policy making, and management of resources are inseparably intertwined with making decisions and the actions following therefrom. Decision-making is also the main activity through which asset management (AM) organizations align their choices, plans, and actions with their objectives in order to realize value from their assets. The achievement of these objectives relies on the coordinated decision-making of those who manage urban drainage (UD) assets within the given political, socio-institutional, and environmental context. This context is dynamic, involves many stakeholders and interconnected systems of technical assets, formal and informal institutions, and their natural environment. The success of an organization within such a context thus critically depends on the collective decision-making skills of those working in the organization and their ability to constructively interact with other decision-makers about decisions affecting the behavior of the system of interest.

Despite its centrality in urban drainage asset management (UDAM), little attention is being paid to understanding decision-making and decision quality in research and practice. Only a few dedicated studies exist, suggesting a strong reliance on intuition and experience in UDAM decision-making (van Riel *et al.*, 2014, 2015). A large number of decision-support guidelines, analytical methods, and tools developed to aid water infrastructure AM (e.g., Tscheikner-Gratl *et al.*, 2019; Zhang *et al.*, 2014) suggests a perceived need for more analytic, evidence-based decision-making based on objective information and models. Obviously, UDAM is not the only human management activity that relies on decision-making and that is facing challenges in achieving them. Decision sciences provide a more nuanced perspective as to what decision-making is, what it entails as well as whether and when relying on intuition or deliberation using decision-support aids is (un)desirable. 'Decision sciences' is here used loosely as an umbrella term for the multi- and inter-disciplinary study of human decision-making and its outcomes, bridging across the social, natural, and applied (technical) sciences. As this research seems to have gone largely unnoticed in UDAM research in practice, with this chapter we aim to:

(1) Introduce and unpack selected concepts, perspectives, and theories that we deem useful for understanding individual and multi-actor decision-making.
(2) Expose known quality issues in multi-actor decision-making settings and illustrate these against examples from UDAM practice in the Netherlands and other countries.

While the suggested associations between the theoretical aspects and the examples presented are selective, based on our own subjective judgment and not independently validated, we hope these will show possible relationships that warrant further study. By this, our aim is to promote cross-disciplinary learning and enhance comprehension of the challenges encountered in UDAM decision-making research and practice. In closing this chapter, we will point to areas for future research, hoping to inspire constructive discussion on decision-making and decision support for quality UDAM.

7.2 UNPACKING DECISION-MAKING

7.2.1 Decision-making perspectives

The Cambridge Dictionary defines decision-making as 'the process of making choices, especially important choices' (Cambridge Dictionary. Last accessed 10 August 2022, https://dictionary.cambridge.org/dictionary/english/decision-making). However, much of the literature and everyday language make use of the term decision-making as a synonym to 'choice', suggesting that the process and outcomes resulting from it are the same – which is not the case. Decision-making scholars have studied human decision-making and its underlying processes from many angles. Three dominant philosophical perspectives include normative, descriptive, and prescriptive views, as illustrated in Figure 7.1 and elaborated upon in more detail in Bell *et al.* (1988).

The normative decision-making perspective is evaluative in nature and studies how humans ought to make decisions, as per some general ideals, values, norms, and standards of what is considered good or bad, right or wrong, acceptable or unacceptable. More specifically, it infers rationality in the strict normative sense whereby 'the rational man of economics is a maximizer, who will settle for nothing less than the best' (Simon, 1978, p. 2). A common behavioral assumption that aligns

Figure 7.1 Three perspectives for studying human decision-making.

with normative rationality is the presumption of intentionality in choice, where someone's actions are believed to result from intentional, deliberate, and self-initiated choices (Goschke & Job, 2023). Another assumption is that the chosen action is deemed better than non-action or other choice options based on some appraisal of expected value.

Much present-day decision-making research on behavioral sciences is based on overwhelming amounts of sound empirical evidence that challenges rational decision making assumptions in the strict normative sense as well as the assumption of intentionality being people's (in)actions. Instead, a descriptive perspective is taken to study and explain human decision-making. Descriptive decision-making deals with analyzing and describing how humans make decisions in the real world under uncertainty and cognitive complexity. From the descriptive perspective, decision-makers are 'boundedly rational' (Simon, 1982a, 1982b), that is, have limited knowledge, time, as well as limited cognitive and information-processing capacities. Therefore, not all potential alternatives and their consequences are considered. Instead, 'satisficing' strategies (i.e., from satisfy and suffice) are used wherein the first alternative that satisfies some minimal acceptability standard is chosen without exploring other options.

Under bounded rationality, intuitive and habitual responses go along with cognitive heuristics and shortcuts that serve to cope with the cognitive complexity of real-world decision-making. These so-called 'cognitive biases' resulting therefrom often refute the normative demands of substantive rationality and are frequently characterized as undesirable deviations that are to be avoided (Gilovich *et al.*, 2002; Tversky & Kahneman, 1974). Other scholars, however, argue that the use of cognitive heuristics is desirable and efficient in many environmental and social contexts, considering them important 'fast and frugal heuristics' that make up our 'adaptive toolbox' (Gigerenzer, 2004; Artinger *et al.*, 2015).

Studies into interventions to improve decision-making generally follow a 'prescriptive' perspective to decision-making. Prescriptive decision-making bridges normative and descriptive perspectives by studying how real people can make better decisions as informed by how people decide in the real world and how they can decide by adopting certain processes, decision aids, and decision-making behaviors leading to better decisions following common normative aspirations. This perspective endorses 'procedural rationality,' or 'the effectiveness, in light of human cognitive powers and limitations, of the procedures used to choose actions' (Simon, 1978). Given the need to make thousands of decisions every day, it makes sense to rely on satisficing, intuition, and automation to reduce the effort rather than taking a maximizing approach to identify and implement the optimal choice. Recognizing when intuitive decision-making approaches may lead us astray and when prescriptive decision-support and intervention are needed to achieve normative aspirations remains a subject of scholarly debate (Kahneman & Klein, 2009).

7.2.2 Individual-level decision-making

Empirical studies show that humans make use of different decision-making approaches, or strategies, reflecting a trade-off between time or effort and accuracy (Payne *et al.*, 1993). The approach is determined as prompted by the perceived situational cues, often without the decision-maker consciously noticing. In line with that, the overly simplistic but useful model of managerial decision-making by Schoemaker and Russo (1993) differentiates four common approaches in managerial decision-making and arranges these in a pyramid. This pyramidal structure reflects the frequency with which approaches are used and the degree of intuition or analysis they entail. Intuition herein refers to the associative processing and quick appraisal of a specific situation as perceived by the decision-maker and the contextual cues they are paying attention to. Accordingly, (1) intuitive judgments from intuition are most frequently used and require the least effort, followed by (2) rules and shortcuts, (3) importance weighting, and finally, (4) value analysis as the most analytical, deliberative, and therewith accurate, yet also most strenuous and least common approach. This pyramid is reflected in Figure 7.2, wherein the four decision-making approaches are mapped against the effort–accuracy trade-off of Payne *et al.* (1993) and another simplistic, yet useful theory of decision-making governed by 'two systems thinking' described by Kahneman (2011).

The effort–accuracy trade-off and the activation of different decision-making strategies by different cognitive processes have been confirmed by other studies that inspired later 'dual-processing theories' such as Kahneman's two systems. Already in the 1970s, Posner and Snyder (1975) presented a model that distinguished between automatic and conscious processing. Thereafter, several other dual-processing theories were developed that distinguish intuitive and reflective processes in learning, cognition, and judgment (e.g., reviewed in Evans, 2008). These theories are widely used to explain the occurrence of mental shortcuts in cognitive processing that underlie human judgment and decision-making and which contradict mathematical logic and economic rationality. They are often referred to as 'heuristics and biases,' following the seminal study of Tversky and Kahneman (1974). The heuristics and biases supposedly arising from dual processing in the brain are attributed to all kinds of 'irrational' thinking features (aka 'system 1' thinking) such as being automatic, quick, low-effort, and uncontrolled as opposed to 'rational' ('system 2') thinking that is conscious, slow, effortful, and controlled (Kahneman, 2011; Stanovich & West, 2000).

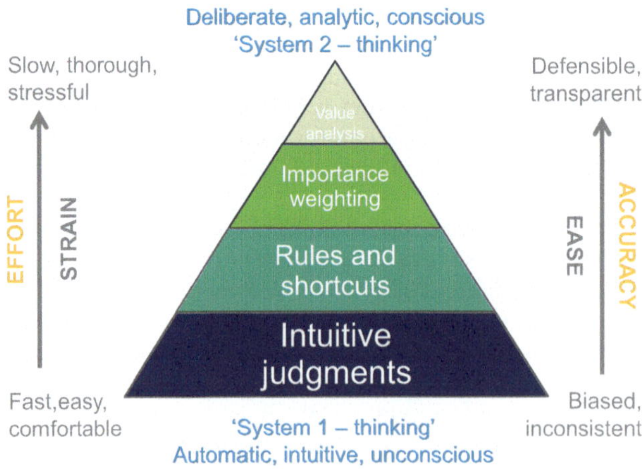

Figure 7.2 Overview of decision-making theories, mapping Schoemaker and Russo's (1993) 'Pyramid of Decision-making Approaches' against the trade-offs of Payne *et al.*'s (1993) effort–accuracy framework, Kahneman's (2011) 'System 1 and System 2 thinking' and attested implications for speed, consistency, affective response, and normative qualities from various 'dual processing theories' (Evans 2008).

When these dual-processing theories emerged in the mid-1970s, functional and structural neuroimaging was still in its infancy (Filler, 2009). In other words, it was not yet possible to link distinct locations of activity in the brain to the associated, abstract mental processes. More recent neuroscientific studies have debunked dual-process theories as too simplistic. As argued by Melnikoff and Bargh (2018), a multi-dimensional typology would better suit the observed interactions between two system features in automatic or deliberate information and emotion processing in the brain to describe, explain, and predict decision-making patterns that rely on both intuition and deliberation. Whether and in which situations reliance on automatic processing and intuition is useful or detrimental depends on the nature of the decision-making task and its context (Myers & Myers, 2002). Jumping out of the way when an object is racing toward you at very high speed will be a well-adapted use of intuition leading to immediate action, whereas rejecting a job candidate based on how well their appearance and demeanors fit with that of the 'in-group' is not. In the first situation, the reliance on intuition instead of pausing to deliberate about whether the object is likely to hit before acting is possibly lifesaving. In the second situation, the decision-maker may be discarding the candidate who is more suitable for the job and end up hiring someone who may look, speak, and act the way others do but is otherwise incompetent.

More recently, the impact of emotions on judgment and decision-making has received significant attention to explain real-world decision-making phenomena. Emotions have been shown to significantly influence decisions via content and depth of thought, as well as goal activation in individual and interpersonal decision-making (Lerner *et al.*, 2015). Another important area of contemporary research concerns 'choice architecture,' which studies the influence of contextual cues, framing, and environmental (re)structuring on automatic processing on behavior (Johnson *et al.*, 2012). Many 'nudging' interventions rely on changing the choice architecture to bring about desired choice outcomes (Münscher *et al.*, 2016). The above reflects only a small selection of well-studied concepts and theories to understand human decision-making. While too simplistic to fully explain and predict decisions, they can be useful to identify and characterize aspects that significantly influence decision-making.

7.2.3 Multi-actor decision-making in and across organizations

Good UDAM requires coordinated action and decision-making of several individuals or groups. We refer to them as 'actors' to reflect their significant interest and leverage to influence the state of UD systems (Hermans & Cunningham, 2013). Within the AM organization, management of UD assets will typically involve actors holding different content expertise, typically, technical planning, technical operation and maintenance, finance, human resources, management, and management levels (operational, tactical, and strategic). The formal relationships within the AM organization are governed by the organizational hierarchy and distribution of tasks, as well as the relationship between these organizations. In the case of UDAM, these relationships are largely determined by the governance context. Governance here refers to the political and institutional frameworks to determine the distribution of roles and responsibilities and within which values and goals are defined and codified into the various political, administrative, legal, social, and economic systems. Many different governance arrangements exist that typically vary between countries (Romano & Akhmouch, 2019). The administrative, functional, and spatial overlaps resulting therefrom make interaction and coordination of activities with other sector organizations necessary such as national/regional/local government oversight bodies (e.g., related to water resources, environmental protection, public health), other infrastructure managing bodies (e.g., municipal road, energy, housing, public greens departments) as well as consultants and contractors. Public authorities have a wide array of policy instruments to influence the management of the various resources in accordance with their goals such as legal and regulatory instruments, economic incentives, instruments for information, and enablement. Management then focuses on specific actions and operational activities to achieve the goals (OECD, 2011).

These formal and informal socio-institutional structures and procedures, along with the broader socio-cultural context, shape individual and multi-actor decision-making behavior across vertical and horizontal decision-making levels. Interactions between individual and groups of actors are influenced by asymmetries in power, knowledge disparities, resource availability, as well as the different interests and action options available in line with their respective roles, responsibilities, and entitlements. It is within this force field that actors strive to identify and solve problems, using the actions and resources at their disposal within the given context. However, no single actor or organization has complete control over all options and resources. Different actors have different roles and responsibilities, such as formulating regulations or monitoring existing assets. The feasibility of these options depends on the availability of necessary resources such as funding, personnel, expertise, mental capacity, and time.

Given this situation, it is not surprising to find misalignment between the actors' system perceptions, needs, and goals, and therewith perceived problems, and possible solutions. Such misalignments in system boundary perceptions are often attributed to a lack of systemic and collective change (Pluchinotta *et al.*, 2022). In such situations, conflicting values, priorities, and demands contribute to a host of political behaviors wherein actors seek to set agenda's, influence the definition of goals and values guiding decision-making and action. Three well-known phenomena affecting multi-actor decision-making in this regard are (1) goal conflicts, (2) action conflicts, and (3) dilemmas (Hermans & Cunningham, 2018). Goal conflicts occur when actors have different desired outcomes for the same goal. Action conflicts arise when one actor's decision negatively affects another actor's goal. Dilemmas arise when an action has both positive and negative consequences for an actor's goals, requiring a trade-off.

The picture would be incomplete without consideration of (inter)personal needs that actors seek to satisfy, resulting in social dynamics that are not often considered in understanding multi-actor decision-making and decision support for UDAM. Social structures have to do with 'the distinctive, stable arrangement of institutions whereby human beings in a society interact and live together' (Wilderdink & Form, 1999). Social structures provide social identity, such as the mission of an organization or work unit and the legitimation for its existence (Smith *et al.*, 2015). Social structures and identity are closely intertwined with social norms that prescribe acceptable behavior within a group and determine who may legitimately wield power or influence over others. Social cognition arises within the context of social structures, it 'refers to a set of neurocognitive processes underlying the individuals' ability to "make sense of others' behavior" as a crucial prerequisite of social interaction' (Arioli *et al.*, 2018). It is within these socially construed structures that actors will seek to realize their personal and group's interests.

Empirical studies have shown the relationships between evolving social values, institutional arrangements, water management paradigms, and technical infrastructure (Hering *et al.*, 2012; Kiparsky *et al.*, 2013; Wong & Brown, 2009). Their impact on decision-making processes and outcomes are often overlooked in UDAM. With changing environmental and social conditions, multi-actor conflicts may arise around the distribution of material and social resources. Dynamics observed in this regard include 'in-group' vs. 'out-group' behaviors, such as mounting commitment to one's own group along with polarization between groups and in extreme cases seeking revenge and retribution (Smith *et al.*, 2015). These factors impact group decision-making performance, including biases in information search and processing, emotion and information cascades, and political behavior aligned with specific group perspectives and interests (Janis, 1982; Sunstein & Hastie, 2015). Changing conditions may also result in intergroup cooperation, wherein exploration of common values and goals leads to the construction of a common social identity. Intergroup cooperation comes with its own challenges in relation to multi-actor decision-making, such as an increased need for efforts upholding the (new) social identity and ensuring cohesion (including attracting and retaining valued members) while avoiding behaviors that undermine group cohesion, morale, and coordination.

7.2.4 Decision-making as a multi-actor problem-solving process

Often, we may only become aware that a decision needs to be made when confronted with an urgent or difficult problem, and especially those we cannot solve or decide on our own. A problem-solving process has been conceptualized in many ways, depending on the foci and units of interest of different fields, and whether it deals with individual or organizational problem-solving (Lang *et al.*, 1978). In the literature the terms 'problem-solving' and 'decision-making' are often used interchangeably, although they are not synonymous but complementary parts of a problem-solving or decision-making process. Problem-solving refers to an analytical process through which solutions to overcome a problematic situation are identified; decision-making is about choosing among different alternatives or choice options to reach a conclusion or judgment. It involves analyzing information, weighing different options, and making a choice based on the available information and the desired outcome.

As a result, most problem-solving and decision-making process models include roughly the same phases that span both problem-solving and decision-making (Harrison, 1996; Keeney, 1982; Lang *et al.*, 1978; Lunenburg, 2010; Nutt, 2008). This similarity between processes and phases also extends to policy making and design processes (Enserink *et al.*, 2010). Moreover, these processes may be described interchangeably as a stepwise process or as a cycle with iterations within and in between steps. These models of problem-solving processes and decision-making processes arguably have more commonalities than differences; for a review and comparison of problem-solving process models see, for example, Lang *et al.* (1978). They are conceptualized as a sequence of steps (also referred to as 'phases') that link problem sensing to the implementation of some solution to the problem. The key differences consist in the terminology used to name the steps, the number of steps into which key problem-solving activities are subdivided and the nuances considered.

Following one of the earlier conceptualizations introduced by Simon, problem-solving has at least four phases: intelligence, design, choice, and implementation (Simon, 1991). Intelligence starts from there being a gap between the desired and existing state as gauged on some goal, norm, or expectation, initiating a problem-solving – or improvement – process. This phase also includes situational analysis and gathering of information. Obvious preconditions for this phase being started is that any gap in the current situation or anticipated future situation has been noticed and attention has been paid to it, next to there being motivation to reduce the gap and being able to do something about it (Lang *et al.*, 1978). In some models, this phase may also include exploration and negotiation between stakeholders to come up with a shared, unambiguous formulation of the problem, its causes, and also the identification of preferences, values, and criteria that matter in addressing the problem. During the design phase, alternative solutions are identified through searches or via a purpose created as a means to address the gap. Then, during the choice phase, the alternatives are compared based on some form of appraisal, possibly involving additional analysis and judgment or evaluation against criteria, and a decision is made whether to adopt any of the alternatives and which. Finally, implementation entails actions to put the decision into effect, involving mobilization and coordination of people and resources, and possibly also monitoring of the effectiveness of the decision. Implementation may be linked back to intelligence, initiating another problem-solving cycle if the effect on the initial state is insufficient or the new situation leads to identification of other problems.

These process models depict a process that is in line with procedural rationality, reflecting normative perspectives. Or, as Lunenburg (2010) puts it, the assumption is one of decision-making under certainty where the alternatives and their outcomes on the decision criteria are known, such that an optimal choice can be made that is subsequently implemented. Obviously, that is not how most individuals and organizations make decisions and it is nearly unachievable in real-world situations. Instead, a more boundedly rational, if not completely intuitive, problem-solving, or decision-making process applies. As illustrated by, for example, Rizun and Taranenko (2014), these processes can still be characterized by the same phases, only that they are incompletely run through and in a way reflect bounded rationality and intuition-driven models. The intelligence phase may be restrained to identifying the problem and decision criteria 'sufficiently,' such as only attending to specific cues and monitoring deviation of an aspired state on a focal goal, without deeper comprehension of the problem faced (Lunenburg, 2010). Different system boundary perceptions may bring about different problem perceptions, which in combination with limited cognitive capacity will influence what information is processed and paid attention to when specifying a problem and identifying solutions. Design based on memory, pattern recognition, and limited searches may yield only those alternatives that are easily accessible and familiar (reflecting 'availability bias,' Tversky & Kahneman, 1973). Choice would then be informed by heuristic judgment and use of simple choice heuristics instead of more elaborate and objective appraisals. Implementation would focus on getting it done to achieve a satisfying improvement of the situation within limited effort.

Various other models exist beyond these so-called 'phase' models of decision-making processes, such as the 'garbage can model' (Cohen *et al.*, 1972), the 'multiple streams framework' (Béland & Howlett, 2016), or the 'rounds model.' For an accessible description of these models, see, for example, Teisman (2000) and Enserink *et al.* (2022). These models do not assume for certain phases to be completed in whichever order, but rather focus on different decision-making instances and dynamics in political and policy decision-making contexts. In these, political alliances and struggles around meanings, power, or influence over scarce resources are inherent to the decision-making process, reflecting different worldviews, interests, and roles come together. Many undesirable outcomes have been associated with political behavior in decision-making: self-serving and strategic behaviors may affect information flows; divisive dynamics are time-consuming and may lead to missing out on opportunities or profits due to delays. A focus on internal matters, power bases, and interests may blind a team or organization. This, in turn, may lead to the inclusion or exclusion of alternatives in line with powerful interests and to the overlooking of critical environmental constraints which may result in underperformance or failure to address the problem. Even when considered objectively 'better' for the organization and its members, establishing more rational decision-making processes is typically met by resistance, as it would involve changing existing mindsets, if not the political and organizational structures and procedures.

Research on the success of strategic decision-making processes in organizations (Nutt, 1999, 2008) found that more than half of these processes fail due to obvious process blunders in decision-making. Although considered more efficient, most organizations relied on top-down 'idea imposition' processes characterized by 'see first' tactics. Conversely, 'discovery processes' that endorse analytical 'think first' tactics were significantly more time efficient, had a 90% chance of decision adoption and resulted in higher, more evenly distributed satisfaction of the involved actors.

In the absence of a well-managed multi-actor process, organizational inertia and resistance are to be expected in decision-making, also in urban water management (Brown & Farrelly, 2009; Marlow *et al.*, 2013). A result is a strong tendency toward incrementalism, where only small adjustments (increments) are made with respect to the existing situation involving limited effort, based on a sequence of limited comparisons and change (Elbanna, 2006). Explanations for the prevalence of incrementalism include structural, cognitive, and political factors at individual, group/multi-actor, and organizational/governance levels. Cognitive factors include limitations in the acquisition, search, and analysis of information due to resource constraints. These may quite simply come down to time limitations and role expectations. For example, studies of the time allocation by decision-makers in leadership positions evidence a strong action-orientation where no more than a few minutes are spent on most activities and only a fraction of activities are being attended to for more than an hour (Lunenburg, 2010). Furthermore, important structural constraints are imposed by the need to adhere to institutionalized processes of planning, approval, and implementation or to act in line with certain self-concepts, institutionalized logics, or social norms. Political behavior may involve not upsetting an important ally or powerful opponent, or exploiting situations of asymmetric information to avoid that certain information may be used to one's disadvantage.

Lastly, the abovementioned constellations of factors that may play out within one particular multi-actor decision-making process are not independent from the earlier history of the organization and the actors involved in earlier decision-making processes. Along with technological path dependency, organizational path dependency exists, wherein various socio-institutional dynamics and self-reinforcing mechanisms may lead organizations to become locked into a situation of structural, procedural, and cognitive inertia (Sydow *et al.*, 2009). To summarize, while there are many models reflecting normative aspirations or prescriptive guidance for decision-making processes that foster rationality in solving problems, decision-making reality is better described as some form of 'muddling through' that may combine elements of rational planning with incrementalism, cognitive limitations, and politics.

7.2.5 In search of decision quality

Many decision practitioners and scientists have prescribed ways to improve decision-making by individuals and groups. While achieving normative ideals of decision rationality remains difficult in the real world, increasing decision quality seems very much possible. A decision-quality framework shown in Figure 7.3 builds upon a set of six principles to satisfy to achieve decision quality (Spetzler *et al.*, 2016):

(1) Appropriate framing that clarifies the problem or opportunity to be tackled and its main stakeholders.

Figure 7.3 Decision quality and what it entails (*source*: adapted from Spetzler *et al.*, 2016).

(2) Creative alternatives reflecting possible actions that can be taken by the people partaking in the decision.
(3) Relevant and reliable information amalgamating what is known or believed about the situation, problem, and solutions.
(4) Clear values and trade-offs concerning what one wants or hopes to achieve as well as how much one is willing to give up achieving one value over another.
(5) Sound reasoning in identifying the best alternative given the values and information at hand.
(6) Commitment to action as good intentions alone do not suffice and actions need to be implemented to make a change.

The framework posits that decision-makers should act in line with all six principles as shortcomings on any of these imply that decision quality is unlikely to be achieved. The most critical is to ensure a clear, appropriate framing, and hence clarity of the problem and its key stakeholders (Ackermann, 2012; Keeney, 2004; Rosenhead, 1996). In complex problem and decision-making situations, it is also common for disagreements about the problems or solutions to consider in decision-making to occur (Head & Alford, 2015; Rittel & Webber, 1973). Even with a clear, shared, and appropriate framing, no decision can be better than the outcomes of the best alternative considered. It is common that decision-makers are not clear as to the full range of the values considered relevant to a decision (Bond *et al.*, 2008). Without clarity of the values to fulfill, decision-making often consists of choosing between readily available alternatives following so-called 'alternative-focused thinking' (Keeney, 1996). There are many examples of solutions that address a wrong, unimportant, or even inexistent problem. A common failure to include non-standard alternatives or to design suitable alternatives as guided by the goals is a main barrier in strategic decision-making (Gregory & Keeney, 1994; Keeney, 1996).

The framing of the problem and the process followed to achieve multi-actor commitment, value identification, alternative creation, and appraisal are key to understanding the decision ultimately made and its odds for successful implementation (Nutt, 2008). Even when several objectives and innovative, suitable solutions are considered, challenges may arise when there is an undue tendency to optimize. Optimizing in decision-making refers to a situation wherein a decision-maker is invested in improving on at least one main goal while avoiding foregoing or 'losing out' on another, wherein the status quo serves as a reference. This avoidance of trade-offs when evaluating alternatives can result in a dilemma when one alternative is desirable in terms of improving on some focal goal, yet at the same time undesirable due to incurring a perceived loss on another. Avoidance of trade-offs and loss aversion are one of the several explanations behind procrastination and failing to take a decision especially when the stakes are high. These are but some of the issues and deviations from substantive rationality which are known to arise when making deliberate choices and weighing trade-offs between several goals (Keeney, 2002).

In addition, there are many 'hidden traps' (Hammond *et al.*, 1998) that may lead to various blunders impacting decision quality outside of the conscious awareness of the decision-makers. Cognitive and resource limitations, the use of heuristics, and motivated reasoning within a predefined decision structure or context are only some out of many explanations for what information is being processed and how. Spetzler *et al.* (2016) highlight five types of biases hindering decision quality in organizational decision-making: (1) protection of mindset, (2) personality and habits, (3) yielding to social influence, (4) faulty reasoning about (complexity and) uncertainty, and (5) relying on automatic associations and relative thinking. First, protection of mindset refers to avoiding changes to pre-existing mental models, assumptions, and preferences constructed in line with memories and lessons learned from experience. This includes biases such as sunk cost, status quo, confirmation, and hindsight bias. Overconfidence and avoidance of cognitive dissonance may arise. Second, personality and habits include habitual framing and preference-based habits, or different decision styles. Third, yielding to social influence involves conforming to group norms to avoid rejection and adjusting views to align with others. This can lead to judgment cascades and group thinking. Suggestibility and conformity may give rise to judgment cascades that prevent dissenting with (majority) judgment and groupthink that discourages diversity in views within a group. Fourth, faulty reasoning about complexity and uncertainty arises when dealing with complexity arising from many interrelated variables that may furthermore be highly uncertain. Confusion and inability to combine many cues into a coherent picture may give rise to selective attention and use of substitution heuristics. Lastly, overreliance on automatic associations in line with a certain framing or context can lead to overweighting certain information while discarding others based on vividness, narrative appeal, ease of recall, or familiarity. Altogether, interaction of these traps may give rise to 'decision mega biases' in organizational and group decision-making that undermine decision quality (Spetzler *et al.*, 2016).

To overcome these decision traps, biases, and judgment cascades, many deliberative and analytic approaches are available (e.g., Eisenführ *et al.*, 2010; Greco *et al.*, 2016; Hammond *et al.*, 2002; Howard, 1988; Keeney, 1982; Mingers, 2011; Mingers & Rosenhead, 2004; Rosenhead, 1996; Roy, 1991; Roy & Słowiński, 2013). These deliberative approaches combine a social process intervention with tools and techniques to support deliberation, analysis, judgment, and reasoning throughout a problem-solving and decision-making process. Although the value and potential of such 'discovery processes' has been clearly demonstrated, they are often perceived as arduous or cumbersome (Nutt, 2008). Processes that are perceived to misalign with organizations' mindset, culture, and habitual ways of doing, are likely to get rejected. Therefore, processes to achieve higher decision quality need to go beyond procedural rationality and suitable methods to tackle structural and cognitive limitations. Instead, given the complex nature of problem-solving in multi-stakeholder environments, the process itself may be contested. This is when moving from 'procedural rationality' toward endorsing 'processual rationality' is required (De Bruijn *et al.*, 2010). This, however, requires social process negotiation, moderation, and facilitation skills that many organizations are unaware of or may lack the expertise to implement.

7.3 DECISION-MAKING IN UDAM PRACTICE

7.3.1 Structural water governance and management gaps in UDAM

The Organization for Economic Co-operation and Development (OECD) suggests that multi-level water governance, regardless of countries' institutional features and organization of the water sector, often faces seven categories of 'gaps' when designing and implementing water policy (OECD, 2011). The OECD multi-level governance framework allows to diagnose vertical and horizontal coordination bottlenecks between levels of government, across policy areas (i.e., ministries and public agencies), and between local and regional actors at the sub-national level. These should be diagnosed and bridged in a systemic way as they are strongly interrelated and may reinforce each other.

7.3.1.1 Administrative gaps

The advantage of top-down task distribution is that decisions can be made routinely, and solutions found quickly. Decisions are relatively simple and straightforward. Yet, because tasks in water resources and infrastructure management are divided and different organizations are responsible for different parts, it becomes difficult to integrate AM activities. In the United Kingdom, all water infrastructure is managed by private-sector water authorities. In the Netherlands, different organizations are involved needing collaboration among drinking water companies, municipalities for the sewage system, and water boards for the wastewater treatment, except for the City of Amsterdam, where the utility company 'Waternet' holds combined responsibility for drinking water supply, wastewater conveyance and treatment, and surface water management. Furthermore, the provinces oversee managing groundwater and together with the national institute 'Rijkswaterstaat,' are responsible for the management of major waterways and (inter)national rivers. Moreover, there are challenges that lie at the interface of the different systems and organizational responsibilities for example, the challenges of aging of centralized infrastructure and climate adaptation of cities. There typically is neither clear authority nor framework for making decisions at this interface and institutional fragmentation affects collaboration and planning.

A study from Sweden suggests that the main challenges a public utility company has to overcome to implement an AM model are (Mårtensson & Rumman, 2019): (1) lack of strategic and long-term planning responsibility, (2) unclear division of asset responsibilities, (3) lack of commitment by senior management toward AM systems, (4) absence of standardized risk management, and (5) lack of information sharing between departments. A big risk when facing these challenges is that measures are taken that do not provide an adequate solution or, in extreme cases, only reinforce the problem. Ineffective interventions can affect trust and acceptance by residents as well as other parties and their willingness to invest (more) into UDAM. Interdisciplinarity also creates issues in terms of available knowledge and capacities. For example, the need for in-house technical capacity in Sweden has led to the creation of inter-municipal bodies (Alm *et al.*, 2021). Differences in water quality are not tolerated and economies of scale are being used.

In the Netherlands, a solution to this challenge is sought in closer cooperation between water management organizations as formalized in 2011 in the National Governmental Water Agreement (in Dutch: 'nationaal bestuursakkoord water,' hereafter BAW-2011). Cooperation between municipalities and water boards has been long-standing and is by now well established. An important starting point was the optimization of wastewater systems to greatly reduce emissions. By combining the knowledge and skills of several municipalities and the water boards in so-called work units, the need for interdisciplinary knowledge and technical capacity is being met. Subsequently, another main argument was cost savings in not letting water rates rise too much, which could be achieved through mergers of management organizations as well as sharing of expertise and machinery in operational tasks (Gerritsen & Sterks, 2004; Oosterom & Hermans, 2013). This also called attention to the quality and vulnerability of many organizations. Looking back, much has been achieved, especially in terms of cost savings. Further analysis revealed, however, that cooperation takes place mainly at the operational level and sometimes at the tactical level. It tends to focus on the coordination of activities to achieve cost savings rather than on determining which actions are possible and jointly arriving at the most appropriate joint intervention. More cooperation at the strategic level could lead to a great improvement in UDAM implementation and its outcomes.

7.3.1.2 Objective gaps

Local administrations face multiple pressures and competing agendas within and beyond UDAM. Climate change, population growth, and urbanization as well as increasing environmental and economic concerns, highlight the limitations of traditional wastewater practices and thereby challenge the management of urban water systems. Both in theory and in practice, it has been widely acknowledged that the challenges of the 21st century require solutions that address problems in a more integrated, systemic way. Although the demand for integration is obvious, implementation has proved challenging given the lack of clarity as to misalignment of objectives.

In the Netherlands, cost reduction is an important goal in sewer maintenance and therefore renovation instead of replacement is often chosen. This makes system change aimed at increasing water retention almost impossible. The emphasis on financial resource efficiency seriously hampers the redesign of public spaces with the intent of climate adaptation. Projects to renew roads and sewers would provide opportunity for climate change adaptation; however, they are often not taken advantage of due to a lack of integrated policy (Bassone-Quashie, 2021).

While current urban water challenges clearly need a more integrated approach, practitioners disagree on what such an integrated approach exactly means in terms of goals and solutions. Integration could therefore be described as a 'wicked' problem (Head & Alford, 2015), with practitioners having a different understanding of the opportunities and challenges they should focus on, of what to achieve as to climate adaptation, and of resource recovery or collective replacement. This lack of consensus challenges decision-making, and thus the implementation of integration (Nieuwenhuis *et al.*, 2022).

7.3.1.3 Policy gaps

The specificity of and scale at which policy is formulated can vary widely between different levels of government and government agencies. In the Dutch situation, municipalities have the most direct link to residents. In the plans and projects of municipalities, issues such as nuisance during work, design and appearance, and management are often directly considered. However, in the plans of the national, provincial, and water boards such criteria are often excluded; instead, the focus is entirely on the aspirations and investments required.

The mutual coordination of policies, and the goals that these should achieve, can obviously cause major problems. In the Netherlands, the tasks of a water board may have primarily a long-term focus, for example, considering not only flooding but also replenishing groundwater, combating drought, and strengthening biodiversity (Waterschap de Dommel, 2021). In contrast, solving flooding is the primary concern for many municipalities. The difference in perspective leads to a gap between the relevant policies at hand.

7.3.1.4 Capacity gaps

Within AM, three main roles may be distinguished: asset owner, asset manager, and service provider. Each of these roles has its own scope. The asset owner's role is to translate the perceived needs of stakeholders into goals. The interpretation of goals and risk preferences of asset owners may be different from those of other stakeholders. The asset managers focus primarily on optimizing the existing assets, whereas service providers are primarily concerned with the availability and reliability of the assets.

In a small organization, one is forced to combine these roles. This can be at the level of the asset owner, where sewer system management is only one of many tasks. At the level of the asset manager, for example, several types of assets that are very different in nature might fall under his responsibility, for example, roads, public lighting, sewers, parks, green space, and so on. Ideally, these three roles are filled independently to ensure that the potentially conflicting interests are well balanced in a decision-making process. However, when these roles are combined, such as the asset manager also being the service provider, then one of the roles – typically depending on the knowledge, skills, and interests of the person involved – can become dominant.

Most municipalities in the Netherlands are understaffed. Qualified personnel are scarce and at the same time, the portfolio of tasks continues to broaden. Sectoral standards distinguish several areas of interest, each requiring its own specialized knowledge and skills (RIONED, 2021). In smaller organizations, only one person tends to oversee UD. Knowledge then focuses on policy, sewer system management, or project management. Sector studies have shown that the other task areas then typically do not receive adequate attention because knowledge or skills are lacking. Attempts to call this to the attention of municipal councils have not yielded much to date. Sewer systems are invisible and the lack of attention at present is likely to manifest itself in a few years down the road. Given the delay in feedback, there is no urgency to act.

7.3.1.5 Knowledge gaps

The question is how decision-making within AM can be advanced given the problems in practice and the search for decision quality. The importance of sufficient competency of UD professionals in terms of knowledge and the associated skills was recognized several years ago in the Netherlands. In addition to calling broad attention to the subject and taking a critical look at existing vocational training programs, it was decided to define a minimum required level of knowledge with associated competencies. This model developed by the RIONED Foundation has been called the 'sector standard': RIONED is the sector organization of urban drainage management professionals in the Netherlands and the umbrella organization in which public bodies, industry, and the educational sector collaborate. It distinguishes 12 knowledge fields related to the general activities of UD professionals: (1) policy, law, and regulations, (2) water management, (3) working methods/basic principles of other

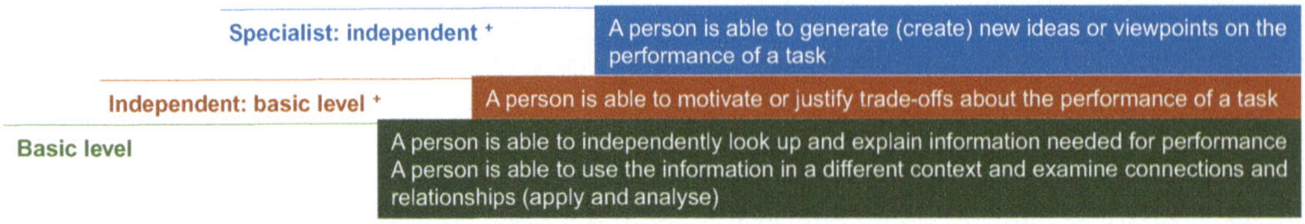

Figure 7.4 Three levels of the Dutch sector standard for UD professional competences.

disciplines, (4) embedding in organization and plans, (5) finance and personnel, (6) research methods and techniques, (7) processing, recording, and validating data, (8) assessing research results, (9) hydraulic calculations, (10) technical knowledge of design, (11) technical knowledge of measures for maintenance, renovation, and replacement of assets, and (12) outsourcing and contracting.

Three competence levels are distinguished (see Figure 7.4). A basic level was determined as the minimum knowledge that would be required to outsource the work and to assess and evaluate the results. This is because many Dutch municipalities – especially the smaller ones – rely heavily on outsourcing to contractors, consultants, or specialized companies. The activities that are most often outsourced include planning and project consultancy, cleaning and inspection, removal of blockages, and rehabilitation works (Oosterom & Hermans, 2013). Billing and troubleshooting services are often still offered by the municipality. A group of experts was asked to indicate the minimum level of knowledge required. In addition to this basic level, the degree of knowledge that must be available to execute the work or tasks independently was also identified, as well as the qualifications required to be considered an expert (see Figure 7.4).

7.3.1.6 Funding gaps

The financial resources at the disposal of organizations are a crucial factor in their ability to manage their staff and assets, which influences the balancing of investment decisions vs. risks of disruption. The use of financial resources for a particular purpose may be time-bound or legally bound to sectoral tasks.

In the Netherlands, there is a dedicated tax or charge for sewerage for which the municipality is responsible and can determine the tax or charge rate (Huisman, 2002). The sewerage charge is an example of a so-called earmarked charge with associated (legal) restrictions as to its use and degree of cost recovery. Weighing of social interests plays an important role in determining the tax or charge rate. Once the tax or charge has been set, asset managers have considerable latitude as to how to use these resources, with operational costs often considered less important than other urgent investments. With limited maintenance and repair budgets being made available for other types of municipal infrastructure as well, collaboration to bridge funding gaps is advisable. This is often achieved using multiple streams of funding; for instance, using road maintenance budgets funded by the municipal mill rate or property taxes and dedicated taxes or charges for water supply and sewer infrastructure (Alm *et al.*, 2021; Bassone-Quashie, 2021).

The Dutch water boards can also levy taxes or charges to fulfill their specific tasks of surface water quality and quantity management, which includes AM of the wastewater treatment plants and the transport of sewage from the urban settlements to the treatment facilities. This 'water treatment levy' is used to finance all the costs of transporting and treating wastewater. The operational costs of treatment are relatively high and therefore the short term (operational costs) and long term (capital expenditures) are treated differently.

Similarly, the availability of funding made available through climate change incentive schemes tends to require submissions for funding or subsidies to be submitted by a certain date. These time limits can act as a major influencing factor barring or enabling UD climate change adaptation projects in the Netherlands (Bassone-Quashie, 2021). Furthermore, as there is no specific levy for the management of urban green space in the Netherlands, it has to be paid from the general municipal tax base which means that the continuity of this funding is not guaranteed but rather subject to the whim of the municipal councils. As these are democratically elected bodies, with inevitable changes in composition and priorities over time, long-term planning of such spaces is highly uncertain.

7.3.1.7 Accountability gaps

An evaluation of the different approaches adopted by the major municipalities in the Netherlands in their approach to 'rainproof cities' illustrates how accountability gaps have become visible in the implementation of climate change adaptation of UD systems (Dai *et al.*, 2018). Under Dutch law, addressing pluvial flooding is a shared responsibility between residents and local government (i.e., the municipality in question). Many different policy instruments exist ranging from local regulations and approval/permitting systems to subsidies and facilitation of participatory projects. Despite common values in terms of equity, flexibility, sustainability, and long-term thinking, the three major municipalities of Amsterdam, Rotterdam, and Utrecht have adopted different approaches to climate proofing via UDAM. On the one hand, the cities of Amsterdam and Rotterdam present climate change and the associated adaptation policies as opportunities to brand their cities or to raise awareness. The city of Utrecht, on the other, is lagging behind without a clear direction that explains or legitimizes its climate change policies (Dai *et al.*, 2018).

In explaining the differences between these three municipalities, Dai *et al.* (2018) identified how clarity about public/private responsibility is important. Urban space in the Netherlands is often privately owned and residents are legally obliged to take care of rainwater on their own property. However, municipalities often act beyond their formal duty of care by collecting runoff from private plots along with runoff from public property to avoid flooding. Consequently, residents are often not aware of their private responsibility and may conveniently shift responsibility to public authorities (Krijnen, 2020). This illustrates how a municipality that is active in addressing pluvial flooding without accompanying public education as to the role of residents on how to protect themselves and their neighbors can result in passivity among its residents, which hampers effective climate change adaptation.

Altogether, the use of policy instruments plays an important role. Binding rules and obligations provide clarity to the division of roles and responsibilities, thus increasing legal certainty. They also make compliance and enforcement easier. Although the soft nature of non-binding policies can yield flexibility to municipalities, accountability mechanisms need to be in place to implement the necessary measures.

7.3.2 Process blunders hindering decision quality and decision implementation in organizations

7.3.2.1 Strategy blunders hindering decision-making success in organizations

Bassone-Quashie (2021) attempted to match decision-making process types and tactics as per Nutt (2008) to interviews conducted about decision-making in 12 sewer system AM and climate change adaptation projects. She found that projects were predominantly initiated by classical UDAM needs such as sewer rehabilitation and neighborhood redevelopments and not due to climate change adaptation needs. Rather, projects targeting standard UDAM works were used as an opportunity for climate change adaptation. According to the studies of Nutt (2008), opportunity-driven processes are on average less successful than discovery processes. In line with that, Bassone-Quashie (2021) observed that the projects involving climate change adaptation generally required bargaining tactics during the selection and evaluation phase (which often involved additional actors), whereas projects addressing rehabilitation projects would rely on analysis. Although half of the projects roughly aligned with decision-making tactics of a discovery process (i.e., the more desirable/successful strategies), there was a strong reliance on less effective tactics for solution development and evaluation in decision-making. This entailed, for example, deciding by edict and persuasion rather than intervention with sound reasoning or structured actor participation (Bassone-Quashie, 2021).

7.3.2.2 Failure to clarify the problem or opportunity to be tackled and its main stakeholders

The first principle that needs to be satisfied to enable decision quality is an appropriate framing that clarifies the problem or opportunity to be tackled and its main stakeholders. The organization of water management influences strongly the framing and to what extent the drinking water supply, UD, and sewerage treatment facilities are managed by one or several organizations. Different organizations all have their own organizational setup, roles, and responsibilities and administrative processes with respective stakeholders. When all UDAM facilities are controlled by one entity, the coordination and decision-making take place at the asset owner level. All interests will be taken into account simultaneously. This coordination does not usually take place when different organizations are responsible for water management. The structure of a decision-making process is different for each of the organization (e.g., drinking water – private, municipal and water board – public/government). The coordination then often takes place at the level of asset managers. Given the interdependence of other tasks in the organizations involved, this coordination becomes difficult, and trade-offs will often be made at the level of the service provider. The tendency is to put the interests of infrastructure (e.g., lifetime, disruption) at the center of the considerations because underlying societal interests are more difficult to weigh at this level.

The situation in Flanders (Belgium) is a good example, where different organizations are discussing coordination of operational efforts (Vlaamse Regering, 2020). For example, the Flemish government has made arrangements between the various managing organizations (i.e., Aquafin, Aquaflanders, de VVSG, VMM, and Vlario) to achieve optimal coordination of AM of the municipal and regional sewerage networks. A characteristic of these differences is the level at which coordination takes place, that is, at the level of the asset owner, the asset manager, or the service provider. The underlying problems and goals cannot be compared and are therefore not discussed.

7.3.2.3 Missing creative alternatives for actions by the people partaking in the decision

In evaluating business cases, the economic book and replacement values of existing assets often play a dominant role. Municipalities, water boards, and drinking water companies have invested enormous sums of money in their infrastructure over the decades to achieve the current service levels. There is thus a high sunk cost, wherein the existing technical infrastructure creates path dependencies that limit changes or new developments (Maurer, 2022). In an economic assessment, rehabilitating existing assets like-for-like is often more beneficial than implementing more innovative technology that requires adapting other parts of the UD system as well.

For example, in the region around the city of Eindhoven in the Netherlands, the Water Board of De Dommel operates one large wastewater treatment plant with a reasonably high treatment performance (i.e., mostly nutrient removal). Over the last century an extensive sewer network was created to connect surrounding municipalities to this treatment plant. The realization of a new smaller treatment plant for one of the furthest small municipalities could not be economically realized. Upgrading or optimization of the existing assets was considered cheaper and hence preferred (Waterschap de Dommel, 2021).

7.3.2.4 Relevant and reliable information and beliefs about the situation, problem, and solutions

The extent to which AM practice has matured within organizations can greatly influence the availability and reliability of information. As a result, the perception of a problem and its possible solutions differ. The data that are available or can be readily obtained tend to receive more initial attention and can prejudice the perception of the problems and required action away from the overarching or more important values and goals. This issue can be exacerbated when the search for information is mostly guided by the available information which may, in fact, be questionable in itself. A case in point is the frequent use of modeling in UDAM practice given the increasing availability of data and the many potential associated benefits (Eggimann *et al.*, 2017). Given the common use of modeling in water management in the Netherlands, one could conclude that all issues require modeling. However, the risk of creating insights based on models is whether or not the decision-making problem lends itself to being modeled and whether it can be sufficiently validated given the available data.

Poor availability and quality of data is often attributed to limited resources, personnel capacity, and knowledge within smaller organizations. Therefore, cooperation between municipalities is strongly encouraged in the Netherlands and formalized via the BAW-2011. Yet, in these cooperations between municipalities operational issues dominate the discussion, reflecting shared information bias and a need for mutual agreement and understanding in decision-making. This hinders the clarification of, discussions about, and balancing of values within and across organizations. A similar problem exists in decision-making between municipalities and water boards. For example, Water Board De Dommel is struggling with its own data management, as different systems are in use and data cannot be validated. As a result, there are difficulties with assessing the condition of pumping stations and pressure mains. The city of Eindhoven, which lies within the catchment area of the De Dommel Water Board has more advanced data management and more accurate condition data. Mutual coordination and decision-making are severely hampered by these differences. In contrast, Waternet (the water utility company serving the city of Amsterdam) has been evolving its AM for years, moving toward an integrated water management approach that considers operational, tactical, and strategic AM aspects.

7.3.2.5 Sound reasoning in identifying the best alternative given the values and information at hand

The concept of a business value and risk matrix is widely used in AM and decision-making in the Netherlands and beyond. An important question pertains to how values are defined. Figure 7.5 shows an example of a typical business value matrix. Three different kinds of potential consequences are presented: finance, safety, and reliability. There are several challenges. First, three values (and their respective goals) are rarely adequate to reflect the various aspirations that organizations take into

	Potential consequences			Likelihood				
	finance	safety	reliability	unlikely	remote	probable	annual	monthly
extreme				moderate	high	very high		
serious				low	moderate	high	very high	
considerable				neutral	low	moderate	high	very high
moderate				neutral	neutral	low	moderate	high
small				neutral	neutral	neutral	low	moderate
negligible				negligible	negligible	negligible	negligible	low

Figure 7.5 Exemplary business value matrix with relevant risks commonly used in Dutch AM.

account in their decision-making. Thus, while tractable, the matrix is often overly simplistic in nature and more appropriate to guide and justify intuition rather than being able to provide tangible decision support in weighing different trade-offs between goals and respective risks.

The inability to explicitly specify all relevant objectives is a known limitation frequently mentioned in decision-making research (Bond *et al.*, 2008). Once the values and goals have been defined, an important question arises as to how to quantify and weigh the different consequences as well as the different risk levels attached to these consequences. Financial consequences are largely quantitative and relatively easy to determine. Reliability is somewhat more difficult but in essence also quantitative. Near misses or unsafe situations are typically of a more qualitative nature. In UDAM these considerations are often handled in a compensatory manner in simplified models or tools, even when they are in fact incommensurate (e.g., a low likelihood of serious injury or death is not commensurate with a high likelihood of a small impact on financial or service outcomes.

In decision-making the consequences that can be quantified with a reasonable amount of effort - generally governed by data that have been routinely collected and processed following current business, operation, and management processes - are over-emphasized. At least, within the technical realm, there is a tendency to rate quantitative information as more objective or valuable than qualitative information. Something that can be measured and displayed as a simple number may receive more attention than a qualitative elaboration of a value. As a result, easily available or accessible quantitative information receives more attention. Depending on the background of the decision-makers, the number of technically educated managers may dominate, which will lead to an even stronger emphasis on the quantitative aspects. To overcome this challenge, attention is needed to define the relevant values that matter. Only then can the appropriate indicators to be used for assessments be selected and decisions be made as to how these will be appraised individually and in concert (Keeney & Gregory, 2005).

7.3.2.6 Commitment to action

The Netherlands is well-known for the 'polder model' way of decision-making (which is even called 'polderen' in Dutch). Problems and possible solutions are discussed by a large group of people with a wide range of stakeholder participation. It takes time to express all different positions and reasoning. The enthusiasm and vigor of individuals can strongly influence decision-making. The impact of selected individuals is not the only problem that may occur. Multiple sessions are typically needed to reach an understanding of the problem, consent on desirable interventions, and a decision for implementation. Decisions are easily revisited again for additional debate at the next session whenever the composition of the group changes. As the respective problem framing, values at stake and potential solutions continue to change, problems tend to drag on and support may even disappear before arriving at a final decision.

7.3.3 Biases in organizational decision-making

7.3.3.1 Protection of mindset

Climate change adaptation efforts may sometimes be at odds with established practice. Appropriate solutions may differ across jurisdictions, which take a lot of effort to identify, design, and implement. When standard solutions or routine measures do not fit, new solutions have to be devised. However, a new, non-standard solution generally is perceived to have a higher risk because it is not well-known, and it is hard to anticipate potential problems and create contingencies for dealing with them. Two examples from the Netherlands include:

- issues with the maintenance of vacuum sewers along with post-construction structural adaptations in houses to meet noise requirements for a novel blackwater collection and treatment system in the city of Amsterdam (Waternet, n.d.), and

- the costly construction of a sewer network in a residential area in the city of Almere-Oosterwold. There, the originally intended nonconventional solution with decentralized treatment, which was based on citizen buy-in, adoption/and action, did not meet the required effluent standards (van Karnenbeek *et al.*, 2021).

Those accountable for decision-making, such as technical asset managers in the case of UDAM, may shy away from novel solutions given the lack of experience and uncertainty about longer-term functioning and associated costs. Pressure on budgets is always present and the unconscious tendency is to avoid risk and keep costs as low as possible. This is unfortunate in that one may thus overlook solutions that may ultimately prove to be attractive and more cost-effective. Small-scale pilot projects where failure scenarios have been thought through and addressed and where post-mortems are executed provide opportunities to overcome the inherent inertia.

Available capacity is a major problem at many municipalities in the Netherlands. Not all proposed works are carried out and programmed investments are only partially realized. Benchmarking of sewerage works showed that, on average, only 60% of all plans had been realized (RIONED, 2005). Management faced with this challenge routinely chooses outsourcing and using external, third-party resources as a solution. The underlying belief is that staff capacity can never be the problem; after all, if there is 'enough' money, this is merely an administrative issue. Reality often tends to be different. Even when external, third-party resources are utilized, data are still needed, procedures must still be adhered to, and coordination with other activities in the public realm must still take place.

7.3.3.2 Personal norms and habits

In a technical environment, for example, UD management, there is a very strong reliance on accumulated experience. A proposed measure is only taken seriously as a solution if it has been successfully applied, in real-world practice within the jurisdiction in question, several times before (Bassone-Quashie, 2021).

In the case of climate change adaptation, this can be a major roadblock. Climate change adaptation requires new infrastructure solutions, or a solution must be sought in other realms, for instance by redesigning public space or adapting the water system. Yet, there may be no clear, standardized, legally 'waterproof' guidelines for designing, building, commissioning, and maintaining new solutions. As long as these guidelines are not in place, implementation of new solutions may require a level of initiative and self-reliance to accommodate unfamiliar practices – often including additional demands – which the staff in charge may not be able or willing to provide.

A second example is the transition to relining and the slow diffusion of its use as an alternative to (open trench) sewer rehabilitation. Standard processes for planning, scheduling, and rehabilitating sewers entailed the open-trench replacement or repair of sewers. Relining required adapting these. The uncertainty about the remaining life span and worries about household service connections were additional reasons not to change the traditional open-trench practice for a long period of time. Successful application elsewhere and the broad sharing of experiences changed this over time as the benefits became more widely apparent and the social norms around viable practices changed.

7.3.3.3 Yielding to social influence

Earlier, when climate change was not yet a major issue discussed in the public realm, government bodies were strongly motivated to relieve their citizens and to take full control of and responsibility for UD. Times have changed and the Dutch law and regulations clearly state that households are responsible for managing rainwater (Dai *et al.*, 2018). Only when households cannot reasonably be required to discharge the rainwater runoff onto or into the soil or into surface waters (e.g., in an old city center) must the government take action by draining stormwater from the boundary of the private property (Dutch Parliament, 2006).

In the mindset of many municipal politicians and the technicians to whom tasks are delegated, the customary social norm of the municipality taking full responsibility for storm water drainage is still valid. Limited organizational capacity in facilitating citizen action while ensuring coordination with public measures is another perceived bottleneck. As a result, approaches seeking to find solutions within public space under full legal, financial, and operational control of the municipality dominate and other solutions are often not even considered.

7.3.3.4 Faulty reasoning about uncertainty

Most of our present-day drainage infrastructure was designed decades ago, when dimensions were determined using relatively simple models and rules of thumb. One of the most basic principles used is that, given certain restrictions, systems can be designed independently. Under extreme conditions of pluvial flooding or prolonged drought, interference between UD and surface waters becomes apparent in terms of water quantity or quality impacts. This complexity of interference is often neglected by continuing to design systems independently. The occurrence of extreme rainfall events is simply taken into account by designing bigger pipes and centralized treatment and pumping capacity, rather than integrating alternative solutions to retain, store, evaporate, treat, and discharge or infiltrate drainage water locally. This may be required within the lifespan of the infrastructure with advancing climate change impacts, yet is given insufficient consideration given limited understanding and consideration of uncertainty and exponentially shifting patterns in drought or rainfall frequency and intensity.

7.3.3.5 Relying on automatic associations and relative thinking

A popular measure for government organizations to introduce or support new policies is to offer financial incentives. To speed up climate change adaptation the Dutch government and several water boards offer municipalities financial incentives through subsidies or co-financing if they realize projects in which climate change adaptation is considered. There are many different arrangements to get funding. Most municipalities are understaffed, and evaluations related to funding are deemed critical, receiving high corresponding levels of attention. Because time is limited, the question often no longer pertains to how much money is needed and where to invest it to achieve the highest return on investment with respect to climate change adaptation, but rather which project can most easily be used to get additional external funding. Getting additional funding is highly rated by many municipal officials and may also reflect positively on their career prospects within the organization.

Furthermore, when looking for solutions, a person's background subconsciously plays a major role in terms of which solutions come to mind and are hence explored. Operations staff who have risen through the ranks to gradually obtain management positions are likely to reason differently than staff having a, say, urban planning or business administration background. Someone who is predominantly familiar with underground infrastructure will likely consider, and trust, such solutions more readily than others who may be familiar with above-ground possibilities to reduce or eliminate nuisance from stormwater flooding.

7.3.4 Outlook

As outlined briefly in Section 7.2.5, there are tried and tested methods that can be used to improve decision-making and decision quality in UDAM. These combine multi-actor decision-making process support along with deliberative and analytic methods and tools to ensure all elements of decision quality are properly addressed. However, only a few such applications have been reported in UDAM.

The state of the art in UDAM thus centers around either 'analytical' measures or (re)enforcement of 'structural' measures. For example, swift change in governance and institutional arrangements including legal, regulatory, and fiscal policies were shown to be highly effective in achieving the development and implementation of proactive, systematic AM efforts in the Portuguese water sector (Alegre *et al.*, 2020). Also, several examples of stepwise processes for indicator identification and assessment by AM organizations (e.g., Cardoso *et al.*, 2012) or oriented toward the application of multi-criteria analysis methods exist (e.g., Carriço *et al.*, 2021; Joubert *et al.*, 2003; Sa-nguanduan & Nititvattananon, 2010; Scholten, 2013; Zheng *et al.*, 2016). They typically include the clarification of goals, of ways to assess these goals, and of ensuring data quality when assessing the current state of the infrastructure and planning into the future. There are, however, only very few documented studies on more processual approaches of how a multi-actor decision-making process can be designed and supported using deliberative methods that are, in practice, capable of addressing the contextual and multi-actor complexity. A few examples include Lienert *et al.* (2015); Alhamed (2020); Scholten (2017, 2018); and Scholten *et al.* (2015).

Of these, the study of Alhamed (2020) has seen continued uptake in practice. During her Professional Doctorate in engineering activities at the Delft University of Technology and Waternet in the city of Amsterdam, Alhamed (2020) bridged the gap between academia and strategic developments in UDAM practice at Waternet. Her research combined stakeholder analysis and multi-actor problem structuring with a tactical, risk-based multi-criteria analysis framework for rehabilitation planning of sewer assets. An important challenge in this study was to align this framework with operational sewer AM needs as well as broader strategic goals of the organization. During the process, 8 decision-makers and 54 subject matter experts were continuously involved. Several group workshops were organized, interlaced with bilateral or small-group discussions with subject matter experts, to ensure a common understanding, consensus, buy-in, and comprehensiveness of the approach. The deliberative methods, and especially problem-structuring using group cognitive mapping techniques, led to the exploration and mapping out of the problems as perceived by different actors and the critical interdependencies. It also helped to facilitate a common understanding and to agree on a common purpose and goals. The identified 5 main strategic goals of the organization were linked to 11 tactical goals for sewer AM, which were either further broken down into sub-goals or directly matched with indicators that could be used to assess performance using operational data. Also, as detailed risk assessments were deemed unviable, a simpler (i.e., 'satisficing') approach was adopted, see Alhamed (2020) for details. Many of the participants praised the inclusivity, adaptability, and usefulness of the approach. By taking part in the process, social learning and alignment at various levels had taken place. After the project was complete, the resulting assessment framework was prepared for operational roll-out to assess all 120 drainage areas in Amsterdam, thereby contributing to improved decision-making in sewer AM.

Building on the above, we believe that both the need and the potential of strengthened efforts in transdisciplinary work across practice and academia is necessary to understand and improve decision-making in UDAM practice.

7.4 CONCLUSIONS AND RECOMMENDATIONS

UDAM is characterized by multi-actor decision-making processes that take place within a complex system setting. The main characteristics are identifiable. To manage the complex system, tasks and responsibilities are divided across different governmental agencies, organizations, or parts of an organization. A number of institutional barriers that result from such fragmentation have been identified both from theoretical literature and empirical examples. Structural solutions such as a different division of tasks and responsibilities may be effective yet will not resolve the fundamental challenge of bridging across fragmented responsibilities and multi-actor decision-making in complex contexts. It is easy to resolve one problem within a complex system with a band-aid while creating others elsewhere in the same or in an adjacent system.

An important explanation is related to the fact that no matter which way the governance and management of water systems are structured, each will imply some form of fragmentation that will serve achievement of some goals better than others. That is because functional fragmentation generally goes along with fragmentation of responsibilities, subject matter expertise, perceptions of the relevant system boundaries and related interests, goals, resources, and legitimate action options. The tendency of managers to avoid complexity can be attributed to the integral picture becoming too complex as system boundaries are broadened beyond the limits of their immediate control and responsibility. Limiting the scope eliminates problems and the need to consider needs and interactions with others. When problem analysis is missing or limited (as it is a 'given' contingent on the perception of the respective actor given their specific context), not all options are considered or not all relevant parties are considered in a decision-making process. Another important consequence of the fragmentation is that UDAM decisions are made such that they make sense and are defensible and doable within a specific political, organizational, and social context. The issue is that the result of these decisions across the organizations or organization units involved in UDAM does not, by definition, translate into a satisfactory solution for the whole. Many barriers resulting from institutional fragmentation can be reduced or even removed within the current division of roles by adapting formal decision-making processes and institutionalized norms, multi-actor arrangements, and the judgment and decision-making tactics used.

In theory, decision-making processes in UDAM can be described by distinct phases to go through to achieve high-quality – if not optimal – decisions that have higher odds of resolving difficult problems. In practice, however, the process plays out differently, where phases are either skipped or cut short, often leading to a less satisfying result. The effects of governance and social structure, political and social dynamics as influenced by social and individual cognition have an important effect on decision-making. Within organizations, there is often an unwritten consensus on how decisions are made, because that is the way things have been done for a long period of time and which reaped satisfactory results earlier. This approach may be suitable for routine tasks and recurring management activities that can be controlled entirely by the organization (unit) without a need for interaction with others. Non-routine problems and decision-making, such as coordination of works for stormwater management and climate change adaptation, however, require a decision-making process that follows its own dynamic and can accommodate the perceptions and needs of several actors to ensure their support and commitment to action. Pushing the system boundaries creates a different picture of the processes driving observed system behavior and issues as well as the players, interests, and possible interventions. In a dynamic natural and institutional environment along with changing social priorities, UDAM is challenged to change as well.

Throughout the process, cognition and judgment underlying decision-making are strongly framed by context and in many cases informed by emotions and gut feeling. Objective reasoning and arguments are often absent or may be constructed in hindsight to align with intuition. A deliberate processing of information in decision-making to align choices and actions toward the achievement of the set objectives is necessary if UDAM strives for a more objective substantiation of decisions. This requires support by data of appropriate quality, and which should be informative to assess progress on the objectives to be achieved through UDAM decisions. Although modeling can provide significant benefits, not all decision-making requires nor lends itself to being modeled in light of the questions at stake and the data situation. Expert knowledge can be systematically elicited to fill the gaps. These can be bolstered by employing decision-making strategies shown to deliver higher quality decisions.

In UDAM, we have long been able to seek more 'optimal' solutions that have direct added value for all parties involved in the decision-making. With climate change adaptation, aging infrastructure, more intensive use of available space, and conflicts around limited resources, negative outcomes can no longer be avoided; instead, trade-offs must be faced and negotiated. Continuing to look for a painless, regret-free solution is delaying decision-making processes and action needed to adapt UD systems. This is short-sighted as it may unduly shift and exaggerate problems into the future or onto other interlinking systems and actors until they can no longer be avoided. Problems that get out of hand typically require more intense intervention that is unlikely to achieve better outcomes than timely, proportionate and adaptive intervention following quality decision-making could achieve.

There are already many examples of possible adaptations in UDAM that reach beyond attempts to merely optimize interventions following the current technical and managerial paradigms. For instance, municipal councils can adapt the norms for climate change adaptation for new developments and redevelopments, which forces adoption of new solutions on both public and private land, while reducing pressure on the system and facilitating the accumulation of experience and precedents for transition within existing infrastructure. In addition, improving interaction with local communities and involving them in earlier stages of the decision-making process, such as the design and development phase, can help commitments to action and reduce the stalling of processes. Other examples include policies to spend a percentage of the UD fees on climate change adaptation measures, providing training and cross-organizational exchanges to facilitate learning and adopt new practices. Behavioral analysis of context and the actors can be employed to better anticipate upon biases and barriers arising from the interplay of formal and social structures with individual and group cognition to adapt policies, technical solutions, and the communication about them in a way to shift social and personal norms and facilitate implementation.

To enhance the understanding of decision quality, it is important to gather practical cases from the sector, analyze them systematically, and discuss them from both a practical and scientific perspective. This will serve as a valuable foundation for developing more effective strategies and processes tailored to specific problems through targeted interventions. This will also promote mutual recognition and comprehension of decision-making challenges, identify areas for improvement, highlight the need for training, and explore methodologies to facilitate multi-actor decision-making processes. There is already a wealth of knowledge, methods, and empirical evidence available from other fields of decision-making research and practice. When combined with a thoughtful consideration of the specific issue at hand, these can greatly enhance decision-making quality in UDAM practice. A systematic study will further provide insights on how to integrate, adapt, and enhance approaches for diverse decision-making situations in UDAM.

REFERENCES

Ackermann F. (2012). Problem structuring methods 'in the Dock': arguing the case for Soft OR. *European Journal of Operational Research*, **219**, 652–658, https://doi.org/10.1016/j.ejor.2011.11.014

Alegre H., Amaral R., Brito R. S. and Baptista J. M. (2020). Public policies as strategic asset management enablers: the case of Portugal. *H_2Open Journal*, **3**, 428–436, https://doi.org/10.2166/h2oj.2020.052

Alhamed H. (2020). Multi-Criteria Decision Analysis for Integrated Risk-Based Asset Management – Demonstration on Sewer Asset Management in Amsterdam. Delft University of Technology, Delft, The Netherlands.

Alm J., Paulsson A. and Jonsson R. (2021). Capacity in municipalities: infrastructures, maintenance debts and ways of overcoming a run-to-failure mentality. *Local Economy: The Journal of the Local Economy Policy Unit*, **36**, 81–97, https://doi.org/10.1177/02690942211030475

Arioli M., Crespi C. and Canessa N. (2018). Social cognition through the lens of cognitive and clinical neuroscience. *BioMed Research International*, **2018**, 1–18.

Artinger F., Petersen M., Gigerenzer G. and Weibler J. (2015). Heuristics as adaptive decision strategies in management. *Journal of Organizational Behavior*, **36**, S33–S52, https://doi.org/10.1002/job.1950

Bassone-Quashie Y. A. (2021). Transitioning to a sustainable urban water future in the Netherlands – How decision-making processes and institutional factors contribute to climate adaptation in urban drainage systems. MSc thesis, Delft University of Technology.

Béland D. and Howlett M. (2016). The role and impact of the multiple-streams approach in comparative policy analysis. *Journal of Comparative Policy Analysis: Research and Practice*, **18**, 221–227, https://doi.org/10.1080/13876988.2016.1174410

Bell D. E., Raiffa H. and Tversky A. (1988). Descriptive, normative, and prescriptive interactions in decision-making.

Bond S. D., Carlson K. A. and Keeney R. L. (2008). Generating objectives: Can decision-makers articulate what they want? *Management Science*, **54**, 56–70, https://doi.org/10.1287/mnsc.1070.0754

Brown R. R. and Farrelly M. A. (2009). Delivering sustainable urban water management: A review of the hurdles we face. *Water Science and Technology*, **59**, 839–846, https://doi.org/10.2166/wst.2009.028

Cardoso M. A., Santos Silva M., Coelho S. T., Almeida M. C. and Covas D. I. C. (2012). Urban water infrastructure asset management – a structured approach in four water utilities. *Water Science and Technology*, **66**, 2702–2711, https://doi.org/10.2166/wst.2012.509

Carriço N., Covas D. and Almeida M. d. C. (2021). Multi-criteria decision analysis in urban water asset management. *Urban Water Journal*, **18**, 558–569, https://doi.org/10.1080/1573062X.2021.1913613

Cohen M. D., March J. G. and Olsen J. P. (1972). A garbage can model of organizational choice. *Administrative Science Quarterly*, **17**, 1–25, https://doi.org/10.2307/2392088

Dai L., Wörner R. and van Rijswick H. F. (2018). Rainproof cities in the Netherlands: approaches in Dutch water governance to climate-adaptive urban planning. *Journal of Water Resources Development*, **34**, 652–674, https://doi.org/10.1080/07900627.2017.1372273

De Bruijn H., Heuvelhof E. T. and Veld R. in 't. (2010). Process Management – Why Project Management Fails in Complex Decision-Making Processes. Springer-Verlag, Berlin-Heidelberg.

Dutch Parliament (2006). Memorie van Toelichting bij de wetswijziging Wet verankering en bekostiging van gemeentelijke watertaken, Hemelwaterzorgplicht (translated title: Explanatory memorandum to the legislative amendment to the Act Anchoring and Financing of Municipal Water Duties, Rainwater Care Duty). Retrieved from https://www.infomil.nl/publish/pages/71309/nvt_hemelwaterzorgplicht_doc.pdf

Eggimann S., Mutzner L., Wani O., Schneider M. Y., Spuhler D., de Vitry M. M., Beutler P. and Maurer M. (2017). The potential of knowing more: a review of data-driven urban water management. *Environmental Science & Technology*, **51**, 2538–2553, https://doi.org/10.1021/acs.est.6b04267

Eisenführ F., Weber M. and Langer T. (2010). Rational Decision-Making. Springer-Verlag, Berlin-Heidelberg.

Elbanna S. (2006). Strategic decision-making: process perspectives. *International Journal of Management Reviews*, **8**, 1–20, https://doi.org/10.1111/j.1468-2370.2006.00118.x

Enserink B., Bots P. W. G., van Daalen E., Hermans L. M., Kortmann R., Koppenjan J. F. M., Kwakkel J. H., Ruijgh T. and Thissen W. A. H. (2022). Policy analysis of multi-actor systems (eleven).

Evans J. S. B. T. (2008). Dual-processing accounts of reasoning, judgment, and social cognition. *Annual Review of Psychology*, **59**, 255–278, https://doi.org/10.1146/annurev.psych.59.103006.093629

Filler A. (2009). The History, Development and Impact of Computed Imaging in Neurological Diagnosis and Neurosurgery: CT, MRI, and DTI. Nature Proceedings.

Gerritsen E. and Sterks C. G. M. (2004). Kostenontwikkeling in de waterketen, 1990–2010. COELO Rapport 04-3. COELO, Groningen.

Gigerenzer G. (2004). Fast and frugal heuristics: the tools of bounded rationality. In: Blackwell Handbook of Judgment and Decision-Making, D. Koehler and N. Harvey (eds), Blackwell, Oxford, UK, pp. 62–68.

Gilovich T., Griffin D. and Kahneman D. (2002). Heuristics and Biases: The Psychology of Intuitive Judgment. Cambridge University Press, Cambridge.

Goschke T. and Job V. (2023). The willpower paradox: possible and impossible conceptions of self-control. Perspectives on Psychological Science, 1271–1574, 17456916221146158.

Greco S., Ehrgott M. and Figueira J. (eds) (2016). Multiple Criteria Decision Analysis – State of the Art Surveys. Springer, New York, Heidelberg, Dordrecht, London.

Gregory R. and Keeney R. L. (1994). Creating policy alternatives using stakeholder values. *Management Science*, **40**, 1035–1038, https://doi.org/10.1287/mnsc.40.8.1035

Hammond J. S., Keeney R. L. and Raiffa H. (1998). The hidden traps in decision-making. *Harvard Business Review*, **76**, 47.

Hammond J. S., Keeney R. L. and Raiffa H. (2002). Smart Choices: A Practical Guide to Making Better Life Decisions. Broadway Books, New York.

Harrison E. F. (1996). A process perspective on strategic decision-making. *Management Decision*, **34**, 46–53, https://doi.org/10.1108/00251749610106972

Head B. W. and Alford J. (2015). Wicked problems: implications for public policy and management. *Administration & Society*, **47**, 711–739, https://doi.org/10.1177/0095399713481601

Hering J. G., Hoehn E., Klinke A., Maurer M., Peter A., Reichert P., Robinson C., Schirmer K., Schirmer M., Stamm C. and Wehrli B. (2012). Moving targets, long-lived infrastructure, and increasing needs for integration and adaptation in water management: an illustration from Switzerland. *Environmental Science & Technology*, **46**, 112–118, https://doi.org/10.1021/es202189s

Hermans L. M. and Cunningham S. W. (2013). Actor models for policy analysis. In: Public Policy Analysis: New Developments, W. A. H. Thissen and W. E. Walker (eds), Springer US, Boston, MA, pp. 187–188.

Hermans L. M. and Cunningham S. W. (2018). Actor and Strategy Models: Practical Applications and Step-Wise Approaches. John Wiley & Sons, Hoboken, USA.

Howard R. A. (1988). Decision analysis – practice and promise. *Management Science*, **34**, 679–695, https://doi.org/10.1287/mnsc.34.6.679

Huisman P. (2002). How the Netherlands finance public water management. Journal of European Water Management Online, **16**, 1–11.

Janis I. (1982). Groupthink: Psychological Studies of Policy Decisions and Fiascos. Houghton Mifflin, Boston, MA.

Johnson E. J., Shu S. B., Dellaert B. G. C., Fox C., Goldstein D. G., Häubl G., Larrick R. P., Payne J. W., Peters E., Schkade D., Wansink B. and Weber E. U. (2012). Beyond nudges: tools of a choice architecture. *Marketing Letters*, **23**, 487–504, https://doi.org/10.1007/s11002-012-9186-1

Joubert A., Stewart T. J. and Eberhard R. (2003). Evaluation of water supply augmentation and water demand management options for the city of Cape Town. *Journal of Multi-Criteria Decision Analysis*, **12**, 17–25, https://doi.org/10.1002/mcda.342

Kahneman D. (2011). Thinking, Fast and Slow. Allen Lane, London.

Kahneman D. and Klein G. (2009). Conditions for intuitive expertise: a failure to disagree. *American Psychologist*, **64**, 515–526, https://doi.org/10.1037/a0016755

Keeney R. L. (1982). Decision analysis: an overview. *Operations Research*, **30**, 803–838, https://doi.org/10.1287/opre.30.5.803

Keeney R. L. (1996). Value-focused thinking: identifying decision opportunities and creating alternatives. *European Journal of Operational Research*, **92**, 537–549, https://doi.org/10.1016/0377-2217(96)00004-5

Keeney R. L. (2002). Common mistakes in making value trade-offs. *Operations Research*, **50**, 935–945, https://doi.org/10.1287/opre.50.6.935.357

Keeney R. L. (2004). Framing public policy decisions. *International Journal of Technology, Policy and Management*, **4**, 95–115, https://doi.org/10.1504/IJTPM.2004.004815

Keeney R. L. and Gregory R. S. (2005). Selecting attributes to measure the achievement of objectives. *Operations Research*, **53**, 1–11, https://doi.org/10.1287/opre.1040.0158

Kiparsky M., Sedlak D. L., Thompson B. H. and Truffer B. (2013). The innovation deficit in urban water: the need for an integrated perspective on institutions, organizations, and technology. *Environmental Engineering Science*, **30**, 395–408, https://doi.org/10.1089/ees.2012.0427

Krijnen K. (2020). Increasing the resilience of urban areas to extreme precipitation: are the residents ready? The receptivity for effective rainproof measures in the neighbourhood De Baarsjes, Amsterdam. MSc thesis, Delft University of Technology.

Lang J. R., Dittrich J. E. and White S. E. (1978). Managerial problem solving models: a review and a proposal. *Academy of Management Review*, **3**, 854–866, https://doi.org/10.2307/257939

Lerner J. S., Li Y., Valdesolo P. and Kassam K. S. (2015). Emotion and decision-making. *Annual Review of Psychology*, **66**, 799–823, https://doi.org/10.1146/annurev-psych-010213-115043

Lienert J., Scholten L., Egger C. and Maurer M. (2015). Structured decision-making for sustainable water infrastructure planning and four future scenarios. *EURO Journal on Decision Processes*, **3**, 107–140, https://doi.org/10.1007/s40070-014-0030-0

Lunenburg F. C. (2010). The decision-making process. *National Forum of Educational Administration and Supervision Journal*, **47**, 7150–7161.

Marlow D. R., Moglia M., Cook S. and Beale D. J. (2013). Towards sustainable urban water management: a critical reassessment. *Water Research*, **47**, 7150–7161, https://doi.org/10.1016/j.watres.2013.07.046

Mårtensson E. and Rumman P. (2019). Asset Management in the Utility Sector: The Challenges of Breaching the gap Between Current Practice and Best Practice. KTH Royal Institute of Technology, Stockholm, Sweden.

Maurer M. (2022). Are hybrid systems sustainable or does winner takes all? In: Routledge Handbook of Urban Water Governance, Routledge, pp. 134–144.

Melnikoff D. E. and Bargh J. A. (2018). The mythical number two. *Trends in Cognitive Sciences*, **22**, 280–293, https://doi.org/10.1016/j.tics.2018.02.001

Mingers J. (2011). Soft OR comes of age – but not everywhere!. *Omega*, **39**, 729–741, https://doi.org/10.1016/j.omega.2011.01.005

Mingers J. and Rosenhead J. (2004). Problem structuring methods in action. *European Journal of Operational Research*, **152**, 530–554, https://doi.org/10.1016/S0377-2217(03)00056-0

Münscher R., Vetter M. and Scheuerle T. (2016). A review and taxonomy of choice architecture techniques. *Journal of Behavioral Decision-Making*, **29**, 511–524, https://doi.org/10.1002/bdm.1897

Myers D. G. and Myers D. G. (2002). Intuition: Its Powers and Perils. Yale University Press, New Haven & London.

Nieuwenhuis E., Cuppen E. and Langeveld J. (2022). The role of integration for future urban water systems: identifying Dutch urban water practitioners' perspectives using Q methodology. *Cities*, **126**, 103659, https://doi.org/10.1016/j.cities.2022.103659

Nutt P. C. (1999). Surprising but true: half the decisions in organizations fail. *The Academy of Management Executive*, **13**, 75.

Nutt P. C. (2008). Investigating the success of decision-making processes. *Journal of Management Studies*, **45**, 425–455, https://doi.org/10.1111/j.1467-6486.2007.00756.x

OECD (2011). Water Governance in OECD Countries: A Multi-Level Approach. OECD Publishing, Paris.

Oosterom E. and Hermans R. (2013). Riolering in beeld – benchmark rioleringszorg 2013. In: S. Rioned (ed.), Stichting RIONED, Ede, pp. 45–46.

Payne J. W., Bettman J. R. and Johnson E. J. (1993). The Adaptive Decision-Maker. Cambridge University Press, Cambridge.

Pluchinotta I., Salvia G. and Zimmermann N. (2022). The importance of eliciting stakeholders' system boundary perceptions for problem structuring and decision-making. *European Journal of Operational Research*, **302**, 280–293, https://doi.org/10.1016/j.ejor.2021.12.029

Posner M. J. and Snyder C. R. R. (1975). Attention and cognitive control. Paper presented at the Information Processing and Cognition: The Loyola Symposium.

Rioned S. (2005). Benchmarking Rioleringszorg 2005. Stichting RIONED.

Rioned S. (2021). Branchestandaard gemeentelijke watertaken. Stichting RIONED. https://www.riool.net/evenement-opleiding/branchestandaard (accessed 31 July 2023).

Rittel H. W. J. and Webber M. M. (1973). Dilemmas in a general theory of planning. *Policy Sciences*, **4**, 155–169, https://doi.org/10.1007/BF01405730

Rizun N. and Taranenko Y. (2014). Simulation models of human decision-making processes. *Management Dynamics in the Knowledge Economy*, **2**, 241–264.

Romano O. and Akhmouch A. (2019). Water governance in cities: current trends and future challenges. *Water*, **11**, 500, https://doi.org/10.3390/w11030500

Rosenhead J. (1996). What's the problem? An introduction to problem structuring methods. *Interfaces*, **26**, 117–131, https://doi.org/10.1287/inte.26.6.117

Roy B. (1991). The outranking approach and the foundations of Electre methods. *Theory and Decision*, **31**, 49–73, https://doi.org/10.1007/BF00134132

Roy B. and Słowiński R. (2013). Questions guiding the choice of a multicriteria decision aiding method. *EURO Journal on Decision Processes*, **1**, 69–97, https://doi.org/10.1007/s40070-013-0004-7

Sa-nguanduan N. and Nititvattananon V. (2010). Strategic decision-making for urban water reuse application: a case from Thailand. *Desalination*, **268**, 141–149, https://doi.org/10.1016/j.desal.2010.10.010

Schoemaker P. J. H. and Russo E. J. (1993). A pyramid of decision approaches. *California Management Review*, **36**, 9–31, https://doi.org/10.2307/41165732

Scholten L. (2013). Multi-criteria decision analysis for water supply infrastructure planning under uncertainty. PhD thesis, ETH Zurich.

Scholten L. (2017). Designing decision processes to overcome barriers to sustainable water systems (PPT). 21st International Conference of the International Foundation of Operations Research Societies, Quebec City, Canada.

Scholten L. (2018). A diagnostic approach to improve the design and evaluation of decision-support interventions. EURO 2018: 29th European Conference on Operational Research, Valencia, Spain.

Scholten L., Reichert P., Schuwirth N. and Lienert J. (2015). Tackling uncertainty in multi-criteria decision analysis – an application to water supply infrastructure planning. *European Journal of Operational Research*, **242**, 243–260, https://doi.org/10.1016/j.ejor.2014.09.044

Simon H. A. (1978). Rationality as process and as product of thought. *The American Economic Review*, **68**, 1–16.

Simon H. A. (1982a). Models of Bounded Rationality/Vol. 1, Economic Analysis and Public Policy. MIT Press, Cambridge, MA.

Simon H. A. (1982b). Models of Bounded Rationality/Vol. 2, Behavioral Economics and Business Organization. MIT Press, Cambridge, MA.

Simon H. A. (1991). Bounded rationality and organizational learning. *Organization Science*, **2**(1), 125–134, https://doi.org/10.1287/orsc.2.1.125

Smith E. R., Mackie D. M. and Claypool H. M. (2015). Social Psychology. Psychology Press, New York.

Spetzler C., Winter H. and Meyer J. (2016). Decision Quality: Value Creation from Better Business Decisions. John Wiley & Sons, Hoboken, New Jersey.

Stanovich K. E. and West R. F. (2000). Individual differences in reasoning: implications for the rationality debate? *Behavioral and Brain Sciences*, **23**, 645–665, https://doi.org/10.1017/S0140525X00003435

Sunstein C. R. and Hastie R. (2015). Wiser: Getting Beyond Groupthink to Make Groups Smarter. Harvard Business Press, Harvard, USA.

Sydow J., Schreyögg G. and Koch J. (2009). Organizational path dependence: opening the black box. *Academy of Management Review*, **34**, 689–709.

Teisman G. R. (2000). Models For research into decision-making processes: on phases, streams and decision-making rounds. *Public Administration*, **78**, 937–956, https://doi.org/10.1111/1467-9299.00238

Tscheikner-Gratl F., Caradot N., Cherqui F., Leitão J. P., Ahmadi M., Langeveld J. G., Le Gat Y., Scholten L., Roghani B., Rodríguez J. P., Lepot M., Stegeman B., Heinrichsen A., Kropp I., Kerres K., Almeida M. d. C., Bach P. M., de Vitry M. M., Marques A. S., Simões N. E., Rouault P., Hernandez N., Torres A., Werey C., Rulleau B. and Clemens F. (2019). Sewer asset management – state of the art and research needs. *Urban Water Journal*, **16**, 662–675, https://doi.org/10.1080/1573062X.2020.1713382

Tversky A. and Kahneman D. (1973). Availability: a heuristic for judging frequency and probability. *Cognitive Psychology*, **5**, 207–232, https://doi.org/10.1016/0010-0285(73)90033-9

Tversky A. and Kahneman D. (1974). Judgment under uncertainty: heuristics and biases. *Science*, **185**, 1124–1131, https://doi.org/10.1126/science.185.4157.1124

van Karnenbeek L., Salet W. and Majoor S. (2021). Wastewater management by citizens: mismatch between legal rules and self-organisation in Oosterwold. *Journal of Environmental Planning and Management*, **64**, 1457–1473, https://doi.org/10.1080/09640568.2020.1829572

van Riel W., Langeveld J. G., Herder P. M. and Clemens F. H. L. R. (2014). Intuition and information in decision-making for sewer asset management. *Urban Water Journal*, **11**, 506–518, https://doi.org/10.1080/1573062X.2014.904903

van Riel W., van Bueren E., Langeveld J., Herder P. and Clemens F. (2015). Decision-making for sewer asset management: theory and practice. *Urban Water Journal*, **13**, 57–68.

Vlaamse Regering (2020). Conceptnota – Optimale afstemming van het gemeentelijke en bovengemeentelijke asset management. Flemish Minister of Justice and Entertainment, Energy and Tourism. Flemish Government.

Waternet (n.d.). Nieuwe Sanitatie: knelpunten oplossen voor Buiksloterham, Waternet. https://www.waternet.nl/werkzaamheden/nieuwe-sanitatie/nieuwe-sanitatie-knelpunten-oplossen-voor-buiksloterham/ (accessed 31 July 2023).

Waterschap de Dommel (2021). Water als basis voor een toekomstbestendige leefomgeving – Waterbeheerprogramma 2022–2027. Boxtel, NL.

Wilderdink N. and Form W. (1999). Social Structure – Definition, Examples, Theories & Facts. Encyclopedia Britannica. https://www.britannica.com/topic/social-structure (accessed 21 July 2023).

Wong T. H. F. and Brown R. R. (2009). The water sensitive city: principles for practice. *Water Science and Technology*, **60**, 673–682, https://doi.org/10.2166/wst.2009.436

Zhang K., Zargar A., Achari G., Shafiqul Islam M. and Sadiq R. (2014). Application of decision-support systems in water management. *Environmental Reviews*, **22**, 189–205, https://doi.org/10.1139/er-2013-0034

Zheng J., Egger C. and Lienert J. (2016). A scenario-based MCDA framework for wastewater infrastructure planning under uncertainty. *Journal of Environmental Management*, **183**, 895–908.

doi: 10.2166/9781789063059_0299

Chapter 8

Data management and quality control

S. Auger[1], J.-B. Besnier[2], M. van Bijnen[3], F. Cherqui[4], G. Chuzeville[2], F. Clemens-Meyer[5], M. G. Jaatun[6], J. Langeveld[7], Y. Le Gat[8], S. Moin[1], G. E. Oosterom[9], W. van Riel[10], B. Roghani[5], M. M. Rokstad[5], J. Røstum[11], F. Tscheikner-Gratl[5] and R. Ugarelli[6]

[1]Integrated Watershed Management, Lake Simcoe Region Conservation Authority, 120 Bayview Parkway, Newmarket, Ontario, Canada L3Y 3W3
[2]Lyon Metropolis, Water Department, 20 Rue du lac CS 33569, 69505 Lyon, France
[3]M. van Bijnen Advies B.V., Drunen, The Netherlands
[4]University of Lyon, INSA-LYON, Université Claude Bernard Lyon 1, DEEP, F-69621, F-69622 Villeurbanne, France
[5]NTNU – Norwegian University of Science and Technology, Trondheim, Norway
[6]SINTEF A.S., Trondheim, Norway
[7]Faculty of Civil Engineering and Geoscience, Department of Water Management, Delft University of Technology, Delft, The Netherlands
[8]INRAE, Bordeaux Nouvelle Aquitaine – ETTIS, Cestas Cedex, France
[9]RIONED Foundation, Ede, The Netherlands
[10]Infralytics, Breda, The Netherlands
[11]Volue Infrastructure, Trondheim, Norway

ABSTRACT

It is often said that data gathering is much more expensive than data-management software (such as geographic information system). Indeed, data are perhaps the most important element of any asset management approach. In this comprehensive chapter, we embark on a journey through digital era, highlighting the pivotal role of data in our contemporary world, emphasizing the importance of data in today's landscape. We delve into the critical question of which data to collect, providing insights into the strategic selection of data based on a cost–benefit approach and the significance of anticipation in data collection. A three-layer approach, encompassing object, system, and urban fabric levels, is proposed as a structure to organize data, elucidating the diverse information requirements at each layer, from descriptive data to performance assessments and requirements. A substantial portion of this chapter is devoted to data models and bias, elucidating the complexities of modeling sewer pipe deterioration and addressing issues such as selective survival and recruitment bias. Quality control emerges as a pivotal concern, clarifying the requirements for data quality, methods to assess completeness, and handling issues such as incompleteness, timeliness, uncertainty, and imprecision. Questions related to data quantity are explored, discussing the data-loop problem, reconstruction methods, and the implications of big data. Practical considerations related to data access and storage are also addressed. The chapter concludes by three enlightening case studies illustrating real-world applications of data models.

Keywords: control, data, quality, quantity, management, model, network, structure, system, requirement.

8.1 INTRODUCTION AND LAYOUT OF THE CHAPTER

8.1.1 The digital era

Since the 1970s, we have entered a new metaparadigm focused on the transformation of information (Figure 8.1).

Human history can be divided into three different long-term metaparadigms (Figure 8.2), following Hilbert (2020): 'The first focused on the transformation of material, including stone, bronze, and iron. The second, often referred to as the industrial revolution, was dedicated to the transformation of

Figure 8.1 'Something somewhere went terribly wrong'. Image Credit: FreeImages.com, Non-copyright for personal and commercial use (https://vector.me/browse/216714/went_terribly_wrong).

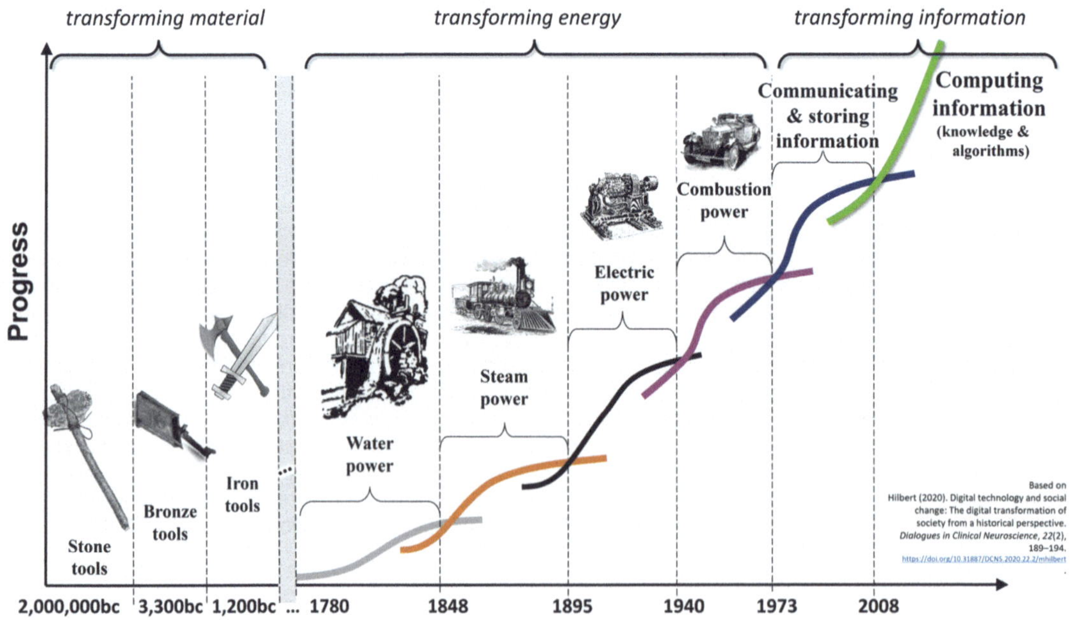

Figure 8.2 Three different long-term metaparadigms of mankind, based on concepts from Hilbert (2020), from https://commons.wikimedia.org/wiki/File:LongWavesThreeParadigms.jpg, CC BY-SA 4.0.

energy, including water, steam, electric, and combustion power. Finally, the most recent metaparadigm aims at transforming information. It started out with the proliferation of communication and stored data and has now entered the age of algorithms, which aims at creating automated processes to convert the existing information into actionable knowledge.'

The last seven decades have seen a strong evolution of our world leading to the so-called 'digital era' (Figure 8.3). The evolution started with computer hardware progress thanks to the military and then gaming demand (computer centric). The second stage (mobile centric) was initiated with the development of communication and social technologies and software. The third stage (data centric), where we currently are, is centered on algorithms and recent technology (such as quantum computing) to make the most of existing data.

Another new development of this third stage is the rise of new artificial intelligence (AI) models to access existing knowledge in a conversational way like, for example, ChatGPT. Their applicability and usefulness for the field of asset management (AM) have yet to be evaluated and improved and its impact is therefore not easily foreseeable. Were one to ask ChatGPT-3 the following question 'How do you see the role of AI in urban drainage asset management for the near future,' the answer is (20.04.2023):

> Artificial intelligence (AI) has the potential to revolutionize urban drainage asset management by improving the accuracy of asset monitoring and predicting asset performance and failures. Here are some ways that AI could play a role in urban drainage asset management in the near future:
>
> (1) Real-time monitoring: AI can be used to monitor water levels, flow rates, and other key parameters in real-time. This can help detect anomalies and potential problems before they become critical.
> (2) Predictive maintenance: AI algorithms can be used to predict when assets are likely to fail based on historical data and real-time monitoring. This can help utilities plan maintenance activities in advance, reducing downtime and maintenance costs.
> (3) Data analysis: AI can help utilities make sense of the vast amounts of data generated by urban drainage systems. This can include identifying patterns in data that are not immediately apparent to human operators, which can help optimize system performance.
> (4) Decision support: AI can be used to provide decision support for asset management activities. For example, it can help utilities prioritize maintenance activities based on asset criticality, maintenance history, and other factors.
>
> In summary, AI has the potential to bring significant benefits to urban drainage asset management, including improved asset performance, reduced maintenance costs, and better decision-making. However, it is important to ensure that AI is used in a responsible and ethical manner, with appropriate safeguards in place to protect against unintended consequences.

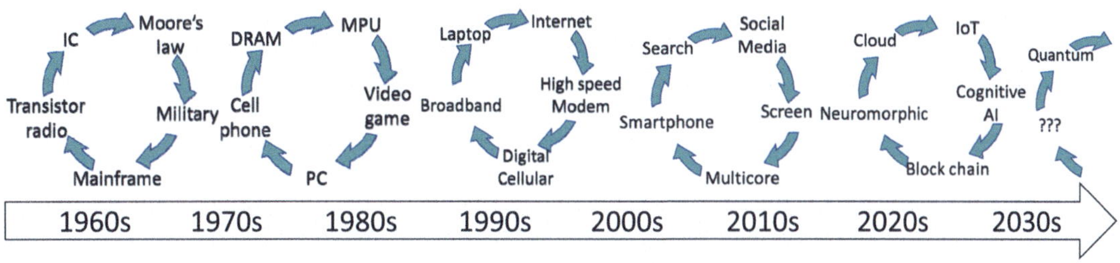

Figure 8.3 Dawn of the Data-Centric Era, adapted from VLSI Research, Industry Strategy Symposium keynote, USA, January 2018, cited on https://semiengineering.com/dawn-of-the-data-centric-era/.

A main problem here is the trustworthiness; on the question 'Can you provide me with a reference book on urban drainage asset management,' the answer provided a reference to a (non-existing) book on urban drainage AM with a (fake) description (20.04.2023):

> Certainly, here is a reference book on urban drainage asset management:
>
> Title: Urban Drainage Asset Management
> Authors: David Butler and Chris Digman
> Publisher: IWA Publishing
> Publication date: 2011
>
> This book provides an overview of urban drainage asset management, including the principles and practices of asset management, data collection and analysis, risk assessment, and maintenance and repair strategies. It also covers the use of advanced technologies such as GIS, remote sensing, and decision support systems for urban drainage asset management. The book is aimed at water utility managers, engineers, and policymakers involved in the planning, design, and operation of urban drainage systems.

8.1.2 The importance of data

There has been an expected increase in the amount and variety of data relevant to AM due to factors such as increased governmental regulations, the emergence of low-cost sensors, the integration of infrastructure management in 'smart cities' concepts, and the digitalization and the capability of social media platforms to report several types of problems. There are both new opportunities and challenges associated with data management created by the diversity of data sources. Due to the need to manage and analyze large volumes of data in near real time, current data-management platforms will require significant changes. Even if intuition is a driver for decision, data provide a solid foundation for informed decision making (see Chapter 7). Data should be used to support decisions, for example, by enabling modeling tools to anticipate the consequences of actions or inactions. In the urban water field, a whole sub-discipline has emerged to provide such models crossing the boundaries between water science, data science, and computer science (Makropoulos & Savić, 2019). Less frequently, data are the center of the decision and human participation consists purely in the validation of the decisions: for example, a computation-based renewal program can rely solely on deterioration models, rehabilitation criteria, and some context parameters (renewal rate or budget), to prioritize assets that need rehabilitation in the short term (year). Utilities are thus increasingly relying on software, in these cases, the outcome is almost fully dependent on the database.

8.1.3 Layout of the chapter

Collecting the data is thus a particularly important (first) step of most decision-making processes in urban drainage AM. It is time consuming and costly, but mandatory as 'the true benefit of collecting data and applying tools for infrastructure asset management is improved decision-making' (Rokstad *et al.*, 2016). To be efficient, data collection must involve medium- to long-term considerations. Among the numerous factors influencing the deterioration of sewer pipes (Davies *et al.*, 2001), gathering information such as the installation date or material could require several years of work with field campaigns, geospatial analysis, and consulting archives. This chapter discusses data collection methods and strategies, along with requirements regarding the use of data models. Several sections focus on specific questions such as data models and bias, the why and how of quality control, and the importance of keeping and securing records (data historian, data archiving, data security).

8.2 WHICH DATA TO COLLECT

Many authors refer to 'minimum data set requirements' for developing an AM strategy (e.g., Carvalho *et al.*, 2018; Rokstad & Ugarelli, 2016) or an 'optimal' dataset (e.g., Ahmadi *et al.*, 2014a; Tscheikner-Gratl *et al.*, 2013). A minimum dataset includes sewer infrastructure characteristics (i.e., pipes and manholes) and management and operational data. Regarding nature-based solutions, a minimum dataset includes information related to the solution itself (e.g., geometry, composition, construction date, etc.) and to its surroundings (i.e., mostly catchment characteristics). It should be noted that the process of acquiring new data is usually associated with interacting with different data owners and navigating varying data quality. In devising data collection plans, this should be considered, as it creates additional challenges.

8.2.1 Targeting the most important data

Data related to water infrastructure such as pipes or nature-based solutions can encompass construction features (e.g., design parameters and objectives, installation date), usage features (e.g., maintenance needs, failure history), external features (e.g., catchment characteristics, rainfall, groundwater, urban planning), or other features (e.g., inhabitant expectations, budget, political consideration). While some data may have already been gathered because of virtuous practices or regulation requirements, not all the data can be used as-is due to data quality issues. Quality and quantity of the available data are often the leading factors to move from a reactive approach (i.e., run-to-failure) to a proactive approach (i.e., anticipate and avoid failures). Efforts to gather and validate data need to target the most promising data, and importance of data will vary from one catchment to another. Data prioritization can be based on utility priorities, experience with respect to the behavior of the infrastructure, and on analytical approaches. Ahmadi *et al.* (2014a, 2014b) proposed an analytical approach to establish the most informative factors for predicting the condition of sewer pipes, using a representative sample of the asset stock.

8.2.2 Cost–benefit approach

Many utilities are experiencing a lack of data availability and/or quality, which limits the quality of information achieved about their assets, hampering their ability to make decisions about their assets that are better than decisions largely based on intuition. This lack of evidence of the potential benefits of investing resources in data collection can, in the worst case, deter the motivation and drivers for data collection, rendering the utility with even less data of lower quality, lowering the quality of information and benefits achieved by systematic data collection even further as time goes on (Rokstad, 2016; Rokstad *et al.*, 2016). This is sometimes referred to as the 'data-loop problem' (Cherqui *et al.*, 2022), as presented in Section 8.6. To counter this potentially vicious spiral, there is a need to develop a data collection and utilization strategy, ensuring that the utility gradually and iteratively improves its data and information quality, rather than allowing to further degrade it (i.e., overturning the vicious spiral). When making a data collection strategy, the utility should consider and anticipate:

- Which criteria will be used to make decisions about their assets in the future?
- The relative mutual importance of the decision criteria applied.
- What information will be needed to rely on to assess these criteria with sufficient accuracy?
- Which types, what amount, and which quality of data are needed to achieve the quality of information needed to achieve this?

The decisions about which data to collect or not, how to collect it, and what quality that data should have, are then to be based on a balance between these points, that is, through balancing the costs of collecting data and the benefit of future improved decision making. In the end, the data collection strategy should comprise an anticipation of future data and information needs, and a data collection and utilization plan that allows one to improve the quality of information gradually/stepwise, as more data and data of higher quality are collected over time.

8.2.3 Anticipation is the key

Significant time (i.e., months to years) may pass between the decision to gather new data and the need/possibility to use these data. Some data can only be collected over time (e.g., condition assessments of sewer pipes or stormwater control measures) because of limited investigation capabilities and the

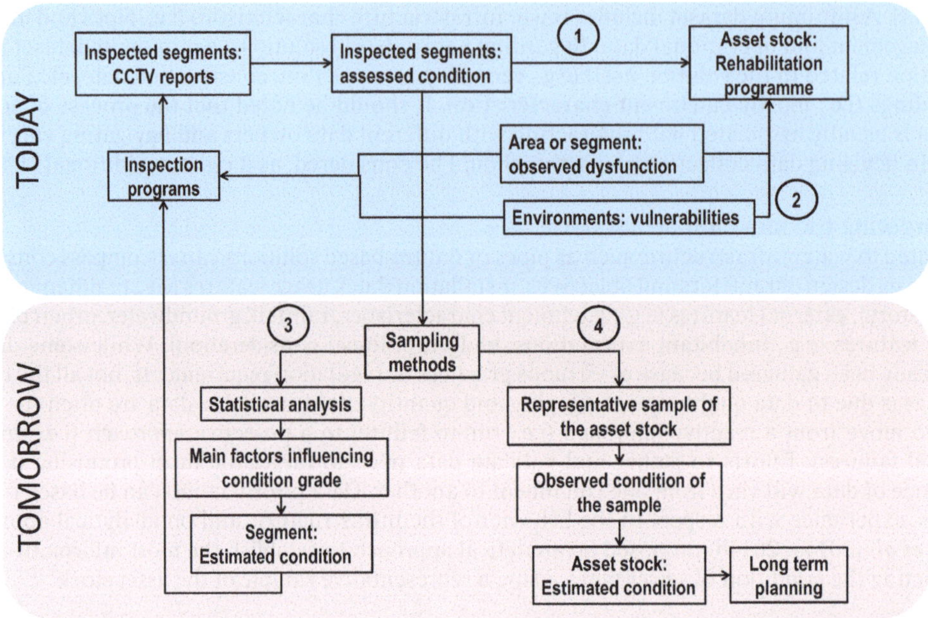

Figure 8.4 Sewer AM practices (*source*: adapted from Ahmadi *et al.*, 2014a, 2014b). The figure shows the different levels of practices, from a simple short-term practice (1) to a long-term approach (4). Utilities often only have the knowledge and data to engage in practices (1) and (2) but they need to start collecting data to evolve to practices (3) or (4).

time-dependent nature of the processes that are monitored (see Chapter 5). Efficient data collection relies on a medium- to long-term strategy involving the planning over the next few years/decades on when to start to collect and use specific data. If it is possible to forecast on technology yet to be developed, this planning needs to consider the development of new algorithms and available software. Another key element is to include training schemes in the planning to ensure efficient data collection and (later) utilization.

Using data as early as possible ensures that it fits the purpose and thus that the collecting process has been tailored to the need. This ensures better data process quality. Contrarily, unused data have usually not been verified and often not been properly stored. Figure 8.4 presents different levels of practice that a utility can have regarding AM. The first level relies only on condition assessments of inspected pipes and does not need additional data. The second level aims at targeting inspection programs where pipe dysfunctions may have major consequences: it requires data on known dysfunctions and on the environment of the pipes in question (e.g., from a geographic information system (GIS)). Evolving to the third and fourth levels requires additional data related to the characteristics of the pipes, a representative sample, and modeling of the deterioration process: the decision to reach these levels 'by tomorrow' (i.e., in the next 5 years) requires 'starting today' the data collection process.

8.3 THE THREE-LAYER APPROACH: OBJECT, SYSTEM, AND URBAN FABRIC

An increasing proportion of the increasing world's population lives in cities (55% in 2018; the United Nations expects this percentage to amount to 68% by 2050: https://www.un.org/development/desa/en/news/population/2018-revision-of-world-urbanization-prospects.html, as a result of ongoing urbanization and urban densification). Urban planning and local topography provide boundary conditions for urban drainage and sewer systems, which provide the foundation for a well-functioning urban area. Traditional 'piped' sanitary and combined sewer systems provide these services out of sight as part of the underground urban infrastructure, whereas dual-drainage stormwater management (SWM) systems including the upcoming stormwater control measures provide these services as an integral part of urban areas, often being shaped as multifunctional infrastructures that cover both public and private space. Consequently, the system boundaries of urban drainage systems are becoming less and less strictly defined, which affects AM both in practical and theoretical terms. Many frameworks can be used to describe the urban drainage network and its environment, but although each framework could have different specifications, in the end, they all should include the same kind of information. For example, Di Sivo and Ladiana (2011) distinguish GIS data for integrated management by form into network, widespread, and punctual structures. Tscheikner-Gratl *et al.* (2015) offer a classification for infrastructure AM into network data, condition data, data related to the environment of the pipe, and model data. At any rate, any used framework needs to be in accordance with the GIS and the local authority departments and must consider information on a larger scale than only urban drainage infrastructure. The Waternet framework is used as an illustrative example here. Waternet, the water cycle company of the municipality of Amsterdam, has developed a conceptual framework comprising three layers, each with its own focus and information needs (see Figure 8.5):

- **Urban fabric:** at the top level, urban planning and urban development determine the layout of the urban area. Living in high-density urban areas is sustained by the underlying urban water infrastructures, such as combined, sanitary and storm sewer networks, water supply networks, and other types of SWM systems, which provide the necessary urban water services. The dominant processes at the level of this first layer are urban planning, such as the mobility transition, energy transition, climate adaptation, ongoing urban densification, and so on. Sewer asset managers need to have information on the urban fabric and urban planning as this provides the boundary conditions for the infrastructures as well as on the required services and service levels.
- **Infrastructure networks:** at the level of the infrastructure networks the focus is on system dynamics and system performance, often expressed in terms of combined sewer overflow (CSO) discharges, receiving water impacts and urban flooding. The tools for an urban drainage asset manager to obtain the required information are continuous monitoring of system performance and modeling to understand system dynamics. Based on the findings, the urban drainage asset

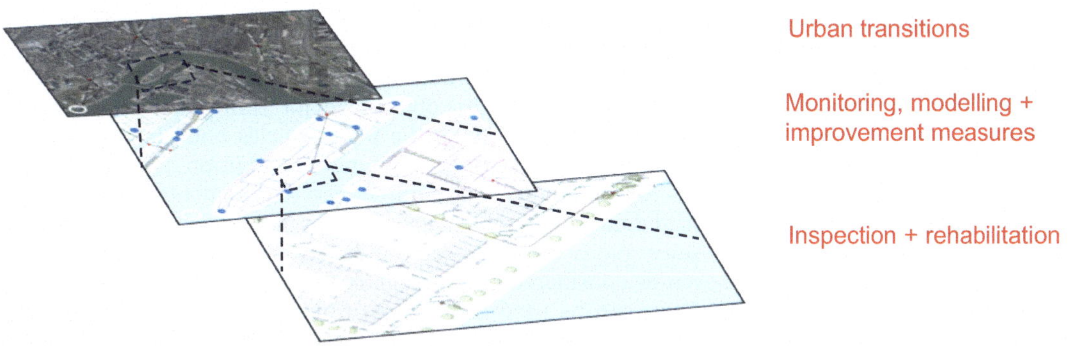

Figure 8.5 Three-layer approach to AM.

manager may decide to upgrade, such as storage tanks and to optimize system controls, quite often by implementing real-time control (RTC) strategies.

- **Network components:** urban drainage networks consist of many individual components, such as pipes, manholes, swales, storage basins, and so on. This object level is the traditional level of AM, where inspections are carried out to obtain information on the condition of individual assets and where rehabilitation strategies can be applied to ensure an adequate condition of assets to fulfill their function as part of the whole network.

In practice, depending on the size of the water utility, the responsibilities for the three layers are divided among several persons and/or departments, which are each optimizing within their scope. In addition, responsibilities for the urban fabric are often outside of the water utility itself. This could easily result in suboptimal decision making. One example is a sewer asset manager responsible for object management who must rehabilitate or replace a pipe that has failed, for example, for structural reasons. The project site is located in a district that is planned to be renovated within the next decade. The current size of the pipe is unfit for the foreseen needs of the changing urban fabric. Should this sewer manager simply replace the existing pipe with the same size, or should the new pipe reflect the hydraulic loadings corresponding to the future demand? This example demonstrates the need for integrated decision making, taking the contents of each of the three layers into account (including adjacent infrastructures), which has of course strong implications for the information needs for AM.

8.3.1 Information required at the object level

8.3.1.1 Descriptive data

The descriptive data that must be available on urban drainage objects have to be sufficient to cater for the needs of the AM at the object level as well as supporting the AM at the network and urban fabric levels as much as possible. Table 8.1 provides a non-exhaustive overview of the required descriptive data related to their use for the most frequently used objects. The table also describes which information needs to be available also in the two other layers. As the data needs vary greatly between utilities such a table can never be exhaustive but rather provides guidance on basic data needs.

One basic feature of any data collection program is the naming of assets and assigning unique IDs. Figure 8.6 presents an example of recommendations, given by ÖWAV-Regelblatt 40 (2010) on how a sewer network should be divided into sections, where every pipe, manhole, house connection, and so on are uniquely numbered and allocated to a certain street section.

Stormwater control measures are typically not a part of a physical pipe network, but rather (adjacent) isolated point or areal assets. They are however an important element of the drainage system as they are all individually contributing to the SWM at the district/urban scale (depending on the contributing catchment to each asset). For stormwater control measures, it is relevant to collect descriptive data necessary for maintenance and rehabilitation, such as location, dimensions (area), and materials used, and data related to the hydraulic performance (storage and discharge rating curves), which is comparable to data collected at the object level for 'traditional' sewer objects. However, it may also be necessary to add data describing the plant species and plant spacing, as these are relevant for the required type and frequency of maintenance. In addition, it is necessary to collect data related to expected services, operation, and maintenance requirements (related to the expected services). All relevant physical data should reflect as-built or as as-planted conditions rather than the design conditions. Unlike sewer pipes, the services provided by these assets are much more numerous and diverse: SWM can even be considered as a secondary function (mobilized only during wet weather conditions): these facilities can be used as a primary function for parking (detention storage or permeable pavement), pathways, roads, green spaces, roofs, and so on. A possible list of required data for a GIS system was collated by Fredriksen (2021) covering a broad range from general information about location, to construction and operation costs, hydraulic parameters, aesthetics and other benefits, construction year, and materials to logging of operation and maintenance interventions.

Table 8.1 Non-exhaustive overview of data required at the object level and information that are required at a larger scale (network or urban fabric).

Type of Asset	Type of Data	Object Level (Focus: Rehabilitation)	Network Level (Focus: System Dynamics)	Urban Fabric Level (Focus: Urban Planning)
Gully pots	Location (x, y, z)	Relevant	–	Relevant
	Material	Relevant	–	–
	Type	Top loader/front loader		
	Dimensions ($l \times h \times d$)	Relevant	Volume	–
	Dimension sand trap	Volume	–	–
	Position connection pipe	Front, side, back	–	–
Pipes	x, y, z (invert level) upstream and downstream	Relevant	Relevant	Relevant
	Diameter	Inner/outer	Inner	–
	Wall thickness	Relevant	–	–
	Material	Relevant	Relevant for hydraulic roughness	Relevant for circularity of pipe material
	Age	Relevant	–	Relevant for integrated projects
Joints	Joint type	Relevant	–	–
	Material	Relevant	–	–
Manholes	x, y	Relevant	Relevant	Relevant
	z (manhole cover)		Relevant	Relevant
	z (bottom)	Relevant	Relevant	–
	Material	Relevant	–	–
Overland conveyance	x, y, z (invert level) upstream and downstream	Relevant	Relevant	Relevant
	Dimensions (cross section)	Relevant	Relevant	Relevant
	Material	Relevant	Relevant for hydraulic roughness	Relevant for circularity of materials
Outfalls	x, y, z	Relevant	Relevant	Relevant
	Material	Relevant		
CSO weirs	x, y, z	Relevant	Relevant	Relevant
	Length	Relevant	Relevant	–
	Free height above Weir	–	Relevant	–
	Material	Relevant		
CSO tanks	x, y, z	Relevant	Relevant	Relevant
	3D internal structure	Relevant	–	–
	Volume	–	Relevant	–

(Continued)

Table 8.1 Non-exhaustive overview of data required at the object level and information that are required at a larger scale (network or urban fabric) (*Continued*).

Type of Asset	Type of Data	Object Level (Focus: Rehabilitation)	Network Level (Focus: System Dynamics)	Urban Fabric Level (Focus: Urban Planning)
	Material	Relevant	–	–
	Cleaning equipment	Relevant	–	–
Pumping stations	x, y, z	Relevant	relevant	Relevant
	Capacity	–	Max cap and (controlled) q-h relation	–
	Pump sump	$l \times b \times h$	V-h relation	–
	3D internal structure	Relevant	–	–
	3D building structure	–	–	Relevant
	Material	Relevant	–	Relevant
Stormwater control measures	x, y, z	Relevant	Relevant	Relevant
	Layout and dimensions	Relevant	V-h relation	Relevant
	Infiltration surface and soil characteristics (if exists)	Relevant	Relevant	
	Storage volume	Relevant	V-h relation	
	Outflow	Relevant	Max outflow and (controlled) q-h relation	
	Vegetation (if exists): type and coverage	Relevant	Relevant for hydraulic roughness	Relevant
	Components description (inlet, outlet, protection, etc.)	Relevant		
	Age	Relevant		Relevant for integrated projects
	Ownership (private/public)	Relevant		Relevant for integrated projects
	Services in charge of the maintenance	Could imply water/road/cleaning/parks departments		Relevant for integrated projects
	Operation and maintenance protocols and frequency	Could imply different departments such as water/road/cleaning/parks		Relevant for integrated projects
	Expected services (see Chapter 6)	Relevant for operation, maintenance, and rehabilitation		Relevant

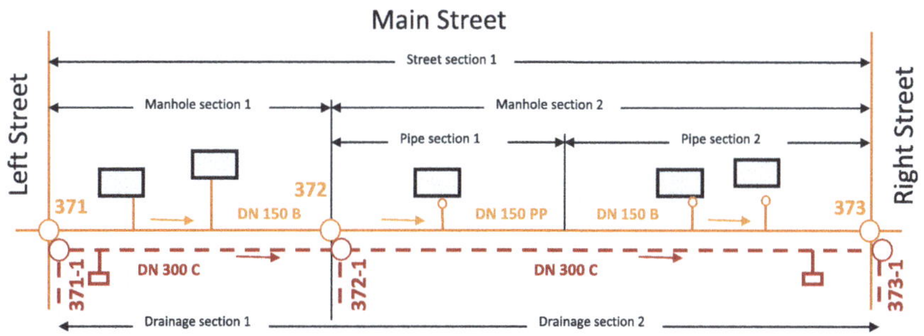

Figure 8.6 Sectioning of wastewater collection network (*source*: revised from ÖWAV-Regelblatt 40 (2010)).

8.3.1.2 Condition information

Ideally, descriptive data are available in AM systems and inspections are used to obtain information on the condition of the assets. However, quite often relevant descriptive data are missing, and inspections are needed to obtain these missing data. This section describes only relevant types of condition data that can be obtained via inspection. Refer to Chapter 4 for a detailed description of inspection techniques. For sewer pipes, condition information can be related to the structural integrity of the assets, which is only relevant at the object level and can be related to the hydraulic performance, which is also relevant at the network level. For stormwater control measures as for sewer assets, condition information is not limited to the structural integrity or hydraulic performance but concerns all expected services and components of the asset. Table 8.2 summarizes relevant condition information for the assets listed in Table 8.1.

With respect to pipes and stormwater control measures, a condition assessment is a vast subject, and the condition is only part of the performance assessment of the asset. This subject is extensively discussed in Chapters 3 (regulation), 4 (inspection), 5 (deterioration models), 6 (service-based approach), and 9 (operation and maintenance as well as rehabilitation techniques). With respect to stormwater control measures, the diversity of the assets (in terms of size, shape, composition, and above all expected services) often requires specific condition information. For example, with respect to green roofs, the structural integrity is related to the substrate bed, its supporting material, and leak tightness of the roofing materials. At the same time, the drain-down time of the roof is a descriptor of its hydraulic performance. For permeable pavements, the structural integrity is to be assessed in a similar manner and with similar parameters as for traditional impermeable pavement, including cracks, roughness, and flatness, whereas the infiltration rate is a key parameter for determining its hydraulic performance. Chapter 9 focuses on the operation, maintenance, and rehabilitation of some of the most used stormwater control measures: ponds and wetlands, permeable pavements, green roofs, bioswales, bioretention systems, and rainwater harvesting systems.

8.3.1.3 Condition requirements and tolerances

The condition requirements for pipes and manholes are well established. Expectations pertaining to the maximum joint width, maximum joint rotation angles, or minimum wall thickness can be found in standards and documentation provided by manufacturers, whereas tolerances as to the dimensions and material properties can be found in standards as well. In addition, inspection results can be coded using for instance the EN 13508 standard (see Chapter 3). The tolerances for pumping stations are less well established. Structural tolerances of the sump and upper structure are clearly defined, for example, for manholes, whereas, in contrast, the hydraulic performance is up to the operator to decide. There is no general agreement on the acceptable tolerance in terms of their hydraulic performance with respect to indicators such as percent capacity loss relative to the design or relative to the actual setpoint.

Table 8.2 Non-exhaustive overview of condition information per type of asset and their relevance per level.

Type of Asset	Condition Information	Object Level (Focus: Rehabilitation)	Network Level (Focus: System Dynamics)	Urban Fabric Level (Focus: Urban Planning)
Gully pots	Blockage of grated inlet	Cleaning	–	Local flooding
	Missing cover	Road safety	–	–
	Filling degree sand trap	Cleaning	–	–
	Blockage of outlet	Cleaning	–	Local flooding
	Misalignment (x, y, or z) with road	Hydraulic performance, road safety	–	–
	Broken shaft	Stability	–	–
Pipes	Wall thickness	Structural integrity		
	Defects and dysfunction, see Chapters 3–5 for detailed information	Structural integrity or hydraulic performance, see Chapters 6 and 9	Hydraulic performance	Leakage
Joints	Joint width	Structural integrity	Leaks	–
	Rotation angle (horizontal + vertical)	Structural integrity	–	–
Manholes	Cover missing or broken	Road safety		
	Wall thickness	Structural integrity		
	Tree roots	Hydraulic performance	Hydraulic performance	
	Holes in wall	Structural integrity		
	Loss of cross section (due to sediment/ Fat, Oil and Grease FOG)	Hydraulic performance	Hydraulic performance	–
	Corroded step iron	Worker safety		
Overland conveyance	Blockage across cross section	Cleaning	–	Local flooding
	Defects and dysfunction, see Chapters 3–5 for detailed information	Structural integrity or hydraulic performance, see Chapters 6 and 9	Hydraulic performance	
CSO weirs: all aspects of manholes+	Blockage of weir	Hydraulic performance	Hydraulic performance	
	Missing parts of weir	Hydraulic performance	Hydraulic performance	–
	Hole in of weir wall	Structural integrity	Hydraulic performance	–
Pumping stations: all aspects of manholes+	All mechanical and electrical items	Hydraulic performance, safety	Hydraulic performance	
	Sedimentation in pump sump	Hydraulic performance	Hydraulic performance	–
	FOG layer in pump sump	Hydraulic performance	Hydraulic performance	–
	Air treatment			H_2S issues, smell, and odor
Stormwater control measures	Defects and dysfunction, see Chapters 3–5 for detailed information	Broad range of performance, see Chapters 6 and 9	Hydraulic performance	Broad range of performance, see Chapters 6 and 9

Similarly, there is no general agreement on the pump failure frequency or pump failure duration. It is thus more straightforward for asset managers to assess the condition of pipes and manholes than to assess the condition and performance of pumping stations. For stormwater control measures, standardization of condition requirements and tolerances (e.g., percentage of green roof area without healthy plants) is still in its infancy. The Low Impact Development Stormwater Management Planning and Design Guide's wiki pages developed by three Ontario, Canada Conservation Authorities under its Sustainable Technologies Evaluation Program (see https://wiki.sustainabletechnologies.ca/wiki/Main_Page) provides an overview of the state of the art of condition assessments and tolerances for stormwater control measures.

8.3.2 Information required at the infrastructure network level

8.3.2.1 Descriptive information

The information required on the infrastructure network level is typically all information required for proper hydrodynamic modeling of any urban drainage system. As such, this requires all information to be able to describe the network topology, or configuration. Effectively, this is all information related to the dimensions, position, and connectivity of all network components as listed in Table 8.1. Next to the description of the network, it is necessary to have information on actuators, system loading, and boundary conditions. The actuator settings may be fixed, that is, switch on/off levels of pumps, or variable with valves and pumps controlled by an advanced RTC system. The system loading typically consists of 'model inputs,' such as rainfall (or net runoff, depending on the system boundaries), dry weather flow, which could be a mix of domestic and industrial wastewater, and extraneous waters, which could be ground water via weeping tiles, infiltration of groundwater, and inflow of upstream surface water, for instance when the original creeks were covered up. All these system loadings should be available at the relevant spatial scale, which is determined by the level of detail chosen in the drainage system model. The boundary conditions could be downstream water levels, or discharge limitations due to hydraulic constraints of downstream pumps or control structures. In case of water-quality modeling, information must be available on the quality of inputs.

8.3.2.2 Performance assessment

The performance assessment of infrastructure networks could be based on (a combination of) monitoring data and results of models, each describing/capturing relevant system dynamics at their relevant scale. Monitoring data should be able to provide, after data validation, high-quality data on the actual system performance. Models could be used to support the interpretation or interpolation of monitoring data. Models could also be applied without monitoring data; however, in that case it is not possible to assess the actual system performance, but rather the potential system performance. The latter is especially relevant for events with a long return period, such as urban flooding (see, e.g., Chapter 6).

8.3.2.3 Performance requirements and tolerances

Performance requirements for urban drainage networks have evolved over time, reflecting interests and state of the art in monitoring and modeling. The following types of requirements have been applied:

- 'Internally focused' requirements related to hydraulic performance, such as:
  - No flooding may occur for a (design) storm with a given return period.
  - No surcharge may occur for a (design) storm with a given return period.
  - Percentage of illicit connections of storm water to a sanitary sewer.
- 'Internally focused' requirements related to environmental performance, such as:
  - Percentage of illicit connections of sewage to a storm sewer system or stormwater control measures.

- 'Externally focused' requirements related to hydraulic performance, such as:
  - The hydraulic load of the downstream wastewater treatment plant (WWTP) cannot exceed a given maximum, typically the design flow.
  - The proportion of extraneous water in the WWTP influent cannot exceed a given percentage on an annual basis; this indicator limits the acceptable infiltration.
- 'Externally focused' requirements related to environmental loading, such as:
  - CSO frequency, expressed in the number of spill events/year. This frequency is often assessed over longer periods of time, for example, a few years, to account for the variability in the precipitation.
  - CSO volumes, which could be measured or modeled.
  - CSO loads (i.e., volume times concentration). Given the large range in concentrations observed in practice and the high expenses related to monitoring water quality, several countries have applied a fixed, theoretical concentration to calculate loads.
  - Sanitary sewer overflow (SSO) loads (i.e., volume times concentration).
  - SSO concentrations.
  - Removal efficiency of CSO tanks/CSO treatment facilities.
  - Removal efficiency of SSO treatment facilities.
  - CSO frequency/CSO volume due to operational issues, such as pump failures.
  - Removal efficiency of stormwater ponds and wetlands or other control measures.
- 'Externally focused' requirements related to urban flooding, such as:
  - Flooding frequency.
  - Flood volume.
  - Flood extent.

8.3.3 Information required on urban fabric

8.3.3.1 Descriptive information

The information on the urban fabric should be sufficient to assess the impact of the network performance on the urban area as well as to assess the impact of urban planning and development on the performance requirements for the urban infrastructure and/or the load on the urban water infrastructure. The sensitivity of the urban fabric to network performance is related to:

- Vulnerability of the urban area for flooding: how much damage will occur given a certain flood depth, or what are the anticipated average annual damages?
- Nuisance, for example, H_2S and other noxious gases may have a negative impact on livability (odor) and public health.
- Impacts of CSOs/SSOs and stormwater discharges on receiving water quality and the potential to use the receiving waters for recreational purposes, drinking water production, irrigation, and so on.
- Benefits in terms of climate change mitigation: resilience against heat island effect or drought, in connection to stormwater control measures.
- Benefits in terms of biodiversity development and conservation, in connection to urban green spaces dedicated to SWM and their connectivity.
- Social benefits for the inhabitants in connection to open-space stormwater control measures.

For each of these impacts, a minimal performance level of the infrastructure is required to be able to sustain a range of urban functions. Impact of the urban drainage system on the urban environment can be negative (odor) but also positive (climate change resilience). Vice versa, the urban fabric affects the urban drainage networks due to:

- Changes in population.
- Changes in degree of imperviousness.

- Changes in housing density.
- Climate adaptation, including urban heating and drought effects.
- Changes in the street networks during urban redevelopment.
- Space constraints above ground and below ground.

8.3.3.2 Performance assessment

The performance assessment at the level of the urban fabric focuses on the contribution of the urban drainage infrastructure to the quality of urban life. This involves criteria that measure the service level provided, which could be framed positively (we are able to flush our toilets 24/7) or negatively (e.g., the damage due to urban flooding in a given year may amount to millions). Methods to assess this are:

- Surveys.
- Assessment of call data or other customer complaints.
- Analysis of public health parameters related to sanitation, such as numbers of patients with gastroenteritis or cholera, life expectancy, or mortality rate of children below an age of five.
- Analysis of damage data of insurance companies.

Chapter 6 discusses risk or impact-based approaches, risk assessment, and multicriteria decision analysis, which are often related to the performance assessment of the urban fabric.

8.3.3.3 Performance requirements and tolerances

At the level of the urban fabric, the urban drainage infrastructure may also be assessed by its contribution to overarching societal issues, such as circularity, sustainability, resilience, adaptability, flexibility, and robustness. These overarching issues are drivers for innovation and experiments with novel sanitation or nature-based solutions for SWM. To be able to compare those 'novel' systems, it is necessary to derive relevant performance indicators that allow for the comparison with the results for 'traditional' urban drainage networks and approaches. These performance indicators should be predefined, where possible, and be comprehensive in scope. For instance, solid waste management criteria should be considered when evaluating black water systems where kitchen waste is suggested as a source of energy. Given its wide potential scope, there is a lack of common performance criteria to be considered, as the system boundaries may change on a case-by-case basis. In practice, the performance criteria should therefore be tailored to the proposed innovative approaches.

8.4 DATA MODELS AND BIAS IN SEWER PIPE DETERIORATION MODELS

Sewer deterioration modeling can help to prioritize sewer segments for annually programming renovation works. It could also be useful, in connection with renovation/decommissioning process modeling, for performing numerical long-term simulations of infrastructure asset management (IAM) strategies. The decommissioning of a pipe, whether followed by its renovation or its abandonment (e.g., in case of network reconfiguration), marks the end of its service life. The calibration of deterioration models should be based on inspection data (see Chapter 5), most often obtained by closed-circuit television (CCTV), and the synthesis of inspection reports in the form of condition notes (Rahman & Vanier, 2004). However, the use of such data requires attention to the following points:

- The data are only available within a limited observation window (see Figure 8.7); this short window of usually less than 20 years does not present a comprehensive history, as the data have only recently become available in digital format (i.e., often after 1990 of later).
- Typically, given the prohibitive cost of CCTV, only a small proportion of sewer pipes get inspected annually.

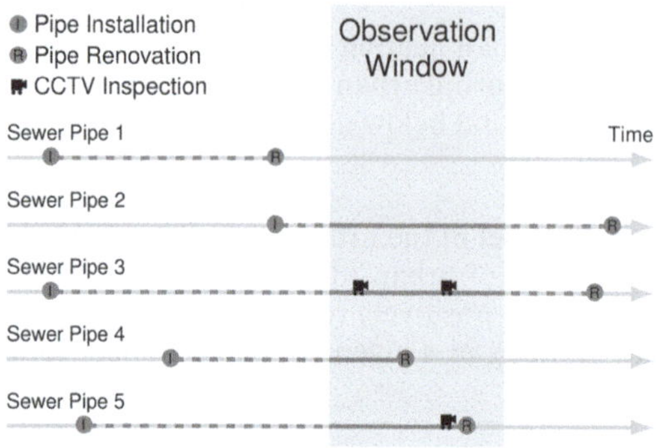

Figure 8.7 Different cases that occur when observing inspection and renovation/decommissioning events for sewer pipelines within a limited observation window.

- Inspections are usually conducted for four reasons:
  - Acceptance or assumption verification of new or rehabilitation works.
  - Diagnosis in the event of an observed malfunction (e.g., presence of odors, overflows, subsidence of the roadway, etc.).
  - Evaluation of the condition of the sewer line before road renovation works.
  - Random sectorial campaign to assess the sewer system condition.
- Exploiting the inspection data also requires that the operational functionality of the pipes, that is, whether they were in service at the start of the observation window, were commissioned or decommissioned during the latter, is fully known.

Consequently, available inspection data are very likely to be subject to three informational imperfections: (1) selective survival bias, (2) recruitment bias, and (3) informative censoring. These data imperfections are said to be 'informative' in the sense that while they occur randomly, their probability of occurrence nevertheless depends on the phenomenon studied, namely the deterioration of the sewer pipes and their inspection and renovation processes. Presenting these data imperfections requires consideration of where the sewer pipes fit as per the five observation cases illustrated in Figure 8.7, which illustrates the relative position of the observation window with respect to the beginning (i.e., initial installation) and the ending (i.e., renovation or decommission) of their service life as well as the timing of inspection activities (if any).

8.4.1 Selective survival (Le Gat, 2008)

Sewer pipes #1 and #3 in Figure 8.7 illustrate a scenario where, due to the limited observation window, the cohort of all pipes which had originally been installed in the same year cannot be completely observed; some pipes of the cohort (i.e., sewer pipe #1) had been renovated or decommissioned before the observation window. But the need for the renovation or decommissioning reflects at least partially the pipe condition as a result, the observed pipes are likely to be the most robust survivors of their cohort. If not accounted for by the calibration method, this 'selective survival' phenomenon may bias the deterioration model underestimating the likelihood of malfunctions. To correct this bias, it is necessary to consider a joint modeling of both the deterioration and the renovation processes, provided the renovation or decommission events had been exhaustively reported prior to the observation window. Particularly, in case of potential renovations based on the data collected within the observation window, sewer pipe data related to the characteristics of the pipes prior to their renovation should be kept in the database, rather than having it completely replaced by information related to the renovated sewer pipes.

8.4.2 Informative censoring (Le Gat, 2008)

Sewer pipes #3 and #5 in Figure 8.7 illustrate a scenario where the observation window may be stopped, either independently of the pipe deterioration (i.e., in case of pipe #3), or prematurely due to pipe needing renovation (i.e., in case of pipe #5), which, in turn, may have been caused by excessive deterioration of the pipe. A major consequence of 'informative censorship' is the potential omission of observations of the most deteriorated objects, due to their renovation having taken place within a short time following their inspection. This informative censoring of the deterioration process by the subsequent renovation process, due to the interdependency between both processes, reinforces the necessity of jointly modeling them.

8.4.3 Recruitment bias

Given the four inspection purposes mentioned above, inspection data are not likely to be based on a random sample of sewer pipes, since a noticeable part of the sample may have targeted malfunctions. It would be necessary, in order to protect the calibration of the deterioration model against a 'recruitment bias' or overestimation bias, to consider the reasons that motivated the inspections. It is then advisable that the inspection purpose be carefully reported as part of the sewer utility information system.

8.4.4 Which data model to allow imperfections to be considered?

The selection of a sewer deterioration modeling approach that allows data imperfections to be taken into account implies that the model input data should be organized in a suitable way. In general, it is advisable to extract from the utility information system three datasets relating respectively to (1) the description of the pipes, (2) the history of inspection results, and (3) the history of operations; these datasets must have in common an unambiguous pipe identifier that allows merging of their contents.

8.4.4.1 Pipe description dataset

The 'pipe description dataset' should contain a single data row per pipe, and at least include the following fields: pipe identifier, installation date or year, decommissioning date or year (to be left empty if the pipe is still in service), material, shape (circular vs egg-shaped vs rectangular, etc.), sectional width, sectional height, pipe length (i.e., from upstream to downstream manholes), sewer function (i.e., wastewater vs combined vs storm), depth, slope, location, nature, and dimensions of any service connections along the pipe (if any), and any additional information considered meaningful to the deterioration process.

8.4.4.2 Pipe inspection dataset

The 'pipe inspection dataset' should contain a single data row per pipe inspection, and at least include the following fields: pipe identifier, inspection date, inspection result (e.g., condition class), inspection purpose (sectorial assessment, reported dysfunction, condition assessment before road renovation works, etc.), if available.

8.4.4.3 Operation dataset

The 'operation dataset' is only needed when the chosen deterioration model is designed to account for operational actions that could potentially influence the deterioration process; this dataset is not needed if the only operation would have consisted of the decommissioning of a pipe. When needed, it should contain a single data row per pipe operation, and at least include the following fields: pipe identifier, operation date, operation type (e.g., flushing, root cutting, etc.).

8.5 QUALITY CONTROL

8.5.1 Data quality requirements

8.5.1.1 Importance of quality control

In many cases it is implicitly assumed that the data on the structure and geometry of a system are well known and are free of errors; however, in practice it turns out that databases on Urban Drainage

(UD) systems fail to live up to these expectations. Deviations between the system as represented in the database and the 'real world' may result in ultimately making wrong decisions (Clemens & van der Heide, 1999; Dirksen *et al.*, 2009). Data validation is a prerequisite to limit the adverse effects of poor data quality, as databases are not static (i.e., even the so-called 'fixed' data such as invert levels, pipe diameters, etc. are subject to mutations); databases used for AM purposes should be subjected to more or less a continuous validation process.

8.5.1.2 Data quantity: incompleteness and how to deal with it?

Data acquisition is a costly and complex process, which is why data are often limited in quality and quantity (Ana *et al.*, 2009). Moreover, it will always be possible to detect some amount of bias and error regardless of the acquisition method. There are also major problems with data such as missing and implausible data, the absence of information about rehabilitation activities and the decisions that underpin them, poorly documented models, and insufficient measurements and calibration data (Egger *et al.*, 2013).

The absence of high-quality and high-quantity data impairs condition-based maintenance and prevents predictive models from being commonly used (Tscheikner-Gratl *et al.*, 2019). According to Scheidegger *et al.* (2011) and Ana and Bauwens (2010), the lack of comprehensive and reliable data is, among others, one of the major reasons that sewer deterioration models fail to provide reliable predictions of sewer conditions. It is necessary to collect information on asset characteristics during their operational lifespan (e.g., date and depth of installation, material, backfill, condition, incidents of failure, and interventions), but such items are often absent from utility databases (UDBs). In addition, problems can arise when there is a discrepancy between design drawings and final constructed assets, which impacts AM solutions (Farrelly & Bach, 2018). This is compounded for nature-based solutions whereas installed information related to vegetation is often missing.

The analysis of various development and operation strategies for urban drainage systems relies increasingly on numerical simulations (Korving & Clemens, 2005). However, the availability of data that is needed for the development of these models is often limited for various reasons. The data about underground infrastructure are often unavailable, confidential, or only available in a 'paper' format. This is particularly true in developing or transitioning countries, where a prerequisite for the numerical description of urban drainage runoff is missing. Furthermore, the number of resources available to improve databases is limited. As a result, there is inherent ambivalence among modelers to accept data scarcity and to try to improve the database by estimating missing data, especially for spatially differentiated systems such as sewer networks (Blumensaat *et al.*, 2012).

To deal with lack/incompleteness of data, some researchers have developed virtual networks to better characterize its consequences in terms of uncertainties. For example, a tree-layout algorithm with predefined boundary conditions was used by Möderl *et al.* (2009) to automatically generate 'virtual' drainage networks. By combining several random processes, a large number of virtual sewer systems were generated, which represented a wide range of cases. In hydraulic dimensioning, rainfall runoff for a design rain was balanced against the cross-sectional flow resulting from it. Based on the simulation results, implemented in a hydraulic sewer modeling platform, it was possible to assess stochastically how hydraulic systems perform under different boundary conditions. An agent-based approach was implemented by Urich *et al.* (2010) in place of the dendritic network generation approach of Möderl *et al.* (2009) to better represent the characteristics and constraints of an urban catchment. Using spatially differentiated, detailed information as input, the method arrived at a sewer network that has a state-of-the-art hydraulic design. The performance of the model was evaluated by comparing a range of performance indicators from a hundred virtual cases with the corresponding values from a real-life catchment. In another attempt, a semi-virtual asset stock (SVAS) was created to evaluate the influence of the available UDB data on inspection programs. This SVAS is based on the asset stock of the Greater Cincinnati Metropolitan Sewer District in the USA (Salman, 2010). The details of the method used can be found in Ahmadi *et al.* (2014b). Virtual (or semi-virtual) case studies can help improve knowledge by providing a dynamic and interactive learning environment, where individuals can engage with realistic scenarios and diverse datasets. These studies offer the flexibility to explore and analyze a wide range of cases and information from different perspectives, enabling learners to gain a deeper understanding of complex topics and enhance their overall knowledge base.

Other studies address datasets lacking historical information by proposing the combination of a sewer deterioration model with a simple rehabilitation model which can be calibrated with these datasets. For instance, in their study, Egger *et al.* (2013) used Bayesian inference to estimate model parameters due to limited information content of the data and limited identifiability of model parameters. A sensitivity analysis indicated whether a model is robust to the uncertainty in the prior. In the analyses, it was revealed that the model's results were sensitive to the means of the priors for specific model parameters, which should be elicited carefully. They concluded that by utilizing this approach, it is possible to calibrate sewer pipe deterioration models even when historical condition records are not available.

8.5.1.3 Data quality: timeliness, uncertainty, and imprecision

Without reliable data, rehabilitation planning is little more than a guessing game. Findings of previous studies confirm that having imprecise data is preferable rather than having no data. For instance, under the hypothesis that a sewer network was constructed at the same time as the surface development above it, Ahmadi *et al.* (2014b) demonstrated that the notion of 'district' could be used effectively instead of precise information on segment age. Data quality requirements include:

- Accuracy – the data correctly describe the real asset with known uncertainties in dimensions and other characteristics.
- Completeness – the data are complete for all relevant attributes.
- Consistency – the data are consistent in terms of format, value ranges, defined rules, and so on.
- Actuality – the data describe the current status of the assets including the management of historical events.
- Uniqueness – each record is unique without duplicates.
- Reliability – The data stem from a known source.
- Validity – the data have been validated, and the methods applied for the validation are described.

Data validation on data used in hydrodynamic modeling is often integrated in the software tools used to this end. Some basis validation steps are described in Clemens (2001):

- Generate a two-dimensional or three-dimensional (3D) plot of the system; obvious wrong relations between elements (e.g., conduits not connected to any other element) become apparent by visual inspection as do incorrect coordinates from, for example, manholes.
- Perform simple checks such as 'no elements above street level,' or the occurrence of unrealistic dimensions (e.g., pipes with a diameter >5 m).
- Check for 'lost storage.' Lost storage is basically a local depression or sag in a system from which no water can be drained. Such depression can be caused by either wrong data or realistic. Incorrect invert elevations, weir levels, or switch on levels of pumping station may cause 'lost storage.' An algorithm for detecting 'lost storage' is found in Clemens (2001).

For further reading regarding data collection and data validation, Chapter 11 (Schellart *et al.*, 2021) of the 'Metrology in Urban Drainage and Stormwater Management: Plug and Pray' handbook is recommended. This chapter proposes seven case studies to serve as illustrations of common challenges as well as lessons learned regarding data collection in urban drainage.

In many organizations managing urban drainage systems, there is a separation of tasks between departments, for example, the departments responsible for the design and maintenance/operations are not in control of the database. Depending on the authorization people have and the efficiency/effectiveness of the internal procedures it may take time between completing the replacement or rehabilitation of an element and the entry of the updated data on this element in the database. In some extreme cases this may take years. In such a situation: this could lead to erroneous actions from other departments not being aware of the replacement. It is thus very important to maintain an up-to-date database and continuous communications between departments. Another example, when a field visit is planned (which can be for any reason), one should supply the visiting employees with the information available in the database and ask them to check this and report back. When implementing such practice over several years the quality of the database can be increased without demanding excessive budgets.

The source of data is also very important as a token of the associated reliability. One may use the following classification (i.e., from doubtful to trustworthy): the data quality management and data storage systems used in existing urban drainage metrology projects are often bespoke. In the quest for a more harmonized approach to data quality management and data storage, the question arises as to what level of harmonization is desirable and how one demonstrates that such an approach should be implemented (Bertrand-Krajewski *et al.*, 2021). With respect to the source of the data, the following hierarchy can be broadly adopted (from best to worst):

(1) Recently measured in the field.
(2) Obtained from 'as-built' drawings.
(3) Construction drawings.
(4) Data from an older database.
(5) Personal information (e.g., obtained from the former construction site manager).

8.5.1.4 Methods to assess the quality and completeness of database

Thanks to advances in sensor technology and automated long-distance data transmission, continuous measurements have become a preferred method for monitoring urban drainage systems. In general, an expert must process the collected or raw data to detect and mark the incorrect data, remove them, and replace them with interpolated data. Data quality assessment or data validation is the first step to detect incorrect, anomalous data. Validation of data consists of three steps: (1) data preparation, (2) validation scores generation, and (3) interpretation of the validation scores. A variety of methods, concepts, and tools are available for data validation (Venkat *et al.*, 2003). There are universally applicable methods (e.g., Rosen *et al.*, 2003; Patcha & Park, 2007) and there are methods specific to urban drainage (e.g., seven simple tests presented in Bertrand-Krajewski *et al.* (2000) and Mourad & Bertrand-Krajewski (2002)). A data quality improvement process is presented in Figure 8.8 (Branisavljević *et al.*, 2009),

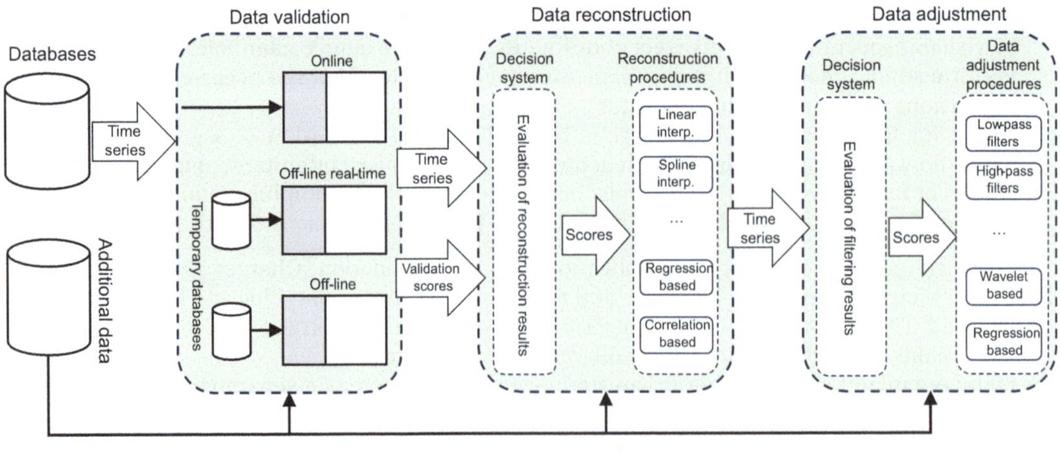

Figure 8.8 Data quality improvement framework (*source*: redrawn from Branisavljević *et al.* 2009).

where the successive steps that follow data validation are data reconstruction and data adjustment. Even after a database has been subjected to this kind of simple tests some uncertainties remain, for example, one can never be sure that all elements in the system in question are represented in the database, errors in, for example, geometrical characteristics or small measuring errors of, for example, invert elevations may go unnoticed, while there may be uncertainties as to the appropriate street elevations.

Clemens and von der Heide (1999) report on the effects of such uncertainties in the database with respect to the results of hydrodynamic calculations and found that these uncertainties may translate into substantial subsequent uncertainties in terms of, for example, flood locations and spilled volumes in CSO events; however, this tends to be case sensitive. A reduction of these uncertainties is only achievable by validating the database against (new) field observations.

When applying models for urban drainage management, for example, in case of predicting pipe deterioration or other performance metrics, it is important to assess the influence the quality of the input data may have on the uncertainty of the resulting predictions and the decisions that are made based upon these predictions (Fugledalen *et al.*, 2021). Depending on the model used and its application, this can be done by scenario analysis, Monte Carlo simulation, or analytically (Rokstad, 2016).

8.6 FROM POOR DATA TO MASSIVE/BIG DATA

In this section, we propose to discuss the contrasting situations regarding the data availability from utilities. These situations illustrate the heterogeneity of knowledge and management methods within the utilities.

8.6.1 The data-loop problem

For small utilities, data applications remain scarce or non-usable in large parts of the water sector depending on the location, size, budget, and motivation of the utilities or organizations. A recent Swedish study (Okwori *et al.*, 2021) confirms the 'low availability, integrity and consistency' for urban water pipe networks, and 'lack of interoperability between asset management tools.' There is a staggering difference between market trends (oriented toward data management and processing software) and the actual data availability and usage within utilities. Data scarcity concerns both the quality and the quantity of data. The problem is not only the incompleteness of data (inaccessible or unavailable) but also the challenges caused by the use of unverified, uncertain, or imprecise data. The data scarcity problem is being progressively dealt with but at a pace that is hampering many AM strategies which will negatively influence our decision making for decades to come.

According to Cherqui *et al.* (2022), the data scarcity problem can be conceptualized as a loop and propose to call it the 'data-loop' problem (see Figure 8.9). Moving from the status quo situation where almost no data are available requires demonstrating the benefits from data usage, which is not possible without data. As investing in data gathering is very often an onerous activity, a utility manager won't consider data gathering without 'immediate' tangible benefits, and the little data that may be available will not have been verified as it is insufficient to run a model. Consequently, the utility manager will not invest in models or tools or will not use data to support decision making. This brings us back to the status quo situation and so on. The 'data-loop' problem is one of the major obstacles to an efficient and long-term AM approach, but this problem is also often caused by the lack of medium- and long-term vision of asset managers (Cherqui *et al.*, 2022). However, vision alone may not be sufficient to overcome all the barriers. These barriers are found at all levels ranging from the individual to the organization (Manny *et al.*, 2021). Some barriers are also directly connected to the organization's characteristics (Sun *et al.*, 2018).

Among the solutions for the 'data-loop' problem, some are already gaining attention and traction: monitoring is more and more based on cheaper and easier to use sensors (Bartos *et al.*, 2019; Cherqui *et al.*, 2021) and models tend to be less data intensive. Okwori *et al.* (2021) proposed a conceptual framework to enable increased data-driven AM in pipe networks. Previous research has also shown

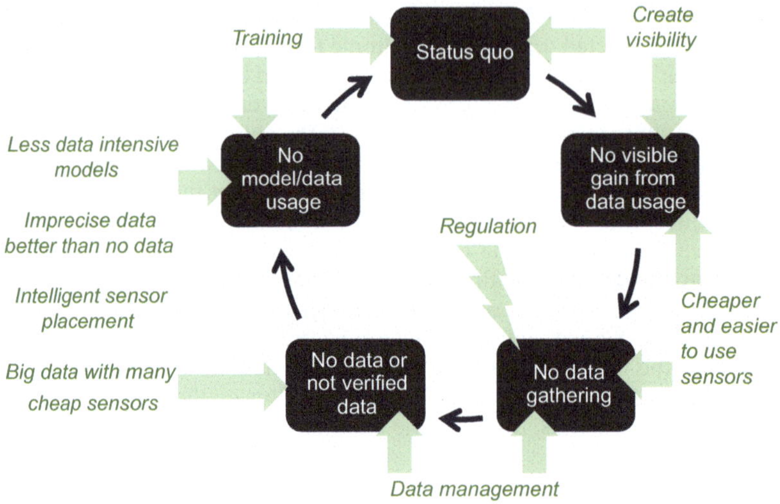

Figure 8.9 How to overcome the 'data-loop' problem: a non-exhaustive list of potential solutions. Data-loop problem (in black) and means to overcome it (in light green) (*source*: adapted from Cherqui *et al.*, 2022).

that, when considering important but unknown variables, imprecision could be better than no data (see, e.g., Ahmadi *et al.*, 2015). Training of utility managers is also a key component explaining the importance and benefits of data and models for more rational decision making. On another level, local or national regulation is also a strong incentive (or a hindrance depending on the regulator) toward UDBs containing a minimum set of data. Similarly, Surbakti *et al.* (2020) have identified seven main themes of factors that may influence effective use of big data: 'organizational aspects; systems, tools, and techniques; people aspects; data privacy and security and governance; data quality; process management; and perceived organizational benefit.' Sharing experience and good practices will help to propose an exhaustive list of solutions.

8.6.2 Data reconstruction

As discussed previously, data reconstruction techniques can help mitigate the 'data-loop' problem. Data collection always relates to data quality issues and gaps, with their size being dependent on the quality of the collection process. In a lot of cases the existing quality (see Section 8.4 for quality control) is not adequate for further utilization whereas measurements or inspections, and so on. will require unavailable time and financial resources. In this case, data reconstruction methods can be applied to fill gaps and enable modeling approaches. These data reconstruction approaches have so far mainly been applied to piped systems. The process is normally started by a data quality check for anomalies. Visualization of the data is helpful and an effective way of identifying anomalies such as missing data, outliers, and implausible values (Robles-Velasco *et al.*, 2020). Missing or anomalous data can then be replaced. It is important to note that filling in the missing values of the most important variables, such as age, with estimations can induce more uncertainties and lower the prediction performance of the models. It is therefore of the utmost importance to highlight the values entered to avoid overconfidence and the misconception of measurement quality for other users. Different techniques of increased sophistication exist:

- Simple input of single data gaps by using average values for numerical information and the highest share for non-numerical ones. This introduces a high uncertainty as this approach emphasizes average and peak values; however, it may be sufficient for modeling at a network level when the data gaps are not too large.

- Regression-based single input fills the gaps using regression-based methods that can include several co-variables. An approach can be found in Graham (2012), but any regression method can be adapted.
- Crosswise reconstruction of available data (Tscheikner-Gratl *et al.*, 2015) taking advantage of available surrogate data (e.g., construction years of buildings and neighboring infrastructure networks that are expected to have been built in the same period) and similarities of neighboring connected pipe segments. Such an approach has been successfully applied for sewer data by Caradot *et al.* (2020).
- Most of the above-mentioned approaches input a single value for the missing data, rather than point estimates and confidence intervals. To account for this, multiple input methods exist (Kabir *et al.*, 2020) introducing uncertainty into the picture. Data requirements and the modeling complexity are the limitations of the multiple input methods.
- Data mining approaches in archives combined with selected punctual inspections will deliver the best quality of data reconstruction, albeit limited by time resources.

All methods have their merits and fields of application. In general, the method of choice will depend on the data quality needed, the available resources, and the timeframe. Data reconstruction should, however, not replace data collection and proper information management in the long term where gaps should be filled with measured data.

8.6.3 Big data

At the same time and for some utilities, the field of urban drainage has witnessed a significant increase in data from various sources (and various data quality), transforming the way cities manage their systems (Ersoy & Alberto, 2019). One notable factor contributing to this surge is citizen participation. With the rise of smart technologies (Zhu *et al.*, 2023) and the Internet of Things (IoT), citizens are actively engaging in data collection through mobile applications and sensor networks, providing real-time information about drainage issues, flooding, and water quality (Etter *et al.*, 2020; Tauro *et al.*, 2018; Yang & Ng, 2017). Additionally, urban drainage systems are increasingly reliant on advanced modeling techniques to predict and optimize asset performance. These models generate vast amounts of data, ranging from hydraulic simulations to long-term condition and performance. Furthermore, the concept of digital twins (Ferré-Bigorra *et al.*, 2022; Grieves & Vickers, 2016) has emerged as a revolutionary tool in urban drainage. By creating virtual replicas of real-world drainage systems, practitioners can analyze and simulate various scenarios, leading to more data generation and insights (Jiang *et al.*, 2021).

Leveraging big data in the field of urban drainage offers a multitude of benefits to city planners, engineers, and decision makers. First and foremost, it enables a more comprehensive understanding of the drainage network's behavior, helping identify potential vulnerabilities and design more resilient systems (Dziedzic *et al.*, 2021; Koo *et al.*, 2015). Big data analytics empower stakeholders to make informed decisions based on real-time data, resulting in more effective responses to emergency situations such as floods and mitigating their impact on urban areas. The wealth of data also enhances predictive capabilities, enabling proactive maintenance and reducing infrastructure downtime (Nie *et al.*, 2020). Moreover, big data foster evidence-based urban planning, facilitating the integration of sustainable water management practices and climate adaptation measures. By harnessing the power of big data, cities can optimize their drainage systems, improve water quality, and ensure the well-being of their inhabitants.

While the abundance of data in urban drainage presents immense opportunities, it also comes with certain risks and challenges. One of the primary concerns is data overload, where the sheer volume of information can overwhelm decision makers and lead to paralysis of analysis. Extracting valuable insights from a massive dataset can be complex and time consuming, potentially hindering timely actions in critical situations. Moreover, data privacy and security become paramount issues when handling sensitive information collected from citizens and urban infrastructure. Ensuring proper data anonymization, storage, and access controls is essential to protect individuals' privacy and prevent potential cyber-attacks. Additionally, the integration of various data sources poses challenges in terms of data interoperability and standardization, as different systems may use disparate formats and structures. Addressing these risks requires careful planning, robust data-management strategies, and the adoption of advanced analytics tools to transform big data into actionable knowledge effectively.

8.7 DATA ACCESS AND STORAGE

Despite the widely accepted value of open data (Carrara *et al.*, 2017), there are relatively few examples of open data in sewer AM. One exception to mention perhaps is the public Sewer-ML dataset (Haurum & Moeslund, 2021) which makes available a dataset of 1.3 million images, from 75 618 videos collected from three Danish water utility companies over 9 years (https://vap.aau.dk/sewer-ml). This dataset allows the training and testing of automatic defect detection algorithms. Apart from the potential benefits of open data, several barriers exist (Janssen *et al.*, 2012). In the case of asset data, some of these barriers may be particularly challenging, as clarity is lacking as to how value can be created, or there is a relatively small number of people interested in and able to use the data.

Embracing open-data practices has multifaceted benefits. First, it facilitates reproducibility, allowing other researchers to verify and build upon existing findings, ensuring the integrity of scientific discoveries. Second, open data promotes transparency, enhancing the credibility and trustworthiness of research and industry outcomes, which is crucial in an era where misinformation and distrust in science can hinder progress. Moreover, by making data openly available, it enhances collaboration among operational and scientists from diverse backgrounds and institutions, fostering interdisciplinary research and accelerating innovation. Additionally, open-data practices encourage data sharing and integration, leading to the accumulation of larger datasets, which in turn can improve the statistical power and generalizability of research results. Utilities can avoid redundant data collection efforts and reduce costs by accessing openly available datasets rather than investing in expensive data collection processes. Lastly, open data allow for data-driven innovations, enabling the development of new tools, algorithms, and insights that can address pressing societal challenges more effectively. By advocating and adopting open-data practices, the scientific community can collectively drive knowledge advancement, facilitate evidence-based decision making, and contribute to the betterment of society as a whole. Interested readers can refer to the FAIR principles (findability, accessibility, interoperability, and reuse of digital assets; https://www.go-fair.org/fair-principles/).

New technologies have become available to improve data accessibility. An example of such recent development is 'cloud data hosting,' which enables utilities to share data more efficiently within their organization and provide stakeholders with controlled access to relevant information. By using this approach, utilities and researchers can create large accessible databases that may be shared among all. Currently, most infrastructure managers believe that the risk of malicious use of infrastructure asset data outweighs the potential benefits of public availability of detailed data. Access to most infrastructure databases therefore usually requires identification and authentication. However, efficient utilization of the data requires that different people need access to different sections of the asset database – the trade-off between data security and flexibility must therefore always be carefully considered.

8.7.1 Password guessing

Guessing passwords has traditionally been an attack method with a frighteningly high degree of success, and regrettably it is probably still the case that most passwords are far too easy to guess. Many hardware products are shipped with pre-configured default passwords (e.g., Simple Network Management Protocol (SNMP) community string in various routers and switches), and all too often nobody bothers resetting these passwords before the equipment is deployed.

8.7.2 Security: access control

Access control is a fundamental mechanism to ensure that the security properties in the CIA triad (confidentiality–integrity–availability) are fulfilled, by ensuring that only authorized users are allowed to read (confidentiality) or change (integrity) specific data, and that authorized users get access to data and systems when they need to (availability). Before any access control decisions can be made, the users must first be authenticated. Access control is normally provided by the underlying operating system.

If a computer system uses password-based access control, then those passwords need to be stored in the system in some way. These passwords need to be protected against prying eyes. An alternative to traditional passwords is the one-time password, which has been quite popular in electronic banking. This can be easily implemented as a list of numbered passwords/codes on a small plastic card, or similar. When the user needs to authenticate himself, he provides his username, whence the bank asks, 'What is code number 13?.' For this to be secure over an open network it is imperative that this exact password is never used again (hence the name 'one-time password'), and it is up to the bank to ensure that the customer receives a new card in good time before the codes have all been used. There are also more technical solutions, such as so-called password calculators that generate an apparently random code when prompted (some versions require that a Personal Identification Number (PIN) is entered first).

Authentication methods that base themselves on determining unique personal characteristics are called biometric methods. Several forms of biometric authentication exist, but the most common is probably the fingerprint scanner. Other popular methods include retina scan, iris scan, voice recognition, and face recognition.

8.7.3 Protecting the data

In addition to using access control mechanisms to protect data, confidentiality and integrity can be further ensured by using encryption and digital signatures. Encryption means transforming data to a seemingly random string of bits using an encryption algorithm and a key. In the case of symmetric encryption, the same key is used to both encrypt and decrypt the data, and it is therefore imperative that the key is kept secret.

A cryptographic hash function will create a fixed-length 'fingerprint' (or hash) of a variable-length file in such a way that it is difficult to find a different file that produces the same hash. Combined with public-key cryptography, this can become a digital signature. User 'A' will sign a file using his private key, and any user 'B' can then verify the signature with 'A's public key (which is freely available). Due to the nature of hash functions and digital signatures, even a single-bit change in a file will be instantly detectable when trying to verify the signature.

8.7.4 The importance of archiving the data and data historian

As the evolution of urban drainage systems is characterized by relatively long timescales (i.e., in the order of decades to centuries) it is of major importance to have access to 'old' data to allow for an evaluation of, for example, the effects/effectiveness of implementation of design/management policies. In many cases, policies on archiving data are poor, for example when a pipe in a sewer system of 90 years old is replaced, the entire record of that pipe (including inspection and rehabilitation data) tends to be deleted from the database. This hampers the option for future generations to learn from past experience. A simple workaround is to use a 'data historian,' which very simply stated, adds, apart from a field 'construction data,' also a field 'removal data' for each element in a system. This allows one to get access to the system as it was on, for example, 20/3/2015 and compare the situation with the situation 5 years later. Such an option allows for a spectrum of information to be extracted. A somewhat more cumbersome manner to achieve the same result is to archive data; however, this is normally done at regular time intervals only, where a data historian offers the option for 'continuous' access. Use of these digital applications will not work for very old data, unless one is willing to digitalize the full history of a system from old, pre-digital sources. More comprehensive information on data archiving can be found in Prodanović and Branisavljević (2021).

8.7.5 Accessing (very) old data

Unlike stormwater control measures that tend to have been built in the last three decades, the construction of sewer networks started more than a century ago (Geels, 2006) in many parts of the world such as, for example, Europe or North America. Figure 8.10 presents an extract from a register which inventoried all sewer pipes installed in the city of Lyon in 1879.

Information related to pipes older than the 1970s was on paper at the time (if it had been stored at all). If the information has not been digitalized, it may remain in the memory of elders or in the archives. Such information may be used to elaborate deterioration models. The most obvious technique to access old information is digging in archives: such an approach could be time consuming depending on how well the archives are accessible and classified. It is, however, a unique starting point when very little information is known. Other techniques are related to machine learning either using partially known information and characteristics of the network (e.g., installation date, material, form, diameter, depth, location, etc.), or to build search algorithms based on pipe connections within the network (Cherqui *et al.*, 2019). In specific cases, understanding urbanization or calling on collective memory could also generate valuable information.

Another example of old but still useful information concerns rainfall series and historic rain events. Flooding associated with historic rain events could help better anticipate vulnerable areas. Long-term rainfall series enable the understanding as to how climate change may impact rainfall patterns. Climate historians investigate ancient meteorological observations (i.e., temperature and rainfall) using past harvesting dates and ancient texts (Ladurie, 2009). Van Gelder (2000) presented an example on how to use historical data in modern extreme value statistical analysis. In old city centers after heavy flooding people sometimes added a marking stone on the wall of the (rebuilt) buildings with a text like 'On the xth day in the month XX of the year YYYY the water level rose to this stone.' In Figure 8.11, the mark is directly engraved in the stone of the house. This information has proven to be of high value when evaluating the statistical properties of extreme events.

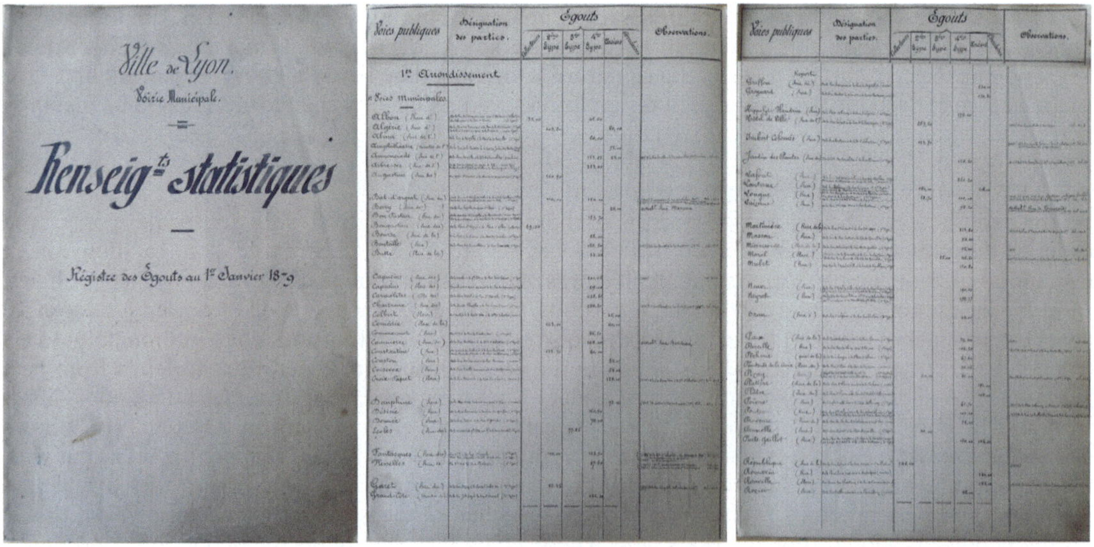

Figure 8.10 Extract from the register of 'statistical information' on the sewer networks installed by the city of Lyon in 1879 (Archives municipales de Lyon, code 923WP434), photographs with permission from F. Cherqui.

Figure 8.11 Photograph of a house in Ousson-sur-Loire (France) showing historical flooding levels for the Loire river: the level reached by each flood is engraved directly in the stone of the house, photograph with permission from F. Cherqui.

8.8 EXISTING APPLICATIONS TO MANAGE DATA

Existing IAM systems tend to be well suited to meet the requirements of individual stakeholders, for example, pavement management or bridge management. Nevertheless, a robust and well-developed data-management solution is essential for multidisciplinary IAM approaches, especially when analyzing diverse and big infrastructure data (Halfawy & Froese, 2007). In an interdependent environment, any error-prone decision on a single infrastructure component could have a cascading effect on all infrastructure components, even resulting in catastrophic consequences under extreme circumstances (e.g., responding to disasters). In this regard, a master data management (MDM) paradigm may provide municipalities with new ways to handle the data quality challenges they have struggled with for years and prevent the dilemma of 'data rich but information poor' situations. It is considered as a 'technology-enabled, business-led discipline that enables business and IT stakeholders within an enterprise to work together to ensure uniformity, accuracy, stewardship, semantic consistency, and accountability for the organization's official, shared master data assets.' In this way, stakeholders will be able to make better decisions with trusted data sources, actionable insights, and a holistic view of their customers, products, and services (White *et al.*, 2006).

MDM is being used in the business sector to orchestrate operational and analytical big data applications. Recently, MDM has attracted increased attention in the field of IAM due to IoT) applications and big data (Walker & Moran, 2017). To develop a suitable MDM solution for smart, sustainable, and resilient city planning, it is important to critically review commercial and open-source MDM systems, smart city standards, smart city concept models, smart community infrastructure frameworks, semantic web technologies, and to gather feedback and requirements from experts who develop smart, sustainable, and resilient city programs (Ng *et al.*, 2017).

Ng *et al.* (2017) designed SIAM-MDM with reference to the open source MDM software products (e.g., Talend MDM), AM standards (e.g., ISO 55000 and IIMM), smart community infrastructures frameworks (e.g., ISO 37152), smart city frameworks (e.g., PAS 181), smart city concept models (e.g., ISO/IEC 30182), domain taxonomy or ontology (e.g., e-COGNOS, see Lima *et al.* (2023)),

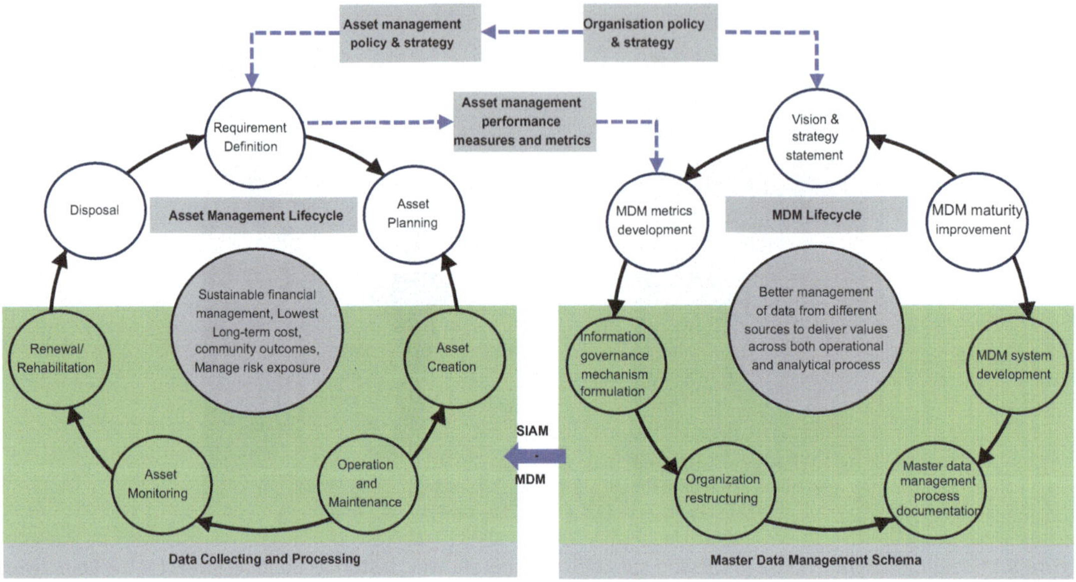

Figure 8.12 Schematic concept of SIAM-MDM (*source*: Ng *et al.*, 2017).

domain-specific knowledge depository (e.g., WATERiD database, see http://waterid.org/), BIM and GIS integration frameworks, and semantic web technologies (e.g., RDF and OWL). SIAM-MDM is piloted through a case study that focuses on the integrated management of building and transportation infrastructure assets of a selected community in Hong Kong. Based on the preliminary results, SIAM-MDM can assist stakeholders in improving infrastructure asset data consistency, accuracy, integrity, timeliness, and transparency, increasing the credibility and reliability of decision making through the analysis of big infrastructure data. The proposed MDM solution is depicted in Figure 8.12.

8.8.1 GIS repositories and inventories

Depending on the municipality, data storage is structured differently (Hansen & Pauleit, 2014). In some municipalities, data storage and management are handled centrally, whereas in others, it is handled in a more decentralized manner. An excellent method to achieve a comprehensive approach is the development of a system in which data management and storage support strategic urban development are integrated into a centrally organized municipal service. Having GIS experts working with the IT unit, which is usually higher up in the municipal hierarchy, may be of benefit. For the development of such structures, it is imperative that data categorization systems are developed so that relevant data are easily found and maintained.

Despite big data technology's potential to store, process, and analyze huge volumes of heterogeneous data, there are still few big data applications in the architecture, engineering, and construction industries (Alavi & Gandomi, 2017). In the absence of a centralized platform for sharing successful experiences and lessons learned, common mistakes, such as inadequate application of technology and unexpected change orders, causing the project to go over budget, are frequently repeated. One of the first initiatives in the United States was the development of a national database of water infrastructure, called WATERiD (http://www.waterid.org), which enabled utilities to communicate among themselves through a web environment and learn from each other's experiences. In addition, WATERiD has a directory that can be used by utilities to identify sister utilities with similar needs and experiences so that they can contact one another and reduce the cost of repeated education efforts. In particular, WATERiD focuses on transferring performance and cost information regarding drinking water and wastewater infrastructure condition assessments and renewal (Jung *et al.*, 2014).

8.8.2 Operational data, supervisory control and data-acquisition (SCADA) systems, and beyond

Besides keeping an eye on new technologies, data infrastructure managers need to evaluate existing tools that can help their organizations better manage their data. It has been shown that by using the IoT to improve infrastructure systems, large quantities of data related to their condition, performance, and citizen behavior can be collected, processed, integrated, and analyzed through cloud-based IAM systems, ubiquitous mobile applications, and big data analytics platforms (Ng *et al.*, 2017).

A collaborative mobile-cloud computing framework was developed by Chen *et al.* (2013) for intelligent civil infrastructure condition inspection and image-based damage analysis. This framework combines advanced mobile and cloud-computing technologies with three innovative features: (1) context-enabled image collection, (2) interactive imaging, and (3) real-time on-demand image analysis (Figure 8.13).

The Internet information services server, which is hosted by Microsoft's Windows virtual machine, is responsible for admitting incoming data. The cloud can be used by many mobile clients simultaneously and can handle high-volume data transmission. An implementation of a GlassFish server combined with MySQL is used in a virtual Linux system to manage imagery data and associated metadata.

8.8.3 Access of data and insights across different vertical digital solutions

There are many challenges facing the water industry and digitalization represents a window of opportunity to solve some of the big problems through data, AI, and sensors (IWA, 2019). Throughout the urban water cycle, a wide range of data sources (GIS, operational data, SCADA, and IoT) exists. Instrumentation and measurements are typically done via a SCADA system with many different vendors active in the water market (e.g., ABB, Siemens, Citect). In addition, a wide range of new IoT vendors are popping up, typically with their own cloud solution for handling the data. The different data sources are often difficult to access and might be stored in different 'silos.' The need for a flexible and robust technology platform which integrates all types of data was one of the recommendations from the European Union (EU) project SmartWater4EU (2017). Regardless of the origin of the data, it is useful for the people in water utilities to have easy access to these data to make good strategic, tactical, and operational decisions.

A conceptual three-step approach can be applied for utilizing real-time data. Step 1 is to get access to different sources of real-time data such as data from SCADA systems, IoT solutions, or from local data

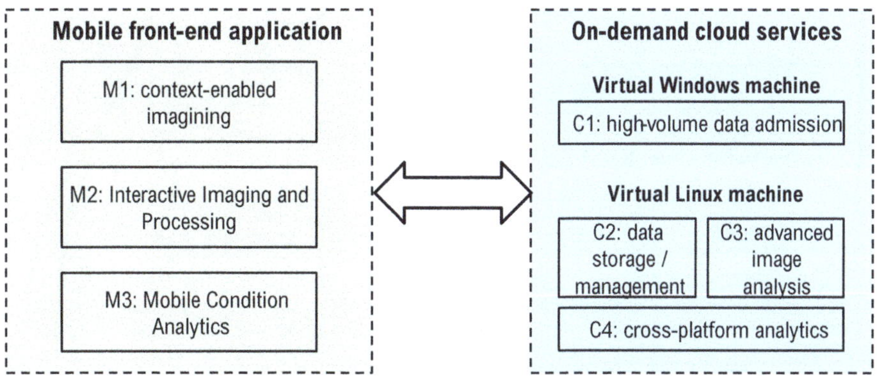

Figure 8.13 Software architecture of the collaborative mobile-cloud computing framework (*source*: adapted from Chen *et al.*, 2013).

lakes. In step 2, different real-time data are presented in the context of the urban water infrastructure, for example, water level sensors at a CSO location should be linked to the relevant wastewater component and viewed in the mapping system the operators and the maintenance teams use at a daily basis. This step enables increased utilization and availability of real-time data to the entire organization. Some utilities have well-defined goals in their strategic masterplans for this. For instance, in Norway, Bergen Water (Bergen Municipality, 2019) states that 'real-time data should be made available on the normal platforms to all employees.' Contextualized real-time data can answer many questions for the operators, but more advanced analysis can yield further insights. In step 3, advanced analytics are applied in addition to the real-time data such as machine learning techniques for identifying events happening in the system, for example, identification of inflow and infiltration (I&I) problems in a wastewater system. In case of I&I, the availability of real-time data provides the potential to both identify where the problems are but also to propose efficient mitigation measures based on the root cause of the problem (Søvik, 2021). Internationally, there is a trend where smart water vendors stream toward platform convergence merging GIS, SCADA, IoT, and operation management platforms into an open 'eco-sphere' which enables access to data and insights across different vertical digital solutions as illustrated in Figure 8.14. Normally, these types of data are not easily available and are locked in different silos because they are separate systems. The key is to bring all these data together and utilize them.

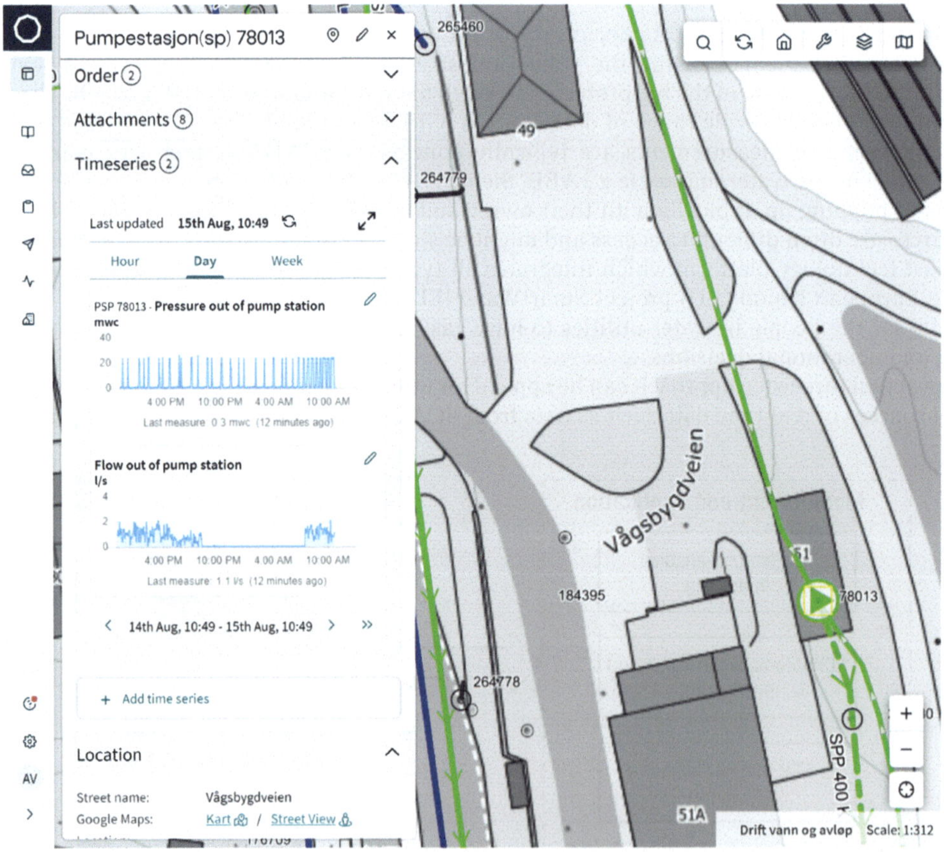

Figure 8.14 Example of a screenshot from a software solution with an 'eco-sphere' approach integrating GIS, SCADA, IoT, and operational management system into one integrated tool.

8.9 CASE STUDIES: DATA MODELS IN PRACTICE

8.9.1 Lake Simcoe Watershed's municipal stormwater management database

In Canada, the Province of Ontario's water, land, and natural resources are managed at a watershed scale by Conservation Authorities (CAs). Protection of Lake Simcoe and its watershed is one of the main objectives of the Lake Simcoe Region CA (LSRCA). Urban stormwater comprises about 31% of the phosphorus loads within the lake (MECP, 2016). While SWM features are in place to reduce pollutant loads, evaluations of the efficiency of stormwater wet ponds to achieve phosphorus removal have shown a significant drop several years after installation (LSRCA, 2010). Therefore, the Province of Ontario's Ministry of Environment, Conservation and Parks (MECP) funded a study in 2016–2017 to evaluate the state of municipal SWM inspections, maintenance, and record keeping within this watershed and provide resources to support improvements. This study identified the need for an AM tool through surveys and workshop series with stakeholders (GHD, 2017).

The LSRCA along with the MECP have provided all members of Lake Simcoe watershed municipalities with AM software (CityWide) since 2018. This software can support organization of the SWM asset inventory (Asset Manager Module) and associated inspection and maintenance programs (Maintenance Manager Module, Figure 8.15). While municipalities may decide to incorporate other AM tools, they are provided with a software license along with support from LSRCA for its adoption. Regardless of adoption of this tool by municipalities, LSRCA collects the SWM asset inventory data and maintains this database as part of its efforts to improve SWM within the watershed. To date, this database contains over 900 SWM assets from 11 member municipalities. Additionally, a User Guide has been developed for municipalities to navigate the application of Citywide for SWM inspection and maintenance programs, including a developed methodology for prioritization of maintenance efforts.

Use of this SWM asset management software has helped the participating municipalities with better organization of inventory and maintenance data in a centralized and digital format. Furthermore, it can help the LSRCA and partner municipalities identify potential inter-jurisdictional issues within the watershed and allow for future evaluation of water-quality benefits of the SWM assets. Potential future plans for the SWM database include improved analytics of SWM data through improvements to the data warehouse along with the broader outreach of these efforts through the Sustainable Technologies Evaluation Program (see https://sustainabletechnologies.ca/).

Figure 8.15 AM modules available to Lake Simcoe Watershed municipalities through Citywide software (printed with permission from PSD Citywide Inc.).

8.9.2 The Dutch 'Urban Drainage Ontology'

8.9.2.1 Introduction

The Dutch 'Urban Drainage Ontology' (UDO) (RIONED Foundation, 2023) is a semantic ontology that defines concepts that are relevant to Dutch urban drainage management practice. The UDO has been developed since 2016 by RIONED Foundation. RIONED Foundation is the branch organization for urban drainage in the Netherlands in which public bodies, industry, and the educational sector collaborate. The term 'semantic' refers to the fact that 'meaning,' that is, metadata and domain knowledge, is added to the extent possible to all elements in the ontology. An ontology is built up from a structured vocabulary (i.e., a uniform set of data and their definitions) with the addition of a hierarchy and explicit relations and semantics. The concepts in UDO include physical objects, object attributes, activities, data exchange formats, mathematical concepts for hydraulic modeling, and all mutually connecting relations. Ultimately, the UDO combined with descriptive data of actual sewer systems forms the knowledge base for real-life applications, data exchange between applications, and automated interpretation.

The development of the Dutch UDO has been driven primarily by the need to create a single and commonly shared source of concepts in order to share data between systems and stakeholders smoothly. A development driven by a similar need is the European INSPIRE Directive (Anon, 2023). As such, software application vendors and urban drainage managers at municipalities, under the wings of RIONED Foundation, have been involved to agree on terms, definitions, formats, and applications. The UDO was gradually developed reflecting the consensus thus reached. The goal of the UDO is to enhance the quality and efficiency of work processes within municipalities and with their contractors. The UDO is expressed in a Resource Description Framework (RDF Working Group, 2014), that is, using the W3C language for linked data, resulting in the data being stored in a directed schematic, which is a collection of nodes connected by links. In the UDO, the nodes are 'concepts' that are connected via relations. All concepts in the UDO are structured in a taxonomy, similar to Linnaeus' taxonomy to classify and structure living organisms by their characteristics. As the UDO is expressed using linked data, it follows four basic principles (Berners-Lee, 2006):

- Use Uniform Resource Identifiers (URIs) as identifiers for elements.
- Use HTTP URIs so that people can look up those identifiers.
- When someone looks up a URI, provide useful information, using standards including RDF and SPARQL (SPARQL Protocol and RDF Query Language) (RDF Working Group, 2013).
- Include links to other URIs so that people and machines can discover more elements.

The UDO has a modular setup, where each module is developed for a specific use-case in practice. Figure 8.16 shows the currently available modules for the Dutch UDO.

8.9.2.2 Usage, data, and applications

A data model without available data has no practical value. Therefore, RIONED Foundation offers users of UDO an entry point: a publicly accessible website: https://data.gwsw.nl.

Second, RIONED Foundation offers Dutch municipalities and water boards a data storage location on a centralized server (the UDO server). This server stores datasets that are setup in line with the UDO. As sewer system management software applications do not themselves store data with the same structure, these applications have implemented data export tools.

Third, the server hosts a range of webapps offering the following functionalities:

- Upload and – if desired by the owner – publication of the instance data.
- Data quality checks, user-case dependent.
- Export to a uniform set of files for hydraulic modeling.
- Export for viewing data on a map via WFS (Web Feature Service, a standard geographical data interface) or a file, such as in QGis or ArcGis.
- An API for retrieving data in a geographical format.
- A SPARQL endpoint for querying ontology and instance data.

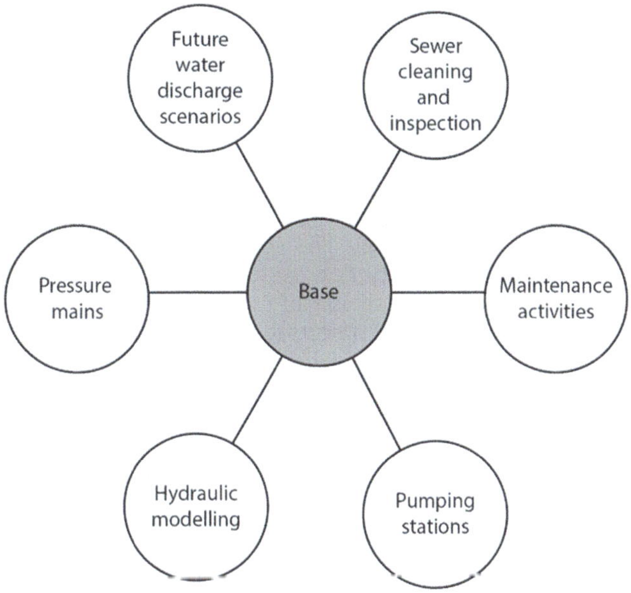

Figure 8.16 UDO modular setup.

Figure 8.17 shows a schematic overview of the different interaction elements surrounding UDO.

Currently, approximately 55% of the 344 Dutch municipalities have uploaded their data in the RDF format to the UDO server, covering 70% of the sewer system in The Netherlands. Once uploaded to the UDO server, the data are publicly available with 'read-only' permission. Data owners use a confidential access key for uploading their data to their pre-allocated environment on the UDO server. RIONED Foundation manages these confidential access keys and environments on the server. In other words, data on the server can be changed only by those who have confidential access to a specific environment. This availability of publicly available sewer data offers opportunities for research.

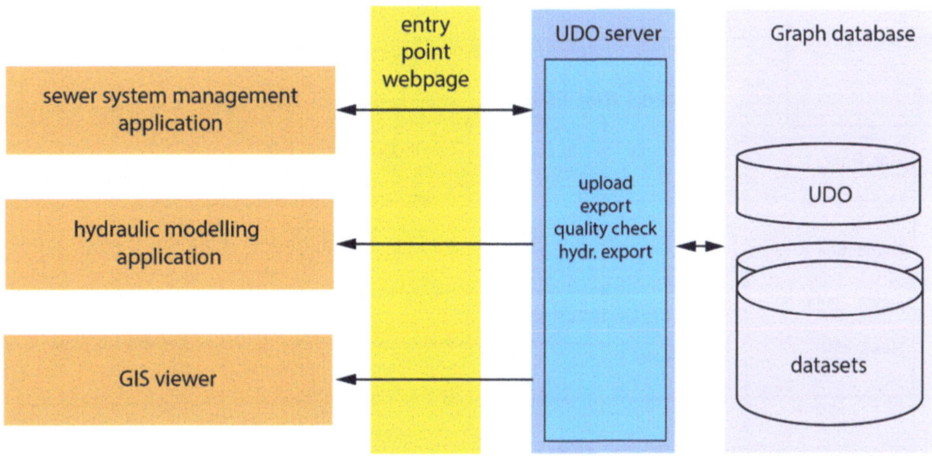

Figure 8.17 Schematic overview of UDO elements.

8.9.2.3 Illustration with the Dutch municipality of Heusden

An example of the implementation of data registration in accordance with the Dutch UDO is demonstrated for the Dutch municipality of Heusden: now all data have been registered in a uniform format and can be shared with and supplied to companies and municipalities. Heusden implemented work processes for sewer maintenance activities and hydraulic modeling (Figure 8.16). The urban drainage network in Heusden, comprising pipes, manholes, gully pots, infiltration basins, pumping stations, sewer overflow structures, and ancillary works (e.g., storage settling tanks), has been registered in various sewer management systems. Contributing areas in terms of roof types and pavement types are stored in a GIS database. The combined and storm water sewer systems are constructed predominantly as looped gravity flow systems, with a total length of 265 km of sewer pipes. Furthermore, approximately 85 km consists of pressurized sewer systems. Figure 8.18 shows an example for a sewer manhole. All physical objects (e.g., overflows, orifices, valves, pumps, settling tanks, and infiltration structures) have been registered as required by the UDO hydraulic modeling module. Although this takes serious upfront effort for the municipality, it has the benefit that the municipality can now utilize a single dataset for multiple purposes (see Figures 8.17 and 8.19).

Figure 8.19 shows the data-management process within the municipality of Heusden. It is now possible for Heusden to publish and share data relatively simply and up-to-date as open data. Every 3 months, the urban drainage data (stored in the management applications and the GIS database) will be converted into the turtle format and uploaded to the centralized server. Additionally, the data will be checked on the national server by using the use-case-dependent data quality checks (e.g., hydraulic modeling). The data from a local water board are also available in this portal. Through WFS, the data are presented in a national online GIS portal, which presents data from many other open public data sources as well. Exchanging data with consultancy firms and contractors for design purposes and sewer cleaning and inspection activities takes place by exporting the data in the standard, uniform UDO format. Consultancy firms that perform the hydraulic modeling will receive a

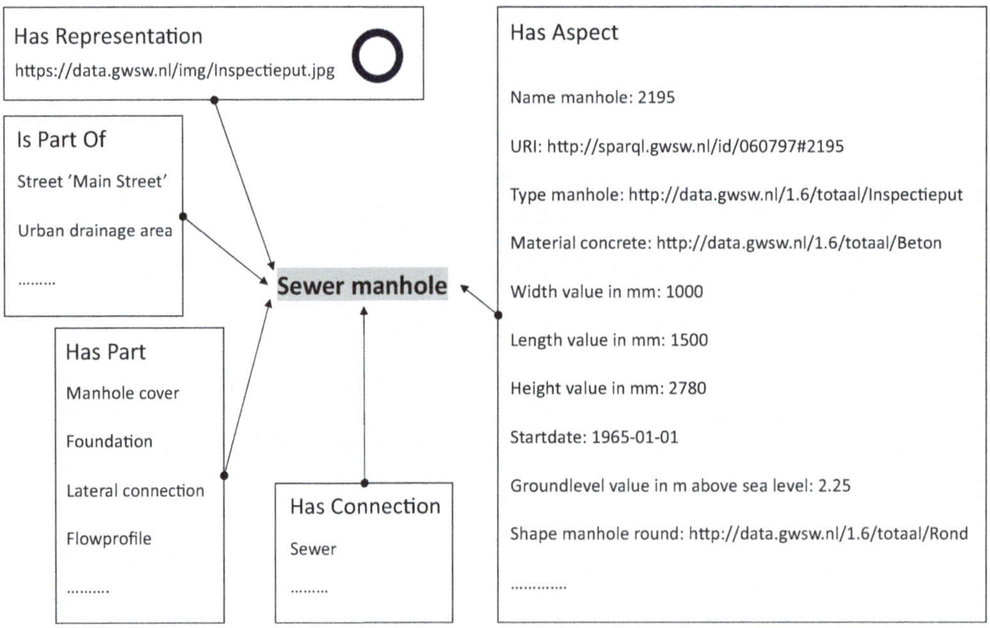

Figure 8.18 Example of UDO attributes of a sewer manhole.

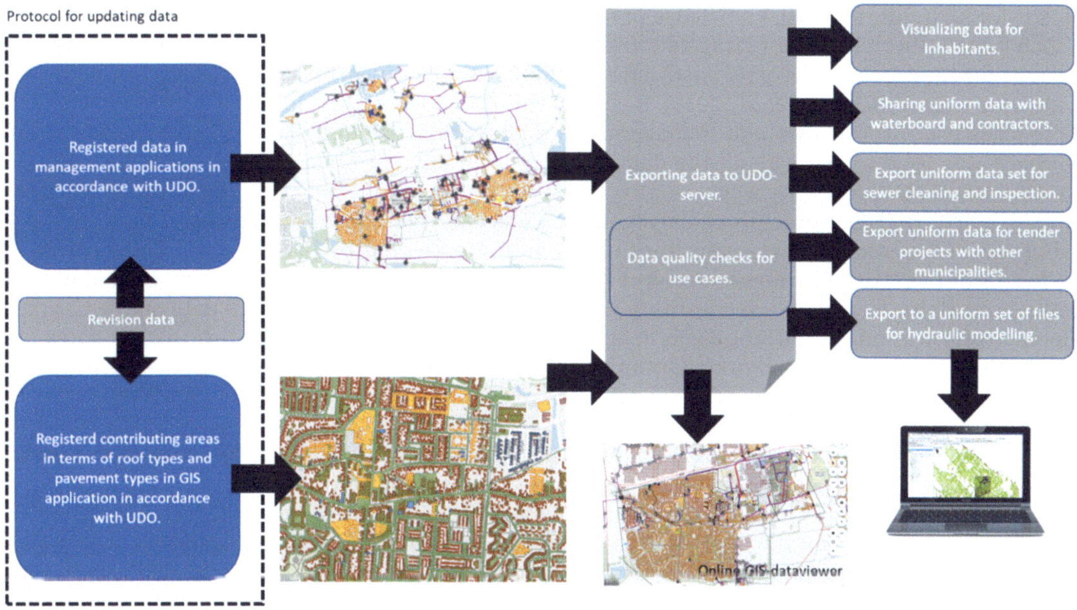

Figure 8.19 Data-management process for municipality of Heusden.

uniform and well-structured set of files from the RIONED Foundation server, which can be imported in hydrodynamic modeling software applications.

8.9.2.4 The future of UDO

The Dutch urban drainage sector and RIONED Foundation have been successfully pioneering the UDO, which now serves a range of use cases. At the same time, the UDO is a work-in-progress with a growing number of modules and extensions. In the future it will serve the management of sewage pumping stations, wastewater pressure mains, measurements, and groundwater management. The growth of the publicly available data of sewer systems, in terms of the number of objects and in 'richness,' allows possibilities for, for example, deep learning, enforcement of permits, and data quality control mechanisms. Effort is currently being put into merging the UDO with other data models for public space management. This is a logical step because managing sewer systems is simply part of managing public space. The RDF-based technology provides the essential building blocks. The challenge here is to align multiple governmental bodies and software application suppliers.

8.9.3 Data reconstruction: installation date and material in Lyon Metropolis

8.9.3.1 Introduction

The Metropolis of Lyon has the objective of reconstructing all the installation dates and materials of the drainage pipes in its heritage. The present communication presents the results of several years of work supported by the Rhone-Mediterranean-Corsica Water Agency. Most French cities have a limited knowledge of the installation year and the material of their drainage pipes, more especially the oldest cities where the network started more than a century ago.

The Greater Lyon Metropolis is responsible for water and wastewater management since it was created as an urban community in 1969. Since the 1990s, Greater Lyon has taken an interest in the management of its network assets. Initially a one-off task, it became a major issue in 2020, leading to the creation of an AM unit within the Water Department. While the pipe network is fairly well known (mostly for hydraulic models), knowledge of the materials and the installation years, as for many other French local authorities, is more fragmentary. In 2016, the HIREAU research project was dedicated to the inventory and assessment of reconstruction methods resulting in an operational guideline dedicated to utility managers (Cherqui *et al.*, 2019). Based on this initial work, this contribution presents various approaches used to estimate and manage missing network attributes. First, a supervised learning approach was used to estimate all the data with an accuracy that remains to be validated. Second, network traversal algorithms were used to find little information, but with certainty. Finally, the addition of external data to improve the algorithms' decision making was discussed.

For the reconstruction, two main lines of research are favored by Lyon Metropolis, inspired by the results of the HIREAU research project. The first axis concerns the implementation of statistical methods (machine learning) to reconstitute the missing information from known data on the heritage. The approach was adapted to the Lyon dataset and provided estimates of the installation dates and materials of the sewerage networks. Optimization work then allowed to improve the prediction by increasing the precision and the recall (fraction of relevant instances among the retrieved instances) for the material, and by reducing the estimation range for the laying date. The second area is the use of knowledge from multiple sources. Exploiting additional data, such as CCTV inspection reports, can provide pipe material directly. The creation of scripts dedicated to constraint propagation of information by graph traversal has also made it possible to fill in missing installation dates. We propose, here, to report on the approaches used and the results obtained. The objective is to inspire other utility managers to obtain this essential information for the management of their assets.

8.9.3.2 Materials and methods

The Lyon metropolitan area's drainage network is around 3200 km long. In 2021, 59% of the materials and 30% of the years of installation (in terms of pipe length) were known. Annual inspection rates are less than 1% of the length of the metropolitan area, and priority should be given to targeting pipes in poor condition. The history of networks, and more specifically of each section of pipe, is a key factor in predicting their condition: installation date but also constituent material, diameter, location (under pavement, etc.), and depth. Increased knowledge of materials and years of laying is therefore essential to better anticipate network malfunctions. To reconstruct the years of laying, the HIREAU report proposes five areas of study, including dating based on computer exploitation or exploration using business rules and/or available data. For example, the use of machine learning is justified for estimating missing parameters in drainage networks. As materials are an important element in determining years of laying, they are estimated first, and these results support the estimation of years of laying. Moreover, these are the most complete data. This means that the estimate of years of laying based on materials is based on more numerous and more reliable field data.

8.9.3.3 Data

Twenty-one classes of materials exist within the metropolitan databases. To create an efficient estimation model, these classes have been grouped by material and structural similarity (e.g., grouping together different types of reinforced concrete). The addition of data consistent with the attributes to be predicted is essential. On the contrary, providing the algorithm with data that have no influence on the determination of materials or years of installation can, at best, do nothing and, at worst, degrade the predictions. One of the most important data points to add is that of spatialization. In a metropolitan context where the center of Lyon is much older than any of the outlying districts, information on the position of the collectors will enable us to identify old or historic districts or recent industrial zones. Working with the teams, business rules have emerged that can be transcribed using attributes. Cast iron pipes are normally those located under bridges and along watercourses. This information can be fed back to the collectors in the form of distance to water. The burial depth of each pipe has also been added. This element characterizes the strength of the material to be used. In all, there are 47 variables describing the collectors in the Lyon metropolitan area.

8.9.3.4 Machine learning

Machine learning consists of two stages. The first involves estimating a predefined model from data. The algorithm will learn from all the data supplied characterizing the sewage collectors. Estimating this same model involves applying the learned model to sewers where information on materials or years of installation is missing. To measure the quality of the prediction, four metrics were used: precision, recall and *f*-score for materials and root mean-squared error (RMSE) for installation years. Precision should be seen as a measure of the quality of the estimates made during training, recall as a measure of the completeness of the results returned and finally the *f*-score is the harmonic mean of the two. The RMSE for years represents the mean-squared difference between the model prediction and the known year. It is used to frame the estimate: the year of a collector is known with an accuracy of ±RMSE years.

The use of a gradient boosting algorithm (Chen & He, 2014) has been justified in several studies on tabular data such as that from collectors (Fernández-Delgado *et al.*, 2014; Olson *et al.*, 2018). Various methods are grafted onto this algorithm to avoid overlearning the model. The aim is for the model to continue to perform well on data on which it has not been trained. The model's hyperparameters were estimated using the Optuna tool (https://optuna.org). Cross-validation and soft-voting methods have also been implemented, with the aim of improving the model's predictions and reliability. These elements also help to resolve the problem of unbalanced support for learning classes. With concrete accounting for more than 60% of known materials, it is important that the model does not overlook other important but under-represented classes. The full method is available in the report on the internship on improving knowledge of the assets of the wastewater and stormwater networks of the Lyon metropolitan area (Besnier, 2022).

8.9.3.5 Information propagation

The aim of estimating attributes using machine learning is to fill a field 100%. However, the more data there is at the outset, the more satisfactory the estimate will be. It is for this reason that a complementary method has been developed upstream, with the aim of filling in attributes reliably: the propagation of information about the year of laying under constraint by graph traversal. Knowing that 40% of the collectors in the Lyon metropolitan area network are less than five collectors away from a pipe where the year of laying is known, it makes sense to try and propagate information from neighbor to neighbor. By topological connection, if a pipe upstream or downstream of a collector has a year of laying, the same material (known or estimated), the same diameter and cross-sectional shape, the same type of effluent and is located in a cone of a given angle, then the year of laying information is propagated. This method yielded 5% more data, which will be used to estimate the missing information.

8.9.3.6 CCTV inspection reports valorization

Pipe materials are recorded in the CCTV inspection reports: this information is digitally coded in accordance with European standard NF EN 13508-2 + A1 (see Chapter 3). These inspections are part of the process of rehabilitating the sewers concerned, since no work of this type can be undertaken before a CCTV has been carried out. They are also part of an inspection program in sensitive areas such as drinking water catchment areas. To extract pipe material from these data, an automatic method has been developed (Besnier, 2022). It was used to add more than 5% of material data to the Lyon databases. Cross-referencing between data known from works and inspected (CCTV) has been carried out. This makes the data more reliable and avoids the transmission of erroneous information. These elements make it possible to train the learning model on more data and thus improve it.

The development of scripts for integrating data from CCTV has also been extended to all the fault codes identified in the inspection reports: 7000 CCTVs have thus been integrated into the city's GIS. This information is fundamental to the implementation of a long-term AM strategy. By improving our knowledge on the network in terms of materials and years of laying, we can estimate its state of health more accurately. As a result, by inspecting the same number of CCTVs, it will be possible to prioritize them and inspect more mains that are damaged beforehand.

8.9.3.7 Main results

Estimation of the materials using supervised learning enables us to go from 58.8 to 100% knowledge (Figure 8.20). The proportion of correct elements among those returned for a class is 82.6% (precision) and the proportion of correct elements returned among those that exist is 87.2% (recall). All material-specific metrics are above 70%. If we look at the distribution of predictions in the different classes of materials, several results are interesting. The estimates produced by the supervised learning model do not follow the proportion of elements from known classes in the estimates. After being presented to the heritage management team, these estimates were validated thanks to their knowledge of the network and its management. These data will therefore be integrated into the GIS in the form of an estimated materials attribute.

Using supervised learning to estimate the installation years, we can increase our knowledge from 30.4 to 100% (Figure 8.21). The years have been grouped in 20 year increments to make it easier to identify urban areas with old pipes. The years correctly show the boundaries of Lyon, as well as the construction of the entire recent network at the confluence of the Rhône and Saône rivers. However, the estimated data are provided with a qualification: the RMSE. This metric provides a confidence interval for the estimated data. With the means presented and implemented, it is 5.8 years. A study of the histogram of the dates of the estimated and known sections (Figure 8.22) shows that most of the known data are from 1990s onward. The lack of knowledge in the years prior to 1960 can be explained

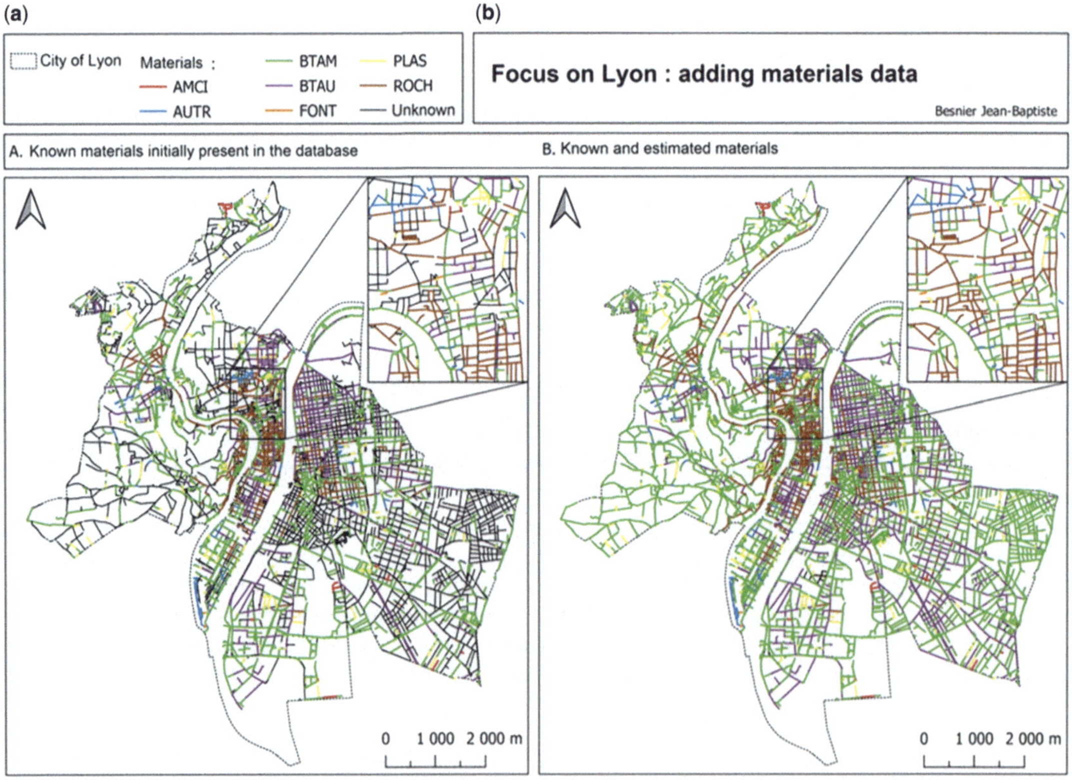

Figure 8.20 Map of known and estimated materials in the Lyon Metropolis: cemented asbestos (AMCI), other (AUTR), concrete with metals (BTAM), concrete other (BTAU), cast iron (FONT), plastic (PLAS), masonry (ROCH), and unknown (*source*: adapted from Besnier, 2022).

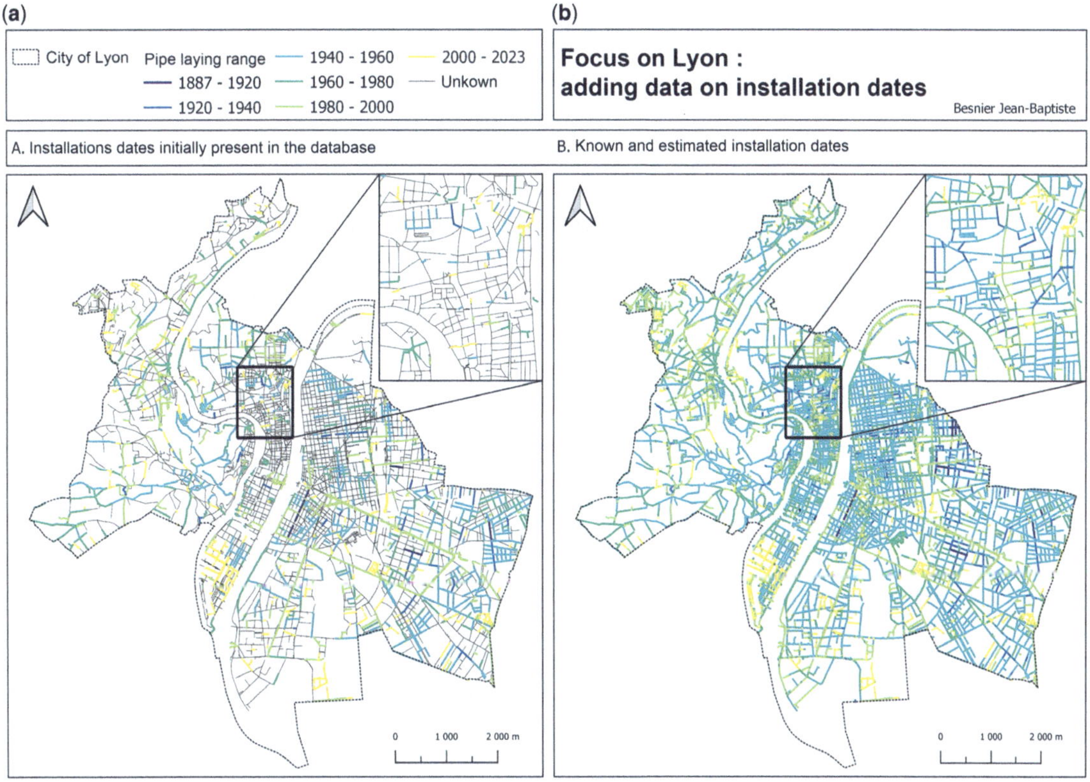

Figure 8.21 Map of known and estimated years for the Lyon metropolitan area, zooming-in on central Lyon (grouped by 20 year age bands for ease of reading) (*source*: adapted from Besnier, 2022).

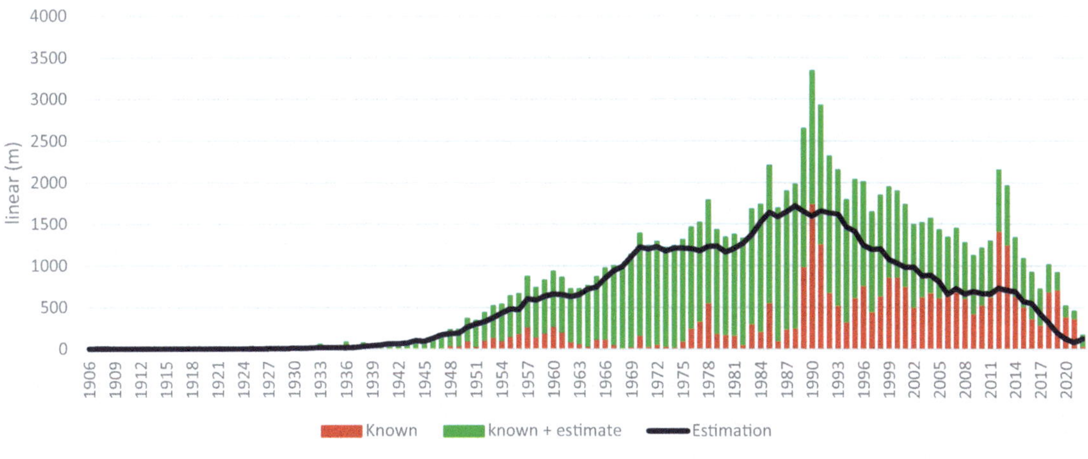

Figure 8.22 Histogram of known and reconstructed installation years for the Greater Lyon Metropolis drainage network (*source*: adapted from Besnier, 2022).

by several hypotheses: (1) these old collectors may have been replaced, (2) data archiving was not computerized, and (3) the information may have been lost when the towns were grouped together to form an urban community. We also note that the predictions correspond well with the expansion of Lyon's urban networks between the 1970s and 1990s. The number of collectors estimated after 1990 is high and may correspond to the ERU-91 Directive, which requires local authorities to collect and treat wastewater. After 2014, the number of estimated collectors falls rapidly, an effect consistent with the fact that the date of new collectors is almost always given.

The information propagation method implemented yielded 2.7% more data for the years of installation, that is, an increase of 8.5% for the training set for the estimation algorithms. This additional data improved the RMSE over the years of installation from 5.8 to 5.4 years.

8.9.3.8 Conclusion

All these methods form part of a hierarchical process in which the data recovered as the project progresses helps to estimate the materials and years of installation. The constrained propagation method yields little information. However, this information is reliable and improves the prediction results. The materials returned by CCTVs are also valuable data that had never been used for this purpose earlier. In the course of this work, the objectives of improving knowledge of the installation years and materials were achieved. Using machine learning algorithms, all the missing attributes were filled in. There has been an increase in the knowledge and reliability of the attributes of the results concerning the methods used. The Lyon metropolitan area now has powerful prediction and estimation tools. The years of installation will be able to be used in other prediction models and to help in the programming of works. The multisource input and analysis of information in the database has made it possible not only to improve predictions, but also to make the most of unused work and data. In terms of future prospects, we need to consider inclusion relationships between material attributes. For example, a collector is initially made of concrete, then a second attribute would specify whether it is reinforced. This method could make it possible to retrieve information from inspections that do not comply with the current data system. During a visit, for example, it is complicated to assess the type of concrete with just a visual analysis of the collector.

REFERENCES

Ahmadi M., Cherqui F., De Massiac J. C. and Le Gauffre P. (2014a). From sewer inspection programmes to rehabilitation needs: research and results related to data quality and availability with the support of numerical experiment. *European Journal of Environmental and Civil Engineering*, **18**, 1145–1156, https://www.tandfonline.com/doi/full/10.1080/19648189.2014.893212

Ahmadi M., Cherqui F., De Massiac J. C. and Le Gauffre P. (2014b). Influence of available data on sewer inspection program efficiency. *Urban Water Journal*, **11**(8), 641–656, https://doi.org/10.1080/1573062X.2013.831910

Ahmadi M., Cherqui F., De Massiac J. C. and Le Gauffre P. (2015). Benefits of using basic, imprecise or uncertain data for elaborating sewer inspection programs. *Structure and Infrastructure Engineering*, **11**(3), 376–388, https://doi.org/10.1080/15732479.2014.887122

Alavi A. H. and Gandomi A. H. (2017). Big data in civil engineering. *Automation in Construction*, **79**, 1–2, https://doi.org/10.1016/j.autcon.2016.12.008

Ana E. and Bauwens W. (2010). Modeling the structural deterioration of urban drainage pipes: the state-of-the-art in statistical methods. *Urban Water Journal*, **7**(1), 47–59, https://doi.org/10.1080/15730620903447597

Ana E. V., Bauwens W., Pessemier M., Thoeye C., Smolders S., Boonen I. and de Gueldre G. (2009). An investigation of the factors influencing sewer structural deterioration. *Urban Water Journal*, **6**, 303–312, https://doi.org/10.1080/15730620902810902

Anon (2023). Inspire Geoportal. Available: https://inspire-geoportal.ec.europa.eu (accessed 1 August 2023).

Bartos M., Park H., Zhou T., Kerkez B. and Vasudevan R. (2019). Windshield wipers on connected vehicles produce high-accuracy rainfall maps. *Scientific Reports*, **9**(1), 170, https://doi.org/10.1038/s41598-018-36282-7

Bergen Water (2019). Masterplan for water and water supply 2019–2028. Bergen municipality.

Berners-Lee T. (2006). Linked data, https://www.w3.org/DesignIssues/LinkedData.html (accessed 1 August 2023).

Bertrand-Krajewski J. L., Laplace D., Joannis C. and Chebbo G. (2000). Mesures en Hydrologie Urbaine et Assainissement. Tec&Doc, Lavoisier.

Bertrand-Krajewski J. L., Clemens-Meyer F. and Lepot M. (2021). Metrology in Urban Drainage and Stormwater Management: Plug and Pray. IWA Publishing, London, UK. ISBN: 9781789060102

Besnier J.-B. (2022). Amélioration de la connaissance patrimoniale des réseaux d'assainissement et eaux pluviales de la métropole de Lyon. Master thesis, 85pp. Available at: https://hireau.files.wordpress.com/2022/09/besnier_jean-baptiste_rapport_tfe_2022.pdf (accessed 1 August 23).

Blumensaat F., Wolfram M. and Krebs P. (2012). Sewer model development under minimum data requirements. *Environmental Earth Sciences*, **65**(5), 1427–1437, https://doi.org/10.1007/s12665-011-1146-1

Branisavljević N., Prodanović D., Arsić M., Simić Z. and Borota J. (2009). Hydro-meteorological data quality assurance and improvement. *Journal of the Serbian Society for Computational Mechanics*, **3**(1), 228–249. http://www.sscm.kg.ac.rs/jsscm/index.php/volume-3-number-1-2009/49-paper-10-2009-1 (accessed 20 April 2023).

Caradot N., Sampaio P. R., Guilbert A. S., Sonnenberg H., Parez V. and Dimova V. (2020). Using deterioration modelling to simulate sewer rehabilitation strategy with low data availability. *Water Science and Technology*, **83**, 631–640, https://doi.org/10.2166/wst.2020.604

Carrara W., Radu C. and Vollers H. (2017). Open Data Maturity in Europe 2017, European Commission, Directorate-General of Communications Networks, Content & Technology. https://www.europeandataportal.eu/sites/default/files/edp_landscaping_insight_report_n3_2017.pdf

Carvalho G., Amado C., Brito R. S., Coelho S. T. and Leitão J. P. (2018). Analysing the importance of variables for sewer failure prediction. *Urban Water Journal*, **15**, 338–345, https://doi.org/10.1080/1573062X.2018.1459748

Chen T. and He T. (2014). Higgs Boson Discovery with Boosted Trees. Proceedings of the 2014 International Conference on High-Energy Physics and Machine Learning, HEPML'14, **42**, 69–80. Available at: http://proceedings.mlr.press/v42/chen14.pdf (accessed 1 August 2023).

Chen Z., Chen J. and Lee Y. (2013). Collaborative mobile-cloud computing for civil infrastructure condition inspection. *Journal of Computing in Civil Engineering*, **29**(5), 04014066, https://doi.org/10.1061/(ASCE)CP.1943-5487.0000377

Cherqui F., Benabdeslem K., Benkabou S. E., Chuzeville G., Erismann J., Gauthiez B., Glandut V., Graja S., Krauss V., Le Gauffre P., Ntissar T., Petit D., Poinard D., Texier F., Sahari S., Sibeud E., Reeb S. and Ursely S. (2019). How to reconstruct the history of drinking water and wastewater networks (in French), 44pp. Available at: https://hireau.files.wordpress.com/2019/10/hireau_guide-operationnel-version-finale-3.pdf (accessed 1 August 2023).

Cherqui F., James R., Poelsma P., Szota C., Burns M. J., Fletcher T. and Bertrand-Krajewski J.-L. (2021). Real-time monitoring of water level at low-cost. The International Conference on Urban Drainage, 25–28 October 2021, online.

Cherqui F., Tscheikner-Gratl F., van Duin B. and Clemens F. (2022). The data-loop problem and what we can do about it. Leading Edge Strategic Asset Management Conference LESAM 2022, 11–14th May 2022, Bordeaux, France.

Clemens F. and von der Heide W. (1999). Effect of geometrical data errors in hydrodynamic calculations in urban drainage. Proceedings of the 8th ICUD, Sydney, Australia.

Clemens, F. (2001). Hydrodynamic models in urban drainage: application and calibration, Dissertation (TU Delft), Delft University of Technology, DUP Science, 372 p. https://www.iospress.com/catalog/books/hydrodynamic-models-in-urban-drainage-application-and-calibration

Davies J. P., Clarke B. A., Whiter J. T. and Cunningham R. J. (2001). Factors influencing the structural deterioration and collapse of rigid sewer pipes. *Urban Water Journal*, **3**(1–2), 73–89, https://doi.org/10.1016/S1462-0758(01)00017-6

Dirksen J., Goldina A., ten Veldhuis J. A. E. and Clemens F. H. L. R. (2009). The role of uncertainty in urban drainage decisions: uncertainty in inspection data and their impact on rehabilitation decisions. In: Strategic Asset Management of Water Supply and Wastewater Infrastructures, H. Alegre and M. d. C. Ameida (eds), IWA Publishing, London, UK, pp. 273–285. ISBN 9781843391869.

Di Sivo M. and Ladiana D. (2011). Decision-support tools for municipal infrastructure maintenance management. Procedia Computer Science – World Conference on Information Technology, **3**, pp. 36–41, ISSN: 1877-0509, https://doi.org/10.1016/j.procs.2010.12.007

Dziedzic R., Amador L., An C., Chen Z., Eicker U., Hammad A., Nasiri F., Nik-Bakht M., Ouf M. and Moselhi O. (2021). A framework for asset management planning in sustainable and resilient cities. 1–10, https://doi.org/10.1109/ISTAS52410.2021.9629158

Egger C., Scheidegger A., Reichert P. and Maurer M. (2013). Sewer deterioration modeling with condition data lacking historical records. *Water Research*, **47**, 6762–6779, https://doi.org/10.1016/j.watres.2013.09.010

Ersoy A. and Alberto K. C. (2019). Understanding urban infrastructure via big data: the case of Belo Horizonte. *Regional Studies, Regional Science*, **6**(1), 374–379, https://doi.org/10.1080/21681376.2019.1623068

Etter S., Strobl B., Seibert J. and van Meerveld H. J. I. (2020). Value of crowd-based water level class observations for hydrological model calibration. *Water, Resources Research*, **56**, 1(17), https://doi.org/10.1029/2019WR026108

Farrelly M. and Bach P. M. (2018). Response to the DELWP Issues Paper for the Improving Stormwater Management Advisory Committee. June.

Fernández-Delgado M., Cernadas E., Barro S. and Amorin D. (2014). Do we need hundreds of classifiers to solve real world classification problems? *Journal of Machine Learning Research*, **15**, 3133–3181. Available at: https://jmlr.csail.mit.edu/papers/volume15/delgado14a/delgado14a.pdf (accessed 1 August 23).

Ferré-Bigorra J., Casals M. and Gangolells M. (2022). The adoption of urban digital twins. *Cities*, **131**, 103905, https://doi.org/10.1016/j.cities.2022.103905

Fredriksen O.-M. (2021). Asset Management for Nature-based Solutions (NBS): A Data Registry to Support Decision Making in Stormwater Management in Norway. https://hdl.handle.net/11250/2787259

Fugledalen T., Rokstad M. M. and Tscheikner-Gratl F. (2021). On the influence of input data uncertainty on sewer deterioration models – a case study in Norway. *Structure and Infrastructure Engineering*, **19**, 1064–1075, https://doi.org/10.1080/15732479.2021.1998142

Geels F. W. (2006). The hygienic transition from cesspools to sewer systems (1840–1930): the dynamics of regime transformation. *Research Policy*, **35**(7), 1069–1082, https://doi.org/10.1016/j.respol.2006.06.001

GHD (2017). Stormwater Inspection and Record Management Best Practices, Data Model Design, and Comprehensive Report. Retrieved from: https://lsrca.on.ca/wp-content/uploads/2023/07/Stormwater-Inspection-Record-Management-Best-Practices.pdf (accessed 1 August 2023).

Graham J. W. (2012). Missing Data: Analysis and Design. Springer Science & Business Media, New York, NY.

Grieves M. and Vickers J. (2016). Digital twin: mitigating unpredictable, undesirable emergent behavior in complex systems. In: Transdisciplinary Perspectives on complex Systems, F. J. Kahlen, S. Flumerfelt and A. Alves (eds), Springer, New York, USA, pp. 85–113, https://doi.org/10.1007/978-3-319-38756-7_4

Halfawy M. M. and Froese T. M. (2007). Component-based framework for implementing integrated architectural/engineering/construction project systems. *Journal of Computing in Civil Engineering*, **21**(6), 441–452, https://doi.org/10.1061/(ASCE)0887-3801(2007)21:6(441)

Hansen R. and Pauleit S. (2014). From multifunctionality to multiple ecosystem services? A conceptual framework for multifunctionality in green infrastructure planning for urban areas. *Ambio*, **43**(4), 516–529, https://doi.org/10.1007/s13280-014-0510-2

Haurum J. B. and Moeslund T. B. (2021). Sewer-ML: A multi-label sewer defect classification dataset and benchmark. 2021 IEEE/CVF Conference on 865 Computer Vision and Pattern Recognition (CVPR), pp. 13451–13462, https://doi.org/10.1109/CVPR46437.2021.01325

Hilbert M. (2020). Digital technology and social change: the digital transformation of society from a historical perspective. *Dialogues in Clinical Neuroscience*, **22**(2), 189–194, https://doi.org/10.31887/DCNS.2020.22.2/mhilbert

International Water Association (2019). https://iwa-network.org/projects/digital-water-programme/, last accessed 12 Nov 2023.

Janssen M., Charalabidis Y. and Zuiderwijk A. (2012). Benefits, adoption barriers and myths of open data and open government. *Information Systems Management*, **29**, 258–268, https://doi.org/10.1080/10580530.2012.716740

Jiang F., Ma L., Broyd T. and Chen K. (2021). Digital twin and its implementations in the civil engineering sector. *Automation in Construction*, **130**, 103838, https://doi.org/10.1016/j.autcon.2021.103838

Jung J. K., Sinha S. K. and Whittle L. G. (2014). Development of a water infrastructure knowledge database. *Journal of Infrastructure Systems*, **20**(1), 04013006, https://doi.org/10.1061/(ASCE)IS.1943-555X.0000163

Kabir G., Tesfamariam S., Hemsing J. and Sadiq R. (2020). Handling incomplete and missing data in water network database using imputation methods. *Sustainable and Resilient Infrastructure*, **5**(6), 365–377, https://doi.org/10.1080/23789689.2019.1600960

Koo D., Piratla K. and Matthews C. J. (2015). Towards sustainable water supply: schematic development of big data collection using Internet of Things (IoT). *Procedia Engineering*, **118**, 489–497, https://doi.org/10.1016/j.proeng.2015.08.465

Korving H. and Clemens F. (2005). Impact of dimension uncertainty and model calibration on sewer system assessment. *Water Science and Technology*, **52**(5), 35–42, https://doi.org/10.2166/wst.2005.0103

Ladurie E. L. R. (2009). Histoire du climat depuis l'an mil: Tome 1. Flammarion, Paris, France.

Lake Simcoe Region Conservation Authority (LSRCA) (2010). Stormwater Pond Maintenance and Anoxic Conditions Investigation Final Report. Retrieved from: https://sustainabletechnologies.ca/app/uploads/2015/01/LSRCA-Stormwater-Maintenance-and-Anoxic-Conditions-2011.pdf (accessed 1 August 2023).

Le Gat Y. (2008). Modelling the deterioration process of drainage pipelines. *Urban Water Journal*, **5**(2), 97–106, https://doi.org/10.1080/15730620801939398

Lima C., El-Diraby T., Flés B., Zarli A. and Ferneley E. (2023). The E-Cognos Project: Current Status and Future Directions of an Ontology-Enabled IT Solution Infrastructure Supporting Knowledge Management in Construction.

Makropoulos C. and Savić D. A. (2019). Urban hydroinformatics: past, present and future. *Water*, **11**(10), 1959, https://doi.org/10.3390/w11101959

Manny L., Duygan M., Fischer M. and Rieckermann J. (2021). Barriers to the digital transformation of infrastructure sectors. *Policy Sciences*, **54**, 943–983, https://doi.org/10.1007/s11077-021-09438-y

Ministry of Environment, Conservation and Parks (MECP) (2016). Lake Simcoe Phosphorus Reduction Strategy. Retrieved from: https://www.ontario.ca/page/lake-simcoe-phosphorus-reduction-strategy (accessed 1 August 2023).

Möderl M., Butler D. and Rauch W. (2009). A stochastic approach for automatic generation of urban drainage systems. *Water Science and Technology*, **59**(6), 1137–1143, https://doi.org/10.2166/wst.2009.097

Mourad M. and Bertrand-Krajewski J.-L. (2002). A method for automatic validation of long time series of data in urban hydrology. *Water Science and Technology*, **45**(4–5), 263–270, https://doi.org/10.2166/wst.2002.0601

Ng S. T., Xu F. J., Yang Y. and Lu M. (2017). A master data management solution to unlock the value of big infrastructure data for smart, sustainable and resilient city planning. *Procedia Engineering*, **196**, 939–947, https://doi.org/10.1016/j.proeng.2017.08.034

Nie X., Fan T., Wang B., Li Z., Shankar A. and Manickam A. (2020). Big data analytics and IoT in operation safety management in under water management. *Computer Communications*, **154**, 188–196, https://doi.org/10.1016/j.comcom.2020.02.052

Okwori E., Pericault Y., Ugarelli R., Viklander M. and Hedström A. (2021). Data-driven asset management in urban water pipe networks: a proposed conceptual framework. *Journal of Hydroinformatics*, **23**(5), 1014–1029, https://doi.org/10.2166/hydro.2021.068

Olson R., La Cava W., Mustahsan Z., Varik A. and Moore J. (2018). Data-driven Advice for Applying Machine Learning to Bioinformatics Problems, arXiv:1708.0507 (q-bio.QM), January, Available at: https://arxiv.org/abs/1708.05070 (accessed 1 August 23).

ÖWAV-Regelblatt 40 (2010). Österreichischer Wasser- und Abfallwirtschaftsverband. Leitungsinformationssystem – Wasser und Abwasser (Pipe information system – water and wastewater), Vienna, Austria, 58pp.

Patcha A. and Park J. M. (2007). An overview of anomaly detection techniques: Existing solutions and latest technological trends. *Comput. Networks*, **51**, 3448–3470, http://dx.doi.org/10.1016/j.comnet.2007.02.001

Prodanović D. and Branisavljević N. (2021). Data archiving and meta-data – saving the data for future use. Chapter 10. In: Metrology in Urban Drainage and Stormwater Management, Bertrand-Krajewski (eds), London, UK, pp. 391–413, https://doi.org/10.2166/9781789060119_0391

Rahman S. and Vanier D. J. (2004). An evaluation of condition assessment protocols for sewer management. Client Report B-5123.6, NRC.

RDF Working Group (2013). 'SPARQL' W3C Semantic Web. https://www.w3.org/TR/sparql11-overview/ (accessed 1 August 2023).

RDF Working Group (2014). 'RDF' W3C Semantic Web. https://www.w3.org/RDF/ (accessed 1 August 2023).

RIONED Foundation (2023). https://www.riool.net/brochure-information-on-rioned-foundation (accessed 12 Nov 2023).

Robles-Velasco A., Cortés P., Muñuzuri J. and Onieva L. (2020). Prediction of pipe failures in water supply networks using logistic regression and support vector classification. *Reliability Engineering and System Safety*, **196**, 106754, https://doi.org/10.1016/j.ress.2019.106754

Rokstad M. M. (2016). Evaluating and enhancing information quality for IAM decisions in urban water systems. PhD Thesis, Department of Hydraulic and Environmental Engineering, Faculty of Engineering, Norwegian University of Science and Technology (NTNU). http://hdl.handle.net/11250/2432168

Rokstad M. M. and Ugarelli R. M. (2016). Improving the benefits of sewer condition deterioration modelling through information content analysis. *Water Science and Technology*, **74**, 2270–2279, https://doi.org/10.2166/wst.2016.419

Rokstad M. M., Ugarelli R. M. and Sægrov S. (2016). Improving data collection strategies and infrastructure asset management tool utilisation through cost benefit considerations. *Urban Water Journal*, **13**(7), 710–726, https://doi.org/10.1080/1573062X.2015.1024692

Rosen C., Röttorp J. and Jeppsson U. (2003). Multivariate on-line monitoring: challenges and solutions for modern wastewater treatment operation. *Water Science and Technology*, **47**(2), 171–179, https://doi.org/10.2166/wst.2003.0113

Salman B. (2010) Infrastructure Management and Deterioration Risk Assessment of Wastewater Collection Systems, PhD thesis, University of Cincinnati, Engineering and Applied Science: Civil Engineering, Cincinnati, USA. http://rave.ohiolink.edu/etdc/view?acc_num=ucin1282051343

Scheidegger A., Hug T., Rieckermann J. and Maurer M. (2011). Network condition simulator for benchmarking sewer deterioration models. *Water Research*, **45**(16), 4983–4994, https://doi.org/10.1016/j.watres.2011.07.008

Schellart A., Blumensaat F., Clemens-Meyer F., van der Werf J., Wan Mohtar W. H. M., Ramly S., Muhammad N., Bonneau J., Fletcher T. D., Costelloe J. F., James R., Burns M., Poelsma P., Ochoa-Rodriguez S., Bourne D., Hancock Z., Wallwork G., Hale J., Nikolova-Peters N., Kroll S., van Assel J., McCarthy D., Shi B., Bloem S. and Ebi C. (2021). Data collection in urban drainage and stormwater management systems – case studies, Chapter 11. In: Metrology in Urban Drainage and Stormwater Management: Plug and Pray, J.-L. Bertrand-Krajewski, F. Clemens-Meyer and M. Lepot (eds), IWA Publishing, London, UK, pp. 415–469, https://doi.org/10.2166/9781789060119_0415

SmartWater4EU (2017). EU-project SmartWater4EU. https://cordis.europa.eu/project/id/619024/fr

Søvik S. R. (2021). Long Short-Term Memory for Detection of Inflow and Infiltration Water in Sewer Systems – A Case Study from Bergen Municipality. Master's thesis, NTNU, Trondheim.

Sun S., Cegielski C. G., Jia L. and Hall D. J. (2018). Understanding the factors affecting the organizational adoption of big data. *Journal of Computer Information Systems*, **58**(3), 193–203, https://doi.org/10.1080/08874417.2016.1222891

Surbakti F. P. S., Wang W., Indulska M. and Sadiq S. (2020). Factors influencing effective use of big data: a research framework. *Information & Management*, **57**(1), 103146, https://doi.org/10.1016/j.im.2019.02.001

Tauro F., Selker J., Van De Giesen N., Abrate T., Uijlenhoet R., Porfiri M., Manfreda S., Caylor K., Moramarco T., Benveniste J., Ciraolo G., Estes L., Domeneghetti A., Perks M. T., Corbari C., Rabiei E., Ravazzani G., Bogena H., Harfouche A., Brocca L., Maltese A., Wickert A., Tarpanelli A., Good S., Lopez Alcala J. M., Petroselli A., Cudennec C., Blume T., Hut R. and Grimaldi S. (2018). Measurements and observations in the XXI century (MOXXI): innovation and multi-disciplinarity to sense the hydrological cycle. *Hydrological Sciences Journal*, **63**, 169–196, https://doi.org/10.1080/02626667.2017.1420191

Tscheikner-Gratl F., Hammerer M., Rauch W., Mikovits C. and Kleidorfer M. (2013). Integrated Rehabilitation Management for Different Infrastructure Sectors, gwf-Wasser, Abwasser, Special 1: 50–56.

Tscheikner-Gratl F., Mikovits C., Sitzenfrei R., Rauch W. and Kleidorfer M. (2015). GIS-Anwendung in der integrierten Rehabilitierungsplanung von urbaner Wasserinfrastruktur. *AGIT – Journal für Angewandte Geoinformatik*, **1**, 308–314, https://doi.org/10.14627/537557043

Tscheikner-Gratl F., Caradot N., Cherqui F., Leitão J. P., Ahmadi M., Langeveld J. G. and … Clemens F. (2019). Sewer asset management – state of the art and research needs. *Urban Water Journal*, **16**(9), 662–675, https://doi.org/10.1080/1573062X.2020.1713382

Urich C., Sitzenfrei R., Möderl M. and Rauch W. (2010). An agent-based approach for generating virtual sewer systems. *Water Science and Technology*, **62**(5), 1090–1097, https://doi.org/10.2166/wst.2010.364

Van Gelder P. H. A. J. M. (2000). Statistical methods for the risk-based design of civil structures. PhD Thesis, Delft University of Technology. ISBN 90-9013452-2. https://repository.tudelft.nl/islandora/object/uuid%3A6a62d6fa-cbcc-4c38-af8a-027c3d191a9d

Venkat V., Raghunathan R., Kewen Y. and Surya N. K. (2003). A review of process fault detection and diagnosis Part I: Quantitative model-based methods. *Comput. Chem. Eng.*, **27**, 293–311, http://dx.doi.org/10.1016/S0098-1354(02)00160-6

Walker S. J. and Moran M. P. (2017). Three Top Trends in Master Data Management Solutions Through 2017. Gartner Inc, Stamford, USA.

White A., Newman D., Logan D. and Radcliffe J. (2006). Mastering Master Data Management. Gartner Group, Stamford.

Yang P. and Ng T. L. (2017). Gauging through the crowd: a crowd-sourcing approach to urban rainfall measurement and storm water modeling implications. *Water Resources Research*, **53**(11), 9462–9478, https://doi.org/10.1002/2017WR020682

Zhu Q., Cherqui F. and Bertrand-Krajewski J.-L. (2023). End-user perspective of low-cost sensors for urban stormwater monitoring: a review. Water Science and Technology, **87**(11), 2648–2684, https://doi.org/10.2166/wst.2023.142

doi: 10.2166/9781789063059_0343

Chapter 9

Operation, maintenance and rehabilitation techniques

Mahdi Bahrami[1], *Floris Boogaard*[2,3], *Bert Bosseler*[4], *Frédéric Cherqui*[5], *Bert van Duin*[6,7,8], *Fabian Funke*[9], *Marcel Goerke*[4], *Francine Kelly-Hooper*[10], *Manfred Kleidorfer*[9], *Magnus Moglia*[11], *Tone Merete Muthanna*[1], *Martin Oberascher*[9], *Franz Tscheikner-Gratl*[1], *Martijn van der Valk*[12] and *Ferry van der Valk*[13]

[1]*NTNU – Norwegian University of Science and Technology, Trondheim, Norway*
[2]*Deltares, Daltonlaan 600, 3584 BK Utrecht, The Netherlands*
[3]*Research Centre for Built Environment NoorderRuimte, Hanze University of Applied Sciences, 9747 AS Groningen, the Netherlands*
[4]*IKT – Institute for Underground Infrastructure, Gelsenkirchen, Germany*
[5]*University of Lyon, INSA-LYON, Université Claude Bernard Lyon 1, DEEP, F-69621, F-69622, Villeurbanne, France*
[6]*City of Calgary, City & Regional Planning, Calgary, Canada*
[7]*University of Calgary, Calgary, Canada*
[8]*University of Alberta, Edmonton, Canada*
[9]*Department of Infrastructure, Unit of Environmental Engineering, University of Innsbruck, Innsbruck, Austria*
[10]*GHD Group, Toronto, Canada*
[11]*Centre for Urban Transitions, Swinburne University of Technology, Hawthorn, VIC 3122, Australia*
[12]*vandervalk+degroot B.V., ABC-Westland 231, 2685 DC Poeldijk, The Netherlands*
[13]*ID-tec B.V., ABC Westland 197, 2685 DB Poeldijk, The Netherlands*

ABSTRACT

Urban drainage operation, management and rehabilitation can be divided into two distinct segments: traditional grey infrastructures (i.e. pipes and associated components) and green infrastructures. For piped systems this boils down to maintaining the operational safety, stability and tightness of the sewers and special structures. However, this chapter provides an overview on both realms and highlights that, while there is a lot of standardization for grey infrastructures, the knowledge on green ones is much more fractured. They are often composed of both engineered and natural elements such as pipes, flow control systems, vegetation, micro-organisms in the soil or growing media, and also deliver a broad range of beneficial services to our communities and their inhabitants. Existing terminology for pipe networks is adapted by defining a similar distinction for green infrastructures based on the severity of the necessary actions. There will be no focus on other special structures and machinery. Adopting these distinctions, this chapter consists of three parts: (1) pipe network operation and maintenance (O&M), (2) structural rehabilitation of pipe networks and the connected manholes and (3) green infrastructure rehabilitation including O&M focusing on some examples. Consequently, this chapter can be used as guidance on available technologies, existing guidelines and research gaps.

Keywords: sewer, repair, renovation, replacement, performance, cleaning, green infrastructure, permeable pavements, swales, wetlands.

The rehabilitation of urban drainage systems is, at its core, a measure to enable continuous fulfilment of the objectives of, for instance, the EN 752 (2017), also discussed in Chapter 3, namely protecting public health and safety by limiting the public's exposure to wastewater, collecting and conveying stormwater at an acceptable risk of flooding and preventing environmental pollution as much as feasible. There is a close relationship and inevitable overlap with operation and maintenance (O&M) activities. In practice, it will also depend on the groups that are involved with the activities and how these activities are funded. Rehabilitation is often associated with capital activities and related financial resources whereas O&M is associated with operational activities. Rehabilitation is defined in EN 752 (2017) as 'measures for restoring or upgrading the performance of existing drain and sewer systems', while 'the purpose of Operation and Maintenance is to ensure that the drain and sewer system performs in accordance with the functional requirements'. Rehabilitation focuses on restoring and improving the condition of an asset that has suffered deterioration, damage or obsolescence, whereas O&M involve ongoing activities to ensure the smooth and efficient functioning of an asset to meet its intended operational goals. There are grey areas, especially when considering repairs. Uncertainty about roles and responsibilities may transform 'O&M needs' into 'rehabilitation needs'. In many cases, poor funding of O&M may lead to the need for rehabilitation which tends to be funded more easily as the activities can be capitalized and hence funds can be borrowed. However, this will likely result in higher expenditures in the long run.

For piped systems this boils down to maintaining the operational safety, stability and tightness of sewers and special structures. The development of major damage and impairments to the surroundings and the environment must be prevented. Operational safety means in concrete terms: the pipeline must reliably continue to do what it had originally been built for. First and foremost, this means that the transported medium flows through pipes without any problems. Pipes must therefore not be blocked, and a sewer must also have an adequate gradient. The two other criteria, stability and tightness, complement this and can also have a direct influence on operational safety. These three criteria, in turn, must be met over the entire planned (remaining) service life of a network at a reasonable cost and in compliance with the locally relevant legal framework. It should be noted that each individual pipe does not contribute to operational safety in isolation, but its performance always depends on the performance of the entire network. For instance, if a sewer is deemed hydraulically too small, it is of little help to enlarge it only at one single point if this exacerbates the problems in other parts of the network (e.g. by increasing combined sewer overflows, or causing pluvial flooding). The capacity and dimensioning of the entire network include pipes, pumps and many other special structures. Nevertheless, while operational safety must therefore always be assessed with a view to the entire network development, tightness and stability can usually be assessed locally for the individual sewer line/pipe section (see Chapter 4).

Accordingly, rehabilitation techniques are to be differentiated as follows:

- **Structural rehabilitation:** This includes techniques that essentially serve to restore the structural integrity and tightness of an asset. As a rule, the original hydraulic condition does not necessarily have to change.
- **Performance based (i.e. mostly, but not only, hydraulic and treatment efficiency) rehabilitation:** This includes techniques or (additional) systems that can improve the performance of the entire municipal drainage system. In addition to a re-dimensioning of the original drainage network, these can also include decentralized measures for the collection, retention, treatment and discharge of runoff. In principle, this includes 'nature-based solutions' (NBS) as well as 'sustainable urban drainage systems' (SUDS). Regarding NBS and SUDS, rehabilitation can target other performance criteria such as visual integration, recreational activities or a complementary function (e.g. acting as a path, road or parking space). This performance or service-based rehabilitation (see Chapter 6) on a network level leads inevitably to a hybrid system that will require a wide array of rehabilitation techniques to maintain the expected current and desired future service.

As SUDS not only consist of technical or structural parts (see Figure 1.3 in Chapter 1) but also often comprise natural components, they will require their own adapted rehabilitation to be able to deliver the range of services attributed to them. Such services are related to the fact that SUDS are often in open air, highly visible to nearby residents and 'share' urban land. Those services are thus not restricted to hydraulic performance, so the need for broadening the term of performance to a wider array is necessary. Also, as structural rehabilitation alone is not sufficient for such infrastructure another point is added to the differentiation of techniques:

- **Environmental rehabilitation:** Any drainage system in poor condition can lead to unintended or undesirable environmental consequences such as spillage or exfiltration of polluted water to water bodies. When considering the specific case of SUDS, rehabilitation can be required for other environmental reasons, namely biodiversity restoration or preservation. SUDS are often vegetated and a main contributor to biodiversity in urban areas.

Another possibility to distinguish between rehabilitation techniques is based on the severeness of the failure and the connected necessary action. This has been well defined in the asset management (AM) of piped infrastructures. The term 'rehabilitation' is considered here with respect to constructional aspects for the field of drainage as per ISO 24516-3 (2017) in conjunction with EN 16323 (2014) (see also Chapter 3).

Accordingly, 'rehabilitation' is defined as 'measures for restoring or upgrading the performance of existing systems, including renovation, repair and replacement'.

The sub-groups are defined as follows:

- **Repair:** Rectification of local damage.
- **Renovation:** Work incorporating all or part of the original fabric of the drain or sewer by means of which its current performance is improved.
- **Replacement:** Construction of a new drain or sewer, on- or offline of an existing drain or sewer, with the function of the new drain or sewer incorporating that of the old.

While we therefore have a well-defined standardization and definition of terms for our 'grey' infrastructure this is not the case for SUDS. In this chapter, we will adapt the above existing terminology by defining a similar distinction for SUDS based on the severity of the necessary action (as presented in Section 9.4).

Adopting these above distinctions, this chapter consists of three parts: (1) pipe network O&M, (2) structural rehabilitation of the pipe network and the connected manholes and (3) SUDS rehabilitation including O&M. Other special structures (e.g. underground detention or concrete retention basins) and machinery (e.g. pumps) are not considered as they should be able to follow either building or machinery maintenance and replacement schemes. The main differentiation will be made on the level of rehabilitation intensity. As there is a lot of knowledge available on the topic, only a brief overview of the main terms and methods is offered in the first part. The opposite is true for the selection of SUDS which will be introduced second. As there is limited data available this part mainly focuses on describing the various parts of the SUDS in question, looking at the long-term performance of SUDS and their needs for rehabilitation and which measures are necessary as a function of the SUDS type in question and at which level of severity. The focus is on environmental rehabilitation, if applicable, as the structural rehabilitation of the 'civil' components is largely similar to 'grey' infrastructure, with some exemptions, for example, for plastic storage chambers. Given the large number of permutations associated with SUDS, it is impossible to paint a comprehensive image; rather, examples and possible research and knowledge gaps are highlighted.

9.1 PIPE NETWORK O&M

This section explains the requirements that a cleaning process and the results of the cleaning of a pipe network should meet, based on international standards (e.g. ISO 9001, 2015; ISO 14001, 2015;

ISO 45001, 2018) and national guidelines (e.g. BRL K10014, 2017; ÖWAV-RB22, 2015). These may slightly change from utility to utility. Besides a well-structured cleaning process, the craftsmanship of sewer cleaning requires all human senses: sight, hearing and smell (within the boundaries of the occupational health and safety guidelines). They are used together by the operator to determine if a sewer is clean allowing the start of inspections. The mastering of the skills for thorough sewer cleaning takes years of experience and training.

A cleaning process consists of:

- Agreement with the client and preparation of the work, whereby all relevant data must be available before the start of the cleaning process.
- Performing the work in accordance with the assignment, including reporting on the work performed.
- Control of the performed work and quality assessment.
- Taking measures to improve the quality of the cleaning.

Before accepting the assignment, the basic principles of sewer cleaning must be laid down between the cleaning company and the client. At least the following aspects should be covered:

- The extent of sewer cleaning.
- The method of cleaning.
- The performance of checks by the cleaning company on the cleaning carried out.
- Location and method of the disposal of the collected solid and liquid waste from the vacuum tank.
- The method of reporting.

For good traceability of the cleaning, identification and, if possible, inventory data must be available. The identification data consist of:

- The name of the municipality or client.
- The name of the district.
- The street name.
- The identification number of the manhole from which cleaning takes place and the identification number of the manhole in which direction the cleaning takes place or the number of the object that is cleaned.
- The cleaning periods.

The inventory data can consist of:

- The diameter or dimensions of the sewer and/or manhole.
- The material of the pipes and/or manhole.
- The inner bottom of the pipe at the location of the manhole relative to the local reference system.
- The height differences between manholes.
- The expected degree of pollution of the objects.

The starting points are defined in the cleaning plan. In the cleaning plan, the cleaning company must in any case deal with the following subjects and record the content:

- which sewers need to be cleaned,
- the order in which the sewers are cleaned,
- the way in which the sewers are cleaned (e.g. using a set-up as shown in Figure 9.1),
- the total length of sewer to be cleaned, indicated per diameter and zone,
- the starting period of the cleaning activities,
- an estimate of the number of working days required for the cleaning activities,
- the periods within which the cleaning activities are carried out,

Figure 9.1 Sewer cleaning set-up with specialized trucks equipped with vacuum suction and high-pressure jetting capabilities. *Source*: vandervalk+degroot BV.

- the disposal of the solid and liquid waste materials,
- draining the process water (including any communication with relevant water quality institutions, if any),
- special circumstances or situation which require specific attention and
- time of reporting.

The cleaning plan must be established by the cleaning company in consultation with the client before the cleaning operations can be commenced unless the client waives this. The cleaning plan or the part of the cleaning plan that is carried out on the relevant day must be present at the workplace during the cleaning activities or the operator must have all the relevant information. Changes to the cleaning plan during the performance of the cleaning operations and any additional or waived work must be agreed upon (in writing) between the client and the cleaning company. If one or more parts cannot be cleaned, it is required to explicitly mention this in the delivery report.

Regarding working safely in sewers, locally relevant health, and work safety (e.g. confined space entry) regulations apply. For example, in the Netherlands the health and safety catalogue 'Regulations for safe working for sewer management' of the Dutch Waste Management Association (Arbocatalogus, 2018) applies. The working conditions catalogue is based on existing (statutory) rules and regulations. Depending on the method of cleaning and the traffic situation at the location of the sewers to be cleaned, plans must be devised to guarantee safety during the cleaning operations. The traffic control measures that must be taken during the cleaning operations must be agreed with the road manager and comply with the boundary conditions.

All collected solid and liquid waste must be transported to a treatment and/or processing facility approved by the client. The treatment and/or processing establishment must have a permit (recognition) for the storage, treatment and/or processing of this waste. The transport of the waste must comply with the Environmental Management regulations.

If sand ingress or other irregularities are observed during the cleaning operations, the work must be stopped immediately, and the client (or its representative) was informed. If it appears that the cleaning activities are hampered, for example, by obstacles, further research will have to be carried out. The immediate removal of fixed obstacles can be done by several methods:

- Removal by hand if entering the sewer pipes is an option.
- Chain cutting (with high risk of damaging the sewer pipe (lining).

- Remotely operated cutting tools.
- Remotely operated ultra-high-pressure tools (see Figure 9.2).

After the cleaning operations are completed, their effectiveness must be checked. When the cleaning work is carried out prior to a sewer inspection, this inspection can be used as a check on the cleaning work carried out. If no sewer inspection is carried out, the cleaning company itself must check whether the cleaning has been carried out properly. The control inspection of the cleaning can be carried out using camera inspections (see also Chapter 4), a mirror or a comparable method. The inspection relates to aspects such as fouling, deposition, and sand or dirt accumulation, and should cover the aspects of root ingrowth, sand ingress, flushing cement and obstacles that could not be removed during the cleaning activities. A check should take place on at least 10% of the cleaned objects/pipes. The presence of water in the pipes should not interfere with the checks. The check should be carried out within 2 working days after the end of the cleaning of the object/pipe.

At the end of each working week, the cleaning company must provide a progress report to the client. This report must include at least the following topics for each day in question:

- The date.
- The weather conditions.
- The type of vehicle/type of machine with number and/or license plate.
- The names of the staff who carried out the cleaning.
- The street names and place where work was done.
- The location and diameter of the cleaned pipes (possibly on a drawing).
- The number of waste deposits with the number of tonnes/m^3 of waste.
- The total daily production of the cleaned sewers, (if applicable) in metres per diameter.

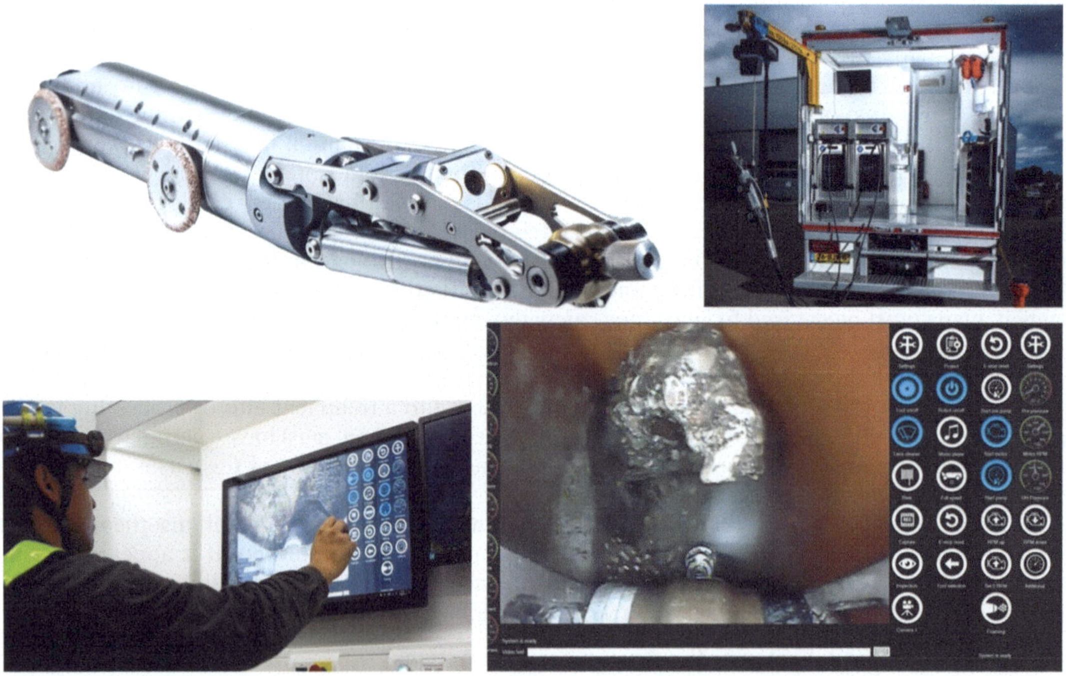

Figure 9.2 Ultra-high-pressure jetting tool with truck and operator controls. *Source*: vandervalk+degroot BV.

- Discovered defects in the sewers.
- Any special circumstances.

Upon completion of the cleaning operations, the cleaning company reports in writing the findings during the cleaning carried out. All relevant data must be stated in the delivery report.

In any case, the following subjects must be mentioned in the report:

- Name of the client.
- Description of the location and location of the cleaning activities.
- The cleaned pipes (including well numbers and strand length), wells and other noteworthy objects.
- The uncleaned sewers, stating the reason.
- The pipe type and material type.
- The diameter of the pipe or dimensions of the manhole.
- The amount of waste deposits with the number of tonnes/m^3 of waste.
- The checks carried out on the cleaning process.

9.2 GULLY POT CLEANING

This section covers the requirements that the cleaning process and the results of the cleaning of gully pots (also known as catch basins) must meet. These requirements are based on international standards (e.g. ISO 9001, 2015; ISO 14001, 2015; ISO 45001, 2018) and national guidelines (e.g. BRL K10014, 2017; ÖWAV-RB22, 2015).

Most of the preparation work for gully pots is the same as for pipe networks. As the cleaning of sewers requires experience and training to do the work properly, the cleaning of gully pots is an entry level work activity (see Figure 9.3 for a standard set-up). But don't be mistaken about required follow-up work and the desk analyses performed based on the data gathered during cleaning operations.

Figure 9.3 Specialized gully pot cleaning truck. *Source*: vandervalk+degroot BV.

Two important conditions will contribute to a proper cleaning of gully pots and the planning of gully pot rehabilitation programme afterwards:

- Proper instruction of operators and taking into consideration that they may be new on the job.
- Providing enough time for operations to clean gully pots and to register proper data and conditions about gully pots.

Carrying out checks during cleaning activities during the work, it should be visually assessed whether:

- The gully pot cover is not broken.
- The gully grates and grids or inlet control devices (if any) are present, in the right place, properly sealed and of the proper dimensions.
- The gully has other defects.
- The gully is clean after the cleaning work. Deviations are detected and recorded.

At the end of each working week, the cleaning company must provide a progress report to the client. This report must include at least the following aspects for each day:

- The date.
- The weather conditions.
- The type of vehicle with number and/or license plate.
- The names of the staff who carried out the cleaning.
- The number of hours worked, start–end, break, travel times.
- The street names and place where work was done.
- The number of uncleaned gullies per street, with reasons and a statement of the location of the gullies.
- The defects found, stating the location of the gullies. These are divided into:
  - Missing and/or broken lid.
  - Missing and/or broken pot grid.
  - Missing, improperly sealed, broken or improperly sized inlet control devices.
  - Other defects in the gully pot.
  - Clogged drainpipes.
  - Other defects.
- The amount of waste deposits with the number of tonnes/m^3 of waste.
- Any special circumstances.

9.3 PIPE REHABILITATION

A high level of performance is expected from all rehabilitation procedures, regardless of their classification as repair, renovation or replacement. For instance, the EN 752 (2017) explicitly demands: 'In principle the performance requirements for a rehabilitated system shall be the same as those for a new system'. This concerns the stability (i.e. the structural integrity), the tightness and the operational reliability of the rehabilitated structure over its entire planned service life. Accordingly, the selection of a suitable rehabilitation method is only possible with a proper assessment of the nature and likely causes of the damage. In the following, the procedures for the rehabilitation of sewer sections are presented. This concerns the repair, renovation, and replacement of sewer pipes, respectively. While the following is a concise review on those techniques, there are more books (e.g. Sterling *et al.*, 2010) and guidelines (e.g. ÖWAV-RB28, 2007) available on these topics in addition to research on decision-making processes as to which methods to select (e.g. Bewszko & Słyś, 2014; Diogo *et al.*, 2018; Marzouk & Omar, 2013; Matthews *et al.*, 2018; Mezhoud *et al.*, 2022; Ochs, 2012).

9.3.1 Repair

The repair of sewer pipes is an essential element for the operation of sewer networks. Trenchless techniques are one way to accomplish repairs. Localized damage scenarios such as leaking joints, missing pipe wall parts and root penetration can be repaired in this way. Typical procedures include injection, flood grouting, grout-injection, trowel-smoothing and the use of sleeves and partial liners. According to the EN 16323 (2014), repair is considered a sub-sector of rehabilitation alongside renovation and replacement and is defined as 'rectification of local damage'. Thus, a repair only concerns individual points in the network and not the replacement or improvement of entire pipe strings or network areas. In this sense, repair is not to be regarded as a fundamental improvement of the substance, but as a localized or individual measure to maintain safe operation. Accordingly, repairs are basically assigned to the operation of the system and are not considered a capital investment. Technically, the question then arises as to which types of damage are considered 'local' in the sense of a repair and to what extent these damages can be repaired using trenchless construction methods.

In the top left of Figure 9.4, we see a typical damage pattern necessitating repair: a transverse crack along the entire circumference of the pipe. This is considered local, because the pipe is only damaged in one location and not along the whole pipe length, thus without any structural implications on

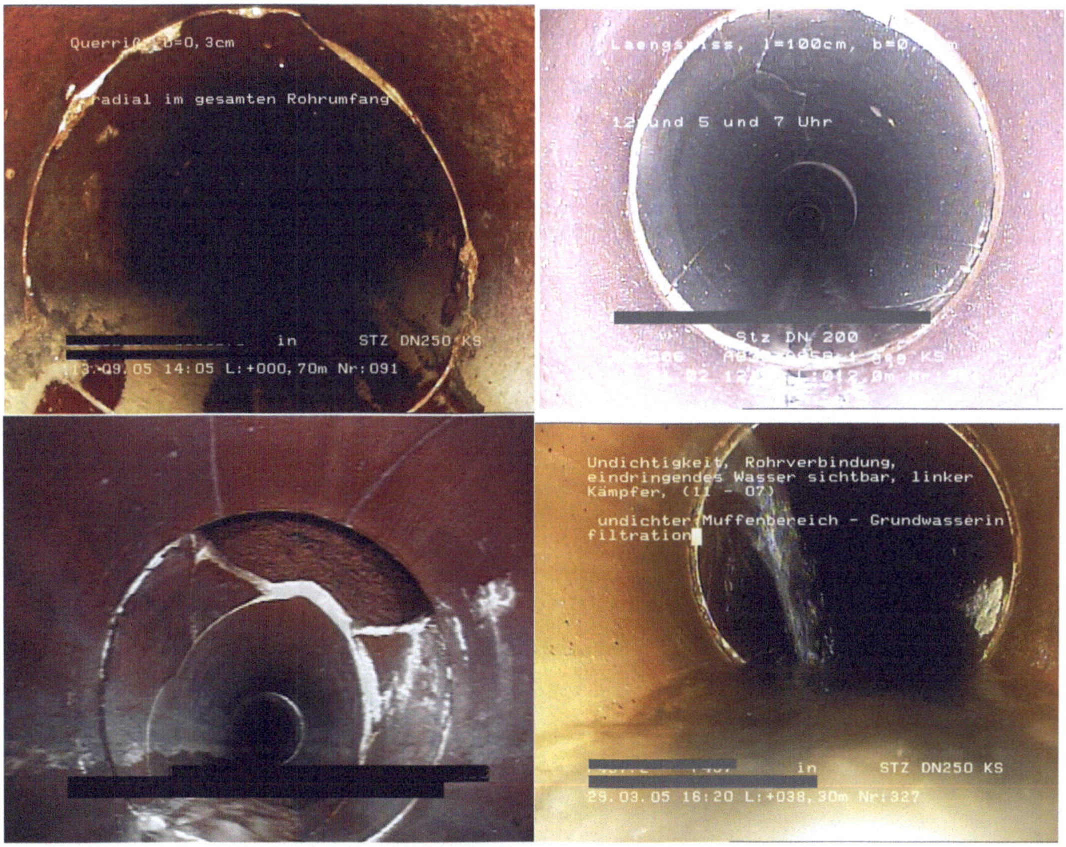

Figure 9.4 Examples of damage patterns and in consequence repair tasks. *Source*: IKT – Institut für Unterirdische Infrastruktur gGmbH.

pipe stability. It can be compared to a leaking joint in terms of failure mode and must be bridged, for example, lined or otherwise sealed.

In Figure 9.4, top right, we see a damage pattern with longitudinal cracks. These longitudinal cracks are always to be regarded as critical regarding the structural safety. Nevertheless, we are still talking about a local measure because longitudinal cracks run exclusively in a single pipe segment. Here, it is possible to replace the damaged pipe or to reinforce it in such a way that it is completely bridged into the undamaged areas.

In Figure 9.4, bottom left, we can also see local damage, namely missing parts of the pipe wall, as the surrounding soil can be seen. In addition, cracks can be detected that propagate from the main point of damage. Such cracks can spread further, sometimes even during a repair measure, for example, due to pressure from repair packers. Again, the entire damaged pipe must be bridged in such a way that the repair measure can connect to a fully intact pipe.

Finally in the bottom right of Figure 9.4, we see how groundwater enters the sewer through a leaking pipe connection. Even if this initially appears to be a simple repair case, that is, local and (still) without major stability problems, the technical demands on the rehabilitation execution are very high. This is because the pipe must be sealed against external water pressure and this is extremely difficult, especially in the case of a strong inflow of water. In areas with changing groundwater levels, it may make sense to wait until a period with lower groundwater levels before carrying out the repair. Also, there is a potential of a void having formed in the surrounding soil or trench for the two bottom damage patterns.

The most common type of damage is shown in Figure 9.5, where roots have grown through a pipe joint thus reducing the flow cross section. In such cases, the cross section is often first milled free, but this is rarely a permanent solution because the roots tend to quickly grow back in. Repairs are also challenging because roots usually press much harder into a pipe joint from the outside than groundwater, for example.

Table 9.1 shows a list of repair methods used in sewers. Some of these procedures are also addressed in the EN 15885 (2018) with similar terms. Basically, three applications can be distinguished:

- Repair from the outside, for example, by excavating, exposing and replacing the pipe, or enclosing the damaged area with a sleeve or cured-in-place pipe.

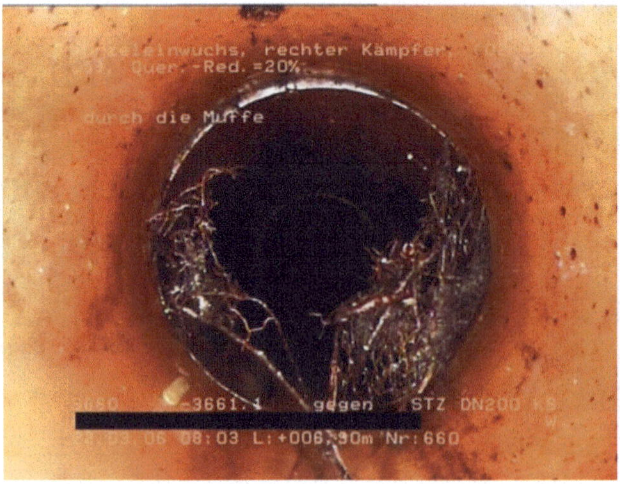

Figure 9.5 Repair task caused by the damage pattern of root ingrowth. *Source*: IKT – Institut für Unterirdische Infrastruktur gGmbH.

Table 9.1 Overview of repair methods.

Pipe replacement	From outside	EN 15885, repair
External sleeve		
Injection		
Injection/trowelled	Accessible	
Injection from inside	Non-accessible	by injection
Repair robots		with trowelled or moulded material
Flood grouting		by flood grouting
Partial liner		with cured-in-place components
Internal sleeve		with internal mechanical devices

- Repair from the inside, by injection of gels, resins or using mortar, when the sewer segments are sufficiently large to allow physical access by staff or contractors. Entry manholes and sewers starting from a diameter of around 1.20 m are considered 'walk-in' accessible.
- Repair from the inside in 'non-man-entry' sewers. Here, injection grouting, flood grouting, so-called robotic methods, internal sleeves and partial liners are used.

The methods used from the inside of sewer pipes, that is, injection, flood grouting, robotic methods, internal sleeves and partial liners, will be discussed next in more detail.

9.3.1.1 Injection methods

The left side of Figure 9.6 shows a typical example of a repair in a walk-in segment. This sewer manhole was sealed from the inside by means of injection grouting. The first pipe of the main sewer, which is obviously a vitrified-clay type, is visible as well. Here, a low-viscosity resin or a mortar was injected from the inside into the points of damage. This material penetrates voids and bonds to the

Figure 9.6 Injection methods: example of manhole repair (left), sewer repair in an overall view (right top) and cross section (right bottom). *Source*: IKT – Institut für Unterirdische Infrastruktur gGmbH.

particles of the surrounding soil, resulting in bodies of grout which continue to adhere to the exterior of the manhole even after its removal.

Injection is also possible in non-walk-in sewers, as can be seen in Figure 9.6 on the top right for diameters of 200 and 300 mm, respectively. The bodies of grout again occupy considerable space and continue to adhere securely to the exterior of the vitrified-clay line even after removal of the soil. Here, the material was installed using a rehabilitation packer, which was positioned above the damaged area. The repair resin was fed to the packer via hoses. Such resins are generally two-component types which are mixed directly at the damaged point and then pumped into it under pressure. The packer then remains in place until the resin has cured, resulting in bodies of grout such as that can be seen in a cross section in the bottom right of Figure 9.6. The resin penetrated here into the pores and bonded with the soil particles into a solid body similar to polymer concrete. Pure resin, completely sealing the pipe joint, can also be seen in the pipe socket.

9.3.1.2 Flood grouting

Similar in nature to injection methods, flood-grouting procedures can be used to fill cavities in soil and thereby to seal cracks and other leaks in the pipe wall from the outside. This method is applied in two stages (see Figure 9.7 with the subsequent steps denoted in green and yellow), using a two-component acrylic or silicate gel. In the first step, the relevant section of the pipeline is filled with one of the two gel components after closing off the line section. This component then penetrates leaks, cracks and cavities. The first component is then pumped out of the line, so that it remains only in the defects and in the soil. In the next step, the second gel component is fed into the line in the same way as the first, and it, too, can penetrate the leaks and into the soil. The two components are then able to mix at these sites and cure to form a body of grout, sealing the line. Flood-grouting methods can be used to rehabilitate poorly accessible sewer and pipeline systems. The duration of flood-grouting procedures depends very much on the surrounding soil since the rehabilitation components must be able to penetrate with certainty into the points of damage and the soil itself. Highly porous soils, such as gravel or sand, will absorb copious quantities of these materials in a short time, whereas cohesive soils appear to make this procedure unsuitable because of the low permeability of cohesive soils. Difficulties occur if greatly differing types of soil, and therefore differing soil permeabilities, are present along the same section of a sewer system. Short filling times will then result in inadequate sealing in areas of low permeability, whereas longer filling times result in severe wastage of materials in the more permeable areas of the soil.

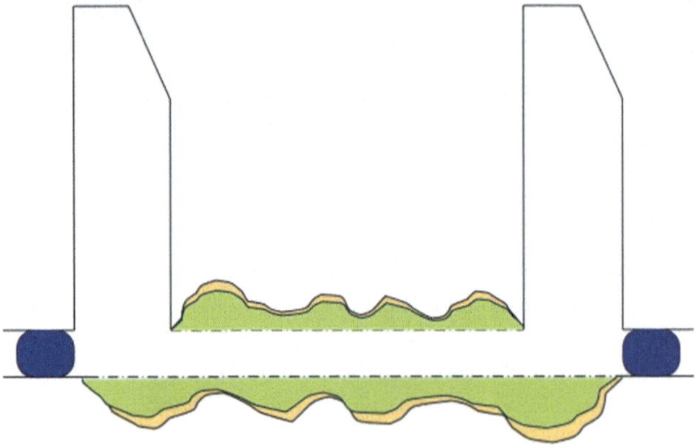

Figure 9.7 Schematic sketch of the flood grouting method. *Source*: IKT – Institut für Unterirdische Infrastruktur gGmbH.

9.3.1.3 Robotics methods

In case of robotic methods, a mechanical tool and a camera are carried on a cart to the point of damage. The damage can then be treated using the tool while the operation is observed via the camera. The process is controlled by a technician working on the surface. Figure 9.8 shows on the top left a robot for milling. A camera is suspended from it at the front which is used to observe how the milling cutter above it cuts away the projecting connection socket. On the top right, Figure 9.8 shows a grout-injection and trowel-smoothing unit. The material is fed under pressure through the hose and into the cavity via the so-called grouting shoe. The whole surface is then smoothed off using a (orange) trowel. The bottom part of Figure 9.8 shows robots usable for cutting, grabbing and grinding, respectively.

Grouting and trowel-smoothing can lead to very different results. Figure 9.9 shows on the left a crack at a laboratory test installation, which has been trowel-smoothed by a robot. Here, the material fills only the defect in the pipe wall, closing it. Cavities in the soil are not filled in so sink-holes in the above constructed road might still occur. On the right, Figure 9.9 shows a body of grout which was cut open. The entire cavity is filled with resin as the resin was extremely viscous during installation, and therefore did not penetrate the voids in the soil, with the soil only adhering to the resin at the interface. In summary, we can ascertain that trowel-smoothing fills the pipe wall, grout injection fills the cavity and injection grouting also fills the voids between the particles of the soil.

9.3.1.4 Internal sleeves

Another repair method is the use of stainless-steel internal sleeves (see Figure 9.10). These sleeves are transported inside the pipe on a packer to the damaged point, clamped there and

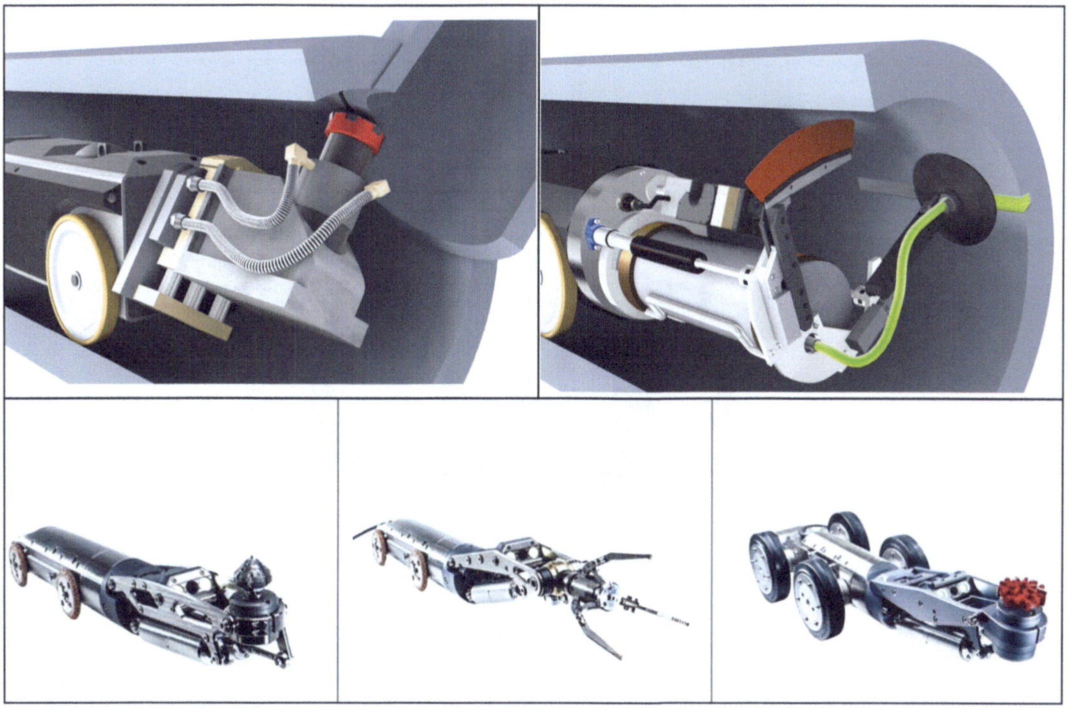

Figure 9.8 Robot for milling (top left) and grouting (top right) damaged areas and for cutting (bottom left), grabbing (bottom centre) and grinding (bottom). *Sources*: UNITRACC.com, Prof. Dr.-Ing. Stein & Partner GmbH, ID-tec B.V.

Figure 9.9 Repair result of trowel-smoothing (left) and grout-injection (right). *Source*: IKT – Institut für Unterirdische Infrastruktur gGmbH.

Figure 9.10 Steel internal sleeves with an elastomer sheath. *Source*: IKT – Institut für Unterirdische Infrastruktur gGmbH.

then permanently fixed by means of a locking gear mechanism. The locking gear engages step-by-step with a toothed rack as soon as the sleeve is expanded. The precondition for this repair is, however, that the old pipe/soil system provides adequate resistance. Sealing is assured by means of embossed sealing studs, which are present on the front and back of the elastomer sheath. Such seals act via compression, and only provide adequate resistance to external water pressure if they themselves are also under pressure. A sealing action is correspondingly less likely if the old pipe is not present as an abutment.

An internal sleeve must therefore always connect to an undamaged pipe at both ends of the repaired pipe. It is therefore particularly suitable for the sealing of leaking sockets, on which the connecting pipes themselves can apply the necessary counterpressure. Not only stainless-steel sleeves with elastomer seals, but also sleeves combined with a so-called 'short liner', or 'partial liner', which are adhesively bonded to the old pipe, are available. An open question remains about the consequences that could result from the change in cross section with the installation of such internal sleeves, as no hydraulic tests are performed for this.

9.3.1.5 Partial (short) liners

To permit the use of 'short' or 'partial' liners (see Figure 9.11 for the methodology), a fabric mat, usually consisting of glass-fibre-reinforced plastics (GRP), is first impregnated with a resin, and then wound onto a rehabilitation packer. The packer is then moved to the damaged point, and the short liner is pressed against the wall of the line until the resin has cured. The aim here is that the liner should be adhesively bonded to the pipe wall across the entire pipe circumference, thus sealing the pipe wall. Typical renovation results from a product test are shown in Figure 9.12.

Such partial liners adhere uniformly to the entire circumference of the pipe and completely line the damaged area of the old pipe. Here, too, the partial liner must connect with certainty to undamaged areas, as the old pipe is exposed to internal pressure via inflation of the packer, and the fine-longitudinal cracks are generally able to spread to the next pipe socket. Special importance needs be given to the preparation of pipe wall surface because the partial liner must adhere reliably to the substrate. Comparative tests have shown that the substrate must always be carefully cleaned and that any glazed vitrified-clay surfaces should first be thoroughly ground or milled off.

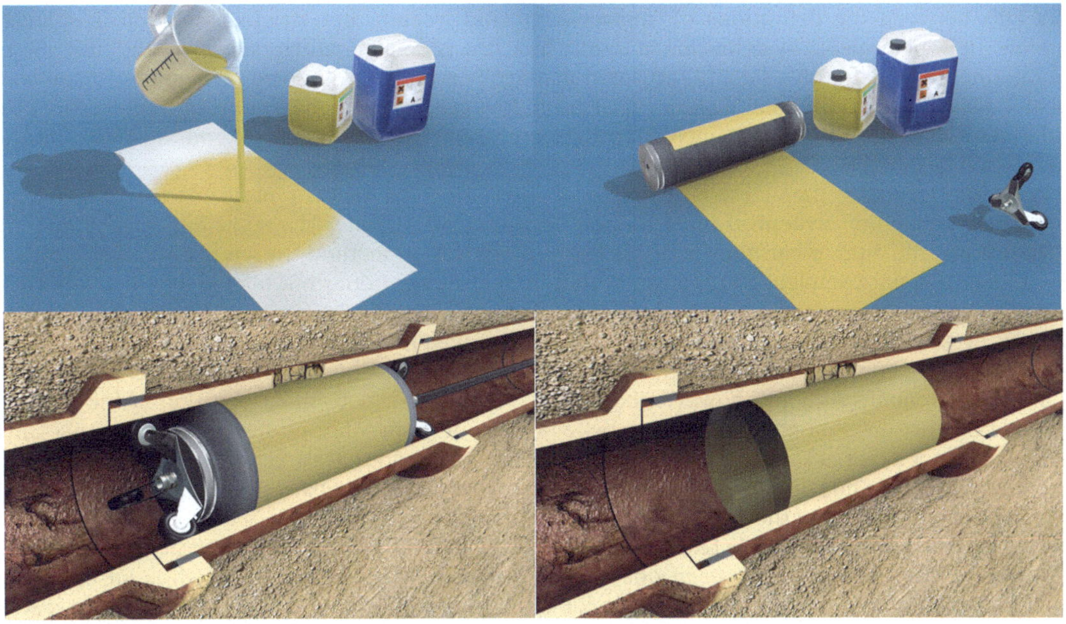

Figure 9.11 Short liner methodology – impregnation of the liner (top left and right), positioning by means of a packer (bottom left), bonding to the damaged area (bottom right). *Source*: UNITRACC.com, Prof. Dr.-Ing. Stein & Partner GmbH.

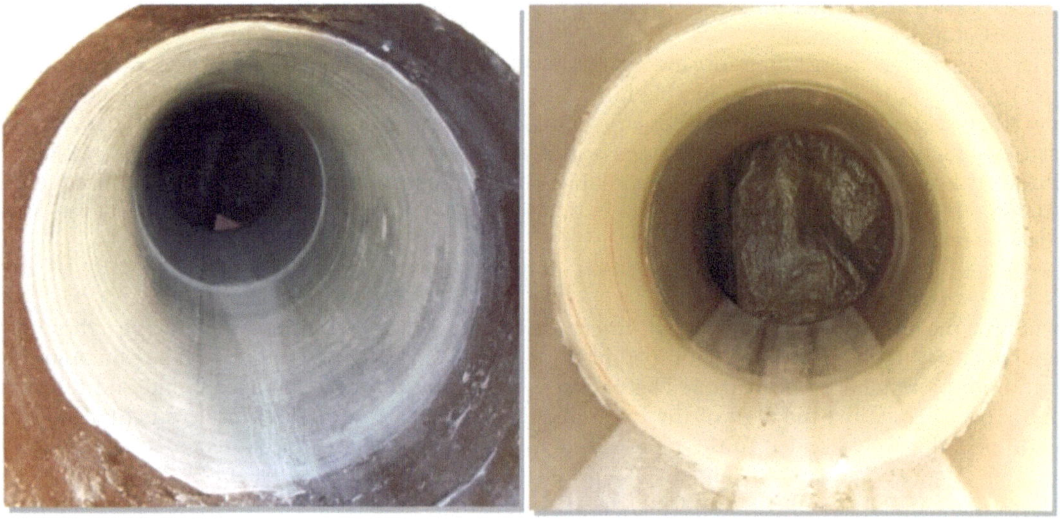

Figure 9.12 Results of the rehabilitation with short liners. *Source*: IKT – Institut für Unterirdische Infrastruktur gGmbH.

9.3.2 Renovation

According to EN 16323 (2014), renovation is considered a sub-sector of rehabilitation alongside repair and replacement and is defined as 'work incorporating all or part of the original fabric of the drain or sewer by means of which its current performance is improved'. The definition of the term 'renovation' thus falls between the definitions of 'replacement' and 'repair', but strict delineation is scarcely possible. In 'replacement', by definition, a totally new line is created which takes over the function of the old line. In 'repair', only locally restricted damage is rectified. The focal principles of 'new' and 'local' address totally different assessment criteria, however, and the definition of 'renovation' also integrates two further aspects, namely 'improvement of current performance' and 'incorporation of the original fabric'.

Renovation should, therefore, be capable of improving the current performance, that is, the operational reliability, structural integrity and tightness. The greater the contribution of a method to these aims, the more it can thus be regarded as a renovation method. In renovation, furthermore, the old fabric is not completely exchanged – that is, replacement does not take place. Instead, while new components are incorporated, the original fabric continues to be used and integrated. We may thus say, by way of conclusion: a sewer line with an improved function is achieved using both old and new engineering components. Differentiation from 'replacement' thus appears possible. Delineation from 'repair' remains difficult, however, because a 'local' repair with 'incorporation of the original fabric' (e.g. short liners bonded to a host pipe) could result in an 'improvement in performance' and thus, strictly speaking, meet both definitions, that of 'repair' and that of 'renovation'. This results in potential disputes in how renovation projects are accounted for. Perhaps, a new fabric is generated as a result of capitalization of the work, as an investment, appears appropriate. The improvement of the current performance usually only results in the restoration of the original functionality, which should therefore be considered largely as an operational measure and thus be classified as an operating expense. In many cases, the argument of prolongation of the service life is therefore used to justify capitalization as an investment. In such cases, renovation will be intended to improve the durability of a sewer or pipeline in such a way that it will remain operationally reliable, structurally safe and tight for a longer time than originally planned. A focus of such methods should be laid also on the house or

Table 9.2 Overview of renovation methods.

EN 15885, Lining	Slip lining (segmental)	Material and shape in factory
with discrete pipes		
with continuous pipes	Slip lining (continuous)	
with close-fit pipes	Close-fit slip lining	Material in factory, shape on site
with spirally wound pipes	Spiral wound lining	
with pipe segments	Segmental lining	
with cured-in-place pipes	CIPP lining	Material and shape on site
by sprayed, trowelled or cast-in-place material	Coating	
with a rigidly anchored plastics inner layer	Grouted-in-place lining	

service connections which can be considered a weak point of sewer systems with a higher failure rate due to the properties of these objects and the associated flows (Post *et al.*, 2016). This will result in a massive increase in the number of such connection pipes eligible for trenchless renovation.

One can find – for example in EN 15885 (2018) – numerous terms for methods which are included in the 'renovation' group. This standard examines in detail the classification and properties of technologies used for renovation and repair of pipes in the urban drainage sector. A list is provided in the left-hand column of Table 9.2.

It is conspicuous that, following EN 15885 (2018), all 'renovation' method names contain the word 'lining'. The middle column shows other frequently used terms in practice, and a sub-division into groups, which is not a feature of the standard. These groupings are intended to assist us in better understanding of the methods and the factors influencing application quality. We differentiate here three groups (see also Table 9.2):

- In the first group, the liner inserts are produced entirely at the factory. This applies both to the composition of the liner and to its ultimate shape.
- In the second group, the material that the liner is composed of is produced at the factory, and then delivered to the site. The liner insert is then shaped on the site to fit the existing pipe.
- In the third group, material components are delivered to the site, and these are only processed to achieve the required material quality and shape on-site.

The essential differences between these methods can thus be found in the extent to which actual liner production takes place in the factory versus on-site. Some examples will illustrate this. The order of these examples corresponds to the list in Table 9.2, starting with the methods in which liner production takes place largely at the factory (i.e. the discrete-pipe relining/continuous-pipe method), followed by factory production of the material and shaping on-site (i.e. the close-fit, spiral-wound and segmental-lining methods), and wrapping up with on-site production of both material and shape (i.e. the cured-in-place pipe (CIPP) lining, spray, trowelling and cast-in-place methods and studded-sheath method/grouted-in-place lining).

9.3.2.1 Discrete-pipe relining/continuous-pipe method

We first examine the discrete-pipe relining and continuous-pipe methods. In both cases, the liner inserts constitute fully formed pipes, with both the material and the final pipe geometry produced entirely at the factory. Both can therefore be tested and dimension-checked before as well as making quality control easier. However, neither the material nor pipe geometry can be changed on-site. Accordingly, pipes featuring socket joints are available in many materials, including plastics, vitrified clay and cast iron.

These methods are illustrated diagrammatically in Figure 9.13. In the first (i.e. left) example, new pipes are pushed forward into the old line. The method for joining these pipes is similar to that for

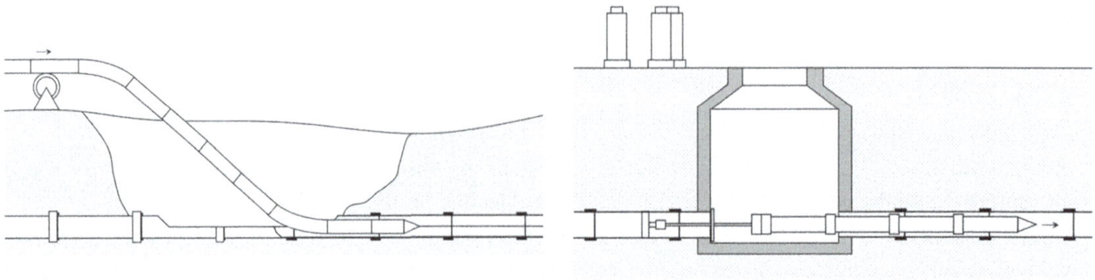

Figure 9.13 Renovation with prefabricated pipes: installation by pushing (left) and lining with continuous pipes (right). *Source*: IKT – Institut für Unterirdische Infrastruktur gGmbH on the basis of EN 15885 (2018).

jacking pipes, which results in the sockets at the joints expand only slightly outwards, and that the new pipe train can slide easily through the old pipe. An abutment for the jacking station is installed in the manhole. The jacking station engages on the rearmost pipe and thrusts the entire train forward. After completion, the outer annular gap is backfilled with an annular gap backfill material, to secure the pipe train in position. In the second (i.e. right) example, the entire pipe train is pulled from the exit shaft. It is possible to pull on the leading pipe provided the pipes are joined in such a way that they can transmit tensile force. Otherwise, pulling is performed against a plate which pushes on the rearmost pipe, and thus thrusts the complete pipe train forward. The pipe train is then subjected to stresses such as those occurring while pushing.

In the continuous-pipe method, the pipes are also produced in advance at the factory and then delivered to the site. In the case of larger-diameter pipes, however, the pipe joint is made only on-site, by means of welding, meaning that only weldable materials can be used. These are generally polyethylene (PE) or polypropylene. A relatively large construction trench is needed for the pulling in of the pipe train because the welding procedure usually takes place on the surface. In the continuous-pipe method, the annular gap between the old and the new pipe is also usually pressure filled with mortar or annular gap backfill material.

In both cases, that is, discrete-pipe relining and continuous-pipe relining, the new pipe must be significantly smaller in diameter than the old one, because the new pipe train will have to pass pipe misalignments, pipe wall surface irregularities and slight curves. A certain clearance between the old and the new pipe is therefore necessary. The resulting reduction in cross section will generally cause a loss of hydraulic performance. The impact of improvements in terms of pipe roughness by such a method on network wide hydraulic performance has proven to be minor (Tscheikner-Gratl *et al.*, 2014).

As incoming sewer laterals are, in both procedures, closed by the installation process, they must be opened again and must be connected by positive locking after insertion of the liner segments. This is generally accomplished using an open-cut method by excavating a so-called 'header pit', that is, the connection between the sewer lateral and the main pipeline is restored from the outside. The use of this method is particularly economically rational in cases in which the rehabilitated length includes only a few sewer laterals.

9.3.2.2 The close-fit method

Figure 9.14 shows several diagrams of the so-called close-fit (i.e. 'lining with close-fit pipes') method. This procedure minimizes two disadvantages of the continuous-pipe method; in particular, (1) the large construction trench needed, to bend the new pipe from the surface down to the old pipe and (2) the severe loss of cross section necessary to permit passage past all obstructions and through curves. The solution provided by the close-fit method is that the cross section of the pipe train has a different geometry when originally pulled in than during operation.

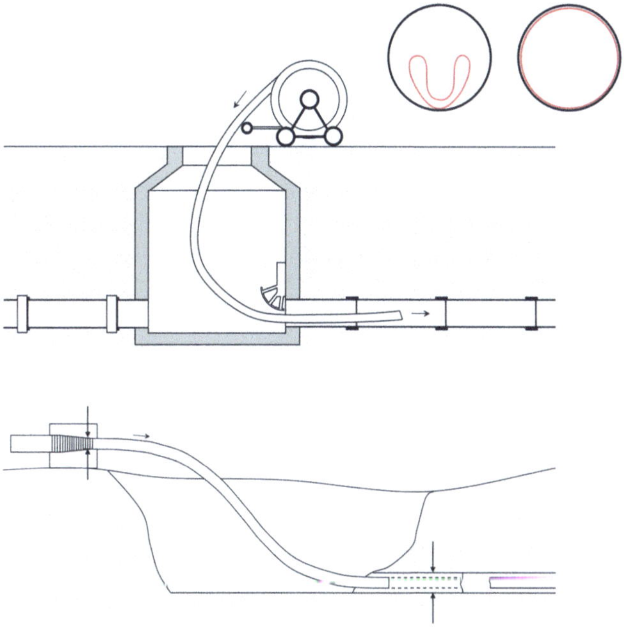

Figure 9.14 Renovation with on-site shaping using the close-fit method: pipe reduced in cross section in the pipe manufacturing plant (method A on top) and pipe reduced in cross section on site (method B on bottom). *Source*: IKT – Institut für Unterirdische Infrastruktur gGmbH on the basis of EN 15885 (2018).

'Method A' is shown at the top. We see to the right of the figure the shapes taken by the liner insert: (1) a 'folded' pipe, with a U-like cross section. The height of this cross section is only about half that of the finished pipe, with the benefit that the moment of inertia of the cross section is significantly reduced. Such a 'folded' cross section has a much lower bending resistance in the longitudinal direction compared to a circular cross section. To the right of this, (2), we can see the new pipe in its ultimate operational state, with a circular section in contact closely with the old pipe. The benefits are obvious: for the pulling operation, the 'folded pipe train' can easily be bent and manoeuvred through a manhole, for example, and then also drawn through the old pipes with adequate clearance. For ultimate operation, the cross section is almost as large as that of the old pipe and enables a similar hydraulic performance.

The actual shaping process consists of the folding and subsequent restoration of the pipe cross section. Pipes are given a U-shape at the factory and are then delivered to the site on drums. On-site, they are pressed back, using hot water, into their proper circular cross section after installation. The higher temperature of the water used increases pipes' deformability, while the internal pressure applied assures the circular cross section. Producers of these pipes also frequently mention a so-called 'memory effect', according to which the material, due to its original circular cross section after production, 'prefers' this circular geometry at a molecular level after reshaping. The application of temperature and internal pressure alone in any case already results in significant 'shape restoration', however.

The bottom section of Figure 9.14 illustrates the second close-fit method, known as 'method B'. Here, pipe's cross section is reduced by means of circular external pressure and longitudinal tensioning before the pulling operation. This is accomplished using rollers and under exposure to higher temperatures. The cross-sectional geometry of pipes remains reduced during the pulling operation, but it expands after the pulling by means of internal pressure in such a way that the pipes press against the wall of the old pipe.

In both methods, A and B, there is scarcely any annular gap, if at all, after completion of the installation, and the new pipes fit 'snugly' (closely) into the old pipe (therefore 'close-fit'). This approach generates several other advantages for non-pressure applications, that is, gravity sewers, in terms of the connection of sewer laterals. Using trenchless methods, the new pipe can be cut from the inside at the location of the sewer lateral and a so-called top-hat section can then be inserted from the inside using packers and then welded.

In summary, for the close-fit method, the composition of the pipe wall has already been fully decided at the factory. On-site, however, the geometry of the pipe train is modified drastically. This is made possible by, inter alia, exposure to compressive and tensile forces, and to temperature. Pipe connections at sewer laterals are subsequently made on-site only using trenchless methods.

9.3.2.3 The spiral-wound method

For the spiral-wound pipe method (see Figure 9.15), a plastic strip manufactured in a factory is unreeled from a drum into the manhole, and then wound to make a new pipe underground. The continuous edge of the strip fits into an interlocking slot which runs spirally along the entire length of the pipe train in the form of a long, continuous joint. The annular gap is generally then backfilled with a backfill material. Challenges can be found here in, inter alia, the interlocking system, which must be permanently tight across the entire length of this joint. Snap-in locking systems, for example, and special welding processes, are used for this.

Several other variants of the spiral-wound lining method exist. In some procedures, for example, the pipe can be retrospectively expanded after installation until it achieves a 'close-fit' with the old pipe. In this case, the strip can move apart at its edges, ultimately resulting in an increase in pipe diameter. In another variant, the pipe-winding machine itself moves forward in the pipe as it progresses, with the strip being added directly at the target location. Here, too, it is possible to work with only a slight annular gap. Pipe geometries with a non-circular cross section, such as rectangular and horseshoe shapes, for example, can also be rehabilitated. In individual cases, the plastic web is reinforced with metal strips, to increase its bending stiffness.

9.3.2.4 Segmental-lining methods

Another method for adjusting prefabricated elements to the shape of a sewer pipe on-site is shown in Figure 9.16, known as 'lining with pipe segments'. When this method is used, the individual elements or 'pipe segments' are assembled in such a way that a new pipe is produced and then permanently

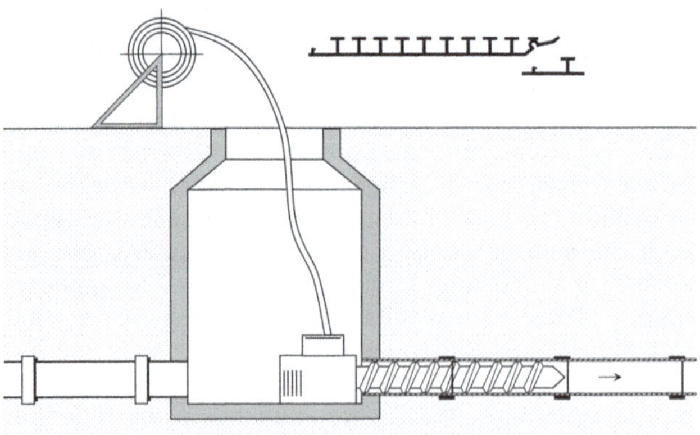

Figure 9.15 Renovation with on-site shaped pipes using the spiral-wound method. *Source*: IKT – Institut für Unterirdische Infrastruktur gGmbH on the basis of EN 15885 (2018).

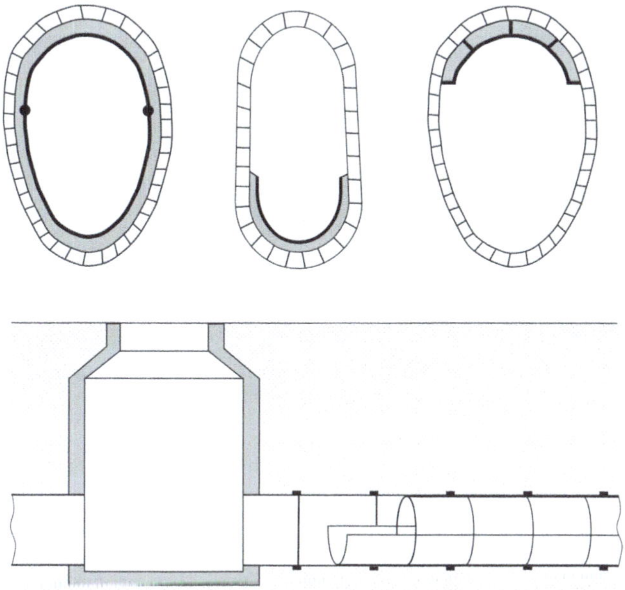

Figure 9.16 Renovation with on-site-shaped pipes using the segmental-lining method (bottom) with full lining (top left), partial lining: invert (top centre) and partial lining: crown (top right). *Source*: IKT – Institut für Unterirdische Infrastruktur gGmbH on the basis of EN 15885 (2018).

fixed to the existing pipe. Two lining variants focusing on the invert and/or the crown are generally used for corrosion-protection purposes. A lining of the invert can also improve hydraulic properties. Significant attention should be paid with this method to the joints between the individual elements. These, too, must be permanently tight and corrosion-proof. The annular gap between the lining elements and the old pipe is generally filled in with grout.

9.3.2.5 CIPP lining

Figure 9.17 shows a schematic view of the CIPP method. In CIPP lining, a flexible tube is delivered to the site, installed in the old pipe and then 'cured' on-site to make a new pipe. The CIPP liner consists of a so-called base material, which is impregnated with synthetic resin. This synthetic resin can be cured on-site when energy, in the form of, for example, heat or ultraviolet (UV) light is applied. The tube is cooled and/or protected against UV radiation during transportation to the site, to avoid premature curing. GRP or needle felt is used as the base, while the resins generally take the form of unsaturated polyester, vinylester or epoxy resins.

Figure 9.17 shows a method variant, known as the 'inversion' or 'slip-in' method. Here, the tube is slipped into the old pipeline by means of water pressure while the liner material is continuously fed into the interior of the liner. The water pressure thrusts the liner against the pipe wall, making it possible for the new pipe to fit into the geometry of the old pipe. In this variant, resin curing is usually accomplished via the application of heat, by feeding hot water into the pipe train after completion of the installation. In another variant, the liner is drawn into the old pipe and then expanded by means of, for example, air pressure. The term 'pull-in method' is also used here. Curing is usually affected by means of UV lamps, which are drawn through the liner. In this case, the UV radiation initiates the curing process in the resin. The base materials also differ, depending on the method variant. In the inversion method, generally needle-felt or textile-fabric liners are

Figure 9.17 Renovation with on-site-produced pipes using CIPP lining. *Source*: UNITRACC.com, Prof. Dr.-Ing. Stein & Partner GmbH.

used because these materials are extremely flexible. With this procedure, no significant friction occurs between the liner and the wall of the old pipe during installation, making it also possible to rehabilitate large-diameter conduits and long lengths. GRP liners with glass-fibre mats serving as the base material are generally used for the pull-in method. Glass fibres have a significantly higher modulus of elasticity than the resins used, and it is thus possible to produce liners with a relatively small wall thickness.

On-site quality assurance is particularly important when using CIPP liner methods since both the ultimate pipe material and the pipe geometry are 'produced' on-site. After installation, such pipes should always be inspected, using a camera, for defects and/or geometrical abnormalities, and pipe specimens should be taken, from within or close to the manhole, for example, to permit subsequent laboratory testing to verify that the material properties were achieved.

9.3.2.6 Spray, trowelling and cast-in-place methods

Another case in which the material and the geometry of the rehabilitation element are created on-site are the 'spray, trowelling and cast-in-place methods', or 'lining with sprayed, trowelled or cast-in-place material'. These procedures are mainly used for manholes and other special structures but seldom in pipes. In pipes, a coating is applied generally only for the purpose of corrosion-protection. In manholes and special structures, such as rainwater retention basins, for example, the aim is usually also to restore their tightness, or even their structural integrity. Several specimen images from manhole rehabilitation are shown in Figure 9.18. At top left, a plastic is being spray-applied, and spread using an automatic spraying system at top right. Mortar can, for example, be 'shot' on automatically (bottom left) or applied manually, using a trowel (bottom right). Mortars are mixed with water on-site and the mixing ratios must meet all the manufacturers' requirements.

Good substrate preparation must be ensured prior to application. The bonding strength achieved between the old and the new materials depends on the strength of the old material and on how the opposing surfaces can be intermeshed with each other. Before the coating activities, the old pipe wall must be cleaned, and layers of unstable substrate (e.g. due to corrosion) removed until a stable layer is established. The bonding strength is usually determined on a random-sampling basis by means of so-called tensile adhesion tests and is then compared against the target data, depending on national guidelines.

Figure 9.18 Lining of a manhole with sprayed, trowelled or cast-in-place material. *Source*: IKT – Institut für Unterirdische Infrastruktur gGmbH.

9.3.2.7 Studded-sheath method/grouted-in-place lining

The studded-sheath method, also known as 'lining with a rigidly anchored plastic inner layer', is a combination of a plastic sheath and injection grout. PE or polyvinyl chloride (PVC) plastic sheath has studs embossed on its rear side, which act both as spacers vis-à-vis the wall of the old pipe and as rear anchoring with the injection grout. The annular gap between the plastic strip and the old pipe is injection-filled with grout, similarly to continuous-pipe relining. Only the cured outer ring of grout generates load-bearing capacity, however. The plastic sheath serves as a sealing element and protects the grout against aggressive fluids in the pipe. We can differentiate two cases in the use of grout. As seen on the left of Figure 9.19, the grout penetrates the cracks (and/or cavities in the bedding) and thereby also directly stabilizes the old pipe and the bedding. There is, however, the risk here that the grout could ingress without control into other cavities, and thus spread out across a large underground area. In the second variant, shown on the right of Figure 9.19, a 'pre-liner' is first injected, which separates the old pipe from the new pipe, thus preventing material from migrating into cracks and/or external cavities. This does not improve the old pipe/soil system itself. It is necessary to decide on a case-by-case basis which functions are more important, and which risks can be tolerated, when selecting this method.

Special importance should be given to the grout quality as part of quality assurance, both for local strength and the homogeneity of the entire mixture. Ultimately, the grout must be pressurized along, in some cases great lengths through an extremely narrow annular gap, distributing uniformly between the studs. All in all, the studded-sheath method appears to be the only procedure in which both the bedding and the old pipe can be improved in a single operation while the pipe train is also sealed simultaneously with an internal plastic sheath.

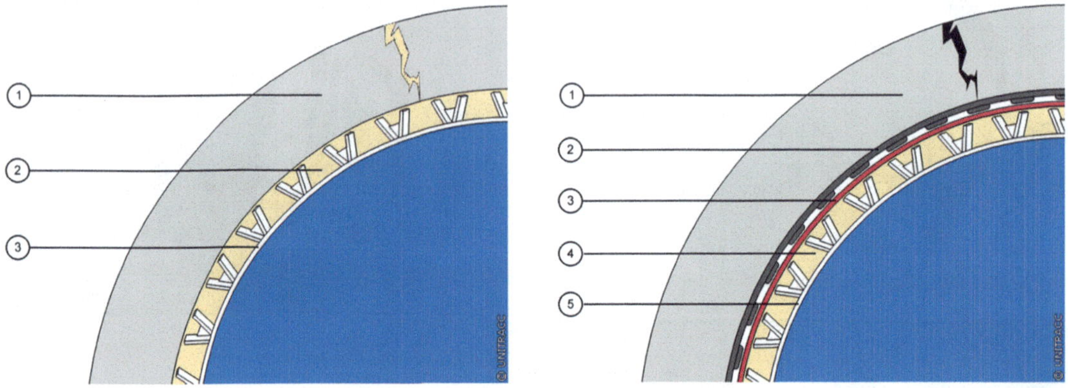

Figure 9.19 Renovation with on-site-produced pipes with two variants of grouted-in-place lining without (on left, (1) existing pipe, (2) injection grout, (3) liner) and with 'pre-liner' (on right, (1) existing pipe, (2, 3) pre-liners, (4) injection grout, (5) liner). *Source*: UNITRACC.com, Prof. Dr.-Ing. Stein & Partner GmbH.

9.3.3 Replacement

According to EN 16323 (2014), replacement is considered a sub-sector of rehabilitation, along with renovation and repair, and is defined as 'construction of a new drain or sewer, on or off the line of an existing drain or sewer, the function of the new drain or sewer incorporating that of the old'.

The function of the original sewers and/or pipelines is thus included and retained, and the old line is correspondingly replaced. The new line, however, is intended not only to perform the original function, but also to be capable of doing more. Replacement also occurs if the new line is constructed at a totally new location and nonetheless performs the function of the old line, because the network as a whole is, in this case, not doing something entirely new, but instead has been structurally rehabilitated and its function, possibly, expanded (e.g. through enlargement of the original cross section). Replacement means the installation of a new line in place of an old line and, as in the case of new construction, two variants are possible in principle in civil-engineering terms: open-cut and trenchless methods.

An open-cut method, in the case of replacement, differs in technical terms only a little from new construction, but there may nonetheless be special challenges when working in an existing network. There may be other intersecting conduits underground, and coordination with other utility providers will then be necessary, in order, among other things, not to have to dig up the road more often than necessary (Daulat *et al.*, 2022).

A trenchless procedure allows for the replacement of lines without the need for digging trenches. It also has advantages regarding indirect costs and carbon emissions (Fuselli *et al.*, 2022). Various methods heavily orientated around underground new-construction methods can be distinguished. These include 'pipe eating', in which the old pipe is excavated at the face, similarly to the procedure in pipe jacking. A new pipe will subsequently be in the same place as the old pipe, but the old pipe would have been eliminated. In a 'pipe extraction' method, the old pipe is also removed. In this case, it is pushed out in front of a pusher head, while a new pipe is drawn in simultaneously. The so-called 'wash-over' method is used for smaller pipelines and cables. Here, the old line is used as a guiding aid for the installation of a new line; the old line is then pulled out after successful installation of the new line. The so-called 'pipe-bursting' method occupies special ranking in trenchless replacement. Here, the old pipe is not removed, but it is instead disintegrated by a bursting head and displaced into the soil. The working principle of this method is explained below in detail.

9.3.3.1 Pipe bursting

Figure 9.20 shows a schematic representation of the bursting procedure, where a 'displacement body' is pulled through the old pipeline. The pulling machine can be located either in the exit shaft, or on the surface, using a deflecting roller in the shaft. The function of the displacement body, or 'bursting head', is that of breaking or cutting open the old line.

As part of a pipe-bursting procedure, the fragments of the old pipe are thrust outwards, and the entire cavity widened sufficiently for the new pipe to be drawn in immediately behind the bursting head. Welded pipes, and trains of short pipes, can be installed. Here, three questions are important:

- How high is the pressure of the displaced soil on the fragments? The displacement process generates a high earth pressure.
- How do the fragments act on the pipe during the pipe-pulling procedure (installation conditions)? Earth pressure thrusts the fragments against the new pipe, which can be subjected to significant abrasion, including scoring ('scratching'), while being forced past the fragments.
- What loads are exerted by fragment pressure on the pipes during operation? Superimposition of tensile stresses from punctiform fragment pressure and design internal pressure, are of importance for pressure lines.

The fragment scenarios in the pipe-bursting method which must, in principle, be anticipated can be determined on actual construction sites or by extensive investigations of pipe/soil/fragment interaction. A typical situation for a concrete pipe disintegrated using the pipe-bursting method is shown in Figure 9.21. The pipe was disintegrated into many individual fragments, which are now located around the (new) pipe. The situation is not always so orderly, and in some cases, fragments are also 'entrained' (carried along) during pipe-bursting, resulting in small accumulations of fragments, and 'fragment-free' areas are formed. Points of contact between the fragments and the pipe always occur, however, which are of particular importance for the loads acting on the new pipe. Scarring, which must not exceed a certain critical extent, occurs on the new pipes at the points of contact during pipe pulling. For this reason, a protective coating is applied to the new pipes, which can be seen at the top of the figure for a drinking-water pipe example. Scarring is also clearly visible here, but it has damaged only the protective coating, and not the actual pipe.

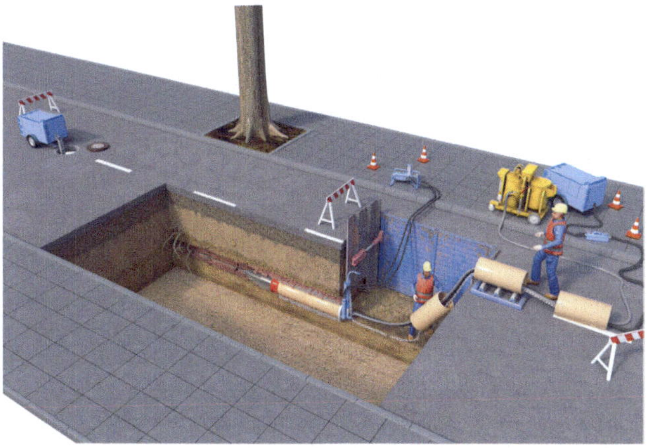

Figure 9.20 Replacement by applying the pipe bursting method. *Source*: UNITRACC.com, Prof. Dr.-Ing. Stein & Partner GmbH.

Figure 9.21 On-site observation of pipe bursting. *Source*: IKT – Institut für Unterirdische Infrastruktur gGmbH.

It is recommendable, for quality control, to collect a sample of the new pipe in the exit shaft, because this sample will already have passed through the entire pulling length and will thus have undergone the maximum exposure to abrasion. If no damage to the new pipe is apparent on the sample, it is highly probable that the rest of the pipe train is also undamaged. Furthermore, when using the pipe-bursting process, it is necessary to ask which stresses the new pipe experiences from the point loads acting at the points of fragment contact, and how one can design and test pipes for these loads. This has been comprehensively investigated with the mathematical and physical phenomena occurring having been described in model concepts (e.g. Liebscher *et al.*, 2003). The aim here is to determine the pressure exerted by the soil on the fragment and ultimately via the fragment on the pipe.

In later operation, the stress pattern in the pipe itself is then decisive. If we apply a point load from the outside, this leads to tensile stresses on the inside of the pipe. For sewer pipes operated without pressure as gravity lines, these stresses do not usually pose a particular problem, even in the long term. However, this is more critical in the case of planned internal pressure loads, as is common in force mains, for example. In this case, the tensile stresses from internal pressure overlap with the tensile stresses from point loads, so that excessive stresses can occur in total. Pressure pipes should therefore be checked regarding their resistance to point loads.

9.3.4 Conclusion

The repair of sewer pipes is an essential element for the O&M of urban drainage networks. Trenchless repair techniques, which minimize nuisance for adjacent property owners or roadway users, can be used for this purpose. Local damage scenarios such as leaking joints, breakouts and root ingrowth can be repaired in this way. Typical procedures include injection, flood grouting, grout injection, trowel smoothing and the use of sleeves and partial liners.

Renovation is of particular importance for the urban drainage sector and is defined by two aspects, the 'improvement of current performance' and the 'incorporation of the original fabric'. Major differences are apparent between the various methods, depending on whether the material and the shape are produced at the factory or on-site. To varying degrees, depending on the method and its qualities, the improvement of the current performance along the entire sewer length, that is, improvement of operational reliability, structural integrity and tightness, is achieved. The fabric of the old pipe is also predominantly incorporated. A basic classification can be found in the EN 15885 (2018). Differences are apparent in terms of the place of production and thus also of quality assurance. Here, we have differentiated between three groups: the first group includes discrete-pipe relining and the continuous-pipe procedure. In both cases, pipes are produced entirely at the factory. This applies to both the material and the ultimate pipe geometry. Both can thus be checked and dimension-inspected at the factory. Neither, however, can be modified on-site. The second group comprises the close-fit, spiral-wound pipe and segmental-lining methods. Here, the pipe material is produced at the factory. Its composition and distribution across pipe wall are decided entirely there. The actual pipe geometry is formed only on-site, from the pipe train, the plastic pipe-winding strip or the wall segments delivered. The material quality is therefore determined primarily at the factory, the only exception being jointing/pointing and internal changes resulting from high energy of deformation, as in the case, for example, of the close-fit liner, with deformation taking place under exposure to internal pressure and temperature. The third group is made up of cured-in-place pipe lining, the spray, trowelling and cast-in-place methods and the studded-sheath method. Here, individual material components are first delivered, either together or separately, to the site. There, they are processed and cured to make a pipe of the required geometry in the sewer itself. Essential parts of the production process thus take place inside the old sewer. This presents a particular challenge for on-site quality assurance since it concerns both the production process itself and the actual rehabilitation result. Methods functioning on-site offer significant advantages, provided quality assurance is correct: the space requirements for the importation of the materials into the sewer are generally lower than when it is necessary to pull in a complete pipe. In addition, the new pipe can be adjusted in situ to conform to the geometry of the old one.

The term ‘replacement’ in principle signifies the replacement of an old line by means of installation of a new line with the same or an expanded function. The engineering methods may be of the open-cut or of the trenchless type. Using trenchless methods, an old pipe can be removed, or disintegrated and left in the soil, as in the case of pipe-bursting procedure. Pipe bursting also makes it possible to install new pipes of larger diameters directly into the route of the old pipe using trenchless methods. The loads exerted by the soil on the fragments of old pipe and, ultimately, on the new pipe, must be assessed in advance, and the new pipes correspondingly tested for resistance to these loads and stresses. When dealing with stormwater pipes and depending on the context, it is worth considering replacement of pipes with sustainable urban drainage solution to infiltrate stormwater at the source (and thus ‘disconnect’ part of the original catchment area from the drainage network).

Current research gaps that have been identified include the question about the uptake and decision process for certain rehabilitation methods, especially with a focus on trenchless technologies (e.g. Rokstad *et al.*, 2022; Skaar *et al.*, 2023), and the life expectancies of the different methods in comparison to a full replacement. Another topic is the link of the rehabilitation method on system behaviour, be it hydraulically (Tscheikner-Gratl *et al.*, 2014) or regarding infiltration and inflow (I/I) and the question how tightening a sewer system may have impact on groundwater conditions. In the end, I/I may simply move to the point of least resistance. The introduced methods are described here for the main, public sewers. A lot of I/I can however be associated with the (private) service connections. Many approaches can be used with slight adaptations to those lines as well, especially the ones suitable for small diameters. In terms of research, little (e.g. Post *et al.*, 2016) is conducted on service connections.

9.4 MANHOLE REHABILITATION

Another important component of a sewer system are sewer manholes. Manholes, at intervals depending on the network type and local guidance and experience, ensure accessibility to the underground sewer system, for example, for cleaning and inspection. In addition, manholes serve to ventilate the entire system and support wastewater or stormwater conveyance, especially at changes in direction or gradient. Manholes are engineering structures that consist of various components (see Figure 9.22). The shape of manholes can be circular or square. For specifically smaller manholes the body and chamber can be integrated without a taper. The possible objects for rehabilitation include (1) the manhole top/cover, (2) the manhole body and chamber, (3) connections of the main line and side inlets and (4) step irons/ladder.

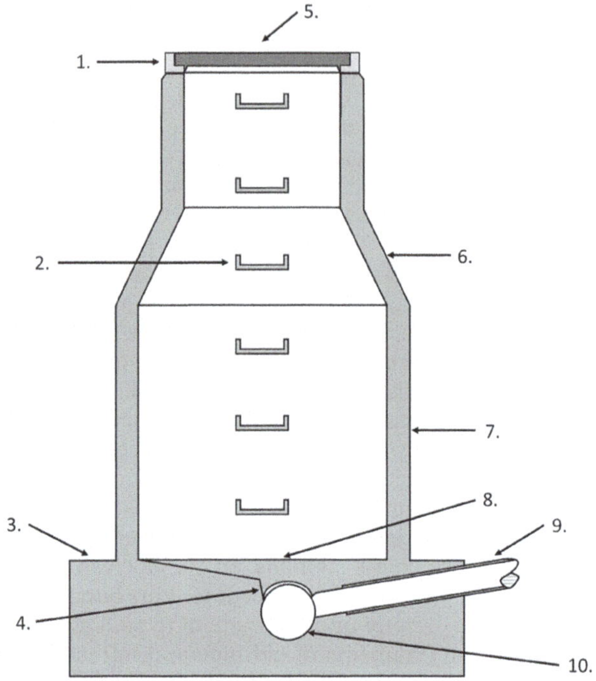

Figure 9.22 Parts of a manhole: (1) ring, (2) step iron, (3) bench/table, (4) channel/chase, (5) cover, (6) cone/straight back taper, (7) shaft/chamber, (8) outlet, (9) inlet, (10) invert. *Source*: IKT – Institut für Unterirdische Infrastruktur gGmbH.

In case of rehabilitation, the quality requirements for manhole rehabilitation are just as high as for the rehabilitation of the pipes. The process of manhole rehabilitation may be divided into the following steps (see an example also in Section 9.4.1):

- Inventory by inspection of the manholes and assessment of the need for rehabilitation.
- Temporary waterproofing, especially in case of external water pressure (e.g. due to rising groundwater).
- Establishment of rehabilitation capability, for example, substrate preparation.
- Rehabilitation execution and monitoring.
- Construction acceptance, if necessary, which also may act as warranty acceptance.

Older manhole structures to be rehabilitated are usually made of concrete or masonry. Typical damages include, for example, cracks, corrosion, leaking manhole ring connections (concrete) as well as damage around the manhole cover.

9.4.1 Example of manhole rehabilitation due to extraneous water intrusion

An important potential rehabilitation task is the sealing of a manhole body under high-groundwater levels because the mitigation of extraneous water intrusion can only be successful with consistent sealing of a manhole. The associated procedural steps include:

- *Preparatory measures/pre-sealing*:
  The aim of pre-sealing is to temporarily seal the manhole wall to stop the groundwater inflow and thus enable lining or coating of the manhole wall in the first place. Pre-sealing can be done

manually or mechanically. The materials used are tamping and surface mortars, crystallizers, cement pastes and suspensions as well as polyurethane resins, epoxy resins and acrylate gels. In the area of pre-sealing, however, a weakening of the substrate (lower adhesive tensile strength) as well as long-term risks for leaks remain.

- *Sealing the connection of the main pipe and side inlets*:
  Similar to pre-sealing, injections are often used here as well, but with higher demands on durability. In addition, inner sleeves, and short liners can be used to bridge leaks. The flexibility of the connection between the manhole and connecting pipes after rehabilitation must be considered.
- *Substrate preparation*:
  A substrate suitable for rehabilitation must be clean, load bearing, solid and undamaged and have a large specific surface area for a good adhesive bond. In addition, the moisture content of the substrate must be adapted to the rehabilitation materials to be used, for example, rather dry for plastic coatings and rather moist for mortar coatings. To achieve this goal, damaged material should be removed, and the substrate roughened and cleaned. The methods used are brushing, grinding, milling, chiselling/needling and the use of high-pressure jets, if necessary, with the use of granulates, with a jet pressure of up to 1000 bar (see Figure 9.23). In principle, one should aim at fully automated use of equipment for substrate preparation to avoid prolonged entry of personnel into manholes during the jetting process.
- *Rehabilitation*:
  Various lining and coating methods can be used for the rehabilitation of the shaft or body of a manhole. These methods are intended to seal the structure and contribute to substance preservation and stability. The following lining methods may be used:
  - Internal shaft systems where a new shaft or body is set into the damaged manhole.
  - Prefabricated segments made of GRP, high-density polyethylene or clay, which are mounted to the manhole wall from the inside.
  - Locally manufactured linings where a GRP laminate is applied or a CIPP liner is placed in a manhole and cured on-site.

Figure 9.23 Corroded surface before (top) and after (bottom) sufficient substrate preparation with high-pressure water jetting (900 bar). *Source*: IKT – Institut für Unterirdische Infrastruktur gGmbH.

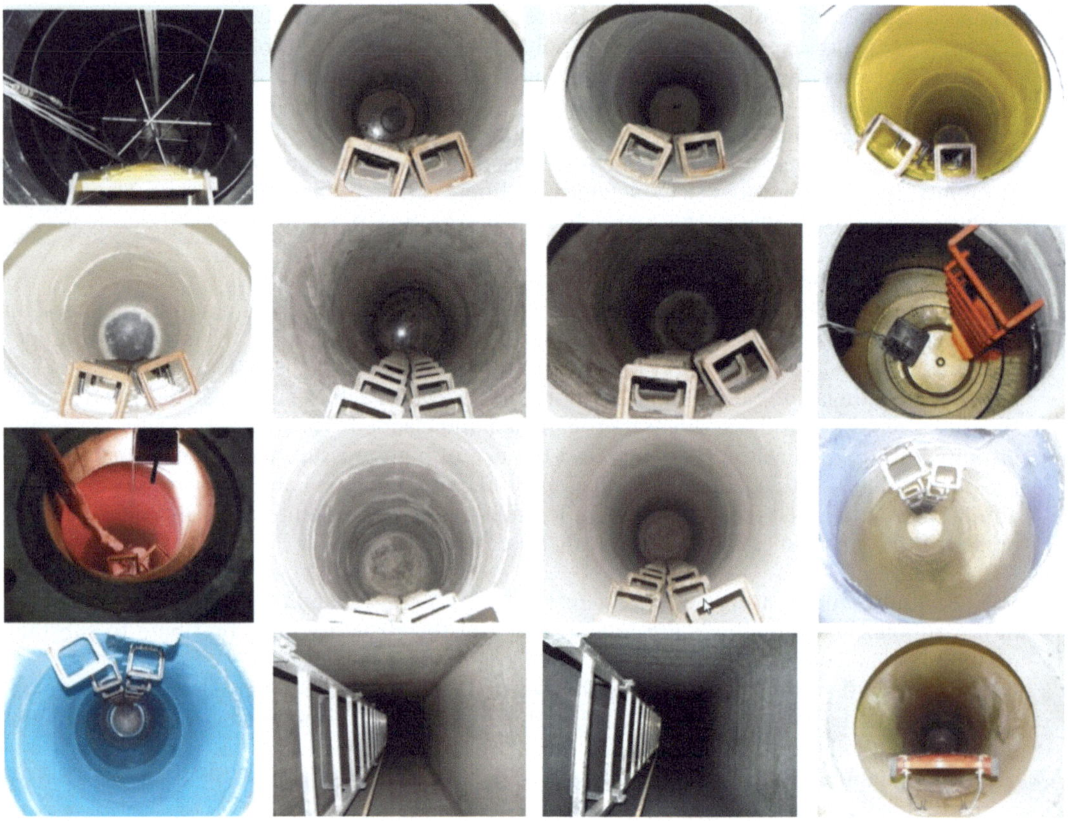

Figure 9.24 Manhole rehabilitation with linings and plastic or mortar coating. *Source*: IKT – Institut für Unterirdische Infrastruktur gGmbH.

Manhole coatings can be applied as mortar or plastic coatings. The preparation of the substrate is of particular importance here. The coating process itself is carried out manually by means of a trowel/spatula or painting, spinning or spraying. Figure 9.24 shows results of rehabilitations with lining and coating methods.

9.5 SUDS OPERATION, MAINTENANCE AND REHABILITATION

As discussed in Chapter 1, SUDS are complex systems. They are often composed of both engineered and natural elements such as pipes, flow control systems, vegetation, micro-organisms in soil or growing media and more. They also deliver a broad range of beneficial services to our communities and their inhabitants. Individual SUDS are unique in terms of composition, context and expectations. O&M and rehabilitation activities are required to maintain the desired services and need to be compatible with the existing system (except when the asset is totally replaced). Rehabilitation actions can target engineered elements, natural elements or both.

Table 9.3 presents a classification of the main and sub-components of stormwater systems including SUDS. The main objective of this table is to illustrate the diversity of SUDS. The classification is provided as an example and not as a unique classification system. Those components that are

Table 9.3 Classification of stormwater system main and sub-components with the one discussed in this chapter highlighted.

Main Stormwater System Components	Sub-components
Overland conveyance	Backyard swales and gutters Streets and pathways Ditches Canals Creeks
Stormwater systems	**Pipe systems (see Section 9.3)** Culverts
Connection between overland conveyance and stormwater systems	Catch basins/gully pots Ditch inlets
Storage systems	Sags in roadways and parking lots Rooftop storage Underground tanks (incl. wells)
Flow controls and appurtenances	Inlet control devices Gates Pumps Weirs and orifices Backflow preventers
Ponds and wetlands (see Section 9.5.1)	**Dry ponds** **Wet ponds** **Constructed stormwater wetlands** **Natural wetlands**
Source control measures	Oil-grit separators Filter systems (incl. gross pollutants) Resilient landscaping including rain gardens **Permeable pavements (see Section 9.5.2)** **Green roofs (see Section 9.5.3)** **Bioretention systems and bioswales (see Section 9.5.4)** Soil cells Infiltration trenches Infiltration basins Infiltration wells **Rainwater harvesting (see Section 9.5.5)** Stormwater capture

discussed in detail in this section are highlighted in bold. A link to an ongoing inventory of guidelines is provided below to complement the present information.

Although SUDS are very different from pipes (as discussed in Chapter 1), a coherent and similar approach for both types of assets is beneficial. Rehabilitation needs can be classified, based on the severeness of the failure and the related necessary action, into:

- Repair or 'low-effort' actions: any rectification of localized damage regarding both the civil-engineering components and the natural elements of the asset. Repairs include mulch replacement, repairs of inlet or outlet pipes or pavement, re-application of waterproofing compounds to retention systems, changing of valves and so on. Some of these repairs can be performed on a regular basis or be triggered following site inspections.
- Renovation or 'moderate-effort' actions: any action on a wider scale or involving the replacement of asset parts, such as replanting, removing clogged layers of soil or growing media, unclogging porous surfaces and so on. The line between repair and renovation actions can be unclear

depending on the asset: renovation is considered when the rehabilitation concerns most of the surface of the asset or when part of the asset must be rehabilitated.

- Replacement or 'high-effort' actions: as for grey infrastructure, actions to remove existing SUDS and replace them with the same types of assets or different ones.

The range of possible actions is wide; the reader can visit the website http://tiny.cc/guidelinesSCMs to discover AM or O&M guidelines from different countries. Many guidelines propose repair, renovation or replacement actions for different SUDS and their components, often reflecting different, local objectives or expected services. For example, the 'WSUD Audit guideline' from Stormwater Victoria and Melbourne Water in Australia (Browne *et al.*, 2017) provides a scoring system organized in a spreadsheet file with a summary report listing all assets with identified recommended maintenance or rectification actions arising from audits or of actions undertaken. The 'Stormwater BMP Guidance Tool' for Louisiana in the USA (James *et al.*, 2010) distinguishes between routine maintenance and major maintenance for a large variety of assets. Finally, the 'Low Impact Development Stormwater Management Planning and Design Guide' of the Toronto and Region Conservation Authority in Canada (Dhalla & Zimmer, 2010) and the 'Minnesota Stormwater Manual' (Minnesota, USA) (https://stormwater.pca.state.mn.us/index.php?title=Operation_and_maintenance_of_green_stormwater_infrastructure_best_management_practices) are excellent wikis that include rehabilitation or repair tasks.

Next, several common SUDS are presented, and for each asset, rehabilitation actions are described. Rehabilitation can be very different when considering engineered or natural elements: each component of SUDS will require its own adapted rehabilitation needs. SUDS AM is not only about maintaining the engineered parts of the asset. Rehabilitation of civil-engineering elements is often 'immediate' in the sense that the function is restored at the end of the construction activities. Rehabilitation of natural elements can take more time and can be dependent on external conditions (e.g. drought). Vegetation cannot simply be replaced but needs to mature (i.e. grow) to achieve its full potential. In the case of SUDS, the main difficulty is that the ultimate desired or expected performance is based on the combination of both engineered and natural elements and often cannot be based only on one type of element.

O&M of SUDS are also included in the following sub-sections. These actions are very similar to the 'low-effort' actions described earlier; however, unlike for repairs, they are often performed periodically and not triggered based on the condition or state of the asset as identified during inspections. Such actions are, for example, litter removal, cleaning of filter media, watering plants (mostly during establishment phase), replanting, weed removal, mulch replacement and so on. The Urban Stormwater Controls Operation and Maintenance Manual of Practice (WEF, ASCE/EWRI, 2022) provides an extensive overview of the O&M requirements of an array of stormwater control measures (SCMs) as do the wiki references provided above.

SUDS are fundamentally different from the traditional grey assets discussed in the previous sections in that they provide additional functionality including 'treatment' of the runoff entering them. In some cases, however, SUDS may leach certain constituents, specifically nutrients, becoming a 'source' of contaminants rather than a 'sink'. This may occur seasonally, for example, in the autumn when vegetation dies off or on ongoing basis. In either case, the origin and nature of the leaching should be examined, and renovation or replacement may be needed in case the leaching would contribute to unintended deterioration of receiving water bodies. In some cases, it may be as simple as maintenance staff applying inappropriate amounts of fertilizers to SUDS. However, in the case of ongoing leaching, this may be a function of how SUDS were originally constructed, for instance:

- The substrate and associated root mass used to construct stormwater wetlands may have been sourced from other, existing natural wetlands in the vicinity of the to be constructed wetland. If the source wetlands had been subjected to agricultural runoff for extended periods of time, the source substrate may already be oversaturated with nutrients.

- While the base materials to create growing media for green roof systems are typically inert, compost is often introduced to support the desired vegetation. The presence and nature of the compost may lead to significant leaching, especially during the early years after initial installation.
- The base materials used for the growing media in, for example, bioretention systems, bioswales and soil cells may be naturally rich in nutrients.

Recent guidance documents (e.g. WEF, ASCE/EWRI, 2022) provide limits with respect to the chemical composition of the materials to be used; however, quality control may have been lacking or absent in the case of older facilities.

Research is ongoing with respect to either amendments that can be mixed in to immobilize nutrients, often reflecting locally readily available low-cost materials, or that treat the effluent or stored runoff in an impoundment before it is being discharged.

As discussed in Chapter 1, SUDS may be installed within either the public realm or on private land. In principle, there are no differences as to where SUDS are located as to needed O&M as well as rehabilitation needs; however, the likelihood of SUDS being properly operated and maintained should be accounted for when selecting and locating SUDS. For instance, residents or tenants on commercial and industrial properties may have little to no clue with even resilient landscaping practices such as rain gardens let alone more sophisticated SUDS, unless a dedicated property manager is present who may call in professional support when appropriate. Municipalities would then be able to focus their educational efforts on these property managers. In theory, property owners should have been provided with O&M manuals. However, these documents do tend to get lost or ignored, especially with the inevitable turnover of staff or after the sale of a property.

Proper O&M of private SUDS are of significance to public entities in that they need to make up for lacking performance. In fact, public entities may even be held liable by regulators in case of spills or loading exceedances to receiving water bodies. As such, an appropriate combination of education and enforcement is needed. While many municipalities may have bylaws that stipulate that property owners should keep their private drainage infrastructure in good operating condition, a lack of monitoring and enforcement tends to jeopardize this. Some municipalities have tied their drainage utility fees to the presence and maintenance of private drainage infrastructure with savings being incurred as a function of the loading reductions (i.e. hydraulic and contamination) to the public system. Regular inspections, as discussed in Chapter 4, by the property owner, third-party professionals or the local jurisdiction are paramount to ensure proper long-term O&M.

9.5.1 Ponds and wetlands

Ponds and wetlands have been around forever, often by default (e.g. from past peat or gravel extraction activities) or created by nature itself. However, their deliberate role in urban drainage has exploded since the mid-20th century. They can be found in many communities built in the western world after the Second World War, but also as retrofit projects in older communities. There are many permutations ranging from dry ponds to wet ponds to constructed stormwater wetlands to natural wetlands that receive stormwater to sustain them. Wet ponds (see Figure 9.24) may be subject to water withdrawal to provide irrigation water, or in some cases even potable water (i.e. after adequate treatment, see also Section 6.4.2). There may be hybrid wet and dry facilities, or facilities that have both surface and sub-surface storage components. Controls may be basic, ranging from a simple pipe or single orifice in a control structure to multiple openings, or active systems incorporating real-time control (Figure 9.25).

Expectations pertaining to the operation and performance of the ponds and wetlands in our communities may have changed since they were first constructed: older facilities were largely implemented from the perspective of providing rate control to minimize downstream flooding. Over time, many other considerations were added, for example, the capture of sediments, nutrients and heavy metals or to address thermal contamination. This trend continues with the increased interest in

Figure 9.25 Water quality research with aquatic drones at constructed wetland in Hoogeveen, the Netherlands. *Source*: https://www.climatescan.org/projects/91/detail.

emerging contaminants and microplastics. In recent years, dam safety considerations (as e.g. laid out by Lacasse, 2022) have started to become considered in an urban context as well.

Expectations by the public have increased as well reflecting the role of these facilities in the urban landscape, for example, necessitating the need to consider aspects such as aesthetics, odours and recreational desires, with the latter moving from passive to active recreation. Some aspects have evolved from how the public interacts with ponds, for instance, the introduction of invasive fish species such as carp by residents or windblown debris and garbage from the surrounding communities. The increased attention to biodiversity, mental health and desire for pollinator habitat has evolved the way ponds and wetlands are viewed. All of this is also upping the ante from a public safety perspective from people accessing the ponds and wetlands, not only during storm events but also during the dry spells in between.

All these considerations and expectations are being influenced by climate change as well, meaning that the demands will even increase, for instance, the desire to add capacity to detention facilities, or store runoff to carry us over droughts or sustain natural water courses.

Given the diversity and different sizes and shapes of ponds and wetlands it is virtually impossible to present a uniform approach to their rehabilitation. However, the one overarching observation is that the expectations should be evaluated and clearly defined upfront, not only as to how these assets were originally intended and designed, but also how these expectations have evolved. There should be recognition that ponds and wetlands contain many different components ranging from the depression itself, to embankments that create the storage impoundments, internal berms, the piping in and out, overland drainage in and out, liners, grates, catch basins, sub-drain systems, control structures, pre-treatment devices, post-treatment devices, electronic controls, aeration systems, algae control systems, pathways and vegetation. All of these will have very different operational needs, and different life spans.

9.5.1.1 Performance over time

Naturally, lakes and wetlands are not static but evolve over the ages as a function of ongoing deposition in them and changes in their upstream catchments. In an urban environment this may be accelerated. However, the functionality of ponds and wetlands within an urban environment should remain more or less unchanged unless wholesale changes are made to the system (e.g. replacing a treatment wetland by a commercially available filter system).

As such, the main aspects that impact the performance of ponds and wetlands over time are:

- The build-up of sediments which may become remobilized when the deposition becomes excessive, and which may reduce the amount of runoff available for use (e.g. irrigation) or water course or wetland sustainment during drought periods.
- The build-up of nutrients, especially in wetland systems that are terminal, which may lead to odour and aesthetic complaints as well as hazards to the public and staff in the presence of blue green algae.
- Lack of water movement due to dead zones in combination with 'unhealthy' habitat conditions which may lead to mosquito breeding problems.
- The maturation and evolution of the vegetation.
- In addition to the inevitable degradation of the structural components.

While unpredictable from a timing perspective, the potential failure of embankments and slopes due to earthquakes should be considered as well.

9.5.1.2 Operation, maintenance and rehabilitation

O&M activities pertaining to ponds and wetlands can be wide ranging and may include (but are not limited to) the following:

- Invasive vegetation and weed control.
- Mowing and pruning of vegetation.
- Annual or seasonal water drawdown, replenishment or flushing.
- Garbage and litter removal.
- Invasive fish species, beaver, muskrat and rodent control.
- Removal of ice hockey nets and Christmas lights.
- Irrigation.
- Maintenance of fountains or oxygenation devices.
- Mosquito control.
- Cleaning of pre-treatment devices and grates.
- Snow and ice clearing from trails and pathways.
- Beaver wiring around trees.
- Water quality sampling.

Key considerations concerning ponds and wetlands include:

- Timing of the activities in relation to the duration of the rehabilitation efforts. In most cases, the facility must remain active which means that efforts may have to take place during the dry season as care of water is crucial. If this is not possible, the area being subject to the rehabilitation activities may need to be isolated to create a safe working environment. The season when the activities are to take place is an important consideration as well from the perspectives of the potential negative impact of frost on structural components (e.g. placement of concrete, asphalt or backfill) and the ease with which vegetation can be restored.
- Temporary protection of replanting until the root structure is fully re-established.
- Staff and public safety issues which can range from getting hit by equipment adjacent to and within the footprint of the pond or wetland, to confined space entry considerations, to contact (e.g. blue green algae), to simply getting stuck in the muck.
- Access for mechanized equipment which may have been overlooked when the facility was originally designed and constructed or may be sub-optimal given current equipment needs.
- The availability of a power supply.
- Preservation of the biophysical aspects, for example, the presence of invertebrates, amphibians and bird nesting.

- The destination of the dewatering water and the need for ongoing dewatering including the intrusion of groundwater, baseflow or runoff during storm events.
- The destination and fate of any sediment removed from the ponds or wetlands.
- The presence of other deep and shallow utilities within the footprint of the ponds and wetlands.

The most difficult rehabilitation efforts are usually associated with the renovation or replacement of

- liners given their spatial coverage,
- control structures given that all flow needs to go through them and
- embankments or retaining walls, given that they hold back part, or all the water stored or to be stored.

In view of the diversity of the components within ponds and wetlands, a multi-disciplinary approach to the rehabilitation efforts is essential. Given the visibility of ponds and wetlands in our urban environment, interaction with and education of the public should be a consideration of any rehabilitation activities.

Repair activities associated with ponds and wetlands are usually relatively simple and include (but are not limited to):

- valves,
- gates,
- electronics,
- grates,
- culverts and weirs,
- railings and fences,
- trails, boardwalks and viewing platforms,
- signage and
- replanting or reseeding of vegetation.

However, repair activities associated with the structural components such as control structures, piping and sub-drains or shoreline and bank erosion may become more elaborate and require more effort.

Renovation activities associated with ponds and wetlands may become more complex and include (but are not limited to):

- Renovation of embankments, side slopes and internal berms, all of which may have sloughed.
- Repair of retaining walls.
- Repairing damage after larger operational cleaning activities, for example, those associated with removal of sediments.
- Clogging of substrates or growing media, especially in the case of sub-surface flow wetlands.
- Transition to naturalization of edge treatment and areas above the high-water line.
- Damaged maintenance access routes and pathways.

Replacement activities associated with ponds and wetlands are often associated with additional demands and expectations, and, as such, will be complex. They include (but are not limited to) the need:

- For additional storage capacity to accommodate increased runoff volumes due to densification and/or climate change.
- To address additional contaminants previously not considered, which may require different (treatment) zones or post-treatment devices.
- To address dam safety considerations which may require replacement of embankments and the provision of proper spillways.
- To address odour and algae complaints, for example, through the provision of recirculation systems, filter systems to remove nutrients or pre-treatment devices.

- To provide space to allow for easy dewatering of removed sediments prior to beneficial use or transport elsewhere.
- To reduce easy access to the wet and mucky areas where people might otherwise get stuck or become easy target for attacks by, for example, alligators or snakes.
- To provide enhanced recreational activities, for example, through the provision of gates and a bypass so that chloride laden runoff would not enter a pond during the winter season.
- To allow the withdrawal of water for, for example, irrigation or other purposes.
- To provide enhanced controls to address hydromodification concerns.

9.5.1.3 Example: 17 year stormwater management pond sediment chemistry and beneficial reuse field study

Engineered stormwater management (SWM) ponds provide flood protection and water quality enhancement for urban environments around the world. Sediment dredging is routinely required to maintain stormwater storage capacity and treatment efficiencies. The results of this 17-year Canadian field study support the beneficial reuse of sediments as a valuable resource rather than costly disposal as a waste. The objectives in this example are to:

- Characterize typical sediment contamination levels within SWM pond inlet, centre and outlet sample zones.
- Determine how many of the 121 SWM ponds studied would meet sediment beneficial reuse regulatory requirements.

Detailed descriptions of the following summarized methods are provided in Kelly-Hooper (2013) as well as in TRCA, CH2M Canada (2016). The 2005–2022 Canadian SWM pond sediment survey included 371 samples collected from 121 SWM ponds. The ponds ranged from approximately 15 to 30 years of age and were located in fully developed residential/commercial catchment areas with no histories of point source release events. A sediment core sampler was used to collect full-depth samples from each basin when conditions allowed. An Ekman dredge was used as an alternative method for shallow and/or sandy sediments. SWM ponds with only one inlet were sampled from three zones, which were identified as the inlet zone, centre zone and outlet zone. One additional sample zone was added per additional inlet. One homogenized composite sample was collected from several discrete samples per zone. Each composite sample was immediately placed into soil jars and submitted to an environmental laboratory for the following bulk soil analysis: trace metals, total petroleum hydrocarbons (TPH), polycyclic aromatic hydrocarbons (PAH), electrical conductivity (EC), sodium adsorption ratio, pH and particle size. Contaminant concentrations were evaluated for exceedances of the Ontario Site Condition Standards for soils (MECP, 2021) and Excess Soil Quality Standards (MECP, 2019). TPH sources were identified by comparing the carbon range distributions (MECP, 2021), gas chromatogram-flame ionization detector chromatograms and PAH ring size distributions to reference source samples (AEP, 2018; Kelly-Hooper, 2013).

Statistically significant differences between sample zones are illustrated in Figure 9.26. Using the non-parametric Dunn's test of multiple comparisons using rank sums it was determined that there were no significant differences between the centre and outlet sample zones. Seventeen analytes were significantly different between the inlet and outlet sample zones, with eight analytes significantly different between the inlet and centre sample zones.

Detailed chemistry data with respect to Regulatory Limit Exceedances is provided by Kelly-Hooper *et al.* (2022). In total, 99 and 88% of the sediments in SWM ponds could not be reused as clean fill due to elevated TPH from asphalt pavement sources and EC from road salt de-icing products, respectively (see Table 9.4). Trace metal concentrations were generally low in most ponds.

In conclusion:

- The statistical analysis results showed a linear shift in contamination levels from the inlet to the outlet. This trend supports the rationale for collecting three composite samples from ponds that have only one inlet, with one additional sample zone per additional inlet.

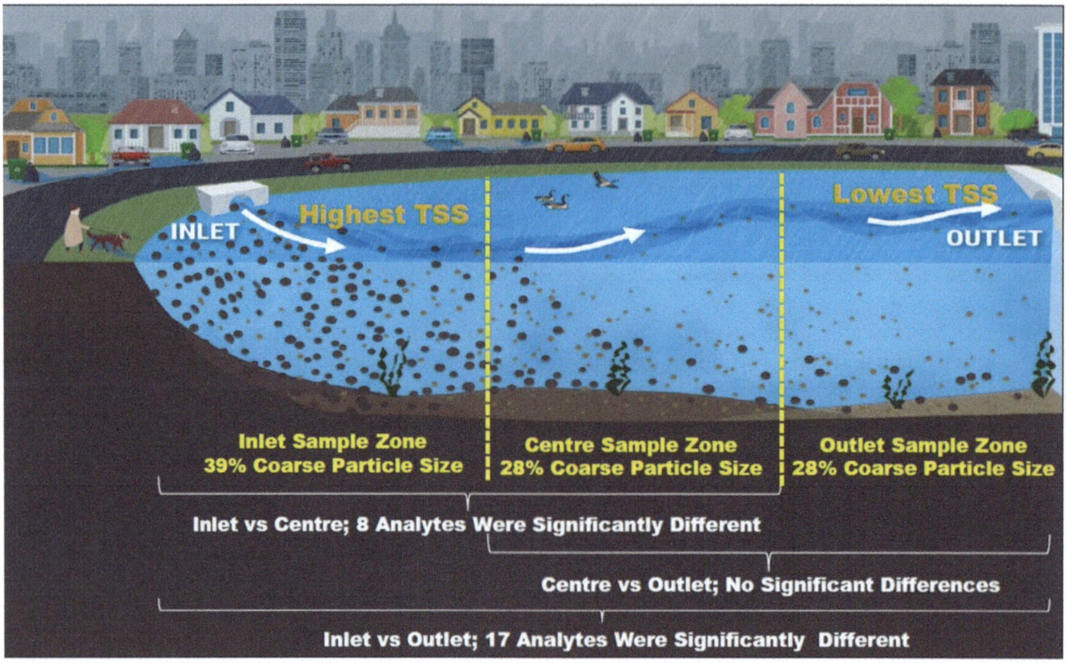

Figure 9.26 SWM pond water quality improvement by gravitational settling of suspended particles; significantly different analyte concentrations between the inlet, centre and outlet sample zones. *Source*: Kelly-Hooper *et al.* (2022).

Table 9.4 Percentages of 121 SWM ponds with sediment samples that exceeded regulatory limits for EC and TPH.

Soil Standards	Land Use	Salt Analyte	Petroleum Hydrocarbons (PHCs)		
		EC	F3 (C16–C34)	F4 (C34–C50)	F4G (Gravimetric)
Table 1 SCS[1]	R/P/I/I/C/C[2]	**88%[5]**	**87%**	**88%**	**99%**
Table 2.1 ESQS[4]	R/P/I	**88%**	**87%**	2%	36%
	I/C/C	**81%**	**87%**	1%	32%
Table 3.1 ESQS[5]	R/P/I	**88%**	**77%**	2%	36%
	I/C/C	**81%**	18%	1%	32%
Number of SWM ponds		58	121	121	69
Number of in situ[6] sediment samples		190	374	374	185

Source: Taken from Kelly-Hooper *et al.* (2022).
[1]*Ontario Regulation 153/04 Table 1 Background Site Condition Standards (SCS) for Soils (MECP, 2021).*
[2]*R/P/I/I/C/C (Residential/Parkland/Institutional/Industrial/Commercial/Community) Land use.*
[3]*Red percentages identify >75% of ponds that exceeded SCS and/or ESQS.*
[4]*Ontario Regulation 406/19 Table 2.1 Potable Groundwater (GW) Excess Soil Quality Standards (ESQS).*
[5]*Ontario Regulation 406/19 Table 3.1 Non-potable Groundwater (GW) Excess Soil Quality Standards (ESQS).*
[6]*Wet in situ sediment samples collected from undrained pond basins.*

- Elevated TPH from asphalt sources was the leading cause of regulatory limit exceedances. Elevated EC from road salt de-icing agents was the second highest cause. However, the Ontario Ministry of Environment Conservation and Parks (MECP) has issued five separate risk-based Environmental Compliance Approvals for the beneficial reuse of asphalt- and salt-contaminated sediments as landscaping material within roadside environments. The benefits of reusing thousands of truckloads of sediment on local lands included disposal savings, conservation of landfill space and reduced greenhouse gas emissions.

9.5.2 Permeable pavements

A wide variety of permeable (or porous) pavements are used around the world to detain, infiltrate and treat stormwater runoff. There are several types of permeable pavements, including concrete pavers with wide joints (Figure 9.27a) and porous concrete pavers, either with or without wide joints (Figure 9.27b), generally referred to as permeable concrete interlocking pavers (PCIPs). In literature, a distinction is made dependent on the block material between impermeable concrete interlocking pavers (IPCIPs, i.e. concrete pavers with wide joints or apertures) and porous concrete PCIPs, (i.e. porous pavers with or without wide joints) with the corresponding system characteristics. Porous asphalt (Figure 9.27c) is another type of pavement which can be used on highways and parking lots. Concrete and plastic grid pavers (CGPs and PGPs) are also often implemented in permeable pavements. The design and function of CGPs and PGPs are similar to that of PCIPs. Stormwater can infiltrate through the openings and gaps in these pavers, which are usually filled with gravel or topsoil planted with grass (Figure 9.27d).

Permeable pavements are specifically designed to promote the infiltration of stormwater through paving and base courses, where it is filtered by various layers (see Figure 9.28). These pavement structures can help reduce discharge rates and runoff volumes from paved surfaces and, in turn, help reduce the risk of downstream flooding. Permeable pavements can also provide considerable water quality improvements by treating and trapping stormwater pollutants (bound to fine particles) and preventing them from reaching downstream receiving water bodies (e.g. Gromaire-Mertz *et al.*, 1999; Lucke *et al.*, 2017; Werker *et al.*, 2011). While infiltrating pavements have a clear beneficial potential to water management in urban areas, they are susceptible to physical and biological clogging problems, if they are not maintained (see Figure 9.29), causing premature degradation and serviceability challenges.

Clogging is a result of fine, organic matter and traffic-caused abraded particles, blocking the gaps and surfaces of permeable pavement systems, due to physical, biological and chemical processes.

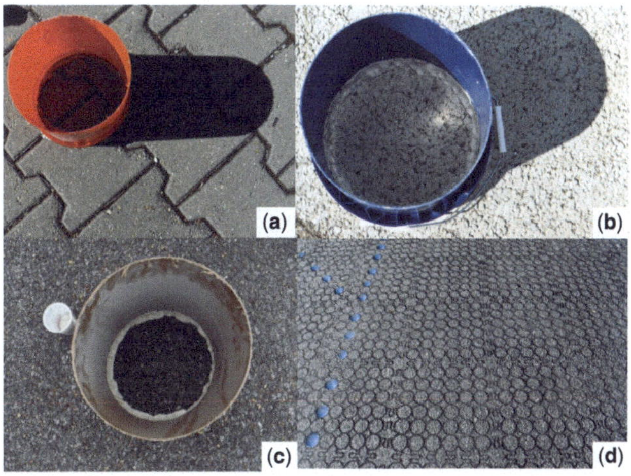

Figure 9.27 Permeable pavement variants in New Orleans during testing: (a) IPCIPs; (b) porous concrete; (c) porous asphalt and (d) PGPs. *Source*: Boogaard *et al.* (2023).

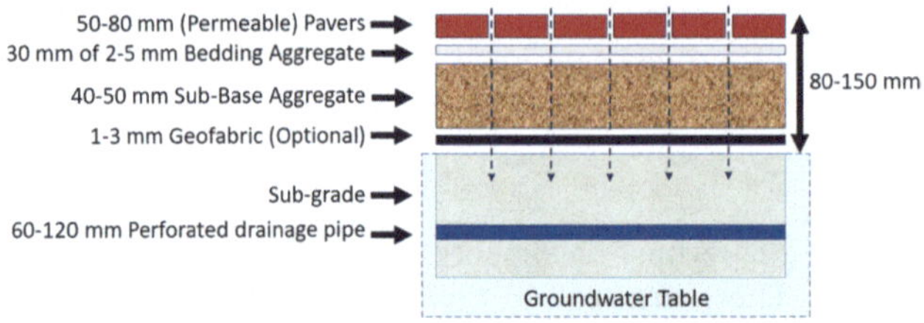

Figure 9.28 Typical permeable pavement structures. *Source*: Boogaard and Lucke (2019).

Figure 9.29 Left: Permeable pavement (stones and aggregate layer). Right: different bedding aggregate layers during construction (butterfly atalanta).

Permeable pavement can be subjected to fine particles during the original construction phases and later in maintenance phases such as winter de-icing operations where fine sand is used. This clogging decreases the porosity/permeability of the paving surface and, hence, the infiltration rate and the effectiveness of the system. It all depends on the structure and location as the type of permeable pavement and the number of implementations depend on the region or country. Nevertheless, permeable pavements have been used in Europe (e.g. the Netherlands and Germany), Australia and the USA for approximately 25 years to treat stormwater runoff and to help recharge water tables, even in low-permeability soils.

9.5.2.1 Performance over time

The performance and maintenance of permeable pavements can vary as a function of

- size of the paving joints,
- different types of bedding aggregates (see Figure 9.28),
- different pavement laying processes and
- the different materials and use of permeable pavement.

Earlier research (e.g. Schmitt *et al.*, 2007; Schönberger *et al.*, 2005) found that construction mistakes in some permeable pavements may affect surface infiltration rates. This could be one of the reasons for the differences in infiltration rates reported in literature (e.g. Boogaard *et al.*, 2023; Veldkamp *et al.*, 2022). In addition, there are distinct variations in the frequency of pavement maintenance (sweeping) procedures among municipalities. Most municipalities conduct occasional street sweeping of their permeable pavements similar to the practice in other districts without permeable pavement. Also, variations in ground water levels may affect the infiltration capacity of the permeable pavements, governing the permeability of the soils where the pavement structure is implemented. In areas with low permeable soils and a high groundwater table, the soil sub-grade underlying the permeable pavement structure may have a higher moisture content. A lack of maintenance of the different parts of the permeable pavement structure (Figure 9.29) can also cause clogging of underdrains resulting in a lower infiltration capacity, thus slowing down the drainage process. This may restrict the infiltration of runoff water into the pavement structure, thereby reducing surface infiltration rates.

When evaluating specific factors that might influence the behaviour of the different types of permeable pavement installations, the following environmental factors should be included in the maintenance and rehabilitation protocols:

- Presence of trees and shrubs.
- Condition of the paving (e.g. rutting, plate formation, loose stones).
- Neighbourhood typology (e.g. traffic and parking intensity).
- Sub soil conditions (e.g. groundwater level, soil characteristics).

In addition to the environmental factors, some characteristics of the permeable pavement structure itself can influence its performance:

- Structure, size and condition of the joints.
- Construction of the street (or top) layer.
- Construction of the foundation; angle of the surface.
- Presence of swirls; system operations.
- Age and type of pavement (PCIPs, IPCIPs; examples are shown in Figure 9.27).

As can be seen in Figure 9.30, tested permeable pavements (most of which are older than 5 years) tend to show a large variation in infiltration capacity. Some permeable pavements show a lower infiltration capacity than the intended water permeability target of $5.4–10^{-5}$ m/s (i.e. when covering at least five paving stones) to enable infiltration of a target rainfall intensity of 270 l/(s ha) or 194 mm/h (Boogaard & Lucke, 2019; SLG, 2020). In these cases, maintenance and if necessary, rehabilitation will be needed. As discussed, low permeable soils and high (ground) water levels can negatively

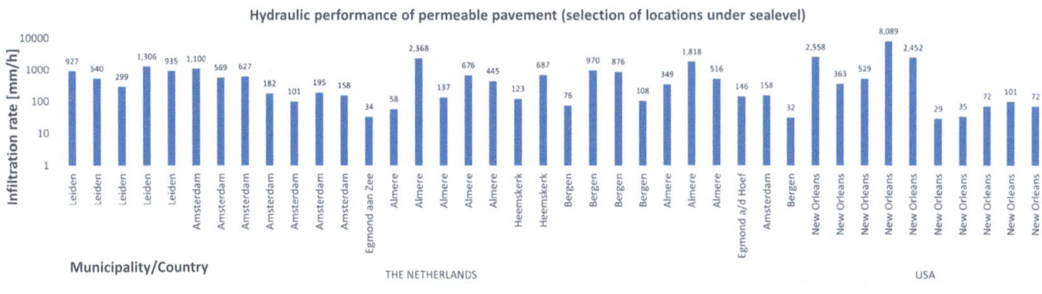

Figure 9.30 Results of full-scale hydraulic test of permeable pavements implemented in municipalities located under sea level (several municipalities in the Netherlands and New Orleans, USA).

Table 9.5 Effects of boundary conditions on permeability.

Influencing Factor	Effect on Water Permeability
Age	With increasing age, dirt particles are introduced into pavements, which can reduce water permeability. However, the degree of contamination does not primarily depend on the age of pavements, but primarily on the use and location (in terms of sunlight and vegetation).
Pollution	Dirt particles, such as dust and moss, can significantly reduce water permeability.
Use	The type of use, such as driving on the permeable surface with vehicles, can lead to the entry of dirt particles and possibly cause settlement of pavements. This can reduce the permeability.
Vegetation and solar radiation	Surfaces underneath trees or shrubs and in shaded areas often showed increased moss growth or soiling by leaves, flowers, pollen dust or needles. As a rule, these surfaces had lower infiltration rates than the adjacent clean surfaces.
Cleaning	The permeability of a pavement can be increased or maintained by cleaning the pavement (including the joints).
Wetting of the pavement	The water content of a pavement (i.e. the weather conditions prior to an infiltrometer test) can significantly influence the infiltration rate.
Installation (own work/ specialist company)	No fundamental differences were observed between installation by a specialist company or a private individual with an effect on the water permeability of a pavement.

Source: Schönberger *et al.* (2005).

affect the infiltration capacity of permeable pavements. Figure 9.30 shows high and low permeable infiltration capacities in municipalities situated under sea level, indicating that this is not the only factor to consider.

Boogaard *et al.* (2023) showed a high variability in infiltration capacities found with both lower and higher infiltration capacities in the City of New Orleans, USA (see Figure 9.27) ranging from 29 to 363 mm/h (constructed in 2011 and 2019, respectively, both tested in 2022). In literature (e.g. Chopra *et al.*, 2010), a lower infiltration capacity of 11 mm/h and a higher capacity of 6000 mm/h were found in studies of pervious concrete cores. Measurements of newly constructed PICPs show infiltration capacities up to 18,000 mm/h, while for pervious concrete and porous asphalt 11,000 and 3583 mm/h, respectively can be observed (Li *et al.*, 2013). Compared to the study in New Orleans, Gilbert & Clausen, (2006), another study showed infiltration rates of PICPs and crushed stone to be in the order of 104 and 96 mm/h concluding that infiltration rates decreased over time due to fine particle clogging. The main findings from testing of water-permeable pavements after several years of operation are shown in Table 9.5.

9.5.2.2 Operation, maintenance and rehabilitation

Infiltration systems such as permeable pavement structures consist of several elements (see Figure 9.28) that may require rehabilitation. As such, the nature of these elements plays a major role in the management and maintenance of the system. Table 9.6 summarizes the elements that are implemented in infiltration systems, including some elements that are not necessarily part of the permeable pavement structures themselves. Also note that the presence of a topsoil layer is restricted to grass-filled grid pavers. Depending on the design and construction of the permeable pavement type one may face different maintenance challenges. Similarly, rehabilitation needs also depend on the details of the system and the different elements present (see Table 9.6).

Relatively little experience has been gained with respect to the long-time behaviour of infiltration facilities such as permeable pavements. Most facilities have been realized in the last few decade(s) and are intended to function for few more decades. The introduction of new innovative systems is still ongoing, resulting in a large variety of systems and elements so there is no 'one-size-fits-all' rehabilitation programme. As a result, there is an increasing need for a systematic, professional

Table 9.6 Elements that may be implemented in infiltration systems and in consequence can be of interest for AM.

Matrix Infiltration Systems and Objects	Details
Topsoil layers	E.g. grass-filled grid pavers
Permeable pavements (porous concrete elements)	E.g. porous pavers or porous asphalt
Pavements (infiltration in open jointed elements)	E.g. Impermeable interlocking pavers
Bedding aggregates	See Figure 9.28
Sub-base aggregates	See Figure 9.28
Sub-grade	See Figure 9.28
Infiltration pipes (perforated)	Sub-surface pipe to distribute the water through the sub-base aggregates
Groundwater drains (perforated)	To prevent standing water in sub-bases
Stormwater sewers	Existing or new stormwater systems as overflow or inflow
Open street gutter s	Surface runoff inflow to permeable pavement systems
Gully pits	Gully from stormwater systems (and sand pit)
Geofabrics/geotextiles	To prevent mixture of bedding aggregate, sub-base and sub-grade
Impermeable liners	Retention for water harvesting or prevention of sub-soil contamination (i.e. no infiltration to environment)
Control units	For monitoring or real-time control
Overflow weirs	Bypass construction
Silt trap/filters	Removal of gross sediment before infiltration

approach to the management of such facilities. Although several monitoring programmes have been executed, in practice, most organizations have little knowledge about the functioning of their facilities and the fate of the infiltrating water. Unfortunately, maintenance tends to be primarily performed when calamities (e.g. flooding) occur or complaints are received. To carry out maintenance measures effectively, insights into the details of the structure and the causes of diminished functioning of the permeable pavement (i.e. the degradation mechanisms) are needed. Diminished functioning of SUDS such as permeable pavement structures can result in:

- reductions of the infiltration capacity,
- reductions of the storage capacity,
- reductions of the discharge capacity (sub-drains and overflow) or
- pollution of sub-soils and groundwater.

For all SUDS with a main function of infiltration the placement in the catchment and its land use are important factors when it comes to the assessment of maintenance needs (Homet *et al.*, 2022). SUDS in the upper parts of the catchment may be designed to capture precipitation where it falls. In the middle parts, the overall catchment generating runoff is larger which, in turn, can increase maintenance needs. In the lower parts, the stormwater is mostly conveyed by pipes, and SUDS are mostly used to capture the remaining runoff. Older residential areas can have more challenges with the accumulation of pollutants (leaching of heavy metals) and leaves (i.e. older trees), which can be a cause of clogging and nitrogen pollution. However, these areas may also be less susceptible to erosion (e.g. better vegetative cover, less construction activities, etc.). A fundamental understanding of the different processes that are responsible for the diminished functioning is needed, relating the direct causes to a proactive, structured management approach instead of a reactive, complaint-driven approach. In Table 9.7, these aspects are summarized in a matrix.

Table 9.7 Diminished functioning of permeable pavements.

Diminished Functioning of Permeable Pavement	Process	Street Litter	Litter from Construction Sites	Falling Leaves	Iron Deposits (From Groundwater)	Sinking of Overflow Weir	Roots in Drainage	Overgrown Overflow Construction	Uneven Subsidence	Polluted Stormwater	Accumulation of Pollution in Topsoil	De-icing Salt
Reduction of infiltration capacity	Silting up	x	x	x	x							
	Compaction								x			
Reduction of storage capacity	Subsidence					x			x			
	Deposits	x										
Reduction of discharge capacity	Clogging	x	x	x			x	x				
	Reduced pressure gradient	x	x	x					x			
Pollution	Sub-soils									x	x	
	Groundwater									x	x	x

The frequency of regular maintenance of streets and parking lots depends on the organization, region or country. In some cases, where clogging of permeable pavements is likely, a higher frequency of litter and leaf removal is utilized ranging from 2 to 52 times a year for litter, 2 to 4 times a year for leaves and 4 to 12 times a year for street cleaning.

In principle, all layers of the superstructure and the sub-soil must ensure sufficient water permeability for long-term operation. However, this is made more difficult by the fact that air pores remain during the infiltration process, which reduce vertical water movement by approximately 50%. To ensure controlled infiltration from the superstructure into the sub-soils, the sub-soil or sub-grade (cf. Figure 9.28) must be >1 m thick. This is not possible in every location (such as high-groundwater tables in the Netherlands where often perforated drainage pipes are installed). To ensure adequate permeability, it must not fall below $k_f > 5.4 \times 10^{-5}$ m/s (approximately 194 mm/h) in the entire pavement structure. If one of the two criteria (Boogaard & Lucke, 2019; SLG, 2020) is not met, supplementary construction measures such as increasing the thickness of the frost or base layer are conceivable to guarantee appropriate water permeability. Drainage pipes can also be installed (see Figure 9.28). To establish long-term functionality in superstructures, the various layers of superstructures must, in addition to the required water permeability, also have sufficient filter stability among themselves. These are subject to the relevant regulations. This is to avoid varying water permeability in the seepage process causing aggregates and fines to be flushed out into the deeper layers by flow pressure.

The need for repair, renovation and replacement depends on all the aforementioned factors and is exemplified in Figure 9.31. On the left, subsided and displaced stones around a waste disposal site due to heavy traffic (i.e. garbage trucks) will require repairs. In the centre, due to an impact from cars as well as street sweeping, the joints within the permeable pavement bricks lost their filling. This lack of grid will affect the performance of the top layer of the permeable pavement and the stability of the road surface. This may result in the displacement and loosening of paving stones. The top layer must be renovated. On the right, a clogged permeable pavement's top layer is shown which causes the need for grid replacement.

Figure 9.31 Examples of the need for repair and replacement.

9.5.2.3 Example from Germany

According to the DWA-A 138 (2005), the condition of surface coverings capable of infiltration should be checked regularly. To prevent and eliminate silting and self-sealing, leaves in particular must be removed from the infiltration area. In the case of infiltration surfaces, it is recommended that an operating plan be drawn up and an operating log be kept. In the event of damage to permeable paved areas where liquids hazardous to water, for example, oil, have spilled onto the pavement structure, for example, from accidental spills or accidents, the responsible water authority must be informed immediately. In the event of a change of land use, for example, in commercial areas, it should be checked whether the existing infiltration-capable surface continues to meet the requirements from a qualitative point of view.

The following measures should be performed to maintain the infiltration capacity of water-permeable surfaces (SLG, 2020):

- Cleaning the surface.
- Removal of damage, subsidence and holes.
- Mowing of grass, removal of weeds.
- Removal of moss layers in joints and possibly renewal of the joint filling.
- Removal of mud deposits between the joints/in the interstices.
- Cleaning by pressure washing/vacuuming of paving stones made of pumice-porous concrete.

Cleaning of permeable pavements by suction sweepers is recommended to maintain the permeability for as long as possible. Special vehicles for this purpose are commercially available. With the vehicle shown in Figure 9.32, the pavement is cleaned by rotating high-pressure water jets, supplemented by suctioning off the water including the dirt particles and the joint material. However, due to their size, these special vehicles may not be suitable on smaller, private areas.

High-pressure cleaners are often used for cleaning private areas, such as terraces, courtyards, driveways, walkways, pathways and so on. Special floor washers are available as accessories, with two rotating high-pressure nozzles mounted on the underside. A circumferential brush rim serves as a splash guard (see Figure 9.33). When cleaning with these devices, superficial dirt on the stones and usually also parts of the joint fillings can be rinsed out. High-pressure cleaners are also offered in combination with suction cups that are connected to the floor washer. This is to prevent the dirt particles from being washed into the stones and clogging them. The cleaning interval specified by the manufacturer or supplier of the

Figure 9.32 Special vehicle for cleaning pavements (using high-pressure water and suction equipment). *Source*: Schönberger *et al.* (2005).

Figure 9.33 High-pressure cleaner and soil scrubber in use. *Source*: Schönberger *et al.* (2005).

pavement structure should be adhered to ensure that the stormwater purifying effects of the pavement structure reflect the conditions under which they had been tested in laboratory experiments during the certification process for construction materials (Goerke & Möllers, 2020).

9.5.2.4 Example from the Netherlands

Historically, after the design and construction of permeable pavement structures, most Dutch municipalities have paid little attention to ongoing management and maintenance. To change this situation, the Dutch RIONED Foundation created the Guideline C3200 (Wentink *et al.*, 2006) for the management and maintenance of infiltration facilities. This guideline states which investigations and which practical management measures are necessary. The guideline reflects knowledge gained from daily operation of these facilities and specifically analysis of the important degradation mechanisms and, as such, presents practical guidelines for the management of infiltration devices. This is illustrated using management diagrams that present desired management activities and frequencies for each object. The management diagrams also include potential investigations such as inspections or monitoring. Table 9.8 shows examples of management diagrams for the investigation and maintenance of the topsoil layer (i.e. in the case of grass-filled grid pavers).

The management diagram about the top layer shows three types of activities:

- The first type is an ongoing process, for instance, providing direction to prevent people or vehicles from accessing the top layer and parking on it (e.g. in the case of grid pavers).
- The second type of activities follows a time-dependent approach. These are repetitive actions such as mowing and removing of litter and leaves. The frequencies in the table reflect initial

Table 9.8 Management diagrams for investigation (I) and measures (M).

No.	Activity	Periodic Frequency
I1	Visual inspection	1× per year
I2	Simple measurement time to empty	1× per year
I3	Detailed measurement time to empty	Based on I2
I4	Measurement permeability topsoil	Based on I2
I5	Geo-survey	1× per 5 years
I6	Taking groundwater samples	1× per 5 years
I7	Taking soil samples	1× per 5 years
I8	Investigation for wrong connections	Based on I1
M1	Ongoing visual inspection (e.g. use of swales as parking space)	Continuously
M2	Lawn mowing	2–26 times per year
M3	Removal of mown grass	2–26 times per year
M4	Removal of litter	2–52 times per year
M5	Removal of leaves	2–4 times per year
M6	Street cleaning	4–12 times per year
M7	Sowing	Based on M1
M8	Fertilizing	Based on M1, M2, M3 and M4
M9	Verticutate	Based on M1, M2, M3 and M4
M10	Fill up of low spots	Based on M1
M11	Removal of silt	Based on M1, M2, M3 and M4
M12	Lowering (scraping) of the verge	Based on M5
M13	Replacing the topsoil layer of the rainwater garden/swale	Based on M6 and M7

advice. Once there is more detailed information about the functioning of the object in question, the manager can adjust the frequency.

- The third type of activities should be executed based on the results of previous investigations. This is a situation-dependent approach.

Good management starts with a good and unambiguous recording of the basic facts of the objects (e.g. location, dimensions, composition). Also, the pertinent items of the design must be recorded, as well as the results of inspections and investigations. For each object, the management diagrams provide an impression which activities are necessary and at which frequency. However, the manager should fine-tune the management reflecting his own local experience. To obtain insight into the functioning of the objects an active role of the manager is necessary. This active role implies:

- to give feedback to other departments after activities have been carried out (i.e. about characteristics and possible challenges),
- to take specific action on reports and complaints (e.g. by regularly visiting SUDS, observing how they function),
- to record clearly performed activities, complaints, observed failures and conducted actions.

Regular visual inspections of the top layer of permeable pavement installations and a simple measurement of the time needed for the water to infiltrate into the object are recommended.

Few research studies are available to date on the effect of maintenance on the performance of permeable pavement. Due to the differences in materials, design, construction, maintenance and specific characteristics, there is not a representative type of permeable pavement. However, in the

Figure 9.34 Infiltration capacity at permeable pavement before (65 mm/h) and after maintenance (184 mm/h) in Werkendam, the Netherlands. *Source*: https://www.climatescan.nl/projects/60/detail.

Netherlands comprehensive studies with full-scale infiltration testing (FSIT) have been conducted at over 100 locations to assess the effectiveness of (different) maintenance methods on the infiltration rate of pavement structures (Boogaard & Lucke, 2019; Veldkamp *et al.*, 2022). The measurements show that the maintenance of permeable pavements generally leads to an increase in the infiltration rate (see Figure 9.34) which is to be expected, or rather desired. However, in some cases it led to fine particles being deposited deeper in the sub-structure leading to an even lower infiltration capacity. Examples of maintenance resulting in a lower infiltration rate include replacement with wrong joint filling, cleaning under unfavourable, that is, wet circumstances and regular street sweeping.

On average, cleaning increased the infiltration rate by a factor equal to three. Based on some studies, the use of a high-pressure air cleaner seems to be the most effective for IPCIPs. If performed properly (i.e. not under wet conditions), this cleaning methodology is able to clean the joints of IPCIPs that increases the average improvement factor to 4.4. All results need to be supported by additional (inter)national experiments and research. However, unfortunately, the maintenance of permeable pavement structures tends to be regarded as expensive and time consuming. As a result, while the initial infiltration rate just after implementation may have exceeded 1000 mm/h, the actual infiltration rates are in practice often considerably lower (Boogaard, 2015). This would suggest that safety factors reflecting clogging should be considered as part of the original analysis and design as is the case in some jurisdictions.

9.5.3 Green roofs

Green roofs encompass a wide range of vegetation solutions used on rooftops. The most common classification (FLL, 2018; NS 3840, 2015) is to divide them into two main groups according to the depth of the vegetation support layers, where intensive green roofs are based on a deep vegetation support layer (from 15 cm up to 1 m), and extensive roofs have a thinner layer (<15 cm) of vegetation support layers. The intensive roofs typically result in high loads on the building structure due to the nature of their composition. They can support a large variety of plants including trees and shrubs often converting them into rooftop gardens. Extensive green roofs are made of lightweight materials and typically with a focus on low maintenance and irrigation requirement. This premise for extensive green roofs results in a plant palette of sedums, herbaceous plants and grasses. Extensive green roofs are by far the most used type as they do not require added roof loads and can be retrofitted onto exciting roofs. Biodiversity has become an increasingly important amenity of green roofs. This has resulted in extensive green roofs can be further divided into extensive sedum roofs and extensive biodiversity roofs

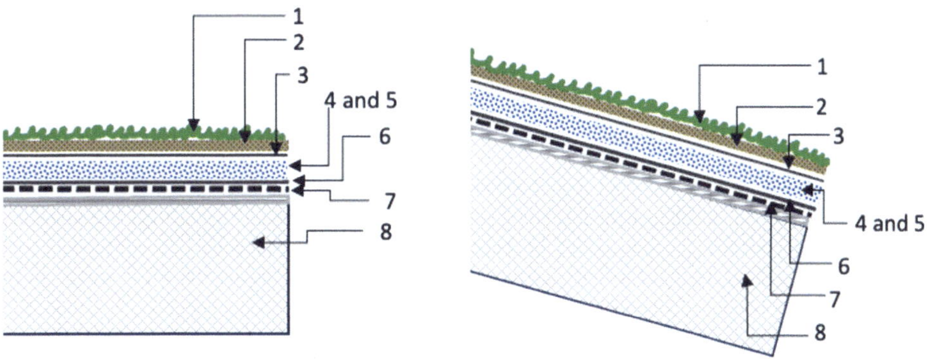

Figure 9.35 Schematic build-up of extensive green roofs for flat and slanted roofs: (1) vegetation, (2) substrate layer on reinforced vegetation mat, (3) separation layer, (4, 5) drainage layer or combined storage and drainage layer, (6) protection layer, (7) root-resistant layer, (8) roof construction.

(https://efb-greenroof.eu/green-roof-basics/). A schematic build-up of the most common layers in an extensive green roof approach is shown in Figure 9.35.

Extensive green roofs consist of a set of different layers. The vegetation is supported by a substrate or growing media layer characterized by low weight, ability to store water and allow for excess water to drain out and contain a low-volume fraction of organic matter (5–20%). The amount of organic matter should be tailored to the local climate and roof composition; compost used with the inert media should be plant-based and well matured to minimize the risk of leaching of nutrients. Amendments to capture nutrients may be considered. The vegetation can be established from seeds, from cuttings, from plug plants or as pre-grown vegetation mats. Pre-grown vegetation mats, widely used, for example in Scandinavia, consist of an around 30 mm thick substrate layer reinforced with coconut or plastic fibres with already established sedum plants for easy installation on roofs. Substrate layer depths larger than 30 mm are normally recommended in central and southern Europe, while the Scandinavian market is dominated by the usage of extensive green roof build ups consisting of only a 30 mm vegetation mat in combination with a choice of drainage layers, separation-, protection- or textile retention fabrics. All the large green roof suppliers recommend the use of a separate drainage layer for relatively flat roofs (i.e. <10% slope) to ensure sufficient drainage of the substrate, while some of the suppliers suggest a solution with only a textile retention fabric under a vegetation mat for sloped roofs. The extensive green roof build-up allows for water storage in substrates, in drainage layers if provided with cuplike shapes, and in textile retention fabrics.

9.5.3.1 Performance over time

Existing literature (e.g. Cook & Larsen, 2021; Jayasooriya *et al.*, 2017) extensively details the performance of green roofs. Green roofs are often said to deliver additional benefits compared to other SUDS because they provide thermal or acoustic insulation, or a protection of the roof waterproofing. In practice, acoustic insulation is only useful in specific areas such as near airports, and thermal insulation benefits exclusively the top floor of a building. Also, green roofs contribute little to energy savings in well-insulated buildings (Andenæs *et al.*, 2018). Roof waterproofing is maybe the most specific and important performance to control over time: such performance highly depends on the proper drainage of the roof. Green roofs are reported to better stand up against hail impact compared to traditional roof structures which should be of particular interest in areas subject to frequent highly damaging hailstorms (e.g. along the eastern slopes of the Rocky Mountains). Regarding SUDS-related performances such as biodiversity preservation, stormwater detention or retention (i.e. through evapotranspiration) or mitigation of urban heat, it strongly depends on the depth of the growing media, the choice of vegetation and its diversity (Dusza, 2017). Intensive green roofs will provide better performance over time. They are of specific interest for infill or redevelopment in older, established parts of our cities where they can significantly reduce the flooding risks from high-intensity summer thunderstorms while mitigating the urban heat island effect.

9.5.3.2 Operation, maintenance and rehabilitation

The first years after initial installation of green roofs are often the most important because vegetation must mature to a level where it will sustain drought or extended wet periods. The vast majority of green roofs require watering after installation, regardless of the type of system installed (i.e. seeds, plant carpet or plant boxes). The duration until maturation is achieved (i.e. when the vegetation has established) depends on the context (e.g. climatic conditions, type of vegetation installed, desired aesthetics, etc.) and on the system installed. A period of 3–6 months is generally recommended for mats/rollers and slabs/plates, 1–2 years for micro-pots, buckets and containers, and 1–3 years for seedlings. These durations should be adapted to the local context and reflect experience gained from earlier installations. After the establishment period, the O&M frequency will be highly dependent on the climatic conditions and the local context reflecting, for instance, the nature, size and use of the roof area, its location and nearby environment, public accessibility or not, type of vegetation, desired aesthetics, and so on.

Regular O&M actions include litter removal, reinstatement of the growing media in case of displacement by wind or rain, manual weeding of unwanted plants, cleaning of drainage systems, watering (in relation to the plant hydric stress or the weather conditions), winterization, purging and refilling of the watering system (if any) and other specific works prescribed and depending on the vegetation installed. Supplementary fertilization should be avoided, if possible, or be judiciously applied to minimize the risk of leaching of nutrients. Extensive or semi-intensive roofs (layer <15–20 cm) may require more frequent watering. Intensive roofs may require mowing or trimming of trees. It is thus paramount to provide easy access to the roof, and a watering point for irrigation. Automatic watering systems (depending on the desired water content of the growing media or the water stress of the plants) are often very useful.

Regarding the vegetation, rehabilitation actions concern the replacement of existing vegetation on a portion or all of the roof. In the event of partial or total failure, actions include additional sowing (seeds or fragments) or planting (of micro-bunch or cups) or installation of pre-cultivated elements. The vegetation survival depends on both maintenance factors and local suitability (Lönnqvist *et al.*, 2021). Plant survival and species variation may be measured by counting the species and numbers in a $1 \times 1\ m^2$ on the roof. In Norway for example, there is a national standard, NS 3840 (2015), developed based on a modification of the European standard EN 12231 (2003). This should preferably be done at different times over the seasons, as in the example shown in Figure 9.36, where four different types of sedum roofs were counted in the autumn for a location in Oslo, Norway.

Careful attention should be given to waterproofing of the roof. Proper initial construction is vital to ensure its longevity. If the installation was correctly performed the main risk comes from unexpected accumulation of water: it is thus paramount to guarantee the correct operation of the drainage system and more specifically the flow-control discharge and overflow system. Rehabilitation of the roof waterproofing often implies the renewal of the entire roof (including the vegetation) unless a leak detection system had been implemented which allows both early detection and pinpointing the location of leakage.

9.5.4 Bioretention systems and bioswales

Large-scale implementation of bioretention systems in urban areas is fairly recent. As such, available information on rehabilitation techniques is still relatively sparse. Although early examples date back to the late 1980s, in most communities, the experience has been limited to the last decade. Given the ongoing evolution of the design practice, the early versions may not reflect current design practice.

Figure 9.36 Different types of sedum roofs in Oslo/Norway showing a box used to count species diversity and ground cover. *Source*: Hanslin and Johannessen (2016).

Bioretention systems include both 'self-contained' bioretention areas and bioswales which, with their linear configuration, also have a conveyance function. The components that make up bioretention systems tend to be similar but there may be differences specifically in the nature of the inlets, the growing media and the vegetation. Older systems may include geotextiles. The main difference could be the presence of sub-drains where sub-soils don't have adequate percolation capability.

The rehabilitation activities associated with bioretention systems requires input from both engineers and vegetation specialists. One should consider the feasibility of harvesting and reusing existing vegetation when these systems are renovated or replaced. The feasibility is a function of the micro-climate and site-specific moisture conditions, together with the location within the bioretention system. The potential for excessive salt (from winter de-icing activities) and other contaminant loadings should be considered as well. The risk of vandalism should be assessed from a vegetation survival perspective.

In view of their relatively small size rehabilitation of bioretention systems can usually be carried out quickly which provides more flexibility from a seasonal perspective, unless more time-intensive structural components need to be replaced. Temporary and biodegradable erosion protection may need to be provided until the root structure of the vegetation has been re-established.

Bioswales (see Figure 9.37) are vegetated surface-drainage system channels that receive stormwater runoff through traditional grey infrastructure pipes or directly from gully pot/catch basin inlets or as

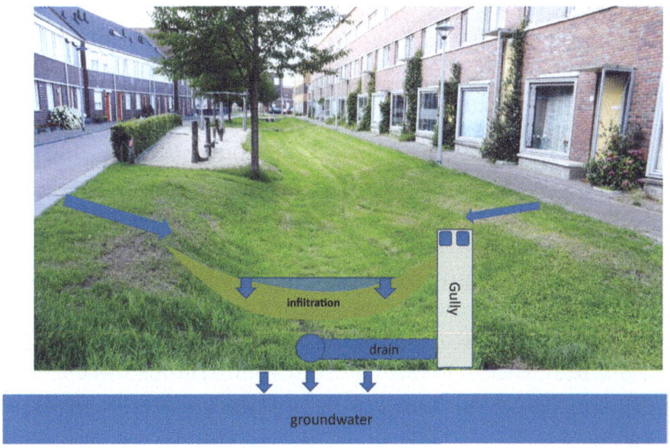

Figure 9.37 Schematic view of a bioswale in an urban area.

sheet flow over gentle side-slopes (Davis *et al.*, 2012a; Stagge *et al.*, 2012; Woods Ballard *et al.*, 2015). Bioswales are an enhanced version of traditional vegetated swales or ditches in that they also include the growing media typically used for bioretention areas as well as check dams/berms to promote infiltration. Water quality treatment in a bioretention area, bioswale and vegetated swale occurs through the process of sedimentation, filtration (infiltration) and biological and chemical interactions with the soil/growing media. Swales have been shown to be efficient in removing sediment particles (with micro-pollutants such as PAH and heavy metals bound to these particles) from urban runoff (Barrett *et al.*, 1998; Deletic, 2005; Winston *et al.*, 2012). Regarding soil composition in general, bioretention areas and bioswales are composed of loamy sands, loams or sandy loams resulting in a wide variability in the infiltration rates of bioswales and biofilters (Le Coustumer *et al.*, 2012). The type of vegetation, such as deep-rooted grass species and proliferation of root and earthworm channels, also improve infiltration (Abu-Zreig *et al.*, 2004; Fort *et al.*, 2013; Le Coustumer *et al.*, 2012). Vegetated swales and bioswales are generally grass-lined, with bioswales in most situations providing more opportunity for diverse vegetation.

9.5.4.1 Performance over time

Bioretention systems may see a decrease in the percolation potential, especially in the first year after initial installation before the root structure of the vegetation has fully established. Some attempts have been made for the installation of gravel chimneys which allow for more rapid drainage while the root structure is establishing; these chimneys clog over time with their functionality taken over by the root structure. However, construction activities, for example, associated with either new sub-division development or infill/redevelopment may result in a sudden overloading of bioretention systems when erosion and sediment control is either poor or overloaded during a severe storm event.

The vegetation in bioretention systems may either take off and flourish given that the growing conditions can be more favourable compared to typical landscaping resulting in the need for thinning or see excessive mortality when watering during the initial establishment period is inadequate or the vegetation is not appropriate for the loss of moisture that may occur during drought conditions given the lean nature of many growing media. Another issue that is important to be addressed in cold climates is snow disposal and management practices in winter (Gougeon *et al.*, 2023), as, for example, excess salt from de-icing practices that ends up in SUDS can result in decreased hydraulic conductivity in soils (Shannon *et al.*, 2020).

Diminished functioning of SUDS including bioretention systems can result in:

- reduction of the infiltration capacity,
- reduction of the storage capacity,
- reduction of the discharge capacity (e.g. overland or sub-drain conveyance capacity or overflow) or
- pollution of the sub-soils and groundwater.

The different processes and causes that may contribute to the diminished functioning are summarized in the matrix presented in Table 9.9.

Given that the performance of swales is spatially and temporally dependent and reflects local climate conditions, a large variability in the hydraulic performance of swales can be expected as a function of the differences pertaining to (1) design parameters (e.g. side-slope, length, type of vegetation, soil or growing media composition, ratio between the footprint of the SUDS and the size of the tributary impervious area), (2) state (e.g. soil moisture content) (Boogaard, 2022) and (3) (human) errors in the design, construction and maintenance phases. International studies have shown a high variability as to the mean volume and peak flow reduction with the presence of check dams improving retention and overall performance (Davis *et al.*, 2012b). Regardless, comparison with national guidelines of the Netherlands, Denmark and Germany suggests adequate performance of swales in those countries. For example, all bioswales in the Netherlands (even those subject to a high-groundwater table being

Table 9.9 Diminished functioning of swales and possible causes.

Diminished Functioning of SUDS	Process	Street Litter	Litter from Construction Sites	Falling Leaves	Iron Deposits (From Groundwater)	Sinking of Overflow Weir	Roots in Drainage	Overgrowing of Overflow Construction	Uneven Subsidence	Polluted Stormwater	Accumulation Pollution in Topsoil	De-icing Salt
Reduction of infiltration capacity	Silting up	x	x	x	x							
	Compaction											
Reduction of storage capacity	Subsidence					x						
	Deposits											
Reduction of discharge capacity	Clogging	x	x	x			x	x				
	Reduced pressure gradient	x	x	x					x			
Pollution	Soil									x	x	
	Groundwater									x	x	x

situated under sea level, see Figure 9.38), met the guideline to drain down (i.e. no more visible standing water) within 48 h. Note that there is a high variability in Figure 9.38 (i.e. using a log scale *y*-axis) with high values corresponding to ideal circumstances (e.g. dry cracked, unsaturated soil) and differences in the composition of swales (sometimes allowing drainage just under the surface giving it a high infiltration/discharge rate).

A low drain down or emptying time can result in less storage capacity being available for future rainfall events, can impede vegetation growth and may result in resident complaints due to nuisance caused by the presence of mosquitoes. A drain down time of less than 5 days will be sufficient to minimize the risk of mosquito nuisance (Boogaard *et al.*, 2021). From Figure 9.38, it can be revealed that not all swales (with a depth of 50 cm) (Boogaard, 2015) will drain down within 1–2 days. If the observed drain down time is more than the required drain down time of a couple of days (Boogaard *et al.*, 2023), maintenance and/or rehabilitation are required as explained in the below sections.

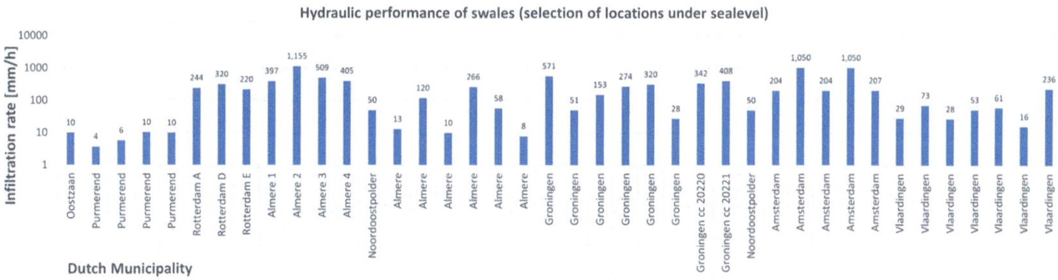

Figure 9.38 Results of full-scale hydraulic tests of swales in the Netherlands.

9.5.4.2 Operation, maintenance and rehabilitation

Typical O&M activities of bioretention systems include:

- mowing or pruning of vegetation,
- weed removal,
- litter removal,
- removal of animal burrows,
- cleaning of gully pot sumps or other pre-treatment devices or
- street sweeping in the contributing catchment area.

Examples of typical O&M activities are shown in Figure 9.39.

Repair is any rectification of localized damages regarding either the engineered or the natural elements of bioretention systems. Repair activities are frequently focused on the composition of the inlets or check dams or focused on re-plantings, either loss due to poor plant selection or vandalism. In case of the former, do not replace the vegetation with the same species. The aim of vegetation selection should be self-sustaining systems with minimal maintenance requirements. The use of native vegetation can contribute to better long-term sustainment and reduced maintenance costs (de la Fuente de Val, 2023). Usually, repair activities are triggered following site inspections.

Renovation activities are often associated with the inlets and may include:

- Widening, if the inlets are too narrow, especially in the case of flow-by situations.
- Corrections to the grades or elevation of the inlets, ensuring there is an adequate drop off from the adjacent roadways or parking lots to the surface of the bioretention systems, minimizing the risk of repeated sediment dams at the inlets.
- Provision of pre-treatment sumps.
- Provision of a continuous plate at the top of the curb so that the blade of mechanized snow removal equipment will not get stuck.

Renovation activities may also consist of the removal and replacement of the top layer of the growing media in case of clogging (see Figure 9.40) or regrading in the case of eroded parts of the system.

As part of **replacement** activities, existing bioretention systems are replaced with the same type of asset or a different one. The need for replacement activities often reflects a poor initial design and construction. This includes but is not limited to the need for:

- Additional or relocated inlets, especially when the grading of the upstream catchments and roadways or parking lots is different from the originally intended grading.

Figure 9.39 From left to right: Mowing swale in Enschede with obstructions, clogged inflow pipe swale in Beek, removal of litter (foul connections) at inlet of the swale in Renkum.

Figure 9.40 Left: Dutch Swale in Enschede (Ruwenbos) littered with leaves. Right: Clogged swale at Haarzicht in Utrecht. *Source*: https://www.climatescan.org/projects/7921/detail.

- Additional or improved pre-treatment devices.
- A reduction in the footprint of the bioretention system if the upstream catchment is too small resulting in not enough moisture being directed into the bioretention system.
- An increase in the footprint if the upstream catchment is quite large, possibly coupled with additional upstream bioretention systems, a redirection of part of the catchment, or the installation of resilient landscaping within the catchment.
- Improving the overflow, possibly combined with changing the vegetation to species that are more resilient to wetter conditions. While the frequency of events may not change, the total moisture introduced may be too much for the species on the side slopes.
- Uncompacting the growing media if it was overly compacted upon the initial installation.
- Changing out the growing media which may need to be done when the original media are not appropriate or may be subject to excessive leaching of nutrients.
- Replacement of geotextiles with a gravel filter layer.
- Installation of replacement of a sub-drain, including appropriate inspection ports and access for cleaning.

Clogging or contamination of the growing media may also be ground for replacement. Examples of renovation activities are presented in Figure 9.41.

Swale rehabilitation can mean different activities with different timelines considering the engineered and natural elements of a system. Rehabilitation of the engineered elements is often 'immediate' in the sense that the function is fully restored at the completion of the construction activities. Rehabilitation of the natural elements, however, can take more time as it takes time for the vegetation to mature and can also depend on external conditions. An example shown in Figure 9.42 reveals a project where replacement of the vegetation was ultimately not needed after the grass managed to grow back after 8 weeks of drought.

Overall, the considerations for replacement activities are usually similar to initial retrofit projects.

O&M and rehabilitation activities can be corrective or preventive. In many cases corrective action is commenced after inspections or to address complaints, often involving the repair or replacement of objects which have failed or broken down. Preventive maintenance includes both condition and life-extending tasks which are scheduled at regular intervals. The choice for one of both methods depends on the following criteria:

Figure 9.41 From left to right: Removing of top layer swale (construction 2000) due to clogged layer in 2012, Nijmegen, replacement of drainage pipes and removing top layer swale (construction 2007) due to clogged layer in 2012, Nijmegen. *Source*: https://www.climatescan.org/projects/4364/detail.

Figure 9.42 Swale at Dalfsen (the Netherlands) showing normal condition on the left in 2017 and on the right the same swale after a 7 week dry period in 2018. *Source*: Boogaard (2022).

- (potential) direct damage and effects of failure,
- insight into causes of reduced function and
- effort for collecting and assessing data.

The (potential) direct damage and effects of failure of the objects ascertain whether preventive or corrective action is needed. To determine which preventive measures to perform, insight is needed into the mechanisms which might cause the objects to fail. As part of a preventative approach, one should differentiate between 'condition' and 'use' (or time) dependent approaches. As part of a condition-dependent approach, investigations take place into the condition or state of the object in question. To do so, criteria for the functioning of the object should have been defined (e.g. 'a drain down or maximum emptying time of 48 hours' or 'pollutant concentrations must not exceed target concentrations'). As part of a use (or time) dependent approach, maintenance only takes place after a certain amount of time has passed or a certain load has been accommodated. The choice between a condition-dependent or use (or time) dependent approach depends on the effort necessary to collect and assess data about the condition or state of the object. As such, to carry out maintenance activities effectively, insight is needed into the degradation mechanisms that lead to diminished functioning of bioretention systems.

Sediment accumulation during construction activities of SUDS themselves or within their catchments or during subsequent operation is a common problem that can lead to clogging of bioretention systems, leading to poor performance and a reduced lifespan. In addition, contamination of the growing media, for example due to excess salt (chlorides) from de-icing practices on adjacent roadways, can result in a decreased hydraulic conductivity (Shannon *et al.*, 2020). Guidelines for good design and operations can mitigate these challenges, but, in most cases, maintenance is required to maintain the infiltration capacity over the long run. Another interesting topic is the need to balance maintenance activities (WEF, ASCE/EWRI, 2022); on the one hand, since leaf and dead vegetation detritus can help maintain the soil structure and preserve macropores that maintain permeability, too much maintenance might lead to underperformance. On the other, a lack of maintenance such as removal of leaves can cause clogging of the growing media. Also, vegetation may die off in those parts of the system (i.e. the lowest parts) where water stands for inordinate amounts of time. The accumulation of the dead vegetation may then decrease the infiltration even further (see Figure 9.43). Appropriate vegetation selection may help in reducing the magnitude of these challenges. Compaction of the top layer, caused by playing children (see Figure 9.43), parked cars or lawn mowers, can also lead to a reduction of the infiltration capacity.

Figure 9.43 Playing children on a swale (top left), swale used as walking route in contrast to regulations (top right), swale with standing water (bottom left) and restricted access (bottom right).

As most countries have only recently started the implementation of green infrastructure, little experience has gained towards the long-term behaviour, structural management and maintenance of bioretention facilities. Virtually no green infrastructures such as bioretention areas and bioswales have made it yet to their expected, or rather desired, life span of 40–100 years. In practice, most organizations tasked with the maintenance of green infrastructure have little knowledge about the performance of their infiltration facilities (Woods Ballard *et al.*, 2015). In most cases, maintenance or rehabilitation is delayed until calamities (e.g. flooding) occur or resident complaints are received. Although the concept of infiltration of stormwater is widely accepted, little practical knowledge has become available so far as to the O&M and rehabilitation needs of bioretention assets. As such, there is an urgent need for a systematic, professional way of managing these facilities, at the same time as we are seeing a dramatic increase in the implementation of this type of green infrastructure to allow us to adjust to a changing climate.

Several studies showed that the performance of SUDS, in general, and vegetated swales, in particular, can be negatively impacted by (human) failures in the design, construction and maintenance of these features (Boogaard, 2015, 2022; Vollaers *et al.*, 2021). Lacking maintenance and monitoring can lead to diminished functioning of SUDS (see Figure 9.44) which can result in flooding, or pollution of ground and surface water. To achieve the desired long-term performance objectives, it is recommended that basic guidelines for the creation of O&M programmes are followed. This will assist stormwater managers with the modelling, planning, testing and scheduling of maintenance activities for vegetated swales with more confidence, so that they will continue to perform satisfactorily over their intended lifespan.

The monitoring and evaluation of SUDS have yielded a wealth of experience which allows us to review and expand our guidelines of SUDS to guarantee their performance in time. Observations pertaining to common failures in the design, construction and maintenance of SUDS can be translated into the following guidance:

- Adequate communication between developers, planners, consultants, architects, engineers, adjacent residents and asset owner is needed from initial design to the O&M phase.
- SUDS and their different elements should be always accessible for maintenance, preferably with mechanized equipment.
- Implement opportunities to adapt the system after construction (e.g. changing weir heights) to overcome uncertainties during the design phase (e.g. uncertainties introduced by hydrogeological, hydrologic or hydraulic models), construction phase (e.g. grading tolerances) and unforeseen developments after construction.

Figure 9.44 Diminished functioning of SUDS. Left: Maintenance with heavy equipment leads to compaction of the soil (*Source*: https://www.climatescan.org/projects/6915/detail). Right: Insufficient hydraulic capacity of a swale and road flooded with stormwater (*Source*: https://www.climatescan.org/projects/210/detail).

Vegetated swales can have different configuration and dimensions depending on the country or even region because guidelines (e.g. DWA-A 138, 2005) for their design, construction and maintenance will vary. In addition, they may have changed over time. Current guidelines should be adhered to after restoring a swale after damage or upgrading an existing swale.

9.5.4.3 Examples from the Netherlands

Bioswales are the most widely implemented nature-based climate adaptation method in the Netherlands (see Figure 9.45). Almost every Dutch municipality has implemented swales. Since its first implementation in 1997, over 5000 swales are estimated to have been implemented. Additional ones are expected in the years to come to address the amplification of flooding, drought and urban heat island challenges due to climate change.

Even in a small country as the Netherlands different types of bioretention have been implemented. Even though a large part of the Netherlands is situated below sea level with high-groundwater levels that are barely 50 cm below street level, bioretention systems have been successfully implemented in clay soils with low permeability. Given the limited storage available within the growing media, the available storage capacity of bioretention systems can be negatively impacted by (1) a too low elevation of the overflow provision or (2) the accumulation of sediments and vegetation residue over time.

Given their nature, vegetated swales are expected to accumulate pollutants. However, this may lead to long-term management challenges in that the growing media tend to become increasingly contaminated with associated rising disposal costs. Toxicity and exposure risks are complex to predict but can be derived from target and intervention values (see Table 9.10).

Most studies of micro-pollutant behaviour such as heavy metal accumulation have been short term, and there is limited understanding as to accumulation rates, remobilization potential or the time required for heavy metals to reach levels of concern in SUDS such as wetlands and bioretention systems. Recent research indicates that plant uptake of heavy metals (phytoremediation) may help prolong their lifespan. However, their long-term ability to retain these contaminants remains unknown. As pollutants are removed through, for example, filtering and sorption, the quality of the infiltrating stormwater largely determines the degree of contamination of the growing media. The presence of salt (i.e. from de-icing compounds) and organic matter (i.e. plant residue) may lead to desorption. Micro-pollutants will display a spatial distribution and are expected to be higher at the inlet and the bottom of vegetated swales (see Figure 9.46).

While the effectiveness of bioretention systems for the retention, infiltration and treatment of stormwater has been repeatedly demonstrated, the uncertainty surrounding long-term operations can be problematic for those responsible for the maintenance of these assets. Proactive planning of the O&M activities to address pollutant accumulation, clogging and plant survival and the associated maintenance budgets are key concerns in AM.

Proactive planning may be facilitated using management diagrams (Boogaard *et al.*, 2006) that present the desired management activities and frequencies for each object. The management diagrams also include inspections and monitoring activities. Table 9.8 can also be used as an example of management diagrams for the investigation and maintenance of the topsoil layer of an infiltration swale. The maintenance management diagram of the topsoil layer shows three types of measures:

- The first type is an ongoing process, for instance, providing direction to prevent people or vehicles from accessing the topsoil layer and parking on it.
- The second type of activities follows a time-dependent approach. These are repetitive actions such as mowing and removing litter and leaves. The frequencies in the table reflect initial advice. Once there is more detailed information about the functioning of the object in question, the manager can adjust the frequency.

Figure 9.45 Aggregated map of the Netherlands with over 1500 locations installed Dutch swales. *Source*: https://www.climatescan.org.

Table 9.10 Dutch threshold values for pollutants in soil.

Metals	Target Value (ppm (mg/kg))	Intervention Value (ppm (mg/kg))
Lead (Pb)	85	530
Zinc (Zn)	140	720
Cupper (Cu)	36	190

Source: Venvik and Boogaard (2020).

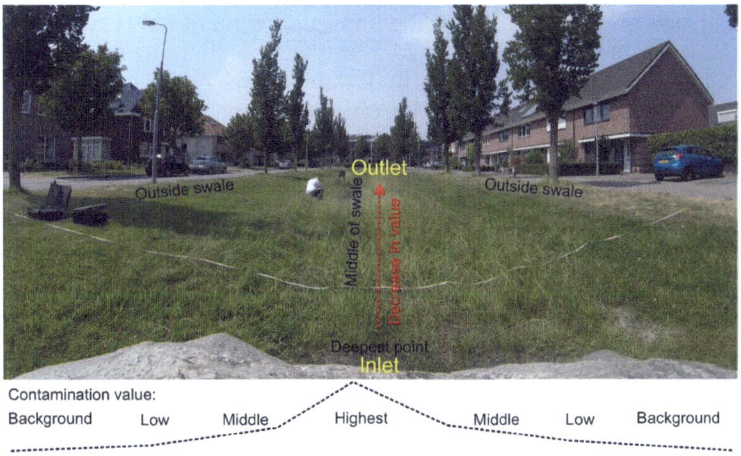

Figure 9.46 Variation of soil quality in a swale.

- The third type of activities should be executed based on the results of previous investigations. This is the situation-dependent approach. For instance, sowing is only necessary if visual inspection (activity I1 in Table 9.8) shows that there are bald spots in the swale. Replacing of the topsoil layer is necessary if samples of the groundwater or the topsoil show an unacceptable amount of pollution.

Good management starts with a good and unambiguous recording of the basic facts of the objects (e.g. location, dimension, composition). Also, the pertinent items of the design must be recorded, as well as the results of inspections and investigations. The management diagrams provide for each object an impression which activities are necessary and at which frequency. However, the manager should fine-tune the management reflecting his own local experience. To obtain insight into the functioning of the objects an active role of the manager is necessary. This active role implies:

- to give feedback to other departments after activities have been carried out (i.e. about characteristics and possible challenges),
- to take specific action on reports and complaints (e.g. by regularly visiting the SUDS, observing how they function) and
- to record clearly performed activities, complaints, observed failures and conducted actions.

Regular visual inspections of the top layer of bioretention systems and a simple measurement of the time needed for the water to infiltrate into the object are recommended. As such, as part of Table 9.8 visual inspection of the topsoil layer combined with a simple measurement of the time in which the water in the swale has been infiltrated, is recommended. The latter consists of a simple visual check that the swale is empty after 2 days. More detailed investigations are only necessary if the initial simple inspection shows that the emptying time of the swale does not meet the target.

9.5.5 Rainwater harvesting

Rainwater harvesting (RWH) systems represent a decentralized water source by retaining rainwater runoff from impervious surfaces for largely non-drinking water applications. Rainwater can sometimes be used for drinking purposes, which can be appropriate when alternative sources are not available (Ahmed *et al.*, 2012; John *et al.*, 2021). However, in this section, the focus is on RWH where water is solely used for non-potable purposes.

The practice has been used since ancient times to meet water demands when natural water sources are scarce or not available. Recently, the implementation of RWH systems has become more prominent because of the increasing demand for water due to climatic, environmental and societal changes (Campisano *et al.*, 2017). This includes both water-scarce regions (e.g. in Australia, Africa and America) and water-rich regions (e.g. Western Europe and Japan). Besides the substitution of drinking water, the retention of rainwater also reduces the frequency, peak rate and volume of urban runoff discharges into the urban drainage network (Campisano & Modica, 2015). In this context, Jamali *et al.* (2020) demonstrated that large-scale implementation of RWH systems may result in a distinct reduction of flood volumes and resulting damage.

Rainwater is typically mainly collected from roofs or other impervious surfaces with minor contamination, for example, pathways, parking spaces and streets with little or no traffic, and stored in a local storage tank. Depending on the quality of the collected rainwater and the intended use, treatment might be necessary (Hamilton *et al.*, 2019). The collected water may afterwards be used for a variety of applications, including external uses such as irrigation and car washing, as well as internal uses such as toilet flushing or washing laundry. In most industrialized nations, rainwater is not used for drinking water applications as this would require specialized treatment and monitoring and may simply not be cost-effective (Woods Ballard *et al.*, 2015).

RWH systems can be installed on residential, commercial and industrial properties. Therefore, the size and design vary greatly, ranging from small-scale and above-ground storage units (e.g. rain barrels for irrigation) to large underground storage tanks (e.g. cisterns for outdoor and indoor usage) with pumps and advanced filter systems. Most RWH systems feature similar designs, consisting of specific components (see Figure 9.47) that are regulated by local, regional or national design guidelines and implementation manuals (ARID, MPMSAA, 2008; TWDB, 2005) (Figure 9.47).

One of the main components of every RWH system is a rainwater tank (cistern) that stores the collected rainwater for later use and inherently treats the collected rainwater through sedimentation. Rainwater tanks are usually made of one of the following materials:

- PVC/PE tanks are made of plastics and have become very popular in the last two decades due to their low cost, lightweight and the many different forms and sizes available. Cracks in PVC/PE tanks mostly happen during winter when the material gets brittle from low temperatures.
- Concrete tanks are state of the art for larger above-ground and underground systems and are characterized by high robustness and long lifespans. Concrete carbonation and reinforcing steel corrosion can affect their durability, resulting in cracks or breaks.
- Metal tanks are made of steel coated with zinc for corrosion protection. The constant contact with water causes corrosion and leakage of zinc into stored water. There are also more expensive alternatives made of stainless steel that do not require a coating.

Due to the weight of the water and to ensure the stability of the rainwater tank, an appropriate foundation is required as a function of the tank design and size. Care should be taken, especially when situated next to walls or fences, to ensure that the heavy weight does not pose a risk to major structures. For underground tanks, particular attention should be paid to possible high-groundwater levels posing the risk of floating and groundwater infiltration as the surrounding ground conditions

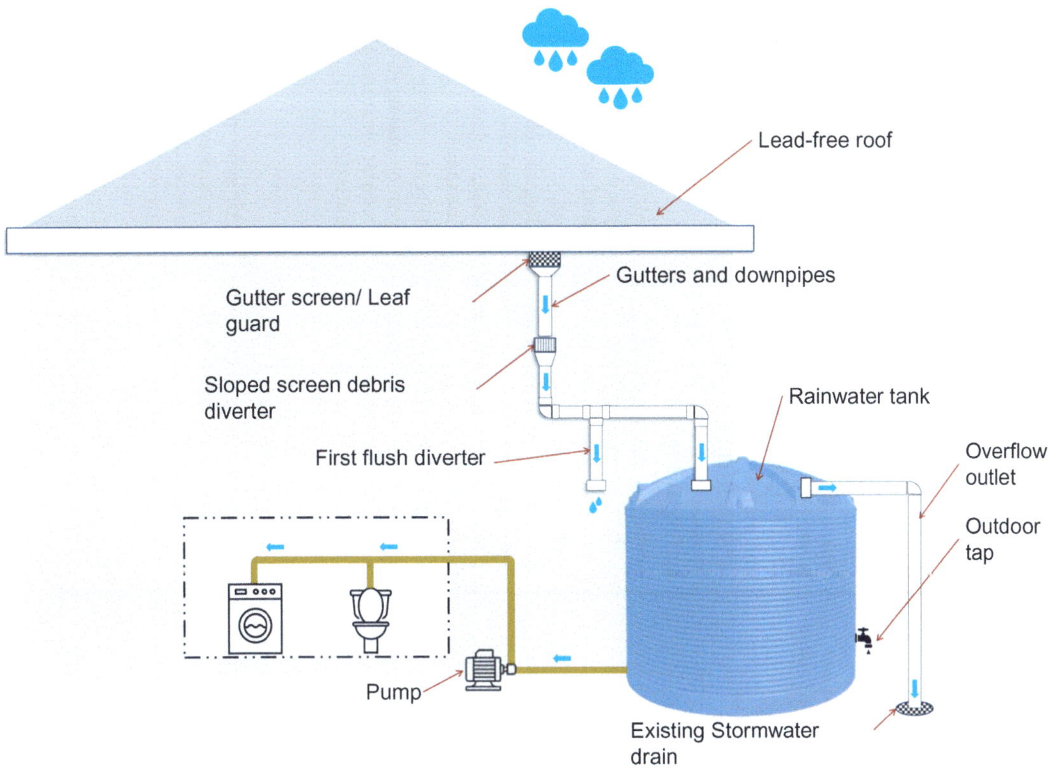

Figure 9.47 Rainwater tank design and components (for above-ground tanks).

may make extra structural support necessary. Additionally, the tank should withstand earth and backfill pressures and additional loads such as from vehicles.

RWH systems with their components may exhibit failures and malfunctions as a function of their age, condition and maintenance intervals. This can result in reduced water availability (Moglia *et al.*, 2016b), or in case of indoor usage, a direct health risk to the connected households, and in geographies with the presence of arboviruses such as Dengue, Japanese Encephalitis or malaria, breeding of mosquitoes (Moglia *et al.*, 2016a). Except for risks associated with arboviruses, the health risk is largely dependent on how the rainwater is used. In rural areas, where rainwater is the only source of fresh water and is also used as drinking water, the consequences of a malfunction or failure of the system may be more serious than in urban areas with alternative water supply. The low risk of failures in urban settings can mean that systems receive less maintenance and care, resulting in that failures sometimes are barely detected and, as such, not repaired. Of perhaps greatest significance of these types of problems, due to its common occurrence and relatively significant consequence, is failures of pumps and switches that can significantly reduce the benefits of the systems. Table 9.11 provides an overview of possible failures in RWH systems and the resulting risks. To restore the functionality of the system, the failures must be remedied.

9.5.5.1 Performance over time

The lifespan of the different components of a RWH system varies, depending on the materials used and the function of the elements in question. For example, prefabricated rainwater tanks (e.g. PVC)

Table 9.11 Risks of urban rainwater tank-related failure events and possible corrective actions.

Failure Events	Consequence (L, M, H)	Likelihood (L, M, H)	Risk (L, M, H)	Corrective Actions
Foundation becoming unstable or leaning	Collapse of walls or fences (H) Collapse of fixing elements (H)	Poor installation (L-M) Incorrect installation of overflow can erode foundation (L-M)	Risk is catastrophic (M-H)	Manageable through good design, installation and maintenance
Faulty switching or top up valves	Failure of switching valves significantly reduce water savings potential (M)	Indications are that this is a very common fault (H)	Severe consequence, limiting any water savings, common fault, high risk issue (H)	Manageable through good installation and maintenance
Broken pump	Pump failure prevents rainwater usage (M)	Problems with pumps are common, replacement every 3–5 years (M-H)	Not noticed unless alarmed (systems with automatic switching between rainwater and mains water), making the tank ineffective, risk can be managed through inspections and repairs (M-H)	Pump repair or replacement restoring capacity of the tank
Blocked gutters	Blocked gutters and screens limit flow into the tank and impact water quality (L)	Very high likelihood, especially with over-hanging trees (H)	Low-consequence event, high frequency, without management little water collection (M)	Manageable through good installation and maintenance
Removal or damage of mosquito meshing	Mosquitoes in urban settings can be a serious issue, spread of diseases (e.g. Dengue fever) (H)	Good design and rigorous installation minimize likelihood (L)	Low likelihood, manageable, consequences can be very high (places where mosquito spread disease) (M)	Regular check of meshing is easy, replacement inexpensive
Poor water quality	Poor water quality can lead to health issues, discolouration of toilets, other fixtures or washing (L-M)	The likelihood is situation-dependent; leaves may contaminate water in wet downpipe systems, small mammals on the roofs, can result in pathogens in the water (M)	Water quality is difficult to manage, alternative water sources should be used for drinking purposes, alternatively, proper disinfection treatment is recommended (M)	Discolouration can be managed by regular cleaning of gutters
Incorrect plumbing	Consequences depend on type of failure; backflow can lead to contamination (L-M)	Common problem; should be identified in post-installation inspections, likelihood is low, subject to such checks (L)	The consequences depend on error, risk is manageable (L)	Air gaps (preferred) or backflow preventers prevent the risk of contamination
Sedimentation in tank	Unless water for drinking purposes only a problem if severe sedimentation, pipes become blocked, water quality affected (L)	Sedimentation is a slow process, manageable with regular desludging (M)	Not a serious issue (L)	Desludging necessary at regular intervals (intervals of minimum 3 years)

The consequences, likelihoods and risks are rated as L (low), M (medium) or H (high), based on Sharma *et al.* (2015).

and household plumbing have a lifespan of up to 35 and 50 years, respectively. In contrast, electronic and dynamic parts reduce the lifespan, which is in the order of 5–15 years for domestic pumps. Furthermore, the lifespan of a RWH system is influenced by the quality of the original installation and by the frequency of subsequent maintenance. Table 9.12 provides an overview of factors that influence the long-term performance of RWH systems.

A reduced functionality or even a complete failure of a RWH system can be caused by a single factor or by a combination of various (minor) factors. For example, poor accessibility of the elements of a RWH system (due to poor placement) reduces the willingness for necessary maintenance works (maintenance efforts), subsequently increasing the risk of operational failures. In this context, Moglia *et al.* (2014) investigated the operational status of 417 RWH systems located in Melbourne, Australia. The operational status was verified by visual inspections and questionnaires. The investigated rainwater tanks had been mainly installed between 2007 and 2010, because of the drought in 2006–2007. In total, 99% of the rainwater tanks had a good or fair condition with only minor degradation, whereas only 49% of the pipe network had been correctly installed (see Table 9.13).

Additionally, while most of the overflows were functional and in good condition, 15% of the investigated rainwater tanks were leaking (see Table 9.14). Further identified problems include leaking taps (9%) and plastic corrosion (2%) (see Table 9.15). Interestingly, 74% of the rainwater tanks surveyed had incorrect household rainwater signage.

Table 9.12 Overview of factors influencing the long-term performance of RWH systems.

Category	Factors	Examples
Quality of the installation	Properly sizing of the elements	Pump capacity Foundation
	Faultlessly and functional elements	Electronic devices working correctly Pipes without cracks
	Correct installation	Tight connections Right placements
Operational failure mechanisms	Structural failures	Leaks Cracks
	Failure of dynamic parts	Pump failure
	Blockages	Sediment Leaves
	Other factors	Insects' penetration
Behavioural factors	Maintenance efforts	Competence Willingness
	Control capabilities	Pump switching

Source: Mukheibir *et al.* (2014).

Table 9.13 General conditions of the investigated rainwater tanks and the corresponding pipe network in Melbourne.

Condition Grading	Overall Condition of Rainwater Tank	Overall Condition of Rainwater Tank (%)	Condition of Pipe Work	Condition of Pipe Work (%)
Good	349	86	177	42
Fair	50	12	28	7
Poor	6	1	1	0
Not connected	N/A	N/A	43	10
Unknown	0	0	168	40

Source: Moglia *et al.* (2014).

Table 9.14 Effectiveness of overflow for the investigated rainwater tanks in Melbourne.

Effectiveness of Overflow	Number of Sites	Percentage of Sites
Good condition/functionality	363	87
Poor condition/functionality	29	7
No overflow	5	1
Slight leak	62	15
Unknown	11	3

Source: Moglia *et al.* (2014).

Table 9.15 Further identified problems of rainwater tanks in Melbourne.

Presence of Issue	Leaking Taps	Leaking Taps (%)	Evidence of Plastic Corrosion	Evidence of Plastic Corrosion (%)	Correct Signage of Rainwater	Correct Signage of Rainwater (%)
Yes	36	9	8	2	107	26
No	320	77	404	97	307	74
Unknown	61	15	5	1	3	1

Source: Moglia *et al.* (2014).

Moglia *et al.* (2013) used a hybrid approach to analyse the impact of inspection intervals on the failure rates of the different components of a RWH system. The analysis was based on an online survey about the expected problems and conditions of water tanks, which was subsequently combined with computer simulations to determine failure rates for blocked gutters (overhanging trees), broken pipes, broken meshing or structural failure of the water tanks. As shown in Table 9.16, RWH systems subjected to frequent inspections have low failure rates, while failure rates increase significantly with a decreasing number of inspections.

9.5.5.2 Operation, maintenance and rehabilitation

Compared to other SUDS, such as green roofs or swales, RWH systems only consists of engineered and no green or vegetated elements. Therefore, rehabilitation techniques are similar to those of grey infrastructure, with a strong focus on the repair of technical components such as pumps and filters, as well as the sealing and renewal of leaking rainwater tanks. In the following, rehabilitation measures from the sub-groups repair (low effort), renovation (moderate effort) and replacement (high effort) are

Table 9.16 Estimated failure rates for different RWH system elements based on frequency of inspections.

Frequency of Inspection	Tanks with Blocked Gutters (%)	Tanks with Broken Pipes (%)	Tanks with Broken Meshing (%)	Tanks with Broken Structure (%)
1 month	17	1	1	0
3 months	39	4	4	0
6 months	54	8	10	1
1 year	67	19	21	2
2 years	75	37	38	5

Source: Sharma *et al.* (2015).

presented. In addition, an outlook is provided on how new technologies (i.e. smart RWH systems) can improve system performance. The boundaries between the sub-groups are often not clearly defined and an exact classification may not be possible for all techniques. In addition, there is a strong overlap with the inspections that should be carried out in conjunction with maintenance activities, which should allow for the recognition of potential failures and the overall condition of the system. In addition, maintenance activities often also include minor repairs. As discussed earlier, there is a clear correlation for most parts of a RWH system between the frequency of maintenance intervals and the general condition.

Repair and maintenance are often strongly interrelated and are therefore here jointly considered. Guidance issued in most countries recommends regular maintenance of RWH systems to maintain their function and satisfactory water quality in the long run, with maintenance intervals of the individual components defined in manuals and guidelines (e.g. ARID, MPMSAA, 2008; Auckland Council, 2016; DIN 1989-100, 2022). Table 9.17 provides an example for maintenance requirements of RWH systems reflecting New Zealand's guidance. Through regular maintenance, minor malfunctions or damages are often detected and either fixed immediately or repaired later by replacing parts. Examples include the reconnection of displaced pipes, cleaning seals, filters, first-flush devices, overflows, roofs, gutters and downspouts. Minor repairs that may require spare parts include replacement or sealing of leaking pipes, damaged hatches, screens, covers, first-flush devices, overflows, pump components and backflow prevention valves. These repair activities usually do not require any special tools and can be carried out either by the owner himself or by specialist personnel. Additionally, the tank should be checked regularly for accumulated sediments and cleaned by qualified personal before the sediments reach the pump inlet, requiring a mud vacuum.

Major damage to RWH systems, which means a medium-to-large repair effort (here referred to as **Renovation**), usually only occurs to the rainwater tank. Leakage of the rainwater tank, as well as structural damage, makes a major repair necessary. Depending on the material of the rainwater tank and the size of the damage, different procedures can be applied. Above-ground tanks, which are common in Australia and the USA, are usually easier to repair than underground tanks, which are mainly used in Europe. In the following, common repair techniques for PVC/PE, concrete and metal tanks are presented.

Plastic tanks made of PVC or PE have become very popular in the last two decades and can be found above or below ground in all different shapes and sizes. Normally, plastic tanks have a long life expectancy of more than 20 years (Shah *et al.*, 2016), but they can be vulnerable to specific external impacts. Failures such as cracks and leaks can occur when the tank's foundation is not level or does not provide adequate structural support, when the tank is exposed to extreme direct sunlight for many years, or when it experiences impacts from livestock or playing children. Especially during winter, when the plastic is more brittle, rainwater tanks are more likely to fail.

Damage to rainwater tanks can be repaired relatively easily using a plastic welding process. The water in the tank should be drained to the point of damage and the area sufficiently cleaned and sanded to provide a good base for the process. There are two different methods for welding plastic. The simpler one consists of melting one or more strips into the spot using a heat gun or soldering irons (see Figure 9.48). To apply the layers of plastic welding strip, the crack is first prepared with a rotary tool and a V-groove is applied along the crack line to achieve optimal bonding between both materials. Finally, the welding strip is heated with a heat gun and fused to the rainwater tank. In a more complex extrusion welding process, the machine extrudes a continual stream of molten plastic which creates a stronger and more durable connection. With this method, even large cracks or holes can be reliably sealed. After cooling, the area is immediately resistant to stress and should provide a secure connection for several years. Both processes are comparatively quick and inexpensive and the welding with strips can be carried out by non-specialists.

Concrete tanks are state of the art for large above-ground and underground systems and are characterized by high robustness and durability. To ensure sufficient stability, the concrete is reinforced

Table 9.17 Maintenance requirements for rainwater tanks in New Zealand.

Component	Device	Recommended Actions	Who	Frequency
Back-up water supply	RT, DP	Inspect all pipes and valves for leaks.	Owner	Monthly
	RT, DP	Float-controlled top-up valves are prone to leakage which may result in constant filling of the tank to overflow which may go unnoticed until the next water meter reading. Check these valves regularly to ensure that they are closing properly and not leaking. Seek professional assistance if necessary.	Owner	
	RT, DP	Check solenoid valve is not switched off	Owner	
Tank hatches and covers	All	Inspect for correct fit and seal. Particularly important for underground tanks to prevent ingress of contaminated surface water.	Owner	
In-line filters	WS, RT, DP	It is advisable to keep a spare filter cartridge. Swap filter cartridge and clean the old one ready for the next replacement.	Owner	Every 3 months
First-flush devices	WS, RT, DP	Inspect and remove any debris and accumulated sediment. Refer to the manufacturer's instructions.	Owner	
Overflow, outlet pipes and orifices	All	Inspect for blockage and clean of necessary. Check flap valves and/or vector screens.	Owner	
In-line leaf and debris diverters	WS, RT, DP	Inspect and clear away any accumulated leaves or debris.	Owner	
Roof, gutters, down pipes and gutter screens	All	Inspect and remove debris and accumulated sediment. Prune any overhanging branches. If the roof and gutters need to be cleaned, the flow into the rainwater tank should be temporarily diverted to prevent ingress of dirty water to the tank.	Owner	Every 6 months
Tanks	All	Inspect the tank for leaks, sediment build up and structural integrity. A professional tank cleaning contractor should be employed to clean out the tank when the sludge level gets close to the pump inlet.	Owner/ professional	Annually
		Most rainwater tanks have a design life of 25 years or more. Tanks older than 25 years should be regularly assessed and replaced if necessary.	Professional	
Backflow prevention valves	RT, DP	Inspection must be carried out by a certified inspector and repaired if necessary. If an air gap is used to provide backflow prevention, no inspection is required.	Professional	
Pumps	WS, RT, DP	Pumps should be professionally inspected and/or serviced every few years. Refer to the pump service manual or contact a local service provider.	Professional	Every 2–5 years
Non-return valves	WS, RT, DP	Check operation when pump is serviced.	Professional	

Source: Auckland Council (2016).WS = water supply, RT = retention tank, DP = dual-purpose, DT = detention tank.

Figure 9.48 Plastic welding of cracked PE tank (for above-ground tanks).

with steel and is thus very robust against external influences. Tanks are designed for a service life of 25 years, although tanks have been known to last more than 35 years. In addition to structural failures due to an inadequate foundation, an undesigned load or insufficient backfilling in the case of buried tanks, decarbonization of the concrete and rusting of the reinforcing steel are the main causes of problems. There are several different methods for repairing damage to concrete tanks as a function of the size of the affected area, the depth of the water and the width of the crack. The methods to address these failures have analogies with the procedures used for the traditional grey concrete infrastructure.

Water tank liners are a popular method to reliably seal leaks in rainwater tanks and additionally improve the water quality affected by concrete carbonation. The tank is completely covered with the liner from the inside so that the original material no longer has any contact with the water (see Figure 9.49). Liners can increase the life expectancy of the rainwater tank. The tank must be emptied and thoroughly cleaned. The liner, which is specifically manufactured to the size of the tank, is then laid out and the inlet and outlet pipes are sealed separately.

If the entire tank does not have to be sealed, three different methods of sealing are possible (DIN 18535-1, 2017; DIN 18535-2, 2017; DIN 18535-3, 2017), differing in terms of maximum water depth, crack width and place of execution (further information can be found in DAfSt, 2001; DIBt, 2020a, 2020b):

- Sheet waterproofing materials (for all water depths, locations and changes in crack width <1 mm):
  - Bitumen and polymer bitumen sheeting (EN 13969, 2005).
  - Plastic and elastomer sheeting (EN 13967, 2012).
  - Combination of polymer bitumen sheeting and plastic or elastomer sheeting.

Figure 9.49 Water tank liners for leaking concrete or metal tanks. *Source*: velacreations (https://www.flickr.com/photos/velacreations/).

- Liquid waterproofing materials:
  - Non-crack-bridging mineral sealing slurry (for all water depths, only tanks not connected to buildings and no change in crack width or formation of new cracks).
  - Crack-bridging mineral sealing slurry (for water depths <10 m, all locations and change in crack width <0.2 mm) (see Figure 9.49, right).
  - Waterproofing systems with liquid plastics (for water depths <10 m, all locations and changes in crack widths <1 mm).
- Liquid waterproofing materials in combination with tiles and slabs (water depths <10 m, all locations and changes in crack width <0.2 mm).

Metal tanks are mostly made of steel coated with zinc for corrosion protection. These tanks were state of the art for decades in the 20th century and can still be found on older buildings or in countryside. To avoid corrosion, steel-lined tanks are mostly used today, consisting of a normal steel tank with a plastic lining, or stainless-steel tanks, where corrosion problems do not exist. However, the cost of stainless steel is much higher than normal steel and both are much more expensive than plastic tanks. For normal steel tanks, manufacturers tend to provide a warranty for 20 years, which can be shortened by heavy rusting. In particular, welds and seams are weak points where rainwater affects the structure of the tank.

There are two possible methods for repairing metal tanks. First, similar to concrete tanks, the interior can be covered with liners. This prevents further rusting and reliably seals all areas. However, care must be taken to ensure that the structure of the tank has sufficient load-bearing capacity. It is also possible to weld holes and cracks in the metal and thus restore the sealing. This method can also be used to close larger holes or restore structural stability. The repair can be complicated, especially if welding is required inside the tank.

If the damage to the rainwater tank exceeds a repairable extent, **replacement** of the rainwater tank is necessary. First, it should be checked whether the dimensioning of the old tank still corresponds to the site-specific requirements and, if necessary, a new dimensioning should be carried out. Second, other materials should be considered for the new tank, with the intent of a longer service life for the new system. The following factors indicate that a repair is no longer sufficient, and replacement is necessary:

- Damaged floor and other irreparable damage: if the bottom of the tank is damaged due to overpressurization, a repair might be not enough and requires a replacement of the whole tank.

- Structural instability: external and internal factors can reduce the structural strength of the rainwater tank, which makes it unsafe to use any longer. This can happen without any signs of leakage, cracks or corrosion.
- End of the expected lifespan: manufacturers of rainwater tanks tend to provide a warranty to their products for 20–25 years which covers the expected lifespan. This requires regular inspection and maintenance. Tanks that are older than the expected lifespan are more prone to failures and need a replacement more often.

9.5.5.3 *Example for updating with modern technology*

The detention performance of 'conventional' RWH systems strongly depends on the demand which is influenced by weather conditions (e.g. higher temperatures increase irrigation demand) and user behaviour (e.g. usage of stormwater for other non-potable water applications) (Quinn *et al.*, 2021). For example, a high frequency of usage in combination with a high amount of rainwater withdrawn results in a faster emptying of these systems, thereby providing more storage capacity for the next rain event. To maximize the use of RWH systems for flood control, recent research has focused on the development of smart RWH systems (Xu *et al.*, 2021). Figure 9.50 shows the key elements of a smart rain barrel concept (Oberascher *et al.*, 2021) as an exemplary smart RWH system: (1) a sensor to measure the actual filling depth, (2) an automatic discharge valve for the release of stored stormwater, (3) a control unit for the exchange of measurement data and control commands and (4) high-resolution weather forecasts for the estimation of potential future rainwater inflow. Using these data, smart RWH systems can release stormwater collected during previous storm events before the next event occurs, thus providing additional storage capacity. The success of these smart systems depends to a large extent on the reliability of weather forecasts; after all, it would be unfortunate if the storm event does not materialize after the tank would have been pre-emptively emptied as the released stormwater would not be available anymore for the desired use.

Figure 9.51 shows the simulation results of a smart RWH system with a forecast period of 1 h for a single-family house located in Austria (roof area of 100 m^2 and green area of 500 m^2). As can be seen, smart RWH systems increases the detention volume (+200% rainwater detention) compared to a conventional (and uncontrolled) RWH system, while providing enough rainwater for the substitution of drinking water. Additionally, the real-time control is mostly effective in case of (partially) filled storage units (Oberascher *et al.*, 2019). Subsequently, smart RWH systems provide a suitable solution for future system extension but will require different maintenance efforts due to the digital components.

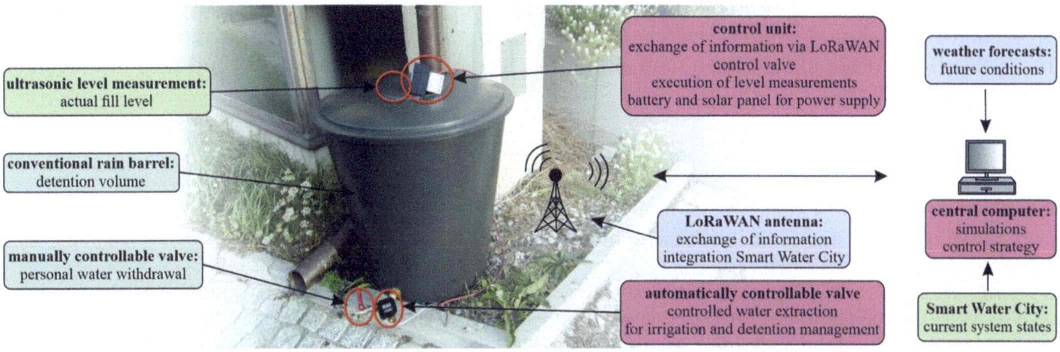

Figure 9.50 Smart rain barrel as an example for smart RWH. *Source*: Oberascher *et al.* (2021).

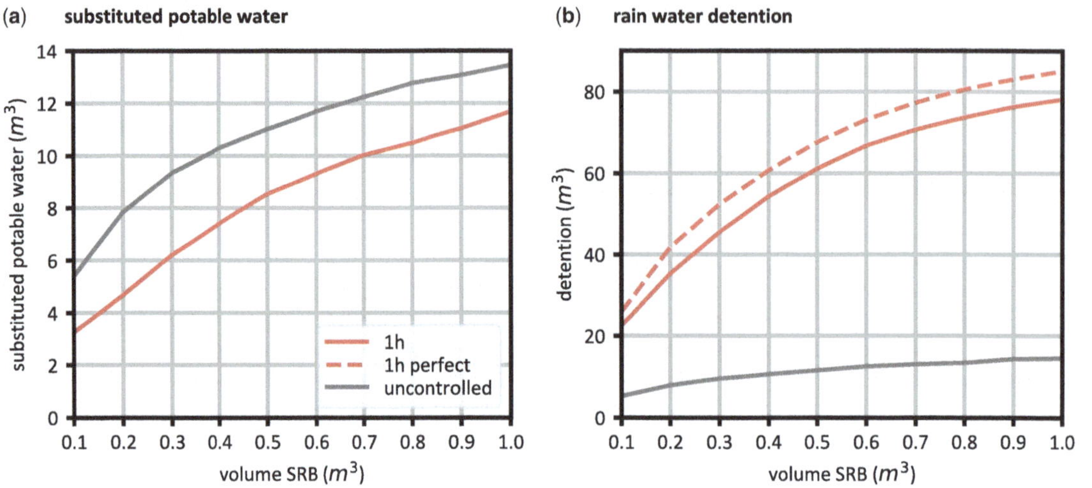

Figure 9.51 Simulation results for (a) substituted drinking water and (b) rainwater detention at household level for different volumes of smart rain barrels (SRBs).

9.6 CONCLUSION

Urban drainage O&M and rehabilitation can be divided into two distinct segments: the traditional grey infrastructure (i.e. pipes and associated components) and the SUDS realm. This differentiation not only originates with the type of assets (e.g. above-ground, accessibility to the public, presence of vegetation, etc.) but also reflects the maturity of the techniques and associated knowledge. SUDS O&M and rehabilitation strategies and actions should build on the knowledge gained for grey infrastructure AM, while considering additional aspects. First, replacement does not necessarily mean identical assets: a swale could be replaced by a trench or permeable pavement, while non-functional drainage pipes could also be replaced by at-source SWM with SUDS. Second, maintenance of SUDS is often done by staff who are used to traditional grey infrastructure: new skills are required for SUDS (especially regarding but not limited to vegetation), and it is important to identify the relevant expectations upfront. For example, when dealing with pipes, the objective is to maintain clean pipes and avoid sedimentation; for SUDS, sedimentation in the way of pre-treatment is often a prerequisite. Combined sewer networks will always have a flow of water, even during dry weather conditions (i.e. wastewater plus possibly an infiltration contribution); in contrast, dry-weather flow directed to SUDS is problematic and should be investigated. Moreover, more interactions are expected in case of SUDS: with residents or other disciplines such as roads departments, property or building management groups or park services.

REFERENCES

Abu-Zreig M., Rudra R. P., Lalonde M. N., Whiteley H. R. and Kaushik N. K. (2004). Experimental investigation of runoff reduction and sediment removal by vegetated filter strips. *Hydrological Processes*, **18**, 2029–2037, https://doi.org/10.1002/hyp.1400

AEP. (2018). BIC Scale for Delineating Petroleum Hydrocarbons in Organic Soils and Compost (AEP Land Policy No. 2018-1), Land Conservation and Reclamation Policy. Alberta Environment and Parks, Calgary, Canada.

Ahmed W., Hodgers L., Sidhu J., Gardner T., Richardson K., Palmer A. and Toze S. (2012). Health Risk Assessment of Roof-Captured Rainwater (Technical Report No. 77). Urban Water Security Research Alliance, Queensland, Australia.

Andenæs E., Kvande T., Muthanna T. M. and Lohne J. (2018). Performance of blue-green roofs in cold climates: a scoping review. *Buildings*, **8**, 55, https://doi.org/10.3390/buildings8040055

Arbocatalogus. (2018). Voorschrift Veilig Werken bij Rioleringsbeheer (Onderdeel van de Arbocatalogus Afvalbranche). Vereniging Afvalbedrijven, s-Hertogenbosch, the Netherlands.

ARID, MPMSAA. (2008). Rainwater Tank Design and Installation Handbook. National Water Commission, Turner, Australia.

Auckland Council. (2016). On-site Stormwater Tanks (Practice Note No. AC2264/2). Auckland Council, Auckland, New Zealand.

Barrett M. E., Walsh P. M., Malina J. F. and Charbeneau R. J. (1998). Performance of vegetative controls for treating highway runoff. *Journal of Environmental Engineering*, **124**, 1121–1128, https://doi.org/10.1061/(ASCE)0733-9372(1998)124:11(1121)

Bewszko T. and Słyś D. (2014). Multi-criteria decision aid for the selection of open trenching technology for modernisation of municipal infrastructure systems. *Tunnelling and Underground Space Technology, Underground Infrastructure Research*, **39**, 94–101, https://doi.org/10.1016/j.tust.2012.11.012

Boogaard F. C. (2015). Stormwater Characteristics and new Testing Methods for Certain Sustainable Urban Drainage Systems in The Netherlands. PhD. Delft University of Technology, Delft, the Netherlands, https://doi.org/10.4233/UUID:D4CD80A8-41E2-49A5-8F41-F1EFC1A0EF5D

Boogaard F. C. (2022). Spatial and time variable long term infiltration rates of green infrastructure under extreme climate conditions, drought and highly intensive rainfall. *Water*, **14**, 840, https://doi.org/10.3390/w14060840

Boogaard F. and Lucke T. (2019). Long-term infiltration performance evaluation of Dutch permeable pavements using the full-scale infiltration method. *Water*, **11**, 320, https://doi.org/10.3390/w11020320

Boogaard F. C., Bruins G. and Wentink R. (2006). Wadi's: aanbevelingen voor ontwerp, aanleg en beheer. Stichting RIONED, the Netherlands.

Boogaard F. C., Boelee E. and Van de Ven F. (2021). Groenblauwe raingardens en muggen. *Vakblad Riolering*, **3061**, 28–29, the Netherlands.

Boogaard F., Rooze D. and Stuurman R. (2023). The long-term hydraulic efficiency of green infrastructure under sea level: performance of raingardens, swales and permeable pavement in New Orleans. *Land*, **12**, 171, https://doi.org/10.3390/land12010171

BRL K10014. (2017). Kiwa Procescertificaat Voor Reinigen van Riolen, Putten en Kolken (Arbeitsblatt). Kiwa Nederland B.V., Rijswijk, The Netherlands.

Browne D., Godfrey M., Markwell K. and Boer S. (2017). WSUD Audit Guidelines. Stormwater Victoria, Melbourne, Australia.

Campisano A. and Modica C. 2015. Rainwater harvesting as source control option to reduce roof runoff peaks to downstream drainage systems. *Journal of Hydroinformatics*, **18**, 23–32, https://doi.org/10.2166/hydro.2015.133

Campisano A., Butler D., Ward S., Burns M. J., Friedler E., DeBusk K., Fisher-Jeffes L. N., Ghisi E., Rahman A., Furumai H. and Han M. (2017). Urban rainwater harvesting systems: research, implementation and future perspectives. *Water Research*, **115**, 195–209, https://doi.org/10.1016/j.watres.2017.02.056

Chopra M., Kakuturu S., Ballock C., Spence J. and Wanielista M. (2010). Effect of rejuvenation methods on the infiltration rates of pervious concrete pavements. *Journal of Hydrologic Engineering*, **15**, 426–433, https://doi.org/10.1061/(ASCE)HE.1943-5584.0000117

Cook L. M. and Larsen T. A. (2021). Towards a performance-based approach for multifunctional green roofs: an interdisciplinary review. *Building and Environment*, **188**, 107489, https://doi.org/10.1016/j.buildenv.2020.107489

DAfStb. (2001). Schutz und Instandsetzung von Betonbauteilen (Instandsetzungs-Richtlinie). Deutscher Ausschuss für Stahlbeton, Berlin, Germany.

Daulat S., Rokstad M. M., Klein-Paste A., Langeveld J. and Tscheikner-Gratl F. (2022). Challenges of integrated multi-infrastructure asset management: a review of pavement, sewer, and water distribution networks. *Structure and Infrastructure Engineering*, **20**(4), 546–565, https://doi.org/10.1080/15732479.2022.2119480

Davis A. P., Stagge J. H., Jamil E. and Kim H. (2012a). Hydraulic performance of grass swales for managing highway runoff. *Water Research*, **46**, 6775–6786, https://doi.org/10.1016/j.watres.2011.10.017, Special issue on stormwater in urban areas.

Davis A. P., Traver R. G., Hunt W. F., Lee R., Brown R. A. and Olszewski J. M. (2012b). Hydrologic performance of bioretention storm-water control measures. *Journal of Hydrologic Engineering*, **17**, 604–614, https://doi.org/10.1061/(ASCE)HE.1943-5584.0000467

de la Fuente de Val G. (2023). The effect of spontaneous wild vegetation on landscape preferences in urban green spaces. *Urban Forestry & Urban Greening*, **81**, 127863, https://doi.org/10.1016/j.ufug.2023.127863

Deletic A. (2005). Sediment transport in urban runoff over grassed areas. *Journal of Hydrology*, **301**, 108–122, https://doi.org/10.1016/j.jhydrol.2004.06.023

Dhalla S. and Zimmer C. (2010). Low Impact Development Stormwater Management Planning and Design Guide. Toronto and Region Conservation Authority, Toronto, Canada.

DIBt. (2020a). Instandhaltung von Betonbauwerken (TR Instandhaltung): Teil 1 – Anwendungsbereich und Planung der Instandhaltung (Technische Regel). Deutsches Institut für Bautechnik, Berlin, Germany.

DIBt. (2020b). Instandhaltung von Betonbauwerken (TR Instandhaltung): Teil 2 – Merkmale von Produkten Oder Systemen für die Instand-Setzung und Regelungen für Deren Verwendung (Technische Regel). Deutsches Institut für Bautechnik, Berlin, Germany.

DIN 1989-100. (2022). Regenwassernutzungsanlagen – Teil 100: Bestimmungen in Verbindung mit DIN EN 16941-1. Deutsches Institut für Normung e. V., Berlin, Germany.

DIN 18535-1. (2017). Abdichtung von Behältern und Becken Teil 1: Anforderungen, Planungs- und Ausführungsgrundsätze. Deutsches Institut für Normung e. V., Berlin, Germany.

DIN 18535-2. (2017). Abdichtung von Behältern und Becken – Teil 2: Abdichtung mit Bahnenförmigen Abdichtungsstoffen. Deutsches Institut für Normung e. V., Berlin, Germany.

DIN 18535-3. (2017). Abdichtung von Behältern und Becken – Teil 3: Abdichtung mit Flüssig zu Verarbeitenden Abdichtungsstoffen. Deutsches Institut für Normung e. V., Berlin, Germany.

Diogo A. F., Barros L. T., Santos J. and Temido J. S. (2018). An effective and comprehensive model for optimal rehabilitation of separate sanitary sewer systems. *Science of the Total Environment*, **612**, 1042–1057, https://doi.org/10.1016/j.scitotenv.2017.08.315

Dusza P. Y. (2017). Toitures Végétalisées et Services écosystémiques: Favoriser la Multifonctionnalité via les Interactions Sols-Plantes et la Diversité Végétale. PhD. Université Pierre et Marie Curie, Paris, France.

DWA-A 138. (2005). Planung, Bau und Betrieb von Anlagen zur Versickerung von Niederschlagswasser (Arbeitsblatt). DWA Deutsche Vereinigung für Wasserwirtschaft, Abwasser und Abfall e. V., Hennef, Germany.

EN 752. (2017). Drain and Sewer Systems Outside Buildings: Sewer System Management (European Standard). European Committee for Standardization, Brussels, Belgium.

EN 12231. (2003). Surfaces for Sports Areas: Method of Test – Determination of Ground Cover of Natural Turf (European Standard). European Committee for Standardization, Brussels, Belgium.

EN 13967. (2012). Flexible Sheets for Waterproofing Plastic and Rubber Damp Proof Sheets Including Plastic and Rubber Basement Tanking Sheet Definitions and Characteristics (European Standard). European Committee for Standardization, Brussels, Belgium.

EN 13969. (2005). Flexible Sheets for Waterproofing Bitumen Damp Proof Sheets Including Bitumen Basement Tanking Sheets Definitions and Characteristics (European Standard). European Committee for Standardization, Brussels, Belgium.

EN 15885. (2018). Classification and Characteristics of Techniques for Renovation, Repair and Replacement of Drains and Sewers (European Standard). European Committee for Standardization, Brussels, Belgium.

EN 16323. (2014). Glossary of Wastewater Engineering Terms (European Standard). European Committee for Standardization, Brussels, Belgium.

FLL. (2018). Green Roof Guidelines – Guidelines for the Planning, Construction and Maintenance of Green Roofs. Landscape Research, Development and Construction Society, Bonn, Germany.

Fort F., Jouany C. and Cruz P. (2013). Root and leaf functional trait relations in Poaceae species: implications of differing resource-acquisition strategies. *Journal of Plant Ecology*, **6**, 211–219, https://doi.org/10.1093/jpe/rts034

Fuselli F., Huber S. and Mambretti S. (2022). Environmental aspects of trenchless pipe rehabilitation methods. *Urban Water Journal*, **19**(8), 879–887, https://doi.org/10.1080/1573062X.2022.2087531

Gilbert J. K. and Clausen J. C. (2006). Stormwater runoff quality and quantity from asphalt, paver, and crushed stone driveways in Connecticut. *Water Research*, **40**, 826–832, https://doi.org/10.1016/j.watres.2005.12.006

Goerke M. and Möllers K. (2020). Versickerungsfähige Pflastersteine: Gestaltung Eines Naturnahen Wasserkreislaufs im Urbanen Raum. Neue Landschaft, 28–33, Berlin, Germany

Gougeon G., Bouattour O., Formankova E., St-Laurent J., Doucet S., Dorner S., Lacroix S., Kuller M., Dagenais D. and Bichai F. (2023). Impact of bioretention cells in cities with a cold climate: modeling snow management based on a case study. *Blue-Green Systems*, **5**(1), 1–17, https://doi.org/10.2166/bgs.2023.032

Gromaire-Mertz M. C., Garnaud S., Gonzalez A. and Chebbo G. (1999). Characterisation of urban runoff pollution in Paris. *Water Science and Technology, Innovative Technologies in Urban Storm Drainage 1998 (Novatech '98)*, **39**, 1–8, https://doi.org/10.1016/S0273-1223(99)00002-5

Hamilton K., Reyneke B., Waso M., Clements T., Ndlovu T., Khan W., DiGiovanni K., Rakestraw E., Montalto F., Haas C. N. and Ahmed W. (2019). A global review of the microbiological quality and potential health risks associated with roof-harvested rainwater tanks. *Clean Water*, **2**, 1–18, https://doi.org/10.1038/s41545-019-0030-5

Hanslin H. M. and Johannessen B. G. (2016). Oppfølging grønne tak fra Fremtidens byer (NIBIO Rapport No. 2/140/2016). Miljødirektoratet, Ås, Norway.

Homet K., Kremer P., Smith V., Ampomah R. and Strader S. M. (2022). Mapping predicted areas of common maintenance impacts to green stormwater infrastructure in Philadelphia, Pennsylvania. *Journal of Sustainable Water in the Built Environment*, **8**, 05022003, https://doi.org/10.1061/JSWBAY.0000986

ISO 9001. (2015). Quality Management Systems – Requirements (International Standard). International Organization for Standardization, Geneva, Switzerland.

ISO 14001. (2015). Environmental Management Systems – Requirements with Guidance for use (International Stakndard). International Organization for Standardization, Geneva, Switzerland.

ISO 24516-3. (2017). Guidelines for the Management of Assets of Water Supply and Wastewater Systems – Part 3: Wastewater Collection Networks (International Standard). International Organization for Standardization, Geneva, Switzerland.

ISO 45001. (2018). Occupational Health and Safety Management Systems Requirements with Guidance for use (International Standard). International Organization for Standardization, Geneva, Switzerland.

Jamali B., Bach P. M. and Deletic A. (2020). Rainwater harvesting for urban flood management – an integrated modelling framework. *Water Research*, **171**, 115372, https://doi.org/10.1016/j.watres.2019.115372

James A., Roberts J. and Ruley J. (2010). Stormwater BMP Guidance Tool. Louisiana Department of Environmental Quality, New Orleans, Louisiana, USA.

Jayasooriya V. M., Ng A. W. M., Muthukumaran S. and Perera B. J. C. (2017). Green infrastructure practices for improvement of urban air quality. *Urban Forestry & Urban Greening*, **21**, 34–47, https://doi.org/10.1016/j.ufug.2016.11.007

John C. K., Pu J. H., Moruzzi R. and Pandey M. (2021). Health-risk assessment for roof-harvested rainwater via QMRA in Ikorodu area, Lagos, Nigeria. *Journal of Water and Climate Change*, **12**, 2479–2494, https://doi.org/10.2166/wcc.2021.025

Kelly-Hooper F. (2013). F2:F3b Ratio and BOC-Adjusted PHC F3 Approach to Resolving False Detections of Crude Oil and Diesel Drilling Waste in Clean Soils and Manure Compost. PhD. University of Waterloo, Waterloo, Canada.

Kelly-Hooper F., Barfoot K., dela Cruz L. and Pike G. (2022). Challenges, opportunities for beneficial reuse of stormwater management pond sediment. *Environmental Science & Engineering Magazine*, **35**, 44–49.

Lacasse S. (2022). Handbook Risk Assessment and Risk Management for Dams. Statkraft, Oslo, Norway.

Le Coustumer S., Fletcher T. D., Deletic A., Barraud S. and Poelsma P. (2012). The influence of design parameters on clogging of stormwater biofilters: a large-scale column study. *Water Research*, **46**, 6743–6752, https://doi.org/10.1016/j.watres.2012.01.026, Special issue on stormwater in urban areas.

Li H., Kayhanian M. and Harvey J. T. (2013). Comparative field permeability measurement of permeable pavements using ASTM C1701 and NCAT permeameter methods. *Journal of Environmental Management*, **118**, 144–152, https://doi.org/10.1016/j.jenvman.2013.01.016

Liebscher M., Sokoll O., El Shahid S. and Bosseler B. (2003). Entwicklung Eines Berechnungsverfahrens für Mittels des Berstverfahrens Verlegter Abwasserkanäle und Leitungen. Ministerium für Umwelt und Naturschutz, Landwirtschaft und Verbraucherschutz des Landes NRW, Gelsenkirchen, Germany.

Lönnqvist J., Hanslin H. M., Johannessen B. G., Muthanna T. M., Viklander M. and Blecken G. (2021). Temperatures and precipitation affect vegetation dynamics on Scandinavian extensive green roofs. *International Journal of Biometeorology*, **65**, 837–849, https://doi.org/10.1007/s00484-020-02060-2

Lucke T., Nichols P., Shaver E., Lenhart J., Welker A. and Huber M. (2017). Pathways for the evaluation of stormwater quality improvement devices – the experience of six countries. *Clean – Soil, Air, Water*, **45**, 1600596, https://doi.org/10.1002/clen.201600596

Marzouk M. and Omar M. (2013). Multiobjective optimisation algorithm for sewer network rehabilitation. *Structure and Infrastructure Engineering*, **9**, 1094–1102, https://doi.org/10.1080/15732479.2012.666254

Matthews J. C., Allouche E., Vladeanu G. and Alam S. (2018). Multi-segment trenchless technology method selection algorithm for buried pipelines. *Tunnelling and Underground Space Technology*, **73**, 295–301, https://doi.org/10.1016/j.tust.2018.01.001

MECP. (2019). On-Site and Excess Soil Management (Ontario Regulation No. 406/19). Ontario Ministry of Environment Conservation and Parks, Toronto, Canada.

MECP. (2021). Records of Site Condition – Part XV.1 of the Environmental Protection Act, R.S.O. 1990. C.E. 19. Ontario Ministry of Environment Conservation and Parks, Toronto, Canada.

Mezhoud C., Berreksi A., Bedjou A. and Bosseler B. (2022). Prioritization of maintenance work in wastewater networks using decision support methods. *Journal of Water, Sanitation and Hygiene for Development*, **12**, 186–199, https://doi.org/10.2166/washdev.2022.165

Moglia M., Tjandraatmadja G. and Sharma A. K. (2013). Exploring the need for rainwater tank maintenance: survey, review and simulations. *Water Supply*, **13**, 191–201, https://doi.org/10.2166/ws.2013.021

Moglia M., Tjandraatmadja G., Delbridge N., Gulizia E., Sharma A. K., Butler R. and Gan K. (2014). Survey of Savings and Conditions of Rainwater Tanks. Smart Water Fund and CSIRO, Melbourne, Australia.

Moglia M., Gan K. and Delbridge N. (2016a). Exploring methods to minimize the risk of mosquitoes in rainwater harvesting systems. *Journal of Hydrology*, **543**, 324–329, https://doi.org/10.1016/j.jhydrol.2016.10.010

Moglia M., Gan K., Delbridge N., Sharma A. K. and Tjandraatmadja G. (2016b). Investigation of pump and pump switch failures in rainwater harvesting systems. *Journal of Hydrology*, **538**, 208–215, https://doi.org/10.1016/j.jhydrol.2016.04.020

Mukheibir P., Boyle T., Moy C. and White S. (2014). Estimating the reliable residential water substitution from household rainwater tanks. *Water Practice and Technology*, **9**, 377–385, https://doi.org/10.2166/wpt.2014.040

NS 3840. (2015). Grønne tak Planlegging, Prosjektering, Utførelse, Skjøtsel og Drift Ekstensive tak (Norsk Standard). Standard Norge, Lysaker, Norway.

Oberascher M., Zischg J., Kastlunger U., Schöpf M., Kinzel C., Zingerle C., Rauch W. and Sitzenfrei R. (2019). Advanced rainwater harvesting through smart rain barrels. *World Environmental and Water Resources Congress, 2019: Watershed Management, Irrigation and Drainage, and Water Resources Planning and Management*, Pittsburgh, Pennsylvania, USA, 75–82, https://doi.org/10.1061/9780784482339.008

Oberascher M., Kinzel C., Kastlunger U., Kleidorfer M., Zingerle C., Rauch W. and Sitzenfrei R. (2021). Integrated urban water management with micro storages developed as an IoT-based solution – the smart rain barrel. *Environmental Modelling & Software*, **139**, 105028, https://doi.org/10.1016/j.envsoft.2021.105028

Ochs C. P. (2012). Multikriterielle Optimierung der Sanierungsplanung von Entwässerungsnetzen. PhD. Technische Universität Kaiserslautern, Kaiserslautern, Germany.

ÖWAV-RB22. (2015). Betrieb von Kanalisationsanlagen (Regelblatt). Österreichischer Wasser- und Abfallwirtschaftsverband (ÖWAV), Vienna, Austria.

ÖWAV-RB28. (2007). Unterirdische Kanalsanierung (Regelblatt). Österreichischer Wasser- und Abfallwirtschaftsverband (ÖWAV), Vienna, Austria.

Post J., Pothof I., Veldhuis M.-C. ten, Langeveld J. and Clemens F. (2016). Statistical analysis of lateral house connection failure mechanisms. *Urban Water Journal*, **13**, 69–80, https://doi.org/10.1080/1573062X.2015.1057175

Quinn R., Rougé C. and Stovin V. (2021). Quantifying the performance of dual-use rainwater harvesting systems. *Water Research*, **10**, 100081, https://doi.org/10.1016/j.wroa.2020.100081

Rokstad M. M., Bruaset S., Skaar B. S., Borgen E. O., Røgenes H. and Tscheikner-Gratl F. (2022). Pardon my trench: Reflections on the uptake of trenchless technologies in the Norwegian water sector. Proceedings of LESAM 2022, Bordeaux, France.

Schmitt T. G., Welker A., Illgen M., Bosseler B., Harting K. and Brüggemann T. (2007). Untersuchung des Abfluss und Versickerungsverhaltens Wasserdurchlässiger Flächenbeläge (Abschlussbericht No. IV-9-042 236 0020). Ministerium für Umwelt und Naturschutz, Landwirtschaft und Verbraucherschutz des Landes Nordrhein-Westfalen (MUNLV), Düsseldorf, Germany.

Schönberger O., Harting K., Gabryl P. and Nolting B. (2005). Prüfung Wasserdurchlässiger Flächenbeläge Nach Mehrjähriger Betriebsdauer. IKT – Institut für Unterirdische Infrastruktur, Düsseldorf, Germany.

Shah K. N., Varandani N. S. and Panchani M. (2016). Life cycle assessment of household water tanks – a study of LLDPE, mild steel and RCC tanks. *Journal of Environmental Protection*, **7**, 760–769, https://doi.org/10.4236/jep.2016.75068

Shannon T. P., Ahler S. J., Mathers A., Ziter C. D. and Dugan H. A. (2020). Road salt impact on soil electrical conductivity across an urban landscape. *Journal of Urban Ecology*, **6**, juaa006, https://doi.org/10.1093/jue/juaa006

Sharma A. K., Begbie D. and Gardner T. (eds) (2015). Rainwater Tank Systems for Urban Water Supply: Design, Yield, Energy, Health Risks, Economics and Social Perceptions, IWA Publishing, London, UK.

Skaar B. S., Stevik T. K. and Johansen A. (2023). Barriers for pipe renewal – a comprehensive study. *Urban Water Journal*, **20**, 248–259, https://doi.org/10.1080/1573062X.2022.2156891

SLG. (2020). Merkblatt für Versickerungsfähige Pflasterbefestigungen aus Beton: Grundlagen, Planung, Ausführung, Erhaltung. Betonverband Straße, Landschaft, Garten e. V., Bonn, Germany.

Stagge J. H., Davis A. P., Jamil E. and Kim H. (2012). Performance of grass swales for improving water quality from highway runoff. *Water Research*, **46**, 6731–6742, https://doi.org/10.1016/j.watres.2012.02.037, Special issue on stormwater in urban areas.

Sterling R., Simicevic J., Allouche E., Condit W. and Wang L. (2010). State of Technology for Rehabilitation of Wastewater Collection Systems (No. EPA/600/R-10/078). United States Environmental Protection Agency, Washington, USA.

TRCA, CH2M Canada. (2016). Inspection and Maintenance Guide for Stormwater Management Ponds and Constructed Wetlands. Toronto and Region Conservation Authority, Vaughan, Ontario, Canada.

Tscheikner-Gratl F., Mikovits C., Rauch W. and Kleidorfer M. (2014). Adaptation of sewer networks using integrated rehabilitation management. *Water Science and Technology*, **70**, 1847–1856, https://doi.org/10.2166/wst.2014.353

TWDB. (2005). The Texas Manual on Rainwater Harvesting. Texas Water Development Board, Austin, Texas.

Veldkamp T. I. E., Boogaard F. C. and Kluck J. (2022). Unlocking the potential of permeable pavements in practice: a large-scale field study of performance factors of permeable pavements in the Netherlands. *Water*, **14**, 2080, https://doi.org/10.3390/w14132080

Venvik G. and Boogaard F. C. (2020). Portable XRF quick-scan mapping for potential toxic elements pollutants in sustainable urban drainage systems: a methodological approach. *Sci*, **2**, 64, https://doi.org/10.3390/sci2030064

Vollaers V., Nieuwenhuis E., van de Ven F. and Langeveld J. (2021). Root causes of failures in sustainable urban drainage systems (SUDS): an exploratory study in 11 municipalities in The Netherlands. *Blue-Green Systems*, **3**, 31–48, https://doi.org/10.2166/bgs.2021.002

WEF, ASCE/EWRI. (2022). Urban Stormwater Controls Operation and Maintenance: WEF Manual of Practice No. 39 and ASCE/EWRI Manuals and Reports on Engineering Practice No. 153. Water Environment Federation and the American Society of Civil Engineers/Environmental & Water Resources Institute, Reston, VA, USA.

Wentink R., Boogaard F., Lemmen G. and Oomens A. (2006). Beheer van Infiltratievoorzieningen Voor Regenwater (Onderdeel van de Arbocatalogus Afvalbranche No. C3200). Ministerie van VROM Bodem, Water, Landelijk Gebied, Houten, The Netherlands.

Werker H., Twardon J., Schmitz S., Wehming U., Hofmann J., Alt K., Ellerhorst S., Ingenhaag S., Bosseler B., Bennerscheidt C., Loos S., Togler R., Overfeld H. and Treunert E. (2011). Dezentrale Niederschlagswasserbehandlung in Trennsystemen – Umsetzung des Trennerlasses (Abschlussbericht). Ministerium für Klimaschutz, Umwelt, Landwirtschaft, Natur- und Verbraucherschutz des Landes Nordrhein-Westfalen, Cologne, Germany.

Winston R. J., Hunt W. F., Kennedy S. G., Wright J. D. and Lauffer M. S. (2012). Field evaluation of stormwater control measures for highway runoff treatment. *Journal of Environmental Engineering*, **138**, 101–111, https://doi.org/10.1061/(ASCE)EE.1943-7870.0000454

Woods Ballard B., Wilson S., Udale-Clarke H., Illman S., Scott T., Ashley R. and Kellagher R. (2015). The SuDS Manual, 6th edn. CIRIA, London, UK.

Xu W. D., Burns M. J., Cherqui F. and Fletcher T. D. (2021). Enhancing stormwater control measures using real-time control technology: a review. *Urban Water Journal*, **18**, 101–114, https://doi.org/10.1080/1573062X.2020.1857797

doi: 10.2166/9781789063059_0421

Chapter 10

Moving forward with creating asset management and operations and maintenance plans

The Urban Drainage Asset Management (UDAM) book is now published, thanks to a huge effort from authors all around the world, but the adventure is not yet complete. After going through all the chapters of this book, theoretically one should be able to set up a comprehensive system for UDAM. It is recognized by the authors however that, while in some places very detailed (scientific) information and introductions to literature for further study are provided, it is doubtful that someone without any experience will be able to do so easily. Therefore, the authors, in collaboration with the members of the UDAM Working Group of the Joint Committee of Urban Drainage envision to publish a 'living' document with relevant checklists that practitioners may consider when creating their own Asset Management (AM) and Operations and Maintenance (O&M) Plans. This living document can be found here: https://udam.home.blog/udam-book-chapter-10/.

This 'living' document is envisioned to provide a comprehensive, though digestible, overview of 'things to be done' when starting from scratch in say, a small municipality or by a private asset owner who wants to create a professional system without relying on commercial products or proprietary software. As pointed out in Chapter 1 of this book, AM essentially boils down to being able to answer three basic questions:

(1) What and where is the 'thing' I am responsible for?
(2) What is the current functionality of the 'thing' I'm responsible for?
(3) In what condition is the 'thing' I'm responsible for?

For each element in the checklists, we intend to indicate to which of the three main questions the action or instrument suggested contributes, and we provide references to the parts of the book that relate to them.

This 'living' document and associated checklists reflect the combined knowledge (and, unfortunately, ignorance) of the authors and members of the UDAM Working Group. With that in mind, the authors welcome feedback from practitioners which can be provided at https://udam.home.blog/contact/.

IWA Publishing's authorised EU representative for General Product Safety Regulations is Diane D'Arras, 15 rue Duret, 75116 Paris, France, e-mail: safety@iwap.co.uk.

Printed and bound by CPI Group (UK) Ltd, Croydon, CR0 4YY
12/05/2026
02108869-0016